HYDROLOGIC FREQUENCY MODELING

HYDROLOGIC FREQUENCY MODELING

Proceedings of the International Symposium on Flood Frequency and Risk Analyses, 14–17 May 1986, Louisiana State University, Baton Rouge, U.S.A.

Edited by

VIJAY P. SINGH

Department of Civil Engineering, Louisiana State University, Baton Rouge, U.S.A.

D. REIDEL PUBLISHING COMPANY

A MEMBER OF THE KLUWER ACADEMIC PUBLISHERS GROUP

DORDRECHT / BOSTON / LANCASTER / TOKYO

Library of Congress Cataloging in Publication Data

CIP

International Symposium on Flood Frequency and Risk Analyses (1986: Louisiana State University, Baton Rouge)
Hydrologic frequency modeling.

Includes indexes.
1. Floods—Mathematical models—Congresses. I. Singh, V. P. (Vijay P.) II. Title.
GB1399.I58 1986b 551.48'9'0724 87-20619
ISBN 90-277-2572-1
ISBN 90-277-2576-4 (set)

Published by D. Reidel Publishing Company,
P.O. Box 17, 3300 AA Dordrecht, Holland.

Sold and distributed in the U.S.A. and Canada
by Kluwer Academic Publishers,
101 Philip Drive, Assinippi Park, Norwell, MA 02061, U.S.A.

In all other countries, sold and distributed
by Kluwer Academic Publishers Group,
P.O. Box 322, 3300 AH Dordrecht, Holland.

Printed in The Netherlands

PREFACE

Floods constitute a persistent and serious problem throughout the United States and many other parts of the world. They are responsible for losses amounting to billions of dollars and scores of deaths annually. Virtually all parts of the nation--coastal, mountainous and rural--are affected by them. Two aspects of the problem of flooding that have long been topics of scientific inquiry are flood frequency and risk analyses. Many new, even improved, techniques have recently been developed for performing these analyses. Nevertheless, actual experience points out that the frequency of say a 100-year flood, in lieu of being encountered on the average once in one hundred years, may be as little as once in 25 years. It is therefore appropriate to pause and ask where we are, where we are going and where we ought to be going with regard to the technology of flood frequency and risk analyses. One way to address these questions is to provide a forum where people from all quarters of the world can assemble, discuss and share their experience and expertise pertaining to flood frequency and risk analyses. This is what constituted the motivation for organizing the International Symposium on Flood Frequency and Risk Analyses held May 14-17, 1986, at Louisiana State University, Baton Rouge, Louisiana.

The objectives of this symposium were therefore (1) to assess the current state of the art of flood frequency and risk analyses, (2) to demonstrate applicability of flood frequency and risk models, (3) to establish complementary aspects of seemingly different models, (4) to enhance interdisciplinary interaction, (5) to discuss practice of flood frequency and risk analyses technology by federal agencies in the U.S., (6) to discuss research needs in frequency and risk analyses, and (7) to determine directions for further research.

We received an overwhelming response to our call for papers. It was indeed a difficult task to select amongst the many excellent papers that were submitted, and we regret that we could not include all of them. The sole criterion for selection of a paper was its merit in relation to Symposium objectives. The subject matter of the Symposium was divided into 20 major topics encompassing virtually all facets of flood frequency and risk analyses. Each topic comprised of an invited state-of-the-art paper and a number of contributed papers. These contributions blended naturally to evolve a synthesized body of knowledge on that topic. Extended abstracts of all the invited and contributed papers were assembled in a pre-Symposium proceedings volume. This helped stimulate discussion and exchange of ideas during the Symposium.

The papers presented at the Symposium were refereed in a manner similar to that employed for publishing a journal article. As a result, many papers did not pass the review and were therefore eliminated from inclusion in the final proceedings. The papers contained in this book HYDROLOGIC FREQUENCY MODELING, represent one part of the Symposium contributions. The other parts are embodied in three separate books, FLOOD HYDROLOGY, REGIONAL FLOOD FREQUENCY ANALYSIS, and APPLICATION OF FREQUENCY AND RISK IN WATER RESOURCES, which are being published simultaneously. Arrangement of these books under four different titles was a natural consequence of the diversity of technical material discussed in the papers. These books can be treated almost independently, although some overlap does exist between them.

This book contains eight sections encompassing major aspects of hydrologic frequency analysis. Each section starts usually with an invited state-of-the-art paper, followed by contributed papers. Beginning with a discussion of hydrologic frequency analysis and its relevance, the papers go on to discuss univariate flood frequency models, mixed distributions, rainfall frequency analysis, application of entropy in flood frequency analysis, methods of parameter estimation, selection of flood frequency models, and multivariate stochastic models.

The book will of of interest to researchers as well as those engaged in practice of Civil Engineering, Agricultural Engineering, Hydrology, Water Resources, Earth Resources, Forestry and Environmental Sciences. The graduate students as well as those wishing to conduct research in flood hydrology will find this book to be of particular value.

I wish to take this opportunity to express my sincere appreciation to all the members of the Organizing and Advisory Committees, and the Louisiana State University administration for their generous and timely help in the organization of the Symposium. A lack of space does not allow me to list all of them by name here. Numerous other people contributed to the Symposium in one way or another. The authors, including the invited keynote speakers, contributed to the Symposium technically and this book is a result of their efforts. The session chairmen administered the sessions in a positive and professional manner. The referees took time out from their busy schedules and reviewed the papers. Graduate students assisted in smooth conduct of the sessions. I owe my sincere gratitude to all of these individuals.

If the success of a Symposium is measured in terms of the quality of participants and presentations then most people would agree that this was a very successful Symposium. A very large number of internationally well-known people, who have long been recognized for their contributions and have long been at the forefront of hydrologic research came to participate in the Symposium. More than 35 countries, covering the five continents and most of the countries of the world active in hydrologic research, were represented. It is hoped that long and productive personal associations will develop as a result of this Symposium.

March 1987
Baton Rouge, Louisiana

Vijay P. Singh
Symposium Director

ACKNOWLEDGEMENTS

The International Symposium on Flood Frequency and Risk Analyses was sponsored and co-sponsored by a number of organizations. The sponsors provided financial support without which the Symposium might not have come to fruition. Their financial support is gratefully acknowledged. The co-sponsors extended their help in announcing the Symposium through their journals, transactions, newsletters or magazines. This publicity helped with attendance at the Symposium, and is gratefully acknowledged. The following is a list of Symposium sponsors and co-sponsors.

SYMPOSIUM SPONSORS

Louisiana State University
- Department of Civil Engineering
- Louisiana Water Resources Research Institute

National Science Foundation
U.S. Army Research Office
U.S. Geological Survey, Louisiana District, Baton Rouge
Woodward-Clyde Consultants, Baton Rouge

SYMPOSIUM CO-SPONSORS

American Geographical Society
American Geophysical Union
American Meteorological Society
American Statistical Association
American Society of Agricultural Engineers
American Society of Civil Engineers
American Water Resources Association
Association of American Geographers
Association of State Floodplain Managers
International Association for Hydraulic Research
International Association of Hydrological Sciences
International Association of Theoretical and Applied Limnology
International Commission on Irrigation and Drainage
International Geographical Union
International Water Resources Association
National Wildlife Federation
North American Lake Management Society
Pan American Institute of Geography
Society for Risk Analysis
Soil Conservation Society of America

TABLE OF CONTENTS

HYDROLOGICAL AND ENGINEERING RELEVANCE OF FLOOD FREQUENCY ANALYSIS

V. Klemeš
National Hydrology Research Institute
Department of Environment
Saskatoon, Saskatchewan S7N 3H5
Canada

"The engineer may be asked to evaluate the risks involved in a certain project and to take effective measures for the protection of lives and property in a catastrophic eventuality. To do this intelligently, the very basis for the design is often missing or hard to obtain: this is an adequate knowledge of the mechanics of the feared phenomena. It is indeed not the engineer's task to study natural phenomena as such: this is the task of the scientist."

Adrian E. Scheidegger (1975)

ABSTRACT. The mainstream of contemporary flood frequency analysis (FFA) qualifies neither as a hydrological science nor as an engineering discipline. It is not a hydrological science because it does not analyze the frequencies of floods - it merely postulates that flood records are random samples from simple probability distributions. It is not an engineering discipline because it pays most attention to, and exerts most energy on, formal polishing of concepts which are crude by their very nature and whose basic assumptions, which have an overriding influence on design parameters, are arbitrary and are dictated by expediency rather than scientific knowledge. At best, much of flood frequency analysis is just a part of small sample theory in disguise, the term "flood" being used merely as a name for the numbers employed; at worst, it is a pretentious game draining resources both from hydrology and engineering research, and a cheap opportunity to satisfy the need of academics to publish papers and supply easy topics for graduate students who know little beyond elementary statistics, probability theory, and computer programming.

1. INTRODUCTION

Flood frequency analysis (FFA) is regarded as a respectable scientific discipline. However, it has the typical features of a pseudoscience, in particular, (1) its most crucial predictions cannot be tested, (2) in response to criticism its exponents often resort to demagoguery.

V. P. Singh (ed.), Hydrologic Frequency Modeling, 1–18.

The first point follows from the basic concept which FFA employs - the return period defined as an average period between two successive attainments or exceedances of a given (maximum annual) flood peak discharge. Thus to verify an estimate of the most popular and basic FFA result, the "100-year flood", one would have to have many hundreds of years of record in order to be able to obtain a meaningful average period of such a length.

The second point is more subtle and often goes unnoticed. The reason is that, as in every well thought out demagoguery, the argument employs commonly accepted standards and beliefs or even facts but uses them to evade rather than meet a criticism. Typically, instead of showing where the challenger's criticism is wrong, the defender will change the issue. We all are familiar with such practices on the scene of international politics where totalitarian governments resort to them whenever challenged on some of their indefensible policies. For example, a critique of the suppression of a free access in a communist country to noncommunist press and literature may be countered by boasting about a universal access to a free medical care in that country, etc. The defense of the practices of FFA has a form of the following circular argument: (a) When a hydrologist criticizes FFA for its assumptions trivializing the complex and inadequately understood processes resulting in extreme floods, and for the consequent lack of realism of its results, the FF analysist will assume the posture of a down-to-earth engineering practitioner and say that the engineer cannot wait until everything is known and needs a result, however crude, now because he must make decisions on flood-related measures now. (b) When an engineer summons his courage and asks why then, given the simplifications and distortions of reality implied in the basic assumptions, should he employ all the sophisticated mathematical machinery to get that crude result, and why not use some simple procedure instead, the FF analysist will immediately change into a meticulous theoretician and extol the merits of mathematical rigour in the quest for extraction of the maximum amount of information from the data. (c) And should a statistician ask him how does he know that his rigorously fitted distribution models give information relevant to flood frequencies, the FF analysist would transform into a seasoned hydrologist and put him to shame for not being aware of the empirically established time-honoured ability of those models to represent hydrological data and of the good service they have rendered generations of hydrologists in such important tasks as flood frequency analysis, etc. At this point the argument returns to item (a) above.

In short, the contemporary FFA is well on its way to intellectual self-sufficiency and immunity to criticisms from any of the three directions from which it originally started; it is becoming an end in itself. It now often appears that streamflow and flood data are no longer gathered to provide statistical information on real floods for real-life engineering needs, but to supply numbers for Flood Frequency Analysis.

Taking cognizance of the few islands of sanity that still exist in the sea of FFA, one may relax its classification as pseudoscience and say that, at best, it is a discipline suffering from the "maladies of cookbookery and mathematistry." This diagnosis has been borrowed from

G.E.P. Box who originally applied it to the state of health of statistics. He described the two maladies as follows:

"The symptoms of the former are a tendency to force all problems into the molds of one or two routine techniques, insufficient thought being given to the real objectives of the investigation or to the relevance of the assumptions implied by the imposed methods...Mathematistry is characterized by development of theory for theory's sake, which, since it seldom touches down with practice, has a tendency to redefine the problem rather than to solve it. Typically, there has once been a statistical problem with scientific relevance but this has long since been lost sight of." (Box 1976).

In our case, the legitimate problem has been the actual frequency of floods which has not only a scientific relevance but also a practical relevance in the context of the economics of flood protection. However, in FFA this legitimate problem has been redefined as an analysis of small random samples from a handful of simple probability distributions, and forced into the mold of curve-fitting techniques. In the following sections, this state of affairs will be examined in the light of Box's criticism, i.e. from the point of view of hydrological relevance and the the relevance to the original engineering problem of flood-protection.

2. SCIENCE, HYDROLOGY, STATISTICS AND MATHEMATICS

The real challenge of statistics is to discover, through analysis of empirical data, patterns in quantitative characteristics of specific objects or events encountered in nature. In this sense, statistics is simply a method of scientific inquiry and it has been through such inquiry that most of the statistical techniques (summarily called mathematical statistics) have been developed. As a rule, a statistical technique has been designed as an aid to the solution of a specific scientific problem, e.g. an assessment of the effectiveness of drugs, the control of the quality of a product, etc. This is why many of the best statisticians have repeatedly emphasized that, in order to make a meaningful analysis, the statistician must have a deep insight into the nature of the scientific problem. Several such statements were quoted in Klemeš (1978). To give just one example, it will suffice to turn to Box's paper quoted above where he says that "...statisticians must learn how to be good scientists."

In the context of flood frequency analysis it would mean that the statistician should learn how to be a good hydrologist. This, however, might turn out to be a difficult task. Those from whom he could learn hydrology are not only very few but seldom hold teaching positions. To his bewilderment, he might discover that, for instance, most courses on "Advanced Hydrology" contain little more than amateurish expositions of elementary linear algebra, probability and statistics. For, paradoxically, it has become almost axiomatic that to be a good hydrologist means to learn how to be a mediocre mathematician or statistician. This attitude is the result of a common misconception about the role of mathematics in science, in particular that, rather than being a method for a logically consistent and parsimonious formulation of scientific insights and exploration of consequences of scientific hypotheses, it is a generator of new scientific knowledge about the outside world; that correct mathematical manipulations and proofs lead ipso facto to correct statements about nature. This misconception has been a permanent source of frustration to scientists as well as to mathematicians who over and again keep encountering it among their students as well as among practicing scientists. Thus, for example, Richard Feynman, the physicist, says:

"Let me say also something that people who worry about mathematical proofs and inconsistencies seem not to know. There is no way of showing mathematically that a physical conclusion is wrong or inconsistent. All that can be shown is that the mathematical assumptions are wrong. If we find that certain mathematical assumptions lead to a logically inconsistent description of nature, we change the assumptions, not Nature" (Feynman, 1971, lecture 13, p. 8).

George E.P. Box, the statistician, expressed his frustration with the same phenomenon as follows:

"In such areas as sociology, psychology, education, and even, I sadly say, engineering, investigators who are not themselves statisticians sometimes take mathematistry seriously. Overawed by what they do not understand, they mistakenly distrust their own common sense and adopt inappropriate procedures derived by mathematicians with no scientific experience" (Box, 1976).

3. HYDROLOGY AND FFA FUNDAMENTALS

3.1. Frequency and Return Period

Frequency of floods, or of some of their quantitative characteristics like peak flows, volumes etc., is a legitimate object of scientific analysis. However, *the fundamental problem with FFA as practiced today is that it does not analyze the observed frequencies of floods.* In fact, the more interesting they are, the less attention they get. The only piece of factual information that FFA provides is that it tells us how many times (or in how many years) a given flood peak occurred or was exceeded during the period of record. However, even this information is presented in a form which implies an interpretation that can often be wrong. Firstly, the exclusive use of relative frequencies tends to suppress the importance of the length of the record from which these frequencies were computed and implies a degree of generality which may be unjustified. There is no a priori reason why, for instance, a flood peak that occurred twice in a specific 20-year period should typically occur five times in any 50-year period or, say, about ten times in the 100-year design period starting in 1990. Secondly, this appearance of generality is reinforced by the neglect of the influence of the actual historic dates of the flood record. Typically, a regional FFA may be based on records which not only differ substantially in their lengths but also in their dating although it is common knowledge that different periods in history often had different hydrologic regimes. Thirdly, the introduction of the concept of return period as an average period between two successive exceedances may further contribute to misinformation, the more so because an average return period is a formally correct number even in cases where it conveys a completely wrong impression, i.e. where the phenomenon has no pronounced central tendency which the average is supposed to characterize. Thus, an example given in Klemeš (1987) shows that one can easily be led to classify as a 2 year flood, i.e. as a flood that was exceeded on the average every second year, a flood which in reality was not reached for 17 consecutive years but subsequently was exceeded in all but one of the following 12 years, only because it was exceeded twenty times within a 40-year historic record.

However, this last example notwithstanding, it is in connection with small and frequent floods where FFA is hydrologically most securely grounded and where it can most likely produce realistic results. This follows from the fact that data on small floods are abundant so that there exists solid empirical evidence regarding their actual return periods, both in different geographic regions and different historic periods. This evidence suggests that there indeed are many rivers where occurrences of small floods have a pattern of random events with a fairly stable frequency which can be meaningfully characterized by the average return period. But this is no excuse for forcing into this mold rivers deviating from this pattern or for automatically postulating it as a general pattern valid for cases arbitrarily far beyond the range of empirical evidence, both of which is routinely done in FFA.

It is important to note that, in order to extend the concept of return period beyond the range where the actual periods between the relevant floods are documented in historic records and thus can be averaged by calculating their arithmetic mean, FFA redefines the return period as the reciprocal of relative frequency. While the two definitions are mathematically equivalent, their "truth content" may be vastly different since the latter definition does not indicate from how many actual periods the (average) return period has been computed. Its most "appealing" feature is that it can produce, without even blushing, and "average" return period of a flood peak based on its one occurrence in a given record, that means *when in reality not a single period between this peak and its closest equal or higher neighbour is documented.* This heroic feat is, so to speak, one of the basic pillars on which the "scientific" edifice of FFA rests.

3.2. Randomness and Probability

In FFA it is assumed that the observed floods constitute a random sample from some mathematically defined probability distribution of floods. This assumption may be reasonable as a null hypothesis. However, since it provides the sole basis for the interpretation of results of FFA as well as for their applications in engineering decisions it should be carefully tested and continuously re-examined in the light of available scientific evidence. Instead, if it is tested at all, (as, for instance, has become a standard practice in Environment Canada in recent years), only crude statistical tests are generally used which are rather ineffective for the small samples usually available. The simple truth is that, while the independence hypothesis often cannot be rejected, neither could be hypotheses of many types of dependencies which suggest themselves on hydrological and other geophysical grounds. For example, there is evidence that maximum annual flood peaks often are related to annual runoff totals or mean flows (e.g. Sangal and Kallio, 1977; Panu et al., 1984) which in turn often exhibit more or less complex patterns of serial dependence traceable to various geophysical causes. However, such leads are seldom seriously explored (a rare recent exception to this rule appears in Booy and Morgan, 1985) and FF analysts are not only too happy to accept the independence assumption but they often postulate or tacitly imply it despite strong evidence to the contrary and justify this attitude by (what else!) the need to perform FFA. Only in the better cases are flood samples that do not pass a randomness test excluded from FFA (e.g. Panu et al., 1984). The philosophy of contemporary FFA is simple: if floods are to be considered for the analysis they *must* be random; or at least they *must be declared* random - otherwise they don't qualify. This might look puzzling only to some frivolous scientists like, for instance, Richard Feynman for whom Nature is just an object for academic nit-picking (see the earlier quotation) and who even seems to rejoice whenever something goes wrong - he is, for example, on record saying that "the thing that doesn't fit is the thing that's the most interesting, the part that doesn't go according to what you expected" (Feynman, 1983). Such a perverse attitude can of course be taken only by somebody divorced from reality, somebody who does not have the heavy responsibility of the FF analyst to aid engineering planning and design.

The crucial point is that, in "design and planning concerned with future events whose time or magnitude cannot be forecast, we *must* resort to statements of probability, or frequency..." (Linsley et al., 1975, p. 338; emphasis added). In order to meet this challenge, floods *must be declared* purely random events since otherwise the quantum jump from (relative) frequency to probability *cannot* be made and it does not follow that "Return period and probability are reciprocals" (ibid., p. 340). Since, while "return period" is at least a mathematically valid notion for every recurring event, be it a win in a lottery, a change in the political party in power, or the appearance of Halley's comet, it certainly is not valid to assume that the occurrences of each of these events in any given year have constant probabilities equal to the reciprocals of their respective return periods. In the case of floods, it simply has been found *convenient* to resort to statements of probability and to calculate flood risk by simple multiplication of elementary probabilities; given this concensus, we then must declare them random events regardless of the actual status of their frequencies in the range between a lottery and a Halley's comet. While this may be quite a legitimate attitude in the context of decision making (which will be discussed later), it certainly cannot be regarded as science. We simply are facing here the old problem identified by Yevjevich (1968) in the context of maximum probable precipitation (and flood), in particular a confusion between "engineering expediency" and "scientific truth" which also is at the root of the continuing controversy about the U.S. Water Resources Council's (1977) endorsement of the log-Pearson III distribution as a basis of FFA in the United States.

3.3. Statistics Used as an Instrument to Justify a Preconceived Idea

Because of the practical importance of adequate flood protection measures, it would be desirable to obtain accurate estimates of the actual probabilities of flood occurrences. Since FFA postulates that floods are random events, it is not surprising that this task is redefined as a necessity "to obtain as accurate an estimate of p as possible" (Linsley et al., 1975, p. 350), where p is the probability of exceedance of a random event whose relative frequency is the same as that of a given flood in a given historic record. This end is pursued with the aid of mathematical statistics which is employed for the justification of the assumptions of randomness, homogeneity, stationarity, goodness of fit and other properties necessary for the accuracy of the above probability p. The justification rests on what may be called "numerical default" of statistical tests which relies heavily on the smallness of available flood samples. Rather than being a handicap, a small sample size has become a godsend which virtually guarantees non-rejectability of any hypothesis. Thus the first half of the success consists in "testing" the right hypothesis. The second half is assured by disregarding the main original purpose of a statistical test which is to arrive at hints of the presence or absence of relationships where these are not known a priori and where the only clue is their possible effect on the numbers.

Statistical test are meant for search in the dark, so to speak, and are pointless in bright daylight. For instance, one hardly needs a statistical test based on the distribution of the sizes of dogs to test a null hypothesis that a large animal is a dog when it is obvious that it actually is a horse. However, the approach of the FF analyst is first to close his eyes and then to say that, for the purpose of Dog Frequency Analysis, the animal can be considered a dog since this hypothesis cannot be rejected at the 5% significance level. On the other hand, if a St. Bernard was a puppy when the dog size distribution model was fitted, the FF analyst would likely exclude him as an outlier when he grew up even if he had a first class pedigree. Because it obviously is counterproductive to look and see if decisions made with closed eyes are both more convenient and can be better justified by rigorous mathematics. Once the status of a random sample from a simply distributed homogeneous population is scientifically "proven" in this fashion, great care is taken to adhere to the rigour of mathematical statistics in order that the estimates of p be obtained with the greatest accuracy, so that the "largest amount of information" (presumably on the size of big dogs or floods, as the case may be) is extracted from the data.

The remarkable feature of FFA is not that it deals with the shadowy world of "*as if* representations of reality" as Willeke (1979) put it in his stimulating paper but that so many otherwise rational people have been taking its mathematical games so seriously for so long, given that they could well have turned their attention to the study of the real features of flood frequencies (if they had a scientific interest in them), or to real engineering problems related to flood protection, or simply to theoretical work in mathematical statistics. What is even more striking is the apparent seriousness with which are regarded the most questionable results of FFA, namely those related to floods with relative frequencies lower than 1/100 and based entirely on wild extrapolations of concepts which are often difficult to take seriously even when only interpolation is involved.

4. EXTRAPOLATIONS OF FLOOD FREQUENCY MODELS

4.1. The General Problem

"With respect to flood frequency analysis, it's all in the tails, so to speak" (Greis, 1983). Extrapolation is the very raison d'être of flood frequency analysis and the assumption that historic floods are a random sample from a simple distribution is the only basis on which the extrapolated "probabilities" of extreme floods can be justified in principle. If this assumption is granted everything else is in order and the whole FFA is just one perfectly legitimate subarea of the small sample theory. And it is fair to say that several valuable advancements of the small sample theory have been stimulated by, and made within the framework of, FFA. In this context, computer simulation has proved extremely useful because it makes possible to test the theoretical predictions in ideal situations that strictly conform to the underlying assumptions. For the study of frequencies of actual floods, simulation results also have been extremely useful, though in a negative sense. They have shown that the estimates of parameters, inferences about population distribution types and about the behaviour of the extremes, etc., face fundamental limitations when attempted on the basis of small samples, even under the ideal conditions of simulation experiments which are not, and never will be, attainable in real flood samples. It is therefore important to give a serious thought to the additional obstacles that may be "superimposed" on simulation experiments by Nature and to re-examine the practices of FFA, notably its cavalier attitude to extrapolation, in this light.

4.2. The Physical Problem

Mathematical models of the behaviour of natural phenomena must fit the empirical data. Once fitted, a mathematical model can easily be extrapolated beyond the range of observations which it was designed to fit. However, there is no a priori reason why its correspondence with the manifestations of the physical prototype should hold beyond the range of observations. Science has documented many examples where an accurate fit in the observed range soon becomes inaccurate beyond it and eventually may become completely wrong; often there even occurs a sudden departure point beyond which the lack of the correspondence is not merely a matter of inaccuracy which can be removed when data in that range become available, but where the fit of the model cannot be repaired because its structure cannot accommodate the manifestations of the phenomenon in that region. Examples of this pattern are so many that they suggest a general hypothesis that *no model developed for a small range of a phenomenon will be valid for its entire range.* This is just another way of saying that different physical forces dominate a given phenomenon in different ranges of its states. To name a few specific cases, one may mention the eventual sudden collapse of Hooke's Law, the transition of laminar flow into turbulent flow, the sudden change of state of most substances with a gradual change in temperature, or the inapplicability of simple vector algebra for dealing with very high velocities.

For this reason alone it seems unlikely that a distribution model that fits the relative frequencies of a few, and mostly small, floods within a 20- to 50-year period, can be used for a meaningful prediction of a thousand- or a million-year flood which are now coming increasingly into fashion.

Another point relevant to such extreme floods is how far it is physically justified to extrapolate any distribution model with an infinite tail. One can sometimes hear that there is no limit since there always could be, say, a 1 mm more rain than there is in any conceivable rainstorm. This is a fallacious argument following from the inherent inability of the human mind to stop extrapolating which, since antiquity, has led to various logical paradoxes, e.g. the paradox of Achilles' inability to evertake the tortoise. This limitation of human mental faculties may easily lead a hypothetical Dog Frequency analyst to calculate the "probability" of a horse-size and even elephant-size dog; and the FF analyst to (scientifically!) estimate the peak flow of a million-year flood without ever thinking of a possibility that during such a deluge the whole area might be submerged under a stagnant pool of water with the flow being close to zero.

4.3. The Hydrological Problem

The main problem is that, in the climate of contemporary hydrology, it has become virtually impossible to discuss the hydrological problems associated with extreme floods. The several decades of FFA indoctrination has blurred most hydrologists' minds to the point that the only images that the notion of "extreme flood" can conjure up are those of distribution tails, quantile estimates, confidence limits, asymptotic distributions, extreme value distributions, outliers, etc. In short, hydrologists now often are able to think about floods only as about numbers to be subjected to statistical computations.

Flood frequency theorists often cannot comprehend that the difference between the occurrence of small and large floods is more fundamental than the difference between drawing a card marked with a small number and with a large one. Unable to see beyond a sample of random numbers each of which contains the same amount of information about the parent distribution, they cannot appreciate that the smallest floods from a given historic period may contain little or no information relevant to the likelihood of a very large flood, that the processes governing the formation of small and large floods may be vastly different. For example, one can often see that the whole shape of a distribution model fitted to a flood peak sample is dominated by the two or three smallest numbers which exert a commanding influence on the upper tail of the distribution and hence on the "probability" of the highest floods - in fact, the bigger the flood and the more hydrologically remote from the nature of the smallest one, the more its "probability" is influenced by the latter. An example given in Klemeš (1986) shows that, in a 25-year sample, almost negligible changes in the three smallest annual peak flows (which can hardly be called floods at all) can easily change the return period of a large flood from 100 to 10 000 years. Taking such an effect seriously - as is the rule in FFA - implies that the analyst believes that, say, a 2 cm difference in snow cover or a few hours difference in the duration of a mid-winter thaw in a small basin in Ontario has a controlling influence on the likelihood of a hurricane originating in the eastern Carribean and finding its way into the Toronto area. It is not argued here that such a connection is impossible in principle but that its implication in the result cited above would normally not even occur to the FF analyst, not to say that it should disturb him or lead him to investigate its substance.

There are other hydrological problems with the extrapolation of flood frequency data by models fitted to a few flood peaks from the lower part of the spectrum. For example, there are cases where, above a certain threshold not reached during the historic period, the flood generating area may suddenly increase and the likelihood of a very large flood may be much higher than the extrapolation of a model fitted to small floods would indicate.

One case in point is the Santa Anna River basin in California where floods from about a third of the basin (controlled by Lake Elsinor) normally do not reach the Santa Anna River but could do so in extreme conditions. Another similar case is documented in Sumas River in the western border area between Canada and the USA where in extreme conditions large overflow into Sumas River may occur from Nooksack River in the US. Such an overflow did not occur during the past 30 years during which records have been kept in Canada. Based on these records, the 100- and 500-year floods can be extrapolated as 52 m^3/s and 64 m^3/s, respectively. However, the U.S. authorities, considering this overflow, quote Q_{100} as about 270 m^3/s and Q_{500} as 540 m^3/s. Several such overflows occurred in the first half of the century and the largest that was actually measured (in 1932) was 127 m^3/s. A complementary effect takes place in the overflowing river where the peaks of extreme floods are correspondingly reduced. Diversion of flow from one basin to another during extremely high flows is quite a likely phenomenon in rivers with flat headwaters controlled by lakes as is often the case in Canada, e.g. the Ottawa and Gatineau Rivers in Quebec, Indian Brook in Newfoundland, etc.

Other aspects that may invalidate the upper tails of flood distribution models include inundation of large detention areas normally out of reach of flooding, floods caused by a collapse of morainic dams as is often the case in alpine streams (e.g. the 1983 flood on Nostetuko and Homathko Rivers in British Columbia), backwater effects of a tributary or a recipient, changes in the natural detention capacity of river valleys by geological processes, climatic anomalies caused by volcanic eruptions, processes in the ocean, anthropogenic climatic and land use changes, etc. Frequencies of extreme floods provide a typical example of *a common situation where the key to a solution of a hydrologic problem lies outside hydrology*. The excellent work of Victor R. Baker and his collaborators (Baker, 1982, and references given herein) demonstrates how, for example, relatively simple geological insights can often be more illuminating in regard of extreme floods than volumes on various maximum likelihood, parametric and nonparametric, and other parahydrologic FF analyses.

In summary, from a hydrological point of view, very extreme floods and their causes tend to be outliers by definition, i.e. very little, if any, information about their likelihood is contained in the frequencies of relatively small floods of which the bulk of a typical flood sample is composed. Extrapolating distribution models fitted to these samples is tantamount to extrapolating the small flood dynamics beyond the range where it can physically function. For example, a 10 000-year flood extrapolated for a small Ontario basin on the basis of, say, a 30-year record of its ordinary snowmelt floods may be a nonsense because, whatever the amount of snow in the basin, there may never be enough energy available during a single day to melt snow at the rate implied in the corresponding flood flow. On the other hand, a summer storm or a hurricane may well cause such a flood quite easily and many times within a single century.

Thus the hydrological problem of a proper FFA is to predict the as yet unobserved outliers by analyzing the plausible dynamics of extreme-flood formation, rather than to worry about exclusion or inclusion of numerical outliers in empirical samples and about their effects on the fitted distribution tails. For it is quite conceivable that for an occurrence of an extreme flood, the tail of Halley's comet may be more important than the tails of all the standard distributions combined.

This may seem like an advocacy of the often ridiculed and presumably long discredited "probable maximum flood". This impression is partly correct, but only as far as the motivation behind the concept is concerned. The procedures for its implementation have many weaknesses and the name of the product is rather unfortunate though its defamation has in part been the result of a snobbery which, impressed by the presumed ability of FFA to quantify the probabilities of real large floods, fails to acknowledge the obvious fact that the term "probable" can legitimately be used in a nontechnical sense and that it may be more honest to do so than to hide arbitrary and often meaningless numbers behind its technical quantitative facade. However, the whole scholastic argument about the "probable" and the "maximum" could easily be avoided if labels like "unusually large flood", or "extreme flood", or "conceivable catastrophic flood" were used. For reasons which will be discussed in the next section, it may often be preferable to give just an idea of the magnitude of a flood that, based on current knowledge of the physical phenomenon involved, could conceivably occur on rare occasions than to base important decisions on seemingly exact values of *probabilities* which *probably* have little or no meaning.

5. BACK TO BASICS

5.1. What the Engineer Really Needs

The basic thing that the engineer needs in connection with flood-related planning and design is protection - not so much from floods but from accusations that his flood protection measures were (1) underdesigned (in cases where flood damage was not completely avoided), or (2) overdesigned (in cases where it was). While this dilemma is not unique to flood-related engineering, it is there where it is exposed in perhaps the most dramatic manner. As a way out of this no-win situation, engineers have always been striving to develop "objective" approaches to design. Their common underlying feature is the replacement of an individual judgement with a collective one. This can be achieved by a variety of means ranging from an informal consensus of a professional organization often presented in the form of guides or handbooks containing recommended procedures, to more formal arrangements like standardization, regulations or even legislation.

The formation of a collective opinion at any level can be achieved only on the basis of some generally accepted principles. Among these, "maximization of public good" is probably the most common one but it is difficult to implement. For practical purposes, the latter has often been divided into a tangible and an intangible component in the hope that at least the first one can be objectively assessed. This has led to cost-benefit analysis, gradually generalized into "economic optimization", which has become perhaps the most important instrument of the planning and design methodology. Its universal appeal derives from the assumption that its result can attain a status of "scientific truth" free of subjective bias. This assumption is reasonable only to the extent to which economic optimization is based on relevant scientific facts and employs scientific methods. In theory, this caveat should naturally lead to a strong commitment to scientific research through which scientific facts can be obtained. In practice, pressures to perform economic optimizations and to maximize the credibility of their results often lead to pseudoscience, typically to a substitution of irrelevant scientific facts for the relevant ones which are difficult to handle or are inadequately understood, or to diverting attention from inadequacies in the substance to methodological subtleties. Flood Frequency Analysis is a perfect example of a result of such pressures.

However, the drift into irrelevance could be easily stopped if engineers did not let themselves be intimidated by pseudoscientific snobbery but exposed it to a scrutiny of sound engineering judgement, of the common sense to which Box appeals in the earlier quotation. It would then soon become obvious that the scientific content of economic optimization is rather modest. This is so even in quite simple and straightforward cases and becomes an illusion in complex situations, especially those involving modification of the environment as is the rule with civil-engineering water-related projects. Here, economic optimization is merely a conceptually appealing rationale for arriving at an acceptable collective opinion. Its appeal comes from its formal resemblance of the straightforward mathematical problem of finding an extremum of the sum of two monotonous functions, one of which is decreasing and the other increasing. The other element of its appeal is the fact that we have been conditioned to homogenizing various non-commensurate utilities by expressing them in monetary equivalents. Thus what we treat as "scientific facts" are often entities that have been made facts by definition or concensus. For example, we can freely mix apples and oranges when we define both as commodities whose only attributes are their unit prices. Similarly, it is we who define what should be included into the "cost" and into the "benefit" brackets and how far their effects should be traced; we treat as objective facts the monetary equivalents while we know that they are a result of many subjective and arbitrary manipulations like taxes, tariffs, interest rates, market fluctuations, speculations, government subsidies, agreements, etc. etc.

In view of this, it is far more desirable to acknowledge explicitly that economic optimization is just an agreed upon rationale and try to standardize it on the basis of simple procedures than to keep pretending that it is a science and try to perfect it by refining the methodology for manipulating its "scientific facts". This would relieve the pressure for the cultivation of the supporting pseudo-sciences such as Flood Frequency Analysis which then itself could be recognized for what it is, namely a rationale useful for performing a cost-benefit analysis, and standardize it within a simple framework. "Bulletin 17" (U.S. WRC, 1977) has been a bold step in this direction and should be viewed in this context. It is unfortunate that the hydrologic community has let itself be tricked into futile arguments about its "scientific" merits and demerits which is a pseudoproblem par excellence.

5.2. Flood Frequency Analysis as a Rationale

"Until this problem of estimating the magnitude of floods with return period of at least 1,000 years [is solved], no satisfactory quantitative hydro-economic decisions based on balancing costs and benefits can be made" (Alexander, 1973). This is the kind of reasoning that has been responsible for pushing FFA into the realm of pseudoscience. The fallacy implied in the above statement is that a "satisfactory" decision is taken to mean an "objective", or "scientifically based" decision. The crucial point is that a satisfactory decision does not necessarily have to be based on scientific knowledge. In many cases such knowledge is simply not available and decisions have to be made regardless. They are judged satisfactory or otherwise according to the degree to which they satisfy, to use a fashionable term, the currently accepted paradigm, not according to their scientific content.

Paradigms may, and necessarily do, have a strong component of myth - in modern times a significant part of this myth component has been a belief that science can alleviate the burden of responsibility in decision making, that "scientific results" may in a way replace decisions. This flight from responsibility and from courage to take a stand then pushes into a "scientific" framework things that have no scientific content thus giving rise to various pseudosciences

(Berlinski, 1976). Facing the ever growing complexities of life and being more and more aware of the multitudes of risks and potentially disastrous consequences of wrong decisions, it is becoming increasingly more "desirable" to be able to wash one's hands and hide behind science and computers. In other words, the subjective element of decisions is pushed one or more steps back, and out of reach of public scrutiny (Bazelon, 1979).

This trend is fundamentally very unhealthy and dangerous and, although in flood-related decisions it may be less dangerous than elsewhere, there is no need for its continuing cultivation. Given the fact that neither the economical nor the hydrological component of flood-related economic analyses have a solid scientific basis, it would be more honest, convenient and practical to anchor the procedures to standards decided upon within the profession than to continue trying to endow the transparent "as if" game with more respectability via more mathematistry. The weaknesses on the economical side have been abundantly discussed by specialists (e.g. Loentief, 1982; Eichner, 1985) and some of the hydrological weaknesses were pointed out earlier in this paper. Some practical reasons for stopping to pretend that FFA is a science are discussed below.

Science is a continuous search for "...order which gives unity to what had long seem unlike... For order does not display itself of itself... [but] must be discovered and, in a deep sense, it must be created. What we see, as we see it, is mere disorder" (Bronowski, 1972). FFA is obviously not in this category. It does not search for patterns in flood occurrences but is only too happy that its basic assumption - that there is no element of pattern involved - "cannot be rejected". *FFA does not use predictions to test the validity of its hypotheses but uses hypotheses to make untestable predictions.* Its only scientific content is the empirical frequencies of floods in a given period of record. Everything beyond and above, including the interpretation of the reciprocal of these frequencies as return periods, equating them with probabilities, representing the latter by various distribution models, etc. is just a superstructure of hypotheses. These hypotheses are not inspired by the study of the process involved but dictated by a need of the decision maker to quantify flood risk in a way acilitating a unique result of cost-benefit analysis of flood related projects.

Thus the driving force behind FFA is not a scientific discovery but a necessity to make a decision. This necessity is an objective reality imposed by society. Given that, the need for some kind of rationalization of the decision is undisputable and quite independent of whether and to what extent the rationalization can be based on scientific knowledge of the problem. Where facts and scientific knowledge are not available the rationalization must be based on other foundations, for example on assumptions, on recourse to a higher authority, tradition, custom, etc. This is perfectly legitimate and beyond challenge. What is not legitimate is to pretend that the latter entities are elements of scientific knowledge, or that they somehow will be transformed into them by being forced into a scientific format either by decree or by subjecting them to treatment by scientific methods. Although this has been pointed out on many occasions (e.g. Hall, 1971; Feynman, 1983), it is still common practice because of the high prestige which the "scientific status" enjoys in the contemporary paradigm.

Once we accept that the present FFA is merely a rationale which accommodates the empirical evidence that large floods occur less often than small floods, and formalizes it in a framework convenient to cost-benefit analysis of design, several things become obvious, in particular: (1) That the whole edifice of models and techniques based on the assumption that floods are random samples from simple distributions is only an "as-if" representation of reality (Willeke, 1979); (2) That there is no theoretical reason why some distribution models should fit the data from our small samples better than others and that the reasons why certain models have been used are purely practical (Moran, 1959); (3) That "the part of the distribution we are interested in is well away from the part where the observations provide some information about the shape. It is therefore easy to construct two different distributions both of which fit the observations closely but for which the tails are of quite different shape" (Moran, 1957, 1959); because this problem arises from the shortness of our records it "cannot be overcome by mathematical sleight of hand" (Moran, 1957). This was well demonstrated by Wallis and Matalas (1973) who showed by a simulation experiment that, for a small sample, surprisingly often the "best fit model" is different than the parent distribution. Not surprisingly, this reference is seldom cited in FFA literature; (4) That, therefore, "... in so far as extrapolation made beyond the range of the recorded floods is concerned, the most elaborate analytical procedure of curve fitting gives results that are no more reliable than those obtained by a simple extension by eye of the flood frequency curve on any kind of probability paper" (Thomas, 1948); (5) That, if we have to extrapolate, then we should pay much more attention to the largest recorded floods than to the small ones since they presumably better reflect the type of conditions influencing the tail behaviour which we want to estimate (Harold Thomas' rationale underlying the Wakeby distribution; also see Reich and Renard, 1981); and finally (6) That, in the light of the above points, the "mathematical prestidigitation" (Moran, 1959) which has become the main preoccupation of contemporary FFA is a preposterous scholasticism that has no value for hydrology or engineering.

This conclusion is further supported by the following:
(a) Flow measurements under flood conditions are very difficult and unreliable so that the upper sections of flow rating curves are subject to large errors. Moreover, flood peak flows are almost never measured but extrapolated from rating curves or by indirect methods. Measured flows are seldom higher than 50% of peak flows. Cases where the maximum measured flow is under 20% of estimated peak flows are not uncommon. For example, on Yakoun River at Port Clements in British Columbia, the annual maxima in the four years 1962-1965 ranged from 396 to 612 m^3/s while the maximum measured flow during this period was 72 m^3/s (for examples from Newfoundland see Panu et al., 1984). (b) Flood peaks are often estimated from single daily staff gauge readings. A

recent study done for coastal streams in British Columbia indicates that errors due to this single cause can be as high as 90%; in 10% of all the cases analyzed it was greater than 50%. (c) However, in FFA, peak flow data are treated as exact deterministic numbers and the only errors recognized are those between these "exact" numbers and their model estimates. These then serve as a measure of "confidence" in the estimates of the various T-year floods which is given in "exact" probabilistic terms. *Few FF theorists seem to question the wisdom of striving for a high confidence in estimating numbers which can claim only a very low confidence level themselves.* (d) From the point of view of the final product, i.e. the result of the cost-benefit analysis, it often makes little difference what distribution model is fitted to the data (Slack et al., 1975). This is the consequence of three things; one, that the influence of the body of the distribution far outstrips that of its tail - an observation made by Moran (1959) in connection with storage design; two, that this body can be equally well fitted with many different models (see item 3 above); and three, that all the fitted models are made to "preserve" the basic parameters of the sample, especially the mean and the variance, which have a dominant role in the standard economic analyses (Klemeš, 1977). (e) It is generally the case that cost-benefit analysis of risk plays an important role only in projects exposed to damage from relatively small and frequent floods. Whenever damages of catastrophic proportions are envisaged - typically involving danger to human life, major disruptions in the functioning of society's infrastructure, danger to strategic defence installations or to disposal sites of highly toxic or radioactive wastes, etc. - flood protection levels are invariably set by policy decisions which are outside the realm of an "objective" cost-benefit analysis. Here, the notions as 100-year, 1000-year or million-year flood have mostly qualitative and public-relations significance and could well be replaced by terms like dangerous flood, extremely dangerous flood, flood of catastrophic proportions, etc. The quantification in terms of the return period can often be even misleading and convey a greater feeling of safety than it actually implies; for example, the flood stage of a "1000-year" flood may often be only a few centimetres higher than that of a "100-year" flood although in the "public-relations" perception a "1000-year" protection seems by an order of magnitude higher than a "100-year" protection. Hence, there is little justification for "mathematical prestidigitation" where the decision is determined largely by the emotional impact of large rounded numbers of the decimal system in very much the same way as in the case of setting the values of lottery prices where a "million-dollar prize" always carries the same fascination which seems to be almost unaffected by inflation.

As a result it can be safely concluded that the chief and only claim in favour of mathematical sophistication in FFA, in particular, of the importance of "exactness" and "accuracy" in engineering decision making, is false. Engineers don't need the "science" of FFA but a simple agreed upon framework in which the rationale of the "as-if" representation of reality can be implemented. Perfection of curve fitting procedures, of estimation and inference problems in small sample theory may be an interesting and legitimate pursuit by themselves - but they are neither *engineering* nor *hydrology* and there is no valid reason why they should continue to drain the much needed resources and talent from these areas.

6. ACKNOWLEDGEMENTS

Many practicing hydrologists provided materials illustrating the points made in this paper, and helped me to sharpen the focus through discussions of the problems raised herein. Special thanks are due to the following individuals: E.G. Smith, and G. Tofte (Environment Canada, Pacific and Yukon Region, Vancouver, B.C.), P.J. Pilon, R. Condie, D.A. Smith and B.P. Sangal (Environment Canada, Ottawa, Ontario), R.R. Wyman, R.Y. McNeil, W. Obedkoff and C.H. Coulson (Ministry of Environment, Victoria, B.C.), M. Miles (Miles & Associates, Ltd., Victoria, B.C.), J.Y. Ding (Ministry of Natural Resources, Toronto, Ontario), J.R. Beck (U.S. Geological Survey, Sacramento, California), and H.E. Kubik (Hydrologic Engineering Center, U.S. Army Corps of Engineers, Davis, California). This does not necessarily imply their endorsement of the opinions expressed herein which are my own.

7. REFERENCES

Alexander, G.N. 1973. 'Estimation of the 10,000-Year Flood.' In: *Proceedings, 11th Congress on Large Dams*, Q.41, R.70, 1327-1350, International Commission on Large Dams, Madrid.

Baker, Victor R. 1982. 'Geology, Determinism, and Risk Assessment.' In: *Scientific Basis of Water Resource Management*, 109-117, National Academy Press, Washington, D.C.

Bazelon, D.L. 1979. 'Risk and Responsibility'. *Science*, **205**, 277-280.

Berlinski, D. 1976. *On Systems Analysis*. MIT Press, Cambridge, Mass.

Booy, C., and Morgan, D.R. 1985. 'The Effect of Clustering of Flood Peaks on a Flood Risk Analysis for the Red River.' Canadian Journal of Civil Engineering, 12, 150-165.

Box, George, E.P. 1976. 'Science and Statistics.' *Journal of the American Statistical Association*, 71(356), 791-799.

Bronowski, J. 1972. *Science and Human Values*. Harper and Row, New York.

Eichner, A.S. 1985. 'The Lack of Progress in Economics.' *Nature*, 313, 427-428.

Feynman, R.P. 1971. *Lectures on Gravitation*. California Institute of Technology, Pasadena, California.

Feynman, R.P. 1983. 'The Pleasure of Finding Things Out.' NOVA Program No. 1002, WGBH Transcripts, Boston.

Greis, Noel P. 1983. 'Flood Frequency Analysis: A Review of 1979-1982.' *Reviews of Geophysics and Space Analysis*, 21(3), 699-706.

Hall, Warren A. 1971. 'Biological Hydrologic Systems.' In: *Proceedings, The Third International Seminar for Hydrology Professors, Biological Effects in the Hydrological Cycle*, 1-7,

Purdue University, West Lafayette, Indiana.
Klemeš, V. 1977. 'Value of Information in Reservoir Optimization.' *Water Resources Research*, 13(5), 837-850.
Klemeš, V. 1978. 'Physically Based Stochastic Hydrologic Analysis.' In: V.T. Chow (Editor), *Advances in Hydroscience*, 11, 285-355, Academic Press, New York.
Klemeš, V. 1986. 'Dilettantism in Hydrology - Transition or Destiny?' *Water Resources Research*, 22(9 Supplement), 177S-188S
Klemeš, V. 1987. 'Empirical and Causal Models in Hydrologic Reliability Analysis.' In: L. Duckstein and E. Plate (Editors), *Engineering Reliability and Risk in Water Resources*. Proceedings, NATO Advanced Study Institute, The University of Arizona, Tucson, May 19-June 2, 1985.
Leontief, W. 1982. 'Academic Economics.' *Science*, **217**, 104 + 107.
Linsley, Ray K., Kohler, Max A., and Paulhus, Joseph L.H. 1975. Hydrology for Engineers, McGraw-Hill, New York.
Moran, P.A.P. 1957. 'The Statistical Treatment of Flood Flows.' *Transactions of American Geophysical Union*, **38**, 519-523.
Moran, P.A.P. 1959. *The Theory of Storage*. Methuen, London.
Panu, U.S., Smith, D.A. and Ambler, D.C. 1984. *Regional Flood Frequency Analysis for the Island of Newfoundland*. Department of Environment, St. John's, Newfoundland; and Environment Canada, Atlantic Region, Dartmouth, Nova Scotia.
Reich, Brian M. and Renard, Kenneth G. 1981. 'Application of Advances in Flood Frequency Analysis.' *Water Resources Bulletin*, 17(1), 69-74.
Sangal, B.P. and Kallio, R.W. 1977. *Magnitude and Frequency of Floods in Southern Ontario*. Inland Waters Directorate, Environment Canada Ottawa.
Scheidegger, Adrian, E. 1975.*Physical Aspects of Natural Catastrophes*. Elsevier, Amsterdam.
Slack, J.R., Wallis, J.R. and Matalas, N.C. 1975. 'On the Value of Information to Flood Frequency Analysis.' *Water Resources Research*, 11(5), 629-647.
Thomas, Harold A., Jr. 1948. 'Frequency of Minor Floods.' *Journal of the Boston Society of Civil Engineers*, **XXXV**(4), 425-442.
United States Water Resources Council, Hydrology Committee 1977. *Guidelines for Determining Flood Frequency Information*, Bulletin 17A, Washington, D.C.
Wallis, J.R. and Matalas, N.C. 1973. 'Relative Importance of Decision Variables in Flood Frequency Analysis.' In: *Design of Water Resource Projects with Inadequate Data*. UNESCO, Studies and Reports in Hydrology, No. 16, 435-442.
Willeke, Gene E. 1979. 'Myths and Uses of Hydrometeorology in Forecasting.' In: *Improved Hydrological Forecasting - Why and How*. Proceedings of the Engineering Foundation Conference, 117-124. American Society of Civil Engineers, New York.
Yevjevich, V. 1968. 'Misconceptions in Hydrology and their Consequences.' *Water Resources Research*, 4(4), 225-232.

STATISTICAL FLOOD FREQUENCY ANALYSIS – AN OVERVIEW

Arun Kumar
Professor of Civil Engg.
Delhi College of
Engineering, Delhi, India

Subhash Chander
Professor of Civil Engg.
Indian Institute of
Technology, New Delhi, India

ABSTRACT. The paper critically examines various issues involved in statistical frequency analysis. Emphasis has been laid on areas where further research is required. It is highlighted that the Water Resources Council modified report of 1981 has largely reiterated their earlier findings thereby ignoring criticism labelled by various researchers on their recommendations. In the light of recent studies, it is advocated that WRC, U.S.A recommendations need reevaluation.

INTRODUCTION

Statistical flood frequency analysis has probably been one of the most active areas of hydrological research for the last thirty or more years. Since fifties almost every issue of journals related to Water Resources research contains papers on this topic. With such a voluminous research material available, one would like to ask whether significant progress has been made in understanding or in arriving at more acceptable methodologies for frequency analysis. The answer to this question is rather difficult. The questions such as (i) which parent distribution the data may follow? (ii) how to account for sampling error while identifying the distribution (iii) what should be the most suitable parameter estimation technique and many other similar questions remain unresolved. It looks that as long as the flood frequency analysis remain within the statistical domain, the problems of this nature will remain persisting. Being aware that statistical flood frequency analysis has inherent shortcoming of not explaining in any rigorous manner the flood phenomenon itself, researchers have tried many approaches other than statistical. For example, the use of geomorphologic unit hydrograph (GUH) in regional flood frequency analysis (Rodriguez-Iturbe and Valdes, 1979, Diaz-

V. P. Singh (ed.), Hydrologic Frequency Modeling, 19–35.

Granados,1984, Bras et al,1985). Hebson and Wood (1982) suggested a methodology of deriving the flood frequency distribution model by assuming certain climatic condition in a GUH model. Gerard Karpuk (1979) derived a method of determining the probability of flood using various historical records obtained from photographs etc. Probably, the use of geologic information through the ring analysis or through dating of land forms and through dating of organics in deposits of alluvium can be incorporated in flood frequency estimation (Costa,1978),Kochel and Baker (1982) have made similar attempts of including geologic information in flood analysis. Notwithstanding the bold attempts of attempting the flood frequency problem as a basic one of understanding the underlying geomorphology or geology of the region (not merely a statistical problem), the solution still remain largely statistical. This is because in evolving a model other than statistical require many assumptions which are as subjective and 'adhoc' as the statistical procedures. In fact, it seems that statistical flood frequency analysis remains the most important means of quantifying flood in a systematic manner. It is with this viewpoint that an overview of statistical frequency analysis gains importance.

There are essentially two different approaches to the problem of flood frequency analysis. One corresponds to the streamflow annual flood series (AFS) and the other to the streamflow partial duration series (PDS). Maximum amount of work has been done on AFS and is the more popular approach of analysing floods, we are restricting this review only on AFS.

ANNUAL FLOOD SERIES

A major component in the flood frequency analysis is the identification of the parent distribution to which the data belongs. It would have been 'ideal' if one could identify the distribution on the basis of physical laws governing floods, as then it could have been possible to relate the parameters of the distribution with the physical process, thereby enlarging the scope of frequency analysis. In that case, for example the effect of nonstationarity and man made modification in the physical process on frequency analysis can easily be studied. Unfortunately, this has not been possible though efforts are being made through GUH. The solution of identifying the 'parent' distribution still remain 'empirical' which is not based on any law other than the principle of the "best fit" to the data. In this regard, U.S. Water Resources Council (1976,1981 revised) issued guidelines recommending Log Pearson type III distribution (LP III) for defining AFS. Following the same guidelines, LP III distribution was also recommended by

Australian Institute of Engineers (AIE,1977) for Australian rivers. The flood studies report (FSR) of NERC,U.K. (1976) recommended the use of general extreme value (GEV) distribution given by Jenkinson (1969). The objective of recommending a specific distribution was aimed to bring uniformity in the analysis and in that respect it has largely been successful. Nevertheless, the preference for LP III distribution triggered lot of criticism on its shortcomings. In fact, after the WRC report, substantial amount of research has been on highlighting the shortcomings of the recommended practices. It is but interesting to observe that though WRC (U.S.A.) modified their report in 1981, none of the research findings between the intervening period of 1967-1981 with the probable exception of Tasker (1978) on the generalised skew coefficient has been incorporated. In fact, it will not be an exaggeration to state that WRC has largely ignored the criticism of LP III distribution made by various researchers and have largely stuck to their earlier studies of 1967. In a recent paper, Wallis and Wood (1985) have strongly advocated for the revaluation of the WRC recommendations in the light of recent studies, failing which, as they forcefully put it, may amount to 'professional negligence'.

PROBLEMS IN LP III DISTRIBUTION

The choice of LP III distribution is primarily based on its ability of 'mimic' the data better than the other distributions studied (log normal, Gumbel, Pearson type III). The LP III which fits the data better is more heavily tailed than the more popular Gumbel distribution.

The Gumbel distribution has a certain theoretical appeal for fitting the flood data, however, it is not obvious, whether the number of flood producing event within a year is sufficiently large for the distribution of annual flood maxima to attain that of Gumbel distribution. It is generally recognised that the Gumbel distribution underestimates flood (Shen et al,1980). From the historical records, it appears that the flood sequences have skews greater than 1.14(the Gumbel Skew), highly kurtotic (larger than comparable GEV distribution) and have coefficient of variation in the range 0.3 - 0.8 (Landwehr et al,1978). This observation prompted study of a more heavily tailed LP III distribution.

There is not much controversy on the 'choice' of the LP III distribution but on the difficulties one experiences in estimating its parameters. One of the important 'statistics' on which the performance of this distribution depends is the estimation of skewness. Since the available records at a particular site is not long enough to give a good estimate of skewness, it was recommended that a weighted combination of regional and sample skew values should be used in the analysis (WRC,1967). The weighted skew was determined as:

$$\gamma_w = w\gamma_s + (1 - w)\gamma_r \tag{1}$$

where, γ_s, γ_r, are the site and regional skew, γ_w is the weighted skew and w is the weighting factor. The weighting factor w is determined empirically based on the N numbers of years of sample record.

$$w = \begin{cases} 0 \text{ if } N < 25 \\ (N - 25)/75 \text{ if } 25 < N < 100 \\ 1 \text{ if } N < 100 \end{cases} \tag{2}$$

Later WRC (1981) modified the procedure on the lines suggested by Tasker (1978) and McCuen (1978) based on the linear combination of regional and station skew as:

$$w = \frac{MSE\gamma_s \;*\; (\gamma_r) + MSE\gamma_r \;*\gamma_s}{MSE\gamma_s + MSE\gamma_r} \tag{3}$$

where $MSE\gamma_s$ and $MSE\gamma_r$ are the mean square error of station and regional skew respectively. $MSE\gamma_r$ is taken as 0.302 and $MSE\gamma_s$ for LP III distribution is adopted from Monte Carlo experiments (Wallis,1974) and is given as:

$$MSE_s = 10\left[\exp\ A - B\ \{\log_{10} N/10\}\right]$$

where

$$A = \begin{cases} -3.33 + 0.08\ |\gamma_s| \text{ if } |\gamma_s| > 0.90 \\ 0.52 + 0.30\ |\gamma_s| \quad |\gamma_s| < 0.90 \end{cases}$$

$$B = \begin{cases} 0.94 - 0.26\ |\gamma_s| \text{ if } |\gamma_s| < 1.50 \\ 0.55 \quad \text{if } |\gamma_s| > 1.50 \end{cases} \tag{4}$$

in which $|\gamma_s|$ is the absolute value of the station skew.

Not going into the arguments for or against this type of approach, it may be better if the $MSE\gamma_s$ is computed using the Jacknife and bootstrap methods as they do not depend on normal distribution theory. Tung and Mays (1961) compared the RMSE of the frequency factor, K_t for different skew coefficeents like the Tasker's variance method, the Jacknife and the bootstrap method. It was observed that both Jacknife and bootstrap methods provide good estimate of the sample variance. Also Tung amd Mays (1981b) observed that both these methods scores over Tasker and other methods in terms of standard error in estimating the weighted estimates of all moments including mean, standard deviation and skew. There is a strong case for the use of nonparameter estimation techniques in reducing parameter uncertainty in hydrologic studies.

Another major problem in using Eq.(3) is the estimation of regional skew γ_r. How to estimate regional skew? The simplest procedure would be to draw regional skew maps in original and log space (for the LP III distribution). This is exactly what the WRC followed . Now the question arises whether one can draw unique skew map of a particular region both in real and log space? Further, can the utility of these skew maps be enhanced by relating the regional skew to the geomorphological or the hydrological factors of the region? Also whether the use of regional skew results in more robust estimation procedures? Probably the answer to all these questions is not very encouraging. For example, WRC offered national regional skew maps for U.S.A. using the site skewness from 3000 gauging sites. This map which was first developed in 1967 and has remained unchanged since then have raised many questions which remain unanswered. Some of the objections are:

1. There is a large degree of subjectiveness in drawing these maps. Using the same data of 2972 station skew, McCuen (1979) constructed another map which has a standard deviation of 0.529 as against the WRC map having a standard deviation of 0.543 (a reduction of 15%) and possibly a map of still smaller standard deviation can be constructed.

2. The use of regional maps in improving the reliability of γ_w is not very significant. The mean and standard deviation of logarithms of 2972 station skews was -0.013 and 0.623 respectively. The standard deviation associated with map values when compared to the sample values is as large as 0.543. This means by using regional maps, the expected standard error gets reduced from 0.623 to 0.543 a reduction of only 13% which is not very significant.

3. Landwehr et al (1978) and Singh (1981) questioned the validity of log skew maps, by showing that given the log space regional skew maps, the real space skew contours cannot be uniquely specified. It was observed that log skew is not uniquely related to the skewness in the log space but also on many other factors like the distribution of flood flows, the coefficient of variation c_v and the kurtosis ck in the real space.

4. Matalas et al (1975) and Wallis et al (1977) showed in nearly all the 14 regions subdividing the United States, a condition of 'separation' existed which can neither be explained by LP III distribution or by drawing regional skew maps. The 'separation' was observed while comparing the historical skewness and the simulated skew values derived using various distributions. It was observed that the natural data consistently displays more inherently unstable skews than the ones derived from various statistical distributions including Pearson III and LP III distributions. Wallis et al (1977) suggested spatial or temporal mixing of skewness coefficient as a possible reason for 'separation'. If that being so then what is the relevance of skew maps? If γ_s changes with time then contour maps of estimate of γ_s would need be constructed as a function of time. How is it to be accomplished remain unanswered. On the other hand if γ_s is spatially distributed within a region there must be a mix of low γ with a high proportion of larger γ. The contour map thus obtained will be subject to large sampling error. Moreover, the identification of subregions of high γ is rather impossible to delineate.

The above discussion only highlights the shortcomings of sample skew or the weighted skew in estimating the parameters of LP III distribution. The WRC recommended practice of the method of moments (MOM) for fitting LP III distribution does exactly like that. As the sensitivity of MOM is based on the estimate of sample skew such an estimation procedure is subjected to large sampling error particularly at short record lengths (Nozdryn-Plotnicki and Watt,1979). It is,therefore, important to stress that if LP III or for that matter any other three or more parameter distribution is to be used, skewness and higher moments should only be used with caution in estimating their parameters. In this respect, the work of Rao (1980) of mixed moments needs fur ther probing. Rao (1980) suggested the use of mean and variance of real data and the mean of the log data (MXM1)as three parameters of fitting of LP III distribution.Compared to the MOM, the MXM1 suggested resulted in lower bias and standard deviation in the quantiles. Such a comparison however, assume that the population follow the LP III distribution with a specific skew coefficient. Obviously, the comparison is affected by the sample size and the parent skew value used to generate the Monte Carlo samples. Also, there does not seem to be any rational argument in choosing the particular MXM1 combination as there can be many more mixed combinations (Phien,1983). Moreover, the properties of standard error cannot be easily obtained as this method is completely distribution dependent.

It seems that an ideal method which does not explicitly include the sample or weighted γ in the analysis and yet a logical is the obe based on the Principle of maximum entropy(POME). The POME has widely been used in telecommunications (Jaynes,1961, 1968,Shannon,1948) and in many other branches of Engineering (Tribus,1968), its use in hydrology has been quite limited (Amorcho and Espildora,1973,Sonuga, 1972,1976, Jowitt,1980, Singh 1985). Probably, POME can be used more effectively not only in estimating the unbiased parameters of a distribution but also in selecting the most 'appropriate' distribution.

The three parameters of LP III distribution given in Eq.(5) using the POME can be computed as

$$f(x) = \frac{1}{a\sqrt{b}} \left(\frac{y-c}{a}\right)^{b-1} \exp\left[-\left(\frac{y-c}{a}\right)\right] \qquad (5)$$

where y = ln x, a,b, and c are the parameters of LP III density function f(x).

The parameters using POME are:

$$ab = \bar{y} - c \qquad (6)$$

$$\psi(b) + \ln(a) = E\left[\ln(y-c)\right] \qquad (7)$$

$$s_y^2 = a^2 b \qquad (8)$$

where $\bar{y}$, s_y^2 are the mean and variance of the log transformed series and $\psi(\ )$ is the digamma function. Parameters a,b and c can be estimated using Eqns.(6),(7) and (8). The advantage of POME is that it is based on rational principle and can be applied to estimate parameters of all the distributions commonly used in hydrology. Jowitt(1981) and Kumar et al(1986) have done comparative study on the relative efficiency of POME for estimating the parameters and quantiles of EV1 distribution and the results are encouraging. Further research is definitely warranted in evaluating the performance of POME for other distributions.

BEYOND LP III DISTRIBUTION

The dissatisfaction with the LP III distribution has led hydrologists in search of other distributions/or the transformation which can be applied to a flood series. Though there are many distributions (Mendal,1972, Singh and Sinclair,1972, Boughton,1980, Rossi,1984). Not much work is available on the applications of these distributions and therefore, we shall be briefly discussing only those distributions/transformation on which we have some experience.

If the objective is to have a procedure which enables 'good' fitting to the data and also 'frees' the analysis from distribution identification then the power transformation suggested by Chander et al (1978) is probably the most appropriate one. This procedure employs Box-Cox transformation (1964) as

$$y_i = \begin{cases} \dfrac{x_i^{\lambda} - 1}{\lambda} & \neq 0 \\ \ln x_1 & = 0 \end{cases}$$

where x_i is the annual flood peak, y_i is the transformed value for the ith year and λ is the power transformation parameter. Chander et al (1978) have shown that there exist an unique value of λ which transforms x_i series in a near normal distribution. The suitability of this transformation in flood frequency analysis has been established by Chander et al (1978). More recently, Hadgraft (1982) compared 12 candidate flood distributions for flood data from 40 gauging station located in Queensland, Australia with records in excess of 40 years; he concluded that the power transformation procedure fits a wider range of data than either the LN or the LP distributions. Also the power transformation is better than many other transformations like the SMEMAX (Bethlabmy,1977, Westphal,1984). The power transformation is similar to LP distribution in the sense that the transformation parameter λ is akin to LP log skew parameter. However, due to better analytically tractable form unlike the log skew in LP distribution the inference of using site and regional flood data can be studied analytically (Kuczera,1982). It is felt that power transformation approach of analysing floods has potential and should be studied more rigorously.

A distribution which has been recommended by NERC (UK) is the General Extreme Value distribution (GEV) of the form:

$$F(x) = \begin{cases} \exp\left[- \{ 1 - k(x-u)/a \}^{1/k} \right] & k \neq 0 \\ \exp\left[- \exp \{ 1 - (x-u)/a \} \right] & \text{for } k = 0 \end{cases} \qquad (10)$$

when k = 0, the GEV distribution is the Gumbel.

Being a three parameter distribution it is more versatile than EV I distribution. U.K. flood data were estimated in the Flood Studied Report (FSR) to be GEV like with derived regional growth curves exhibiting skews ranging from 1.9 to 52.8 with regional c_v to be less than 0.6. Many a times, the original FSR algorithm has been criticised at the accuracy of the estimated regional flood hazard as measured by regional growth curves at very large return periods which may yield quantile estimate to be either unreasonably high or unreasonably low (Johnson, et al 1981). Hosking et al (1984) have shown that the estimation procedure suggested by the FSR for regional analysis is not efficient especially for small sample sizes. It is shown that GEV distribution with parameters estimated through probability weighted moments (GEV/PWM) yield significantly better results than the presently followed FSR algorithm. PWM can be used for estimating the parameters of such distributions which are expressed in an inverse form and is defined as:

$$M_{ijk} = E\left[x^i\ F^j\ (1-F)^k\right] \tag{11}$$

where x is a random variable distributed as $F = F(x) = P(X < x)$ and parameters, i,j,k are real numbers. PWM treats outliers in a special fashion thereby improving its performance over existing methods. An application of PWM technique of parameter estimation in regional flood frequency analysis is proposed by Wallis (1982). It combines regional and site specific information by simple averaging and scaling. Moments for regional analysis are obtained by averaging the probability weighted moments at each site. The regional quantile estimate x(ij) is then obtained by scaling the atsite estimate w_j by a regional estimate of the mean M(0,0)

$$x_{(0,0)} = M_{(0,0)} w_j \tag{12}$$

The PWM method works well for situations where records are extremely short and streamflow samples are highly skew- and kurtotic. This situation more often arises when the sample size at each site is small. The major constraint of PWM is that it cannot be used on distributions like LP III which are not amenable in an inverse form. In such a situation, the POME definitely scores over PWM. POME gives comparable results to PWM and yet can be used for all the distributions whose pdf is defined. PWM has gained predominance due to some exceptionally good work done by researchers like Wallis, Matalas, Greenwood, Landwehr,Greis and Wood. POME has not received that much attention and there is a need for critical studies on this approach.

Recently, a distribution which has aroused interest is the Wakeby distribution (Houghton,1978,Wallis and Wood,1985). It is a five parameter distribution and is expressed in an inverse form as:-

$$x = a\left[1 - (1-F)^b\right] - c\left[1 - (1-F)^{-d}\right] + e \tag{13}$$

where a,b,c,d and e are the parameters of the distribution and F is a uniform (0,1) variable. There are two distinct advantages of this distribution.

1. The right and the left hand side of the distribution can be modelled separately. The parameters and b govern the left hand (low flows) tail while the parameters c,d and e

govern the right hand (high flow) tail.

2. This distribution can explain some of the 'separation effects' (Matalas et al,1975) which many other distributions cannot.

Though, the Wakeby distribution is a five parameter distribution, its parameters value can be obtained without going into the higher moments such as the skewness coefficient. The essential difference between Wakeby and GEV is that for a particular skew, the former can be many times more kurtotic than the latter. For example a Wakeby distribution with a = 24, b = 0.2, c = 8, d = 0.2495 and e = 6.5 (c_v = 0.53, skew = 3) may have a kurtosis equal to 890.5 compared to a corresponding value of 29.4 for the GEV distribution. Sample functionsderived from this Wakeby, while they most look like those taken from GEV with identical three moments do exhibit a tendency towards occasional 'outliers' and also exhibit the phenomenon of 'separation', both of which phenomenon appear to exist in annual flood series (Hosking et al,1984).

The five parametess of the Wakeby distribution can be estimated using the incomplete mean algorithm suggested by Houghton (1978) which involves iterating on assumed values of b and d until what is defined as 'best fit' is achieved. Greenwood et al (1979) suggested more elegant PWM approach in estimating the Wakeby parameters. Since Wakeby has five parameters, their estimation using single site data is likely to be poor and it can be done only through regional analysis. One way of carrying out regional analysis can be by using the index flood method involving the following steps:

(i) pool the data from each site,

(ii) scale the data by dividing the 'at site' data by 'at site' mean so that the regional flood curve will have a mean equal to unity,

(iii) fit the Wakeby distribution using PWM to the scaled data, and

(iv) obtain the flood quantile at each site by scaling the regional estimates at the 'at site' mean.

Wallis and Wood (1985) showed that WAK/PWM has virtually zero bias with very close confidence bands and the maximum uncertainty for the WAK/PWM arises from the error in estimating the 'at site' mean which is used to standardize the data prior to regionalisation. These findings are very much in contrast to the WRC procedures which show a near exponential increase in the width of the confidence band with increasing return periods.

One major difficulty in regional estimation technique arise when the catchments in the region are hydrologically hetrogeneous and in particular if the heterogenity takes place in the form of varying coefficient of variation amongst the basin of the data base. For example,if the set of catchments has flood sequences with varying c_v then the regional analysis will give positively biased quantile estimate for same site and the negatively biased estimate for the other and it is rather impossible to establish which individual catchments are likely to have quantile estimate with a positive bias and which with a negative bias.

Being aware of the basic uncertainty in the choice of proper distribution from a small sample, the parameter estimation choice has shifted from 'good' estimation procedure to robust estimation procedure. Regional estimation of parameters is one example of robust estimation procedures. Robust estimates are those estimates which are relatively in sensitive or robust to deviations from ideal assumptions. For example, if the flood peaks are assumed to follow LP III distribution by mistake, whereas in reality Gumbel distribution is the true distribution then by robust estimation we ensure (as far as possible)that T year recurrence estimate under LP III assumption does not vary very much from Gumbel distribution. Similarly a robust estimation procedure does not lead to a large variation in the estimated design flood with small changes in the data like the addition or removal of one or two observations. Statistically robust estimator seek to provide this assurance by treating the outliers in a special way. Fiering and Kuczera (1982) have shown that good estimation procedure will be of less use in flood analysis if small perturbation in the parameter, no matter how well estimated produce large deviation in the probability density function. In fact the use of regional parameters related to topographic or climatic factors in the analysis is the first step towards robust estimation procedures.

Probability weighted moments as used in estimating GEV or WAK distribution is another form of robust estimation procedure. The other robust estimator procedure which have been applied in flood frequency analysis are the James Stein estimator (Efron and Morris,1975) and the empirical Bavesian approach (Kuczera,1982a, Stedinger,1983,1985). Landwehr and Matalas (1984) discussed the applicability of James Stein estimator attached to the quadratic loss function in regional hydrologic studies which entails the estimates of n-dimensional location parameter. The basic concept about the James Stein estimator is that an improved estimate can be obtained by shrinking the desired estimate, for example, the at site quantile event towards some ensemble average, for example, the regional mean. When the number of elements in the ensemble average is graeter than two, the James Stein estimator can be shown to provide estimates with smaller total risk. The loss function according to which the estimation procedure is measured is a composite loss function in the sense that increased loss on some estimate on the region is exchanged for decreased loss in other regions (Greis and Wood,1981). It was observed that the James Stein estimator has the maximum advantage over the traditional estimators in the case of independent random variables. However, one big difference is observed. In the traditional method of regionalisation,the cross correlations amongst the sites are used to improve the estimates of the parameter. In the case of James Stein estimator, relative advantage over traditional estimator disappears if the correlation between sites increases. Thus although there is advantage in using James Stein estimator in water resources studies, the relative advantage decreases appreciably because to high degree of cross correlation amongst the variables and the heterogenity present in the hydrologic region.

In the Bayes analysis, the assumption is made that there is an unknown super population of flows from which the parameters of individual basin are drawn. The parameters of individual basin are then known realisation of the super population which can be pooled to form some estimate of the regional population. By combining the regional or superpopulation information with the site specific information using Bayes theorem, it is possible to arrive at a posterior distribution of quantile events. Mathematically:

$$f(q_j \mid w_j;\ \theta_j) \quad \xi(\theta \mid x_j, \beta) = \xi(\theta \mid q_j;\ w_j\ x_j\ \beta) \qquad (14)$$

The posterior distribution is proportional to the product of the marginal at site density function and the super population density functions. Using the method of ordinary least square, Kuczera (1982a) demonstrated improved performance over at site procedures in estimating distributions of 100 year flood using record lengths of 10 to 30 years per site. As expected, the difference between the empirical Bayes and site specific densities for 100 year flood were very significant particularly at shorter record lengths. Kuczera (1983) also used the empirical Bayes analysis on the power transformation parameter λ used by Chander et al (1978) in frequency analysis. It is shown that the regional analysis on the basis of Bayes approach is better than the recommended WRC practices using LP distribution. In Bayes case, knowing the posterior distributions of the parameters, $\bar{x}$, σ_x and λ the posterior distribution of T year flood can be obtained. This enables the construction of probability limits of the population T year flood (Stedinger,1983). If one specifies a loss function, a point estimate of T year flood can be obtained (Wood,1978).The ability of Bayes estimator to compute an associated risk which can in turn be related in economic terms makes it an important tool in regional analysis.

REFERENCES

1. Amorcho,J. and Espildora, B., 'Entropy in the Assessment of Uncertainty in Hydrologic Systems and Models', Water Resources Research,1973, 9, pp.1511-1522.

2. Bethlahmy,N., 'Flood Analysis by SMEMAX Transformation', Journal of the Hydraulics Division, American Society of Civil Engineers, 1977, 103, pp.69-80.

3. Bras, R.L. and Gadboury, D.R., 'Spatially Varying Rainfall and Flood Risk Analysis,'Journal of the Hydraulics Division, American Society of Civil Engineers,1985,III, pp.754-773.

4. Boughton,W.C., 'A Frequency Distribution for Annual Floods', Water Resources Research,1980,16, pp.347-354.

5. Box,G.E.P. and Cox,D.R., 'An Analysis of Transformation', Journal of the Royal Statistical Society,1964,26, pp. 211-252.

6. Chander,S., Spolia,S.K. and Arun Kumar, 'Flood Frequency Analysis by Power Transformation',Journal of the Hydraulics Division,American Society of Civil Engineers,1978, 104, pp.1495-1504.

7. Costa,J.E., 'Halocene Stratigraphy in Flood Frequency Analysis', Water Resources Research,1978,14, pp.626-632.

8. Diaz-Granados,M.A., 'A Physically Based Flood Frequency Distribution',Water Resources Research,1984, 20 pp. 995-1002.

9. Efron,B. and Morris,C., 'Data Analysis Using Stein Estimator and Its Generalisation', Journal of the American Statistical Society, 1975, 68, pp. 311 - 319.

10. Fiering, M.B. and Kuczera, G., 'Robust Estimators in Hydrology',Scientific Basis of Water Resources Management Studies in Geophysics, 1982, National Research Council.

11. Gerard,R. and Karpuk, E.W., 'Probability Analysis of Historical Flood Data', Journal of the Hydraulics Division, American Society of Civil Engineers,1979,105,pp.1153-1165.

12. Greis,N.P. and Wood,E.F.,'Regional Flood Frequency Estimation and Network Design, Water Resources Research,1981, 17, pp.1167-1177.

13. Greis,N.P., 'Flood Frequency Analysis: A Review of 1972-1982', Reviews of Geophysics and Space Physics,1983,21, pp.699-707.

14. Greenwood,J.A., Landwehr,J.M., Matalas,N.C. and Wallis,J.R., 'Probability Weighted Moments:Definitions and Relation to Parameters of Several Distribution Expressable in Inverse Form', Water Resources Research, 1979, 15, pp.1049-1054.

15. Hadgraft,R.G., 'An Empirical Study of Flood Frequency Distribution', Hydrology Report 000716, P.R.Queensland Water Resources Commission, 1982,Australia.

16. Hebson,C and Wood,E.F., 'A Derived Flood Frequency Distribution Using Horton Order Ratios', Water Resources Research, 1982, 18, pp.1509-

17. Hosking, J.R.M., Wallis,J.R. and Wood,E.F., 'Estimation of the Generalised Extreme Value Distribution by the Method of Probability Weighted Moments', MRC Technical Report, Mathematics Research Centre,University of Wisconsin,1984, pp.24.

18. Houghton,J.C., 'Birth of a Parent; the Wakeby Distribution for Modelling Flood Flows', Water Resources Research,1978, 14, pp.1105-1110.

19. Houghton,J.C., 'The Incomplete Means Estimation Procedure Applied to Flood Frequency Analysis', Water Resources Research, 1979, 14, 1111 -

20. Institution of Engineers,Australia,'Australian Rainfall and Runoff', Flood Analysis and Design, ACI,Canberra, 1977, pp.150

21. Jaynes,E.T., 'Probability Theory in Science and Engineering',McGraw Hill Book Co., New York,1961.

22. Jaynes,E.T., 'Prior Probabilities', IEEE Transactions, Systems Management and Cybernetics', 1968, 3,pp.227-241.

23. Jenkinson,A.F., 'Estimation of Maximum Floods',Chapters, World Meteorological Technical Note,1969,98,pp.183-257.

24. Johnson,F.G., Jarvis,R.M. and Reynolds,G., 'Discussion of Papers 7-10, in Flood Studies Report -Five Years on', Institution of Civil Engineers,London,1981,pp.119.

25. Jowitt,P.W., 'The Extreme Value Types Distribution and the Principle of Maximum Entropy', Journal of Hydrology, 1979,42, pp.23-28.

26. Kochel,R.C. and Baker,V.R., 'Paleoflood Hydrology', Science, 1982, 215, pp.353.

27. Kuczera,G., 'Combining Site Specific and Regional Information Bayes Approach',Water Resources Research,1982, 18, pp.318-324.

28. Kuczera,G., 'A Bayesian Surrogate for Regional Skew in Flood Frequency Analysis', Water Resources Research, 1983,19, pp.821-832.

29. Kuczera,G., 'Robust Flood Frequency Models', Water Resources Research,1982, 18, pp. 318-324.

30. Kumar,A.,Parida,B.P. and Rema Devi, 'Entropy Principle in the Estimation of Gumbel Parameters', International Symposium on Flood Frequency and Risk Analysis,U.S.A. 1986, (preprints).

31. Landwehr,J.M.,Matalas,N.C. and Wallis,J.R., 'Quantile Estimation with More or Less Flood Like Distribution', Water Resources Research,1980,16, pp.547-555.

32. Landwehr,J.M.,Matalas,N.C.,and Wallis,J.R., 'Some Comparisons of Flood Statistics in Real and Log Space', Water Resources Research,1978,14, pp.902-920.

33. McCuen,R.H , 'Map Skew?', Water Resources Planning Division,American Society of Civil Engineers,1979,105, pp.269-

34. Matalas,N.C., Slack,J.R. and Wallis,J.R., 'Regional Skew in Search of a Parent', Water Resources Research, 1975,15, pp.1361-1379.

35. Mendel,H.G., 'A New Proposal for Flood Frequency Computation',Proc.Second International Symposium in Hydrology, Fort Collins,Colorado,1972,pp.325-335.

36. National Environmental Research Council, Flood Studies Report,1975,1-5, 50 NERC,London, pp.550.

37. Nozdryn Flotnicki,M.J. and Watt,W.E., 'Assessment of Fitting Technique for the Log Pearson type III Distribution Using Monte Carlo Simulation',Water Resources Research,1979,15, pp.714-718.

38. Phien,H.N. and Hira,M.A., 'Log Pearson type III distribution Parameter Estimation', Journal of Hydrology, 1983,64, pp.25-37.

39. Rao,D.V., 'Log Pearson type III distribution - a Generalised Evaluation', Journal of the Hydraulics Division, American Society of Civil Engineers,1980, 106, pp.853-872.

40. Rao,D.V., 'Log Pearson type III distribution,Method of Mixed Moments', Journal of the Hydraulics Division, American Society of Civil Engineers,1980, 106, pp.999-1019.

41. Rodriguez-Iturbe,I., and Valdes,J., 'The Geomorphologic Structure of Hydrologic Response', Water Resources Research, 1979,15, pp.1409.

42. Shannon,C.E., 'The Mathematical Theory of Communication', Bell System Technical Journal,1979,27,pp. 379-428.

43. Shen,H.W., Bnyson,M.C. and Ochoa,I.D., 'Effect of Tail Behaviour Assumption on Flood Predictions', Water Resources Research,1980, 14, pp.902-920.

44. Singh,K.P. and Sinclair,R.A., 'Two Distribution Method for Flood Frequency Analysis', Journal of the Hydraulics Division, ASCE, 1972,98, pp.29-44.

45. Singh, K.P., 'Map Skew? A Discussion', Journal of the Water Resources and Planning, American Society of Civil Engineers,1981,107, pp.298.

46. Singh,V.P., 'Derivations of Some Frequency Distributions Using the Principle of Maximum Entropy',1985, (Private Communication.

47. Sonuga, J.O., 'Principle of Maximum Entropy in Hydrologic Frequency Analysis', Journal of Hydrology, 1972,17, pp.177-191.

48. Sonuga,J.O., 'Entropy Principle Applied to Rainfall Runoff Process', Journal of Hydrology, 1976, 30, pp.81-94.

49. Stedinger,J.R., 'Estimating a Regional Flood Frequency Distribution', Water Resources Research,1983, 19, pp.503-510.

50. Stedinger,J.R., 'Design Events with Specified Flood Risk', Water Resources Research,1983, 19, pp.511-522.

51. Tasker,G.D., 'Flood Frequency Analysis with Generalised Skew Coefficient', Water Resources Research,1978, 14, pp.373-376.

52. Tribus, M., 'Rational Descriptors, Decisions and Design', 1969, Pergamon Press, New York.

53. Tung,Y.K. and Mays, L.W., 'Generalised Skew Coefficient for Flood Frequency Analysis', Water Resources Bulletin,1981, 17, pp.261.

54. Tung,Y.K. and Mays, L.W., 'Reducing Hydrologic Uncertainty', Journal of Water Resources and Planning Division, American Society of Civil Engineers, 1981, 107, pp.245-262.

55. Wallis,J.R., Matalas, N.C. and Slack,J.R., 'Just a Moment!' Water Resources Research, 1984, 14,pp.626-632.

56. Wallis,J.R., Matalas, N.C. and Slack,J.R., 'Apparent Regional Skew', Water Resources Research, 1977,13, pp. 159-182.

57. Wallis,J.R. and Wood, E.F., 'Relative Accuracy of Log Pearson III Distribution', Journal of the Hydraulics Division,American Society of Civil Engineers, 1985, 111, pp.1043-100 .

58. Water Resources Council, U.S.A., 'Guidelines for Determining Flood Flow Frequency', Bulletin No.17B, Washington,D.C.,U.S.A.,1981.

59. Westphal, J.A., 'Smemax? Caution!', Journal of the Hydraulics Division, American Society of Civil Engineers, 1984, 110, pp.209-218.

60. Wood, E.F., 'Analysing Hydrologic Uncertainty and Its Impact Upon Decision Making in Water Resources', Advances Water Resources, 1978, 1, pp.299-305.

DETERMINISTIC NATURE OF FLOOD FREQUENCIES: SOME OBSERVATIONS

Ivan Muzik
Department of Civil Engineering
The University of Calgary
Calgary, Alberta, Canada T2N 1N4

ABSTRACT. The subject of the paper is the deterministic role that physical characteristics of watersheds have in the transformation process of probabilistic rainfall inputs into probabilistic peak runoffs. Analytical solution of kinematic wave flow over a single runoff plane and experimental hydrographs from a laboratory catchment were the methods of investigation.

Results indicate that the dispersion and skewness of peak runoff series increase with increasing resistance to surface runoff and with decreasing bifurcation ratio, the mean basin slope, and the duration of rainfall. The results also indicate that when the rainfall intensity distribution is of two parameter type, such as normal or gamma distribution, with rigid relationship between its coefficients of skewness and variation, the peak runoff distribution does not retain that rigidity.

1. INTRODUCTION

Effects of physical properties of a fluvial system, such as the roughness, slope and geometry of runoff surfaces and channels, on a flood movement, characterized, for example, by the mean flow depth and velocity, are intuitively obvious and can be described by appropriate equations. How physical features of a watershed effect statistical characteristics of floods, however, is not so easily understood(Eagleson, 1972; Klemes, 1978). Yet, every watershed has its unique physical characteristics of more or less stationary nature which may be expected to affect runoff outcomes due to rainfall inputs in a unique way. In this respect, the series of runoff outcomes may be compared to outcomes of experiments conducted with biased dice. Every catchment is somewhat biased in its own peculiar way, so that even if the rainfall inputs were exactly identical for various catchments, the frequency distributions of outcomes will differ due to the biases introduced by individual catchments. In nature, the deterministic influence of catchments on flood probabilities is obscured by superimposed stochastic effects produced by

V. P. Singh (ed.), Hydrologic Frequency Modeling, 37–47.

nonstationary watershed properties such as the antecedent moisture conditions and seasonal changes, and by the uncertainty of the rainfall input.

The objective of the paper is to demonstrate how the probability distribution function of peak discharges from a laboratory catchment depend on the rainfall and catchment parameters. Two approaches are employed in the research, analytical and experimental approach, respectively. In the analytical approach a kinematic wave model is used to derive the probability density function of peak discharges from a uniformly sloping plane when the rainfall input intensity has a normal probability density function. The experimental approach is used to study the effect of two physical parameters of a laboratory watershed on the probability distribution function of peak discharges. While the probabilistic structure of the rainfall input was kept constant the bifurcation ratio and mean slope of the laboratory watershed were varied in order to study their deterministic influence on the probabilistic structure of the peak discharge series.

Experimental data were obtained by means of a computerized laboratory rainfall-runoff facility capable of generating rainfalls of predetermined duration and intensity over an area of approximately 2.5 m by 2.5 m. The entire study is limited to impervious runoff surfaces.

2. ANALYTICAL APPROACH: THE KINEMATIC WAVE MODEL

The kinematic wave equations of overland flow form a suitable mathematical model by means of which it is possible to link deterministically the probability density functions of rainfall excess intensity and peak outflow from a single runoff plane, respectively. Duration of the rainfall excess is constant. Transformation of a single rainfall event into runoff by the kinematic wave model is briefly reviewed, followed by development of the probability density transformation equation for the entire population of the rainfall-runoff events.

2.1. Single Event Transformation

The kinematic wave equations of overland flow due to rainfall, the momentum and continuity equations, respectively, have the following form

$$q = \alpha y^m \tag{1}$$

and

$$\partial y/\partial t + \partial q/\partial x = i \tag{2}$$

in which q is the discharge per unit width, y is the depth of flow, t is the time, x is the distance in the direction of flow and i is the rate of rainfall excess.

When Manning's equation is used to solve for the velocity of flow, the exponent m is equal to 5/3 and α is equal to

$$\alpha = S^{\frac{1}{2}}/n \tag{3}$$

where S and n is the slope and Manning's roughness coefficient of the plane, respectively. The coefficient α describes the physical characteristics of the basin, a runoff plane in this case, which govern the runoff process.

For the case of rainfall having a constant intensity uniformly distributed over the runoff plane and of a duration t_o longer than the time to steady state t_s, kinematic wave equations have the following analytical solution (Henderson, 1964)

$$\text{for } t < t_s: \qquad q = \alpha(it)^m \tag{4}$$

$$\text{for } t_o > t \geq t_s: \qquad q_s = Li \tag{5}$$

where L is the length of the runoff plane, q_s is the steady state discharge and the time to steady state is given by

$$t_s = (L/\alpha i^{m-1})^{1/m} \tag{6}$$

The recession part of the hydrograph is of no interest since only the peak discharge is required.

Consider a runoff plane characterized by certain length L and coefficient α. Rainfalls of various intensities i but of a certain constant duration t_o are applied uniformly to the runoff plane and the corresponding peak discharges are computed by eq. (4) or by eq. (5). The peak discharge q is computed by eq. (4) for those intensities for which eq. (6) yields $t_s < t_o$. For intensities yielding $t_s \geq t_o$ the peak discharge is computed by eq. (5). The changeover occurs when $t_s = t_o$ and the corresponding intensity is designated by i_{cr}. The transformation relationship between i and q is shown graphically in Fig. 1 for L = 10 m and $\alpha = 10\ m^{1/3}$/sec.

2.2. Population Transformation

Individual pairs of i and q are functionally related by the kinematic wave equations. When i is a random variable (the duration of rainfall, t_o, remaining constant) the peak discharge q becomes also a random variable. It is of interest to see if a functional relationship can be found between the probability densities of i and q. Such relationships would allow transformation between entire populations of i and q in terms of their statistical parameters.

The probability density functions f(i) and f(q) of the two functionally related variables i and q, respectively, can be related by the equation

$$f(q) = f(i)\ |\, di/dq\, | \tag{7}$$

where di/dq is the reciprocal of the derivative of q with respect to i,

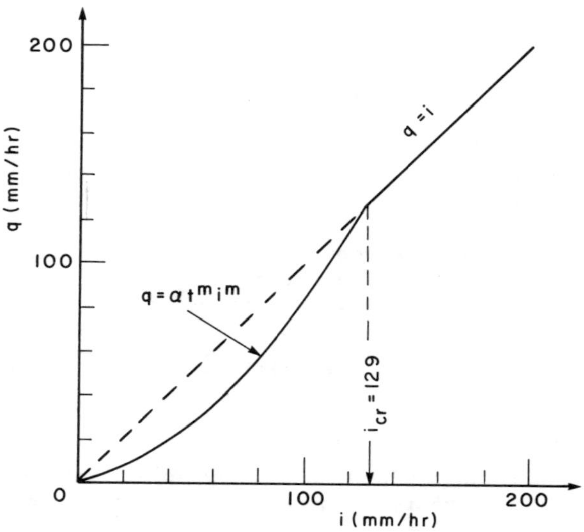

Figure 1. Transformation relationship between intensity i and peak discharge q for a runoff plane with L = 10 m and α = 10.

which can be evaluated by eqs. (4) and (5), respectively, as follows:

$$\text{for } i < i_{cr}: \qquad dq/di = \alpha m t_o^m i^{m-1} \tag{8}$$

$$\text{for } i \geq i_{cr}: \qquad dq/di = 1 \tag{9}$$

When i, for example, is normally distributed with a mean $\bar{i}$ and a standard deviation s_i the probability density function of q is given by the following equations.

$$\text{for } i < i_{cr}: f(q) = (2\pi)^{-\frac{1}{2}} s_i^{-1} \exp[-\tfrac{1}{2}(i-\bar{i})^2/s_i^2]/(\alpha m t_o^m i^{m-1}) \tag{10}$$

$$\text{for } i \geq i_{cr}: f(q) = f(i) = (2\pi)^{-\frac{1}{2}} s_i^{-1} \exp[-\tfrac{1}{2}(i-\bar{i})^2/s_i^2] \tag{11}$$

Dependence of the probability density of peak discharge on the probability density of rainfall and on the catchment characteristics is clearly demonstrated by eq. (10). To illustrate the effect of catchment characteristics, quantified by the coefficient α, three probability density functions of the peak discharge were computed for the following values of α: 2.5, 5.0 and 10.0, respectively. The rainfall intensity was assumed to be normally distributed with the mean of 100 mm/hr and the standard deviation of 25 mm/hr. The duration of rainfall was 60 seconds and the length of the runoff plane was 10 metres. The computed probability density curves are shown in Fig. 2. It can be seen that as the coefficient α decreases, that is, the slope of the runoff plane decreases and/or the Manning's n increases, the mean of q rapidly decreases as compared to the mean of i, while the density function is becoming very peaked and positively skewed. On the other hand, as the coefficient α increases, the critical intensity i_{cr}, computed by eq. (6) for $t_s = t_o$, decreases so that eq. (11) applies to progressively larger range of intensities. The discharge q follows then a mixed probability distribution given by eqs. (10) and (11). This case is shown in Fig. 2 for a runoff plane with α value equal to 10. The critical rainfall intensity for this plane is 129 mm/hr. For all intensities lower than 129 mm/hr the probability density function of q is given by eq. (10). For intensities greater than 129 mm/hr the density functions of q is normal and identical to that of i as given by eq. (11). It should be noted that similar results will be obtained when the rainfall intensity is kept constant and the duration of rainfall is allowed to vary.

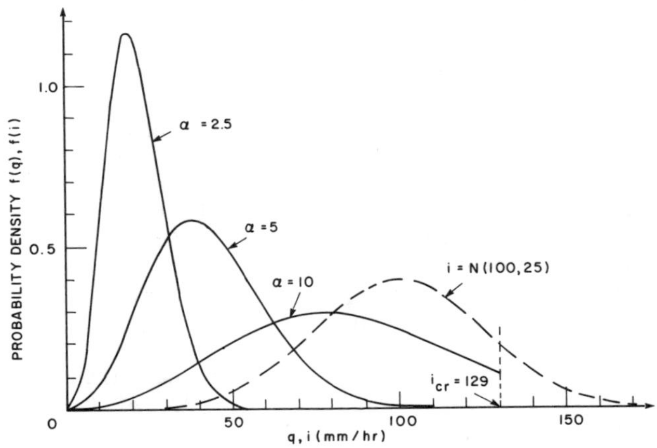

Figure 2. Effect of the physical characteristics of a runoff plane, represented by $\alpha = S^{\frac{1}{2}}/n$, on the probability density function of peak discharge q.

3. EXPERIMENTAL APPROACH

3.1. Experimental Equipment

Experiments with controlled rainfall applied over laboratory catchments were carried out to verify and extend the results obtained analytically for a single runoff plane to more complex catchment geometries.

The laboratory rainfall-runoff facility used for the experiments consisted of a rainfall generator, a catchment area, and of measuring instruments recording the catchment outflow. A complete description of the facility is provided by Beersing (1983). The simulated rainfall was produced by means of a large number of small diameter polyethylene tubes mounted in the bottom of overhead pressurized plexiglass boxes. The area covered by the simulated rainfall was 6.35 square metres. The rate and duration of rainfall was controlled by an HP 9845B microcomputer. The microcomputer also acquired discharge data measured by a small flume at catchment outlet.

The experimental catchment consisted of two planes draining laterally into a main channel or a combination of up to ten planes and six lateral channels draining into the main channel. The planes were made of galvanized sheet metal, the main channel was made of plexiglass. The slope of the planes and the main channel were adjustable.

3.2. Experimental Method

Aim of the experimental work was to obtain a regression equation between the rainfall intensity i and the corresponding peak discharge q for a given duration of rainfall and a particular catchment configuration. This empirical relationship should be similar to the q-i relation for the kinematic wave model shown in Fig. 1 except that the steady state relation, given by q=i, will be achieved gradually for the observed flows. Such empirical relationships established for a number of catchment configurations can be used to study the effect of catchment characteristics on the transformation of the probability distribution of rainfall intensity into the probability distribution of peak discharge. Choosing a probability distribution of i, a random series of q can be generated by means of the regression equations between q and i. Statistical parameters of the q series can be determined and compared to the known parameters of the i series.

The rainfall generator was calibrated to generate uniform rainfall intensities of magnitudes 64.8, 77.5, 111.8, 142.2, 171.4, 177.8, 200.7 and 221.0 mm/hr over the catchment area. Two durations of rainfall were chosen, 25 and 35 seconds, respectively. Four catchment prototypes were used. The characteristics of the prototypes were the bifurcation ratio, the mean ground slope, the mean stream slope and the total stream length. The bifurcation ratio varied between 1 and 6, a range which occurs in natural basins. (The bifurcation ratio equal to 1 is used here to designate a basin configuration consisting of two planes contributing runoff to one channel). The catchment characteristics of the four prototypes are listed in Table I.

3.3. Analysis and Results

Peak discharges, q, were obtained for eight rainfall intensities, i, ranging from 64.8 mm/hr to 221.0 mm/hr, for 16 catchment configurations listed in Table I, and for two rainfall durations of 25 sec and

35 sec, respectively. For each catchment and rainfall duration the i vs. q data were fitted by a regression equation of the type

TABLE I - Basin parameters used in experiments

Basin prototype	Bifurcation ratio	Mean ground slope	Mean stream slope	Total stream length (m)
1(a)	1*	0.026	0.017	2.44
1(b)	1*	0.042	0.017	2.44
1(c)	1*	0.060	0.017	2.44
1(d)	1*	0.079	0.017	2.44
2(a)	2	0.053	0.018	3.05
2(b)	2	0.067	0.021	3.05
2(c)	2	0.082	0.025	3.66
2(d)	2	0.102	0.029	3.66
3(a)	4	0.099	0.018	3.66
3(b)	4	0.108	0.030	3.66
3(c)	4	0.119	0.042	3.66
3(d)	4	0.131	0.053	3.66
4(a)	6	0.146	0.019	4.27
4(b)	6	0.149	0.038	4.27
4(c)	6	0.153	0.057	4.27
4(d)	6	0.161	0.077	4.27

*Bifurcation ratio of 1 designates a basin with one channel and two planes.

$$q = a(b_o + b_1 i + b_2 i^2) \qquad (12)$$

The regression equations can be used to determine parameters of the probability distribution of q for known parameters of the i distribution. A method developed by Beersing (1983), allows direct calculation of statistical parameters of the q series from the coefficients of eq. (12) and the known parameters of i. The mean $\bar{q}$, the coefficient of variation, C_{vq} and skewness, C_{sq}, of the generated q distribution are given by the following equations

$$\bar{q} = a(b_o + b_1 \bar{i} + b_2 \underline{i}^2) \qquad (13)$$

$$C_{vq} = [(\underline{q}^2 - (\bar{q})^2)^{\frac{1}{2}}]/\bar{q} \qquad (14)$$

$$C_{sq} = [\underline{q}^3 - 3\bar{q}\underline{q}^2 + 2(\bar{q})^3]/[\underline{q}^2 - (\bar{q})^2]^{3/2} \qquad (15)$$

where $\bar{i}$ is the mean of the probability distribution of i. The remaining terms are defined in the appendix.

Values of C_{vq} and C_{sq} were computed for the sixteen experimental catchments and two rainfall durations used in the experiments for the case of rainfall intensity distributed normally with the mean of 140 mm/hr and two standard deviations, 40 mm/hr and 20 mm/hr, respectively, and plotted on a C_{vq} - C_{sq} plane in Figure 3.

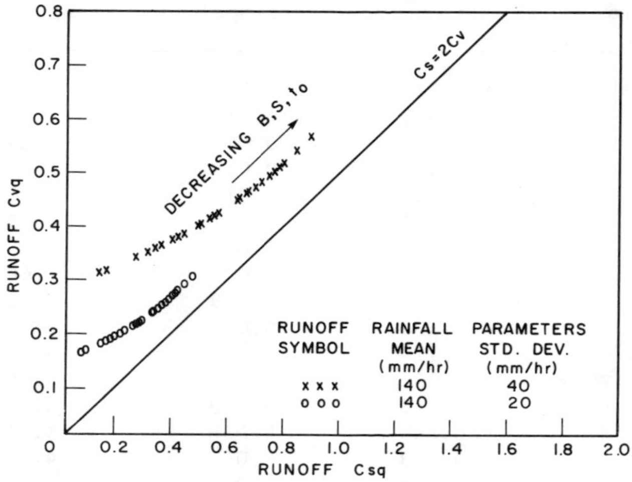

Figure 3. Relationship between C_{vq} and C_{sq} for laboratory catchments when the rainfall intensity is normally distributed.

The two curves, corresponding to the two standard deviations, plot between the lines corresponding to a normal (C_s = 0) and gamma (C_s = 2 C_v) distributions. As the bifurcation ratio B, the mean slope, S, and the duration of rainfall t_o decrease, the coefficients C_{vq} and C_{sq} increase in comparison to coefficients C_{vi} and C_{si} of the rainfall intensity. Thus, the probability distribution of q approaches the normal distribution of i for fast responding catchments on one hand, and becomes more skewed for slower responding catchments, on the other hand. The results are in agreement with those obtained analytically for a single runoff plane.

The analytical and experimental results support a probabilistic argument presented by Klemes (1978) in favour of a flexible relationship between C_{sq} and C_{vq}. A probability distribution which has a flexible relationship between its coefficient of skewness and variation is the Pearson type III distribution. For every set of catchment parameters and rainfall durations used in the experiments the Kolmogorov-Smirnov goodness of fit test was performed. In each case it was found that the hypothesis that the observed peak discharge distribution was a Pearson type III distribution could not be rejected. Typical plots of the observed values of q and the fitted Pearson type III distributions are shown in Fig. 4 for the catchments 1(b) with t_o = 25 seconds, and 4(d) with t_o = 35 seconds. The rainfall intensity was normally distributed with the mean of 140 mm/hr and standard deviation of 40 mm/hr. The plots again clearly demonstrate the tendency of the skewness of the peak discharge distributions to increase as the bifurcation ratio, slope and duration of rainfall decrease, and the associated decrease in the distribution mean.

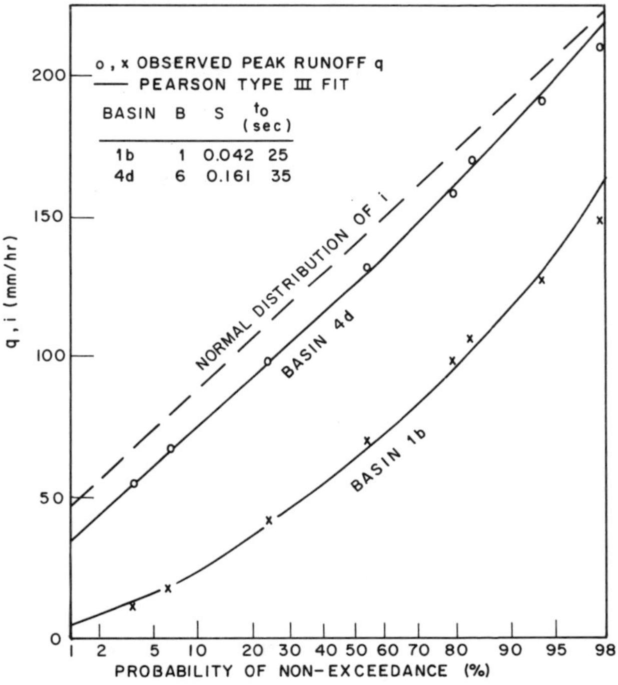

Figure 4. Pearson type III fit to observed peak discharges generated by a normally distributed intensities with mean of 140 mm/hr and standard deviation of 40 mm/hr.

SUMMARY AND CONCLUSIONS

The objective of the paper was to demonstrate the existence of physical basis for probabilities of peak discharge. Numerical and laboratory simulations of overland flow were the means by which the transformation of the probability density function of rainfall intensity, f(i), into the probability density function of peak discharge, f(q), was analyzed. The following assumptions were made: the runoff surface is impervious, the rainfall inputs consist of intensities i, uniform both in time and space, the duration of rainfall, t_o, is constant for a given f(i), and f(i) is normal.

Runoff from a single plane was simulated by the kinematic wave equation linking the peak discharge q to the causative rainfall intensity i. Laboratory rainfall-runoff experiments were used to establish empirical relations between q and i for catchments of various bifurcation ratios, B, and mean slopes, S. Based on the analysis of the simulation results the following conclusions may be made.

1. When there is a unique deterministic relationship between q and i, f(q) is a derived distribution which can be completely specified for known f(i).
2. Since the (q, i) relationship depends on physical characteristic of catchments, this dependency also effects the transformation of f(i) into f(q).
3. In case of a single runoff plane the f(q) depends on the parameters of f(i) and on the slope and roughness of the plane. The steeper and smoother the plane is the closer f(q) approaches f(i). The milder the slope and the rougher the surface of the plane, the lower the mean and the greater the skewness of f(q).
4. Theoretically, f(q) is a mixed distribution consisting of a certain f(q) which is valid for non-equilibrium flows, and a second f(q) identical to f(i), which applies to equilibrium flows.
5. Experimental results are in agreement with the numerical simulation results. In general it can be stated that as the response time of a catchment to a given rainfall input decreases (B and S increase), f(q) approaches f(i). On the other hand, as the response time increases (B and S decrease), the mean of f(q) decreases and the skewness of f(q) increases.
6. The duration of rainfall, t_o, effects the parameters of f(q). As t_o approaches the time to steady state flow f(q) approaches f(i). As t_o decreases, the mean of f(q) decreases and the skewness of f(q) increases.

Based on the last conclusion one may reason that flood frequency analysis should be performed on samples homogeneous with respect to the duration of rainfall excess. This requirement may in practice severly limit the sample size and thus the accuracy of the analysis. The alternative is to study the relationship of f(q) and the joint probability distribution of i and t_o, f(i, t_o).

The study has been limited to normally distributed rainfall intensities. It seems reasonable to assume that the conclusions made here will be valid also for other distributions of i.

ACKNOWLEDGEMENTS

This research has been supported by the Natural Sciences and Engineering Research Council of Canada Research Grant No. A9320. Thanks are due to A. Beersing who assisted in data processing.

REFERENCES

Beersing, A.K. 1983. Experimental Study of the Statistical Relationships Between Rainfall and Runoff. M.Sc. thesis presented to the University of Calgary.

Eagleson, P.S. 1972. ' Dynamics of Flood Frequency.' Water Resources Research, Vol. 8, No. 4, pp. 878-897.

Henderson, F.M., Wooding, R.A. 1964. ' Overland Flow and Groundwater Flow from a Steady Rainfall of Finite Duration.' Journal of Geophysical Research, Vol. 69, No. 8, pp. 1531-1540.

Klemes, V. 1978. ' Physically Based Stochastic Hydrologic Analysis' in: V.T. Chow (Editor), Advances in Hydroscience, Vol. 11, pp. 285-356, Academic Press, New York, NY.

APPENDIX

The following expressions define the terms in eqs. (13), (14) and (15), respectively.

$$\underline{q}^2 = a^2[b_0^2 + 2b_0b_1\bar{q} + (b_1^2 + 2b_0b_2)\,\underline{i}^2 + 2b_1b_2\underline{i}^2 + b_2\underline{i}^4] \quad (16)$$

$$\underline{q}^3 = a^3[b_0^3 + 3b_0^3b_1\bar{q} + (3b_0b_1^2 + 3b_0^2b_2)\,\underline{i}^2 + (b_1^3 + 6b_0b_1b_2)\,\underline{i}^3 + (3b_1^2b_2 + 3b_0b_2^2)\,\underline{i}^4 + 3b_1b_2^2\underline{i}^5 + b_2^3\underline{i}^6] \quad (17)$$

$$\underline{i}^2 = (\bar{i})^2(1 + C_{vi}^2) \quad (18)$$

$$\underline{i}^3 = (\bar{i})^3(1 + 3C_{vi}^3 + C_{si}C_{vi}^3) \quad (19)$$

$$\underline{i}^4 = (\bar{i})^4(1 + 6C_{vi}^2 + 4C_{si}C_{vi}^3 + k_iC_{vi}^4) \quad (20)$$

$$\underline{i}^5 = (\bar{i})^5(1 + 10C_{vi}^2 + 10C_{si}C_{vi}^3 + 5kC_{vi}^4 + M_5C_{vi}^5) \quad (21)$$

$$\underline{i}^6 = (\bar{i})^6(1 + 15C_{vi}^2 + 20C_{si}C_{vi}^3 + 15kC_{vi}^4 + 6M_5C_{vi}^5 + M_6C_{vi}^6) \quad (22)$$

where k is given by

$$k = 1/N \sum_{j=1}^{m} (i_j - \bar{i})^4/s_i^4 \quad (23)$$

and

$$M_5 = 1/N \sum_{j=1}^{m} (i_j - \bar{i})^5/s_i^5 \quad (24)$$

$$M_6 = 1/N \sum_{j=1}^{n} (i_j - \bar{i})^6/s_i^6 \quad (25)$$

where m is the sample size and s_i is the standard deviation of i.

REVIEW OF STATISTICAL MODELS FOR FLOOD FREQUENCY ESTIMATION

Conleth Cunnane,
Department of Engineering Hydrology,
University College,
Galway, Ireland.

ABSTRACT. The relation between flood frequency estimates and the economic decision-making process is briefly discussed and selected developments in frequency analysis are traced in tabular form. The main types of frequency analysis procedures and models are surveyed including at-site, at-site/regional and regional only cases. The at-site/regional cases include Index Flood, Bayesian and TCEV methods. Criteria for selecting a frequency analysis procedure are discussed under two headings, descriptive ability and predictive ability. The former relates to a model's ability to preserve statistics of observed flood series while the latter relates to quantile estimating ability in a robust manner. The relative merits of different types of flood quantile estimators are discussed, beginning with at-site estimators and the search for a robust at-site estimator. This is followed by the principal results available about at-site/regional quantile estimators. Further topics considered are regional homogeneity of flood statistical behaviour, the effect of spatial and temporal interdependence of flood magnitudes, and the detection and use of historical floods.

INTRODUCTION

The objective of flood frequency analysis is to obtain an estimate of flood quantile magnitude Q_T, for one or more locations on a river system. Such an estimate must be accompanied by a statement about its reliability. The error or lack of reliablity of such a quantile estimate stems from both model error and sampling error. Quantile estimates are used in a planning or decision process relating to hydraulic works or flood alleviation programmes. Since many of these decisions have economic implications, flood frequency analysis procedures should ideally be sensitive to the requirements of the decision-making process. However, in current practice the production of a flood quantile estimate is frequently divorced from the economic decision-making process for which it is intended.

Past and current research into methods of flood quantile estimation have concentrated for the most part on the flood statistical aspect of the problem. This is largely because it is relatively easy to conduct investigations into specific aspects of statistical flood frequency analysis in isolation, based on easily-available annual maximum flood series data, while the necessary economic data for research into the decision-making process tends to be application-specific and therefore difficult to assemble in large quantities for research purposes.

For practical purposes, models for flood frequency analysis should be

V. P. Singh (ed.), Hydrologic Frequency Modeling, 49–95.

compared on the basis of economic loss functions which express the loss incurred due to mis-specification of a design flood by a particular model. However the form of the loss function is application-specific and researchers have not been inclined to investigate model performance over a range of reasonable loss function types. As pointed out by Stedinger (1980) "It is necessary, therefore to use surrogate loss functions to evaluate alternative fitting procedures [and models] so as to identify those which appear to be most efficient and most likely to minimise design losses when used for different purposes."

One such surrogate loss function is the symmetrical quadratic one which is proportional to mean square error of the quantity (flood quantile of specified return period in this case) being estimated. In all recent studies rmse, and its components se and bias, has been used as a criterion of model performance.

Davis et al (1972) introduced Bayesian decision theory for a flood alleviation design which involved calculation of estimated expected flood costs and expected flood alleviation benefits while Wood and Rodriguez-Iturbe (1975a) combined flood cost and damage data with Bayesian inference of flood magnitudes, using a range of three distributional assumptions, to choose a flood protection design which maximises expected future monetary benefits. While these and associated works form a framework for decision related flood frequency analysis, research papers of the past decade (with some notable exceptions, such as Kuczera (1982a,b)) have concentrated on other aspects of flood frequency estimation. The usefulness of doing so is brought clearly into focus by Marin (1986) who draws the distinction between "optimal actions" and "optimal parameters [estimates]".

It would seem that hydrological researchers have largely ignored the advice of Fiering (1967) who wrote "The hydrological literature contains many references to the propriety of one statistical distribution or another in empirical flow records. The statistical significance of one distribution or another, the degree of resolution of the tail of one distribution or another, the plotting position or the return period associated with a particular event represent operationally meaningless questions unless they are framed within the context of making decisions. It is evidence of statistical immaturity to argue whether the flows at a particular site are distributed normally, lognormally or gamma if the subsequent design decision is not materially affected".

Historical Development.

A survey of the historical development of frequency analysis is not explicitly attempted here but selected developments under a number of different headings are traced in Table 1. Moss and Kirby (1985) have traced the principle aspects of American practice since the 1860's while at the other end of the time scale Greis (1982) reviewed the early 1980's and Lettenmaier (1985) has reviewed very succinctly recent developments in regional flood frequency analysis.

Dissatisfaction with statistical flood frequency analysis has been voiced in the past and usually arises from the use of single short samples of annual maxima from highly-variable populations for the purpose of estimating large return period quantiles. Flood populations, from which such samples arise, are quite variable in nature having high values of coefficient of variation and skewness and it is inevitable that some samples from such populations will contain either low or high outliers, which in turn effect quantile estimates in a manner which may not always be plausible.

Another source of dissatisfaction occurs in regions where the flood population might be considered to have two components, one of which (for example hurricane-caused floods) occurs very rarely and in a spatially-scattered pattern. In such cases the length of hydrological record available may be quite insufficient to enable plausible flood quantile estimates to be obtained which are in any way reconcilable with regional experience. It will be suggested here that some of the sources of dissatisfaction with statistical flood frequency analysis may be eliminated by adopting regional flood frequency analysis procedures.

PROCEDURES FOR FREQUENCY ANALYSIS

Procedures may be grouped under different types of headings such as

(i) Amount and type of data used (a) at-site data.
(b) at-site / regional.
(c) regional only without at-site data.

(ii) Type of model (a) Annual maximum (AM) series.
(b) Peaks over threshold (POT) series.

(iii) Form of distribution and estimating procedure used, of which there have been many possible combinations suggested.

The distinctions between the items in the first group are of the greatest importance even though the greatest amount of published literature relates to various subdivisions of the last group.

Classification of Procedures by Amount and Type of Data Used.

At-site data only. An AM or POT frequency model is proposed and calibrated from the N-year record of AM or POT flood peaks at a site. The quantile Q_T is then estimated as

$$\hat{Q}_T = \hat{\theta}_1 + \hat{\theta}_2 \, y_T(\hat{\theta}_3) \qquad \text{...(1)}$$

in the AM model where $\hat{\theta}_1, \hat{\theta}_2, \hat{\theta}_3$ are estimates of location, scale and shape parameters of a selected distributional form $f(q;\theta)$ and $y_T(\theta_3)$ is a standardised variate value of return period T from the f(.) distribution. Alternatively if a POT model is assumed

$$\hat{Q}_T = \hat{\beta}_1 + \hat{\beta}_2 \, y_T(\hat{\beta}_3, \lambda) \qquad \text{...(2)}$$

where again $\hat{\beta}_1$, $\hat{\beta}_2$ and $\hat{\beta}_3$ are estimates of a distribution's parameters and λ represents the average number of independent flood peaks per year exceeding the specified threshold.

At-site plus regional data. In a given geographical region let there be M hydrometric gauges having Nj, j = 1,2....M years of record at each. Let

$$L = \Sigma_1^M \, Nj \qquad \text{...(3)}$$

be the total number of station years of record in the region. If a flood quantile estimate is required at one of these sites there are two options open, namely.

(i) use the at-site data as in case (a) above and ignore the existence of the other (M-1) gauges, or

(ii) use a regional estimating procedure which avails of all the L years of record in the region. This requires an assumption about regional homogeneity of catchments with respect to flood statistics which may or may not be justified.

Under the latter heading there are three important types of procedure, Index Flood Method, Empirical Bayes and Two Component Extreme Value (TCEV).

<u>(i) Index Flood Method.</u> The simplest regional procedures estimate the distribution of a dimensionless flood variate X, where $X = Q/Q_I$. Q_I is called an index flood and is usually taken as the mean annual flood μ, whose at-site sample estimate is denoted by $\bar{Q}$. Then x_T is estimated from the M data sets and μ is estimated by $\bar{Q}$ from the at-site data and

$$\hat{Q}_T = \hat{x}_T \cdot \bar{Q} \qquad \text{...(4)}$$

A graphically implemented version of this index flood approach was suggested by Dalrymple (1960) while an objective numerical method, based on regionally averaged, standardised PWM's was suggested by Wallis (1980). The index flood method was also used by NERC (1975) for British and Irish conditions using regionally averaged, standardised order statistics in a graphical procedure to estimate portion of the X distribution. This procedure was later implemented for New Zealand (Beable and McKercher, 1982, p.24) and Nepal (WEC, 1982).

The distribution of dimensionless variate X could also be estimated from regional average values of C_v (Nash and Shaw, 1965) and if necessary, g. A two-parameter X distribution would then have mean = 1 and $C_v = \bar{C}_v$ while if a three-parameter X-distribution were chosen $\bar{G}$ would also have to be used. Use of such a method would have to take note of the bias in C_v and g (Wallis et al. 1974). A variation of this approach is to estimate two moments of the logarithms m_z and s_z from at-site data and adopt a regional average value of g_z or a weighted average value of regional and at-site estimates of g_z (USWRC, 1981).

Basically the USWRC estimator assume that floods are distributed as LP3 and quantile estimation is achieved by moments estimation in the log-domain. m_z, s_z and g_z are estimated from logarithms of at-site AM data (USWRC,1967). The skewness estimator was modified in USWRC (1977) to

$$\hat{g}_z = \bar{G}, \qquad N < 25, \qquad \text{...(5a)}$$

$$= \bar{G} + (g_z - \bar{G})[(N-25)/75], \qquad 25 \le N \le 100, \qquad \text{...(5b)}$$

$$= g_z, \qquad N > 100, \qquad \text{...(5c)}$$

where $\hat{g}_z$ is the log-skew value to be used in LP3 quantile calculation, $\bar{G}$ is regional skew obtained from neighbouring catchments or from a published "map of generalised [regional] skew" and g_z is the skewness calculated from the logarithms of the at-site data. USWRC (1981) has again modified $\hat{g}_z$ and re-expressed it as a weighted sum of $\bar{G}$ and g_z for all sample sizes. The published recommendations also include procedures for dealing with years having zero flood values, incomplete years, low outliers, high outliers and historical flood data.

While there is no guarantee that AM floods are LP3 distributed, the above "base procedure" was adopted for reasons of institutional uniformity for use by US Federal Agencies (Benson, 1968). LP3 was chosen on the basis of a goodness of fit test on eight long AM records in USA i.e. on the basis of descriptive ability. This was not a very satisfactory test as it later appeared that it was sensitive to plotting position formula (NERC, 1975, Vol 1, p158). NERC also found that three parameter distributions are always favoured by such tests.

(ii) Empirical Bayes method. Use of Bayes' Theorem allows regional and at-site data to be combined in a manner different from the index-flood approach. Let $f(q;\theta)$ be the pdf of AM floods dependent on parameters θ. Let $\mathbf{c}$ denote a vector of catchment characteristics such as area, slope, mean annual rainfall.... It is possible to imagine several catchments having the same value of $\mathbf{c}$ but having different values of θ. If the different values of θ are assigned randomly to the several catchments having a common $\mathbf{c}$ value, the corresponding distribution $\phi(\theta/\mathbf{c})$ is called the prior distribution of θ. The population of such θ values may, in Kuczera's (1982) terminology, be called a superpopulation.

The dependence of θ on $\mathbf{c}$ is usually expressed by regional regression relations

$$\theta = \mathbf{c}^T\beta + \varepsilon \qquad \text{...(6)}$$

where β are the regression coefficients and ε is a random vector, or by the corresponding type of relation in the log-domain. For instance, based on the data of 57 catchments in Britian, with $\theta^T = \{\mu, C_v\}$ and $\mathbf{c}^T = \{A,\overline{R}\}$, Nash and Shaw (1966) found that the dependence between θ and $\mathbf{c}$ could best be described by

$$\ln \mu = 0.85 \ln A + 2.2 \ln \overline{R} - 4.67 + \varepsilon_1 \qquad \text{...(7a)}$$

$$\ln C_v = -0.5 \ln \overline{R} + 5.39 + \varepsilon_2 \qquad \text{...(7b)}$$

where ε_1 and ε_2 are the residuals of the regression relation or those random components of μ and C_v which do not depend on A and $\overline{R}$. Cunnane and Nash (1971) showed that the observed sequences of residuals were found to satisfy the hypothesis that each was normally distributed with insignificant cross-correlation. Thus $\phi(\theta/\mathbf{c})$ could in this case be considered a bivariate lognormal density function. Relations between mean and variance, in both Q and Z domains, and catchment characteristics derived from data of 36 New England basins were used by Wood and Rodriguez-Iturbe (1975) to define prior distributions of model parameters. Kuczera (1982a) related s_z^2 to four catchment characteristics using data of 19 catchments in New England while Lettenmaier and Potter (1985) related m_z and s_z^2 to catchment area for use in the Kuczera (1982a) linear Bayes regional flood model.

The prior distribution $\phi(\theta/\mathbf{c})$ expresses the regional information about θ. The at-site information is expressed by the likelihood

$$\mathbf{L}(\mathbf{Q}/\theta) = \Pi_1^N f(q_i;\theta) \qquad \text{...(8)}$$

These two are combined in Bayes' Theorem to give a posterior distribution of parameters $\psi(\theta/\mathbf{Q},\mathbf{c})$

$$\psi(\theta/\mathbf{Q},\mathbf{c}) = \phi(\theta/\mathbf{c}) \,.\, \mathbf{L}(\mathbf{Q}/\theta) \Big/ \int_\theta \phi(\theta/\mathbf{c}) \,.\, \mathbf{L}(\mathbf{Q}/\theta) d\theta$$

$$= k\,\phi(\theta/\mathbf{c}) \,.\, \mathbf{L}(\mathbf{Q}/\theta) \qquad ...(9)$$

where k is a normalising constant.

Point estimates $\hat{\theta}$ of the parameters θ may be obtained by taking the expectation of the posterior distribution. Hydrologically, however, the parameters themselves are not of immediate interest but some function of them $h(\theta)$ which represents the T-year flood is required. Using the distribution theory of transformations the posterior pdf of $h(\theta)$ can also be obtained and its expectation may be taken as one Bayesian estimate of Q_T. However, this procedure fails to make use of available decision-theory methods which enable the estimator $h^*(\hat{\theta})$ to be obtained which minimises some specified loss function on either an at-site basis or on an aggregate regional basis (Kuczera 1982a, p.308). The Bayesian approach opens the way for such choices but uses of such procedures have not been widely reported in flood hyrology, apart from Davis at al (1972) , Wood and Rodriguez-Iturbe (1975a) and Kuczera (1982a).

Kuczera (1982a) refers to the Bayesian approach as a model for "borrowing strength" from similar basins to estimate some hydrologic quantity for a specific basin. It is not alone in this, however, as other regionalisation models also "borrow strength" for the same purpose and not necessarily in a less efficient manner. The Bayesian method does not have to assume perfect regional homogeneity. In fact specifying a prior distribution, equivalent to acknowledging the super-population concept, acknowledges heterogeneity. The index flood method assumes regional homogeneity but it does not need to be discarded in the presence of moderate heterogeneity. The Bayesian method, in giving a posterior distribution of parameters, allows legitimate subjective probability statement to be made about parameters and quantiles and this holds even if a non-informative prior distribution (one which is not based on region flood information, in this context) is used. This is one of its major advantages.

Bayesian flood estimation studies which have used informative prior distributions to date have all used regional regression models which express parameters θ in terms of catchment characteristics **c**. Since these models are quite imprecise they do not provide low variance prior distributions. Nash and Shaw (1965) showed that $\theta_1 = \bar{Q}$ estimated from catchment characteristics is only as good as $\bar{Q}$ obtained from one year of record or less. Their result holds for a catchment located at the centroid of the catchment characteristic space. For other catchments, the conclusion is much worse (Hebson and Cunnane,1986, Table 7). Thus prior distributions add very little precision to the at-site estimates.

A prior distribution based on index flood estimates of standardised flood quantiles could well be more informative than those used before now. Index flood methods give more precise quantile estimates than at-site estimators (Lettenmaier et al., 1985, Wallis and Wood, 1985) which in turn are more precise than regional regression estimates. Hence index flood based prior distributions or TCEV based prior distributions ought to be worth investigation.

<u>(iii) TCEV Regionalisation procedure.</u> The TCEV method is based on a four parameter distribution for annual maximum (AM) floods (Rossi et al. 1984) whose parameters are estimated by ML from the combined regional data set of standardised AM values. The model is consistent with assuming that flood peaks do not all come

from a single parent distribution but rather that the most extreme events and the more ordinary ones can be considered to come from different distributions. This allows separate consideration of different physical mechanisms of flood production as was done successfully by Waylen and Woo (1982). In the application to Italian flood data, quoted by Rossi et al (1984), the different physical mechanisms could not be identified uniquely with season unlike the Waylen and Woo (1982) case, and the parameters of the separate component distributions could not be estimated independently of one another.

Two independent sequences of iid floods are considered to occur in any year on a catchment $\{Q_{1i}, i = 1, 2...K_1, \; Q_{2j}, j = 1, 2 ...K_2\}$ where K_1 and K_2 are random Poisson variates with expectations $E(K_1), = \Lambda_1$ and $E(K_2) = \Lambda_2$ respectively. The total number of floods in a year is also a Poisson variate, $K = K_1 + K_2$, with expectation $\Lambda = \Lambda_1 + \Lambda_2$, while the distribution of peak magnitude Q regardless of which sequence it comes from is a mixture

$$F_Q(q) = p F_{Q1}(q) + (1 - p) F_{Q2}(q), \qquad q>0 \qquad \text{... (10).}$$

where $p = \Lambda_1/\Lambda_2$. One of the components Q_1, is regarded as a "basic" or ordinary one while the other component is more variable and generates rarer and larger floods. Q , the annual maximum, is the maximum of a compound Poisson process with df

$$F_Q(q) = \exp\{-\Lambda_1 [1 - F_{Q1}(q)] - \Lambda_2 [1 - F_{Q2}(q)]\}, \quad q>0$$

$$= F_{Q1}(q) \, . \, F_{Q2}(q) \qquad \text{...(11)}$$

where $F_{Q1}(q)$ and $F_{Q2}(q)$ are the df's of the maxima of the two component compound Poisson processes. In the event that Q_1 and Q_2 are exponentially distributed with $E(Q_1) = \theta_1$, $E(Q_2) = \theta_2$ then

$$F_Q(q) = \exp\{-\Lambda_1 e^{-q\theta_1} - \Lambda_2 e^{-q/\theta_2}\} \qquad \text{...(12a)}$$

$$= \exp\{-e^{-(q-\varepsilon_1)/\theta_1}\} . \exp\{-e^{-(q-\varepsilon_2)/\theta_2}\} \qquad \text{...(12b)}$$

where $\varepsilon_1 = \theta_1 \ln \Lambda_1$ and $\varepsilon_2 = \theta_2 \ln \Lambda_2$, which represents the product of two EV1 distribution functions which suggests that annual maxima in this model can be considered to be the maxima of two independent EV1 variates - subject to the modification required in acknowledging that a zero Q value may occur with finite probability, $\exp(-\Lambda)$.

As expressed above, the model applies to a single site. In such a case its parameters could be estimated from POT data, in a simple fashion if separation into recognisable components is possible, or failing that, θ_1, θ_2 and p of eqn (10) can be jointly estimated from the aggregate POT data by ML (Hasselblad, 1969). In the absence of POT data the parameters $(\theta_1, \theta_2, \Lambda_1, \Lambda_2)$ can also be estimated iteratively by ML from AM data as described by Rossi et al (1984).

However, at-site estimation can lead to unstable parameter estimates which give rise to physically unrealistic flood frequency predictions. Thus Rossi et al (1984) recommend that TCEV be used in a regional estimation context and also that validation of the model as a suitable parent for AM floods can also only be done on a regional basis. As a first measure in regional parameter estimation some standardisation of the data is required so that data from different sized catchments can be jointly used. Secondly some kind of simplifying assumption is needed to keep the problem manageable (as for instance in the index flood method $X = Q/Q_I$ is assumed to have the same distribution at all sites i.e. constancy of C_v and g). In this case standardisation is achieved for each site by

$$Y' \quad = \quad (Q - \varepsilon_1)/\theta_1 \quad = \quad Q/\theta_1 - \ln \Lambda_1 \qquad ...(13)$$

where $(\varepsilon_1, \theta_1)$ are ML estimates of EV1 parameters obtained from the at-site AM data after rejection of outliers, if any, in a specified way. Y' is distributed in a similar way to Q (eqn 12a)

$$F_{Y'}(y') = \quad \exp\{ - \Lambda'_1 e^{-y'/\theta'_1} - \Lambda'_2 e^{-y'/\theta'_2} \} \qquad ...(14)$$

with modified parameters

$$\theta'_1 = \Lambda'_1 = 1, \quad \theta'_2 = \theta_2/\theta_1, \quad \Lambda'_2 = \Lambda_2/(\Lambda_1^{1/\theta'_2}) \qquad ...(15)$$

The simplifying assumption then introduced is that θ'_2 and Λ'_2 are constant over the region i.e. the rate of occurrence of extraordinary floods and the mean standardised magnitude of them is constant for all sites.

The $L = \Sigma_1^M N_j$ values of standarised Y' so obtained are then considered to be a single random from a TCEV distribution and the four parameters are estimated iteratively by ML from it. Note that the values θ'_1 and Λ'_1 are estimated also rather than retaining the unit values already mentioned in eqn 15.

Rossi et al. (1984) applied the above method to 39 Italian AM series, totalling 1525 station years while Beran et al. (1986) applied it to 2334 station years of AM data from 57 sites in Britian (UK). By simulation experiments the TCEV model was validated in both cases in regard to the regional distribution of skewness. Random samples of size comparable to the observed series used for parameter estimation gave skewness values which were as variable as the variability of skewness observed among the observed AM series, thus accounting for the condition of separation of skewness (Matalas et al.,1975) for Italian and UK data. Rossi et al.(1984) had earlier shown that the regional distribution of AM skewness of Italian data could not be satisfactorily modelled by EV1, Power Extreme Value (PEV) or lognormal hypotheses.

Rossi et al. (1986) in replying to Beran et al. (1986) suggest that a hierarchical approach to use of TCEV can be considered in which θ'_2 and Λ'_2 might be estimated at regional scale, Λ'_1 on a subregional scale and lastly $\varepsilon_1 = \theta_1 \ln \Lambda_1$, which is a type of index flood, is estimated at a basin scale. A weakness in the current implementation of the model is the "station year" type of assumption that L values of Y' form a random sample. An improvement, if it could be effected, would be to average estimated quantities across sites, as in most index flood methods, thus being less sensitive to the effects of inter-site correlation.

Regional Data Only (i.e. at-site data not available). In this case an index flood approach may be used with x_T estimated by one of the methods mentioned above. $\overline{Q}$ is estimated using a relation between $\overline{Q}$ and catchment characteristics, calibrated from the available (M-1) neighbouring data sets. Such relations (Benson (1962), Nash and Shaw (1965), Thomas and Benson (1970), NERC (1975)) usually take the form

$$\overline{Q} \quad = \quad c.\, A^a R^r S^s \ldots\ldots \qquad ..(16a)$$

The ability of such equations to predict $\bar{Q}$ satisfactorily leaves much to be desired (Hebson and Cunnane, 1986).

As an alternative to using $\bar{Q}$ from eqn 16a with an index flood approach the regional average values of C_v and g could be used, along with $\bar{Q}$ from eqn 16a to estimate by method of moments the parameters of two parameter or three-parameter distributions respectively.

Another alternative is to estimate, using at-site methods, Q_T, T = 2, 5, 10, 25, 50, 100 at all guaged sites in a region and then to estimate parameters of the relations

$$Q_T = c.A^a R^r S^s \ldots\ldots \qquad (16b)$$

for each T. Such a technique has been used in many flood frequency studies in United States of America (Roeske (1978), quoted by Tasker and Moss (1979), Eychaner(1984)).

CRITERIA FOR SELECTING FREQUENCY ANALYSIS PROCEDURES

A procedure in this context comprises three things, viz. model type, distribution type and method of parameter/quantile estimation. The latter refers also to the amount of data (at-site and /or regional) to be used.

Evaluation criteria can be divided into two categories:

(a) Criteria relating to ability of a chosen model to describe/reproduce chosen aspects of observed flood peak hydrology i.e. descriptive ability.

(b) Criteria relating to a procedure's statistical ability to achieve its assigned task, with minimum bias and maximum efficiency and robustness i.e. predictive ability.

Descriptive Ability

AM flood series in general are positively skewed with observed average regional C_s values in the range 0.5 to 3.0 ((NERC,1975, Vol.1, p122), Matalas et al. (1975), Beable and McKerchar (1982)). Matalas et al. (1975) also found that C_s values calculated from M annual maximim flood records of length N (=10,20,30) in U.S.A. are more variable than C_s values calculated from M random samples of size N from any of the distributions commonly used for AM flood series. They named this discrepancy the "condition of separation". This condition has also been noted in Italian AM data by Rossi et al. (1984) and in British data by Beran et al. (1986). The performance of GEV samples in this context, not tested by Matalas et al.(1975), is shown in Figure 1 where it is seen that GEV offers some improvement over the other distributions, but nevertheless fails to account for the variability in C_s observed among abserved AM data.

AM flood series have C_v values which vary from 0.1 in tropical rainforest climates (Atkins, 1982) to around 1.0 in semi-arid areas (Wallis, 1982 b) and sometimes in less arid but low rainfall areas such as the south east coast of New Zealand (Beable and McKerchar, 1982).

It should be noted that the population values of C_v and C_s are necessarily higher than the observed average values because of the known small sample bias in these statistics (Wallis et al 1974, Kirby, 1974). Some, but not all, AM series contain extra-ordinarily large flood values, known as outliers (Kottegoda 1984, Costa 1985).

These observations taken together suggest that AM floods come from populations with positively skewed distributions and that these distributions are relatively thick-tailed. This excludes immediately symmetrical distributions such as normal for AM flood populations but nevertheless leaves a large number of positively skewed distribution families as possible candidates, with EV1/GEV, Gamma/Pearson Type 3, lognormal/log Pearson Type 3 and recently Wakeby and TCEV being the most popular.

Choosing between such candidates is by no means trouble free as the sampling variations inherent in small samples from skewed, heavily tailed populations tend to mask true between - distribution differences. Goodness of fit indices such as χ^2 calculated on the histogram, Kolmogorov-Smirnov index based on the cumulative histogram and sums of squares of deviations on probability plots (Benson 1968, NERC 1975, Vol.1, pp 150-160) all tend to be inconclusive. This is in part due to the fact that each distribution being tested has its parameters calculated from the same data which forces a strong resemblance on the very distributions between which the tests seek to discriminate. Use of moment-ratio diagrams (Wu and Goodridge, 1976, McMahon and Srikanthan, 1981) to discriminate between distributions is unsatisfactory because of bias and large sampling variance of the plotted quantities Wallis (1982a). However, if the condition of separation of skewness were to be taken seriously , then only the Wakeby (Houghton 1978) and TCEV (Rossi et al. 1984, Beran et al ,1986) distributions would be considered as suitable.

Houghton (1977) plotted robust measures of skew and scale for forty-six 60 year high quality records and then subdivided the scatter plot into four quadrants with origin at the centroid of the plot. A separate Wakeby distribution was fitted to the records of the sites identified by the four quadrants. By simulation experiments these four regional Wakeby distributions were shown to be able to generate random samples whose statistics agreed well with those of the observed data and which also accounted for the Matalas et al. (1975) condition of separation of skewness.

Rossi et al. (1984) suggest that discriminiation between models or distributions should be based on regional behaviour of statistics, such as skewness and standardised largest value, which reflect more the upper tail behaviour of observed data and of the proposed model. The use of regional sets of such statistics, although implying regional homogeneity, can be more effective than single highly variable values. Single values of skewness have such high standard error as to be almost valueless in this context, as remarked fifty years ago by Slade (1936). However, the regional distributions of C_v, C_s and C_k as well as the standardised largest value $Q_{max}/\overline{Q}$ behave quite regularly and are comparable with corresponding distributions observed among samples simulated from theoretical distributions, see Figures 2 and 3.

Tests of hypotheses relating to GEV parent flood distributions have been conducted by Njenga (1985) on such regional sampling distributions of C_v, C_s, and C_k but the power of these tests against other parent distribution types has not yet been established. The hypothesis of GEV parent for the data of Figure 2 was not rejected by these tests on regional distributions of C_v, C_s and C_k, but these data come from what may be described as a benign flood region , whose flood C_s values do not display the condition of separation . For a less benign flood region in Italy

Rossi et al.(1984) found that neither EV1, lognormal nor PEV distributions could model the regional distribution of skewness but that TCEV could do so.

Predictive Ability

In this category, apart from the fact that consideration is restricted to "flood-like" distributions, observed data facts are temporarily ignored. Then the statistical properties of estimating procedures are considered in isolation. Properties of quantile estimates, $\hat{Q}_T$, are more appropiate to consider for flood hydrology than those of individual parameter estimates. These properties, obtained either algebraicly or by Monte Carlo methods, are

$$\text{bias} = E[\hat{Q}_T - Q_T] \qquad ...(17)$$

$$\text{se} = \{E[\hat{Q}_T - E(\hat{Q}_T)]^2\}^{1/2} \qquad ...(18)$$

$$\text{rmse} = \{E[\hat{Q}_T - Q_T]^2\}^{1/2} \qquad ...(19)$$

Use of rmse as a criterion of quantile estimator performance is equivalent to using a symmetrical quadratic loss function for expressing the consequences of under- or over - specification of a design flood quantile. Likewise, use of bias is equivalent to a symmetrical linear loss function.

Since there is interaction between distribution type and method of estimation (i.e. moments estimators are in general not as inefficient for two parameter distributions as for three parameter ones) they cannot be treated separately. Hence such studies can span the following topics simultaneously:-

- choice of model or distribution,
- choice of estimation method,
- choice of data series.

Robustness (Kuczera 1982) is an estimator property that is now receiving well merited attention from hydrologists. In this context it is best to consider assumed parent distribution and estimator (D/E) jointly. A D/E is robust in the flood frequency context if it estimates medium and high return period quantiles with low bias and low standard error even if the sample on which it operates is drawn from a distribution other than the assumed one. It is acknowledged that the distribution from which the sample is drawn is in the words of Landwehr et al. (1980), reasonably"flood-like".

Suprisingly, Slack et al.(1975) found on the basis of Monte Carlo experiments that "in the absence of information about the distribution of floods and the economic losses associated with the design of flood reduction measures, the use of the normal distribution to represent the distribution of floods is generally better than either the Gumbel, lognormal or Weibull distributions". This study involved a between-distribution robustness aspect and in the absence of better information on flood losses assumes a symetrical quadratic loss function (analogous to rmse of eqn 19 above) to discriminate between competing models. However, this need not be interpreted as recommending the normal distribution for flood frequency analysis since the wealth of empirical knowledge about floods is completely ignored in such a statement. The study goes on to conclude that given prior knowledge about flood skewness a "considerable reduction in the expected opportunity design loss results". It does not seem reasonable to ignore the available empirical knowledge on flood distributions and the latter finding confirms this.

Kuczera (l982b) used mse as a criterion for adjudicating the relative robustness of competing LN2, LP3, EV1, log EV1 and Wakeby models for at-site quantile estimation from samples drawn from each of four Wakeby populations, which had earlier been advanced by Houghton (1977) as being representative of flood experience in U.S.A. . He concluded that LN2 was the most resistent estimator for return periods up to 200 years, EV1 being a close second while Wakeby performed better than LP3 despite having more parameters and hence seemingly being more susceptible to sampling error.

Descriptive Versus Predictive Ability

Having separated these two model properties it is valid to ask whether one or other aspect should be allowed to over-ride the other. The answer is **no**. Study of predictive ability helps to exclude non-robust and inefficient models but the range of realistic populations on whose data the predictive abilities of models must be tested depends on choices made as a result of study of descriptive ability of models. The two aspect complement one another.

Notwithstanding Gumbel's (1966) remark, "It seems that rivers know the [EV] theory. It remains to convince the engineers..........", theoretical arguments about model type, such as those advanced by Gumbel (1941) in favour of EV1 and by Chow (1954) in favour of lognormal are not readily accepted by hydrologists without strong empirical support. Incidentally EV1 and LN2 as models for predicting Q_T, $T \leq 200$, have both been supported by Kuczera's (1982b) predictive ability study, which in turn is conditional on Wakeby populations which had earlier been selected on the basis of empirical descriptive ability. On the other hand EV1 and LN2 models were rejected by Rossi et al.(1984) on the basis of descriptive ability of regional skewness.

Influence of other Theoretical Results.

Theoretical results about estimator bias and efficiency influence choice of estimation method from among maximum likelihood (ML) , moments (MOM) , probability weighted moments (PWM) , least squares (LS) and graphical procedures. Theoretically, since ML is asymptotically most efficient and consistent it ought to be first choice but the fact that it may be biased in small samples and computationally tedious lessens its popularity. Corrections for bias can be made (Fiorentino and Gabrielle (1984), Hosking (1985)) and computation should no longer pose a problem. PWM, where applicable, and LS tend to be unbiassed because they involve linear functions of the observed data.

Theoretical results (obtained algebraicly or by Monte Carlo methods) relating to censored-sample estimation methods may, in future, have a bearing on the treatment of data series prior to quantile estimation. Some data series contain low outliers which, if estimation is carried out in the log-domain, can have undue effect on the estimated upper quantiles. Omission of the lowest 20% of the sample data, noting only their existence in the likelihood function and making use of censored sample estimating techniques, Leese (1973), leads to upper quantile estimates in the simple EV1 case which are less biassed than full sample estimates and only slightly less efficient (H.N. Phien, 1986, Personal Communication). Thus, censoring from below as a matter of course might be advantageous. It may be that smaller sample values have only a nuisance value in the context of upper quantile estimation and also in model form testing and verification. On a related point, inclusion of more and more flood peaks in peaks over a threshold (POT) series, by lowering the threshold, frustrates the choice of distribution type and may lead to underestimation of upper quantiles (NERC, 1975, Vol.1.2.7.). Senaratne (1986) has found that (POT) estimates of upper quantiles increase, on average when taken region-wide, with increase of threshold level.

RELATIVE MERITS OF FREQUENCY ANALYSIS ESTIMATORS

Comparison may be made internally among at-site estimators and among at-site/regional estimators and on another level between at-site, at-site/regional and regional only estimators.

Comparisons of At-site Estimators

At-site estimators may be compared on the based of model type, AM or POT, and separately on the basis of distribution type. AM and POT models do not define the same Q-T relation at low return periods (Langbein (1949), Chow (1950), Beran and Nozdryn-Plotnicki (1977), Takeuchi (1984)) but this is not a major factor at high return periods. In order to be statistically as efficient as an AM quantile estimator from N annual maxima, a POT series has to include at least M = 1.65N peaks (Cunnane (1973), Taesombut and Yevjevich (1978), Tavares and da Silva (1983), Rosjberg (1984)). This means that the annual exceedance series (i.e. M = N or equivalently $\lambda = 1$) is less efficient than the AM series.

Comparison of quantile estimating ability. Of greater importance are the comparisons which may be made of a particular model's ability to estimate flood quantiles. Different estimators applied to the same real data set usually provide numerically different results. Such results cannot of themselves provide any information on which of the methods is best. Estimators are best compared on the basis of their performance at estimating Q_T from samples which are randomly drawn from a known statistical population (the parent) for which the true value of Q_T is known. For such comparisons to be validly applicable to flood frequency analysis it is necessary that the parent distribution have statistical characteristics representative of observed AM data sets. While it is desirable that such comparison be achieved analytically it can much more easily be achieved by simulation (Monto Carlo) methods.

Effect of log transformations. Many AM flood series have negative log-skew values which are especially sensitive to low outliers. If the LP3 model is used, negative skew in the log domain (Z) implies an upper bound to the Z distribution and hence to the Q distribution. Sometimes this calculated upper bound can be unrealistically low (Reich, 1977, Fig 6) which suggests that log transformations should not be used lightly. Other inconsistencies between log- and real-domain statistics are highlighted by Landwehr et al (1980).

Robustness. Because the true distribution of floods cannot be known with certainty, each estimator's robustness to change of parent distribution must also be examined

(Kuczera 1982b). For any given parent distribution of AM floods, competing estimators can be compared with respect to bias, se and rmse. It may happen that an estimator based on a particular distributional assumption enjoys small bias and se when tested on samples drawn from the same distribution, but it may suffer from a large bias when the samples are drawn from different populations, especially those having skewness markedly different from that of the distribution assumed by the estimating procedure. Such simulation experiments involve the following broadly specified steps.

(1) Select the flood-like parent distributions.
(2) Select the estimators (distributional assumption and method of parameter estimation).
(3) Select the region size (No. of stations, M , and length of record at each station, N_j, $j = 1,2 \ldots M$).
(4) Select the parameters of the parent populations so as to produce homogeneous or heterogeneous regions as required.

These selections determine the range of applicability of the results.

(5) For each parent population

(a) Generate a region-full of data,
(b) For each estimator calculate $\hat{Q}_T$, for each site. Repeat for different selected values of T,
(c) Repeat steps (a) and (b) a large number of times,
(d) For each estimator calculate bias, se, rmse.

(6) Compare the different estimator on the basis of results of 5(d).

Such experiments yield a very large amount of numerical results which require care and patience to assimilate and digest. Rarely does a single estimator emerge as best when taken over all stations, all return periods, all possible region sizes and record lengths and all parent populations. Some form of compromise summarisation of the results is usually adopted. For instance each estimator might be characterised by its average rmse value taken over all parent populations. That estimator having the minimum average value might be selected as the most robust estimator or alternatively that estimator having the minimum value over all parents of maximum rmse for each parent (Kuczera 1982b) might be selected.

Many results have been published relating to standard error and bias of quantile estimates in AM models, see Kaczmarek (1957) and others listed in Table 2. Some of these refer to cases where the parent population and the assumed model have the same distribution and can be classified as within-distribution studies.

Between Distribution Comparison of At-site Estimators.

Of greater importance are between-distribution studies where the parent and model distributions differ. It is important to assess how each model performs over a range of parent distributions since in reality nature's distribution is unknown to the hydrologist. A model's ability to estimate Q_T over a range of parent distributions is related to its robustness (Kuczera, 1982b). The range of parent distributions to be considered is a matter of choice but all reasonable "flood-like" (Landwehr et al. 1980) parent distributions have to be considered.

Choice of parent distribution for robustness studies. Landwehr et al (1980) used six Wakeby distribution (Houghton, 1977) parents

$$q \quad = \quad m + a[1 - (1 - F)^{b}] - c[1 - (1 - F)^{-d}] \qquad \text{...(20)}$$

having skewnesses varying from zero to 4.14 as the basis of a robustness test on three quantile estimating techniques. These are EV1 (with MoM, ML, and PWM estimation), LN3/MoM and WAK/PWM which vary from the simplest to the most complex methods available. They state in support of Wakeby as a suitable parent distribution for this type of test:

> "Empirical evidence, in relation to the condition of separation [see Matalas et al., 1975], suggests that the distributions of floods are more nearly Wakeby-like with b > 1 and d >0 (i.e., long stretched upper tails) than like any of the other more commonly suggested flood distributions [see Houghton, 1977; Landwehr et al., 1978]. That the Wakeby distribution can satisfy the condition of separation does not imply that indeed floods are distributed as Wakeby. However, the Wakeby distribution provides a plausible description of flood sequences, and it also provides a means for representing the seemingly long, stretched upper tail structures of flood distributions, as well as the tail structures of the distributions of other hydrologic phenomena. Thus the Wakeby distribution provides a convenient analytical and a reasonable hydrologic basis for assessing the relative performances of alternative techniques of estimating the unknown quantiles of the distribution of hydrologic phenomena."

Landwehr et al. (1980) found that the flexible WAK/PWM estimator is less biased than EV1 or LN3. If parent skewness $<<1.14$ or $>> 1.14$ the EV1 estimators give precise (small se) but inaccurate (large bias) quantile estimates. When parent skewness is high both EV1 and LN3 underestimate quantiles. Since flood population skewness is larger than that measured in observed samples (Wallis et al., 1974), Landwehr et al. suggested that WAK/PWM, although having larger se than the other methods, should be used in preference to EV1 or LN3 if flood estimates are required for high skew locations. However, it must be noted that rmse $(\hat{Q}_T)$ is very large for WAK/PWM applied to single records and it must be acknowledged that WAK/PWM is more suitable for use in the regional context.

In this context, Kuczera (1982b) used the Wakeby distributions estimated regionally by Houghton (1977) from long records in four regions in U.S.A. as parent distributions. He considered a selection of competing model distributions (Normal, LN2, LP3, EV1, log-EV1 and Wakeby) fitted by a variety of methods. He found that LN2 was the most resistant at-site estimator of those tested, followed closely by EV1, for return periods up to 200years and sample sizes 15 and 30. Thus the two-parameter distributions LN2 and EV1 outperformed their more general 3-parameter relatives LP3 and log EV1. The latter two, because of their more flexible shapes were better able to reproduce the parent distribution shapes i.e. they displayed small bias. However, they experienced very large standard error so that their rmse values exceeded considerably the rmse of LN2 and EV1 estimators, despite the bias experienced by the latter two. Thus the increased efficiency of two-parameter models more than compensates for their possible inherent bias.

Stedinger (1980) also found that LN2 and LN3 distributions estimate quantiles from LP3 populations quite satisfactorily.

The poorer performance of LP3 and log-EV1 as at-site estimators is not solely due to their having three parameters as the four-parameter Wakeby model performed better than either of them showing that method of parameter estimation is important also. PWM was used for the Wakeby model while ordinary moments were used for LP3 and log EV1. Quantiles based on ordinary moments perform poorly for three parameter models (Matalas and Wallis, 1973) but they are not nearly so bad for two parameter distributions (Lowery and Nash (1970), Kuczera (1982 b)).

Kuczera's (1982b) results are qualitatively similar to results obtained within the confines of GEV parent populations defined by

$$F_Q(q) \quad = \quad \exp\{ - [1 - k(q - u)/a]^{1/k} \}. \qquad ...(21)$$

This distribution was presented in this form by Jenkinson (1955,1969) and was recommended by NERC (1975) for use as a flood distribution for Britain and Ireland.

Lettenmaier et al (1985) show that at-site procedures based on the 2-parameter EV1 distribution are likely to suffer from quantile bias if the parent distribution is not EV1 while those based on 3 parameter GEV distribution are likely to suffer from unacceptably large standard error of quantile estimates. The worst consequences of these effects were intended to be avoided by NERC (1975, Vol.1, p.292) which recommended use of EV1 rather than GEV with short records.

In general, if the parent population has a lower/higher skewness than an assumed fixed skew model, then upper quantile estimates will be biassed upwards/downwards. For instance, if parent population has a skewness of 3.0 and EV1 is the assumed fixed skew model, Q_T will on average be underestimated. On the other hand, use of a variable skewness model with an unbiassed parameter estimation method such as WAK/PWM (Landwehr et al. 1980) or GEV/PWM (Hosking et al., 1985b, Lettenmaier et al., 1985) will tend to give unbiassed quantile estimates. This can be done more efficiently if an independent unbiassed estimate of population skew can be obtained, a priori, leaving only scale and location parameters to be estimated from the regional or at-site data.

It should be pointed out that Kuczera (1982b), having found the most robust at-site estimator then concluded that its accuracy was insufficient and he looked to regional estimating methods which would improve the accuracy of flood quantile estimation (Kuczera 1982 a).

Comparison of Regional Estimators.

A number of at-site and at-site/regional procedures, namely,

(i) Flood Studies Report (FSR) index flood procedure based on regionally averaged, standardised order-statistics combined with a station year component for the upper quantiles (NERC, 1975, Vol. 1, 2.6),

(ii) Use of regionally averaged, standardised PWM's to estimate the parameters of an assumed distributional form, denoted WAK/PWM or GEV/PWM, (Wallis 1980), also an index flood procedure,

(iii) U.S.W.R.C. use of log Pearson Type 3 (LP3) distribution in conjunction with either regional average log skew or a map of "generalised (log) skew".

(iv) Empirical Bayes (EB) procedures (Cunnane and Nash 1971, Wood and

Rodriguez-Iturbe 1975, Kuczera, 1982a) in which regional information is used to define the prior distribution (super-population) of the flood distribution's parameters and this in turn is modified by the at-site likelihood function to give a posterior distribution of the distribution's parameters from which the quantile estimates are obtained,

(v) TCEV distribution adapted for regional estimation of the parameters (Rossi et al 1984, 1986),

have been the objects of investigation and intercomparison by Kuczera (1982 a,b), Hosking et al (1985a), Lettenmaier et al. (1985), Wallis and Wood (1985), Lettenmaier and Potter (1985) and Arnell and Gabrielle (1985). Table 3 gives an indication of the scope of these intercomparisons.

These works, apart from the last, have been synthesised by Lettenmaier (1985) who concludes that

> "regionalisation is the most viable way of improving flood quantile estimation. The performance of regional PWM estimators for GEV and WAK distributions, in particular, are so superior to the currently used institutional methods that no viable argument for the continuation of current practice is evident. Particularly where the flexibility of using a three-parameter distribution is required, the reduction in the variability of flood quantile estimators achieved by proper regionalisation is so large that at-site estimators should not be seriously considered. The lone exception will be those where no physically defensible region can be identified. In such cases, it will be necessary to use a two-parameter distribution, such as EV1 [or LN2], and accept large estimation bias, or a three parameter distribution such as GEV [or LP3] and accept large estimation variability".

Even where the assumption of regional homogeneity is violated Lettenmaier et al. (1985) and Lettenmaier and Potter (1985) have shown that at-site/regional procedures are still preferable to at-site procedures.

The range of parent distributions considered in the works reviewed by Lettenmaier (1985) was extended by Arnell and Gabrielle (1985) who included TCEV as a parent. In particular, they used two TCEV parents whose parameters were estimated from Italian and British data as well as GEV and WAK parents which match the statistics of each of the two TCEV parents i.e. six parents in all. They compare regional estimators only of growth curves, $Q_T/\bar{Q}$, and re-scaled site quantiles, Q_T.

In this study WAK/PWM performed very well over all parents and was found to be the optimum algorithm for estimating growth factors. TCEV estimates had downward bias and low variability when the parent was GEV but had small bias and large variability, particularly at large T, with the other parents. The GEV estimates performed very well with GEV parents but noticeably less well with the other parents. This inflexibility is also noticed in the results of Hosking et al. (1985a). It is imaginable that TCEV parents especially those applicable in semi-arid and arid zones would be very difficult to model. In such a case the TCEV model, despite its higher variability in the above tests may well prove very valuable and it deserves further consideration.

LP3 regional procedure. The quantile estimating ability of the LP3 procedure has been found to be poorer than other readily available methods. As stated above,

Kuczera (1982b) found that LN2, EV1 and Wakeby are more resistent at-site quantile estimators than LP3 when the parent distribution is Wakeby, while Wallis and Wood (1985) have found that even when LP3 is the parent distribution the LP3 estimator is much less precise than regional GEV/PWM or regional WAK/PWM. The LP3/regional skew estimator (eqns. 5,a,b,c) is better than the LP3 /at site skew estimator. The same type of conclusion is true when other parent distributions are considered but to a greater degree, that is the LP3 procedure's performance relative to WAK/PWM and GEV/PWM is even worse than when the parent is LP3. Thus, as well as experiencing high standard error, when used both at-site and regionally, LP3 is also lacking in robustness.

RELATED TOPICS

While the importance of the use of regional estimation techniques for gauged catchments as well as ungauged ones emerges as the main finding of recent research there are a number of other topics which are of either direct or indirect importance in all flood estimation schemes. Some of these are now discussed in different degrees of detail. They are

(i) Regional Homogeneity of Flood Statistical behaviour.
(ii) Spatial and Temporal interdependence of Flood Magnitudes.
(iii) Detection and Use of Historical Floods.

Regional Homogeneity of Flood Statistical Behaviour.

A basic assumption in regional flood estimation is that the variate $X = Q/E(Q)$ has the same distribution at every site in the region. This is what is understood by regional homogeneity of flood frequency behaviour. Departures from this assumption would weaken the procedure. Since $E(X) = 1$ at all sites, invariance of the entire distribution of X could be approximated by invariance of σ_x and g_x over the region ($\sigma_x = c_v(Q)$ and $g_x = g_Q$). Alternatively it could be approximated by the invariance of selected quantiles x_T, say T = 10 and T = 50.

It is to be expected that C_v (X) and perhaps g(X) decrease with increase in catchment size. Hence combining small and large catchments into common regions, as can happen in geographical regionalisation, leads to mixing of heterogeneous populations.

Nevertheless if this heterogeneity is not great its negative effect may be more than balanced by the advantage of estimating the X-distribution parameters from M heterogeneous sites rather than two or more separate estimations each based on M/2 or M/3 sites, especially if the typical site record length is short.

A number of questions may be asked when considering regionalisation problems:

(i) is flood frequency behaviour of any one of M sites in a region, with AM records available, inconsistent with that of the remainder of the group?
(ii) are geographically defined regions better or worse than regions obtained by partitioning the catchment characteristic data space?
(iii) how can a large group of catchments be divided into homogeneous sub-groups or regions?
(iv) how can an ungauged catchment be allocated to one of a number of pre-selected homogeneous regions?
(v) what degree of departure from regional homogeneity can be tolerated in a

flood quantile estimation procedure?

Identification of non-homogeneous catchments. Dalrymple (1960) used a test of homogeneity, attributed to Langbein (1947). In it a regional distribution of $X = Q/\bar{Q}$ is derived and $\hat{x}_{10}$ is read off. Then for the site under test the at-site/regional estimate $\hat{Q}^*_{10} = \hat{x}_{10} \cdot \bar{Q}$ is obtained, where $\bar{Q}$ is at-site mean. An analysis of the N years of at-site data gives a Q-T relationship and the T value attributed to $\hat{Q}_{10}{}^*$ in this relationship is determined. If this T value falls outside theoretically determined confidence limits, dependent on N and available in graphical form, the site is rejected as non-homogeneous with respect to the region. The properties of this test have been discussed by Benson (1962) but do not seem to have been further evaluated since then.

Geographically-defined regions. A simple method of allocating catchments to regions for index flood estimation is the geographical one. This was adopted by NERC (1975) for Britain and by Beable and McKercher (1982) for New Zealand.

In general geographical proximity of catchments is no guarantee that their X distributions are identical, or almost so. Consequently such geographically-defined regions are not necessarily homogeneous in X distribution. Its adoption by NERC (1975) is not, however, entirely implausible as the geographical division adopted reflects changes in rainfall pattern and topography which in turn affect the distribution of X. A disadvantage of geographically-selected regions is that physically-similar catchments which are adjacent but nevertheless in different regions are ascribed different X distributions, a fact which is not considered reasonable.

Wiltshire (1985, 1986 a,b,c) and Acreman and Sinclair (1984, 1986) have examined the consequences of choosing regions geographically for Britain and Scotland respectively. In general, their conclusions are that geographical regions do not display as much internal homogeneity of flood frequency behaviour nor as much between region heterogeneity as regions selected by other methods proposed by them. It would appear that of the ten geographical regions proposed for Britain, by NERC (1975), some are too similar to one another to justify the use of separate X-T relations (growth curves) for them. However, Wiltshire (1986a) points out that geographically-defined regions still "may retain a role in a basin (catchment) characteristic-based grouping scheme" by providing clues as to which characteristic axes to use".

Subdivision of catchments into homogeneous regions. In this section the most recent publications (Wiltshire, (1985, 1986 a,b,c), Acreman and Sinclair (1984, 1986)) are surveyed. Earlier related works (De Coursey,(1972) Farhan, (1984), White (1975), Cole (1966), Biswas and Fleming (1966), Mosley (1981)) are acknowledged but are not explicitly surveyed here. In addition to examining a specific example of flood regionalisation in New Zealand, Mosley (1981) gives a rigourous account of the principles that ought to underlie any regionalisation scheme.

Wiltshire (1985) divides the entire group of catchments into two regions, initially, "on the basis of a single partitioning value of one chosen basin characteristic". For instance, they could "be divided into large and small basins on the basis of basin area, A, or into wet and dry basins on the basis of annual average rainfall, R". He also used a soil runoff index (SOIL) in a similar manner. Initially single partitions on one variable (A, R or SOIL) were performed and subsequently four-way partitions on two variables were performed. The X distribution in the kth

such selected region is estimated by ML by considering ($x_{ij} = Q_{ij}/\bar{Q}_j$, i=1,2...$N_j$, j = 1,2,...$M_k$) to be a random sample (station year assumption) from a GEV distribution.

Measures of the variability of the observed AM data from the regional X distribution are devised and evaluated in the Q-domain, such as

$$SSQ5 \quad = \quad \Sigma_1^p \ SSQ5_k \quad = \quad \Sigma_1^p \ \Sigma_1^{M_k} \ (q_{5j} - \hat{q}_{5j}) \qquad ...(22)$$

where

k = k^{th} of p regions

q_{5j} = distribution-free estimate of Q_5 from at-site data at station j.

$\hat{q}_{5j}$ = $\hat{x}_{5k} . \bar{Q}_j$ is the at-site/regional estimate of Q_5 at site j.

Three other measures, one of which is a log-likelihood, in addition to SSQ5 are also considered. A plot of any of these measures against the partitioning point (A_p, R_p or $SOIL_p$) yields a definite minimum value for some A_p*, R_p* or $SOIL_p$* which can then be chosen as the optimum partitioning point. In the multiple partition case contours, on a two-dimensional space, help to define the optimum partitioning points.

Acreman and Sinclair (1984, 1986) have proposed a method of searching for homogeneous regions of catchments from within a larger, possibly heterogeneous, group of catchments. Their procedure is a two stage one,

(a) use of a NORMIX clustering algorithm which allocates the sites to regions according to their catchment characteristics (such as area, slope, mean annual rainfall, lake index...) and

(b) use of a likelihood ratio method to test whether the estimated GEV distribution of X within one selected region differs significantly from that of another selected region. The selected catchments in a region are adjacent in the catchment characteristic space but they need not be geographically adjacent.

Ultimately 168 catchments in Scotland were classified into five regions "corresponding to physically-reasonable basin types and all but one of which yield homogeneous distributions" of standardised variate X.

Wiltshire (1985) and Acreman and Sinclair's (1984, 1986) proposals have some apparent similarities even though they differ in philosophy. Wiltshire (1985) partitions the group of catchments by a single partitioning of axes in the catchment characteristic space while Acreman and Sinclair (1984, 1986) achieve (multiple) partitioning by a clustering algorithm. Then, both procedures test the efficacy of the partitioning by likelihood ratio methods (inter alia in Wiltshire's case) which test whether the estimated flood distributions differ significantly between the selected regions. Both procedures use the station year assumption that (x_{ij}, i = 1,2 ... N_j, j = 1,2... M_k) forms a random sample of independent values which may not be entirely realistic.

However, a difference between them is that Wiltshire's procedure conducts a systematic search through the catchment characteristic data space while Acreman and Sinclair's procedure seems to make a once-off partitioning. It can be argued that clustering methods produce regions which reflect gaps in the catchment characteristic data space rather than reflect gaps in flood frequency behaviour.

Acreman and Sinclair (1986) argue that their procedure obviates the possibility of misallocation that can occur when regionalisation is based on observed sample quantities such as C_v or g. Such misallocation is quite a possibility in such a scheme. Consider a group of catchments each of which comes from one of two homogeneous regions, as measured by C_v, regions A and B. If population $C_v(A) = 0.4$ and $C_v(B) = 0.8$, it is quite possible that some A catchments have sample C_v values sufficiently large, say 0.6 - 0.7, to make them appear to have come from region B and vice versa.

The quantity C_v features prominently in Wiltshire's (1986a) regionalisation scheme. The procedure performs a systematic search along catchment characteristic axes, as in Wiltshire (1985), but uses C_v instead of the previously used SSQ5 and other measures, to assess the success of any partitioning of the catchment characteristic data space. The partitioning point is successful if

$$F = \frac{\text{between region variance of Cv values}}{\text{within region variance of Cv values}} \qquad ...(23)$$

is significantly large, while obviously that partitioning point among successful ones that maximises F is selected. A second independent measure, S, of within-region homogeneity of C_v was also developed which should not, in a homogeneous region, exceed a critical value. The sampling distributions of F and S were established by Monte Carlo methods, critical points F_c and S_c were established, and partitioning values sought for which $F > F_c$ and $S < S_c$ simultaneously. While partitioning values in the A, SOIL and R axes which satisfied $F > F_c$ were readily found, the condition $S < S_c$ seemed almost impossible. In other words, between-region heterogeneity was easily achieved but within-region homogeneity was not.

Wiltshire (1986a) applied this procedure to the data of 376 sites in Britain and found five significantly different regions, one of which is internally homogeneous with respect to C_v at 5% significance level, three of which are likewise homogeneous at 1% level while the fifth one is hetereogeneous. Wiltshire argues that the regions so obtained "accord with physical reasoning". He also examined ten previously-defined geographic regions from the point of view of F and S statistics and found that some of the regions were redundant and only five of the ten were found to be internally homogeneous. However, Wiltshire (1986a) points out that geographical regions "may retain a role in a basin characteristic-based grouping scheme, either by providing clues as to which axes to use and how to partition them, or explicitly as regions which are defined independently of the (catchment characteristic) data space groups".

Wiltshire (1986b,c) further develops the use of C_v as a measure of regional homogeneity and introduces yet another statistic R for the same purpose as follows. If F(x) is the distribution function of the variate X, then F(x) considered as a function of x, the probability integral transform, has a uniform distribution. If random sample values $(x_1, x_2...x_N)$ are transformed to values $(G_1, G_2... G_N)$ in the (0,1) interval by the probability integral transform, then the G values should be approximately uniformly distributed, the departure from uniformity being solely due to sampling effects. On the other hand if random sample values $(z_1, z_2 ... z_N)$ from another variate Z, $Z \neq X$, are so transformed their distribution would be different from uniform by more than random amounts. For instance if $\sigma_z < \sigma_x$ (or $C_{vz} < C_{vx}$) the G values obtained from $(z_1, z_2 ... z_N)$, would tend to concentrate in the middle of the (0,1) interval, giving a unimodal distribution of G. On the other hand if $\sigma_z > \sigma_x$ (or $C_{vz} > C_{vx}$) the G values tend to concentrate at the extremeties of the (0,1) interval, giving a U-shaped distribution of G.

Thus, if F(x) is the regionally estimated d.f. of the variate $X = Q/\bar{Q}$ and if $(z_{1j}, z_{2j},...z_{ij}...z_{Nj\,j})$ are a standardised sample $(Q/\bar{Q})$ from site j in the region let G_{ij}, i

$= 1,2 \ldots N_j$, be the F(.) transform of z_{ij}. Under the hypothesis that z is distributed as x (i.e. homogeneous region) the G_{ij} values are approximately uniform. If the hypothesis is not true, the difference can be amplified by the transformation

$$G'_{ij} = 2|G_{ij} - 0.5| \qquad \ldots(24)$$

The statistic R is then defined for the M sites in the region as

$$R = \Sigma_1^M (G'_{\cdot j} - G'_{\cdot\cdot})^2 / U_j \qquad \ldots(25)$$

where $G'_{\cdot j}$ and $G'_{\cdot\cdot}$ are site and regional average values of G' and U_j is the sampling variance of $G'_{\cdot j}$, defined by Wiltshire, (1986b). The sampling distribution of R is approximately χ^2, values of R exceeding χ^2_{crit} leading to rejection of the homogeneity hypothesis.

Wiltshire (1986b) further examines the statistical power of this test by Monte Carlo methods using EVl distributed parent populations. Naturally it was found that the power of the test increases with increasing record length N and with increasing number of sites M (region size). A comparison of power curves for regions having the same number of station years, NM, namely

N = 20 years at M = 50 sites and N = 50 years at M = 20 sites

N = 10 years at M = 20 sites and N = 20 years at M = 10 sites

"shows that the information content of the data is greater when deployed in a small number of long records rather than in a large number of short records". Furthermore he concluded that "if the series available to a regional analysis are typically ten years or less in length then there will rarely be sufficient information present to detect heterogeneity even when there are gross differences amongst the constituent flood series". Finally for a typical British region, with 50 sites of 20 years record each, the R-test is only of moderate power when the correct form of estimating distribution is used and is even less powerful when more flexible 3 and 5-parameter estimating distributions are used. He states that with more parameters with which to characterise the data " the GEV and Wakeby distributions are more tolerant of heterogeneous effects in the data and so the region is more likely to appear homogeneous".

Allocation of unguaged or poorly-guaged catchments to regions . All the above methods deal with testing homogeneity of flood frequency behaviour among gauged catchments, for each of which an AM series is available. They do not deal specifically with allocation of an unguaged catchment to one of a number of pre-selected (homogeneous) regions. Geographically defined regions pose no problem in allocating an ungauged catchment but users never feel fully at ease when the unguaged catchment is close to the regional boundary, fearing a mis-allocation. Wiltshire (1986a) suggests calculating a flood quantile estimate for an ungauged catchment as a weighted sum of estimates obtained from contiguous regions, the weighting factors being inversely proportional to the catchment characteristic data space distance of the catchment to the centroid of each region. These procedures allow quantile estimates to vary smoothly across the data space. It avoids the abrupt changes which would otherwise occur at region boundaries and it also reduces the error which results either from misallocating a catchment to a region or from uncertainty in the optimum positions of the region boundaries.

A further method of allocating "fractional membership" in several regions to a particular catchment, based on discriminant analysis of clusters, is presented in Wiltshire (1986c). In this approach catchments which are hydrological similar, as expressed by proximity in the two-dimensional ($\bar{Q}/A$, C_v) space, as suggested also by Mosley (1981), are identified and the resulting similar flood frequency regions are subjected to examination in the catchment characteristic space using multivariate discriminant analysis. The discriminant analysis allows a probability of membership of region k, P_k, to be calculated for any catchment whose catchment characteristic vector is given. The standardised quantile value for that catchment is then given by

$$x_T = \Sigma\ P_k \hat{x}_{Tk} / \Sigma\ P_k \qquad ...(26)$$

where summation is over all regions and $\hat{x}_{Tk}$ is the region k estimate of $x_T = Q_T/\bar{Q}$.

Effect of regional heterogeneity on quantile estimate. Hosking et al (1985a) examined the effect of regional heterogeneity of C_v and g on quantile estimates obtained by an index flood regional method in which the X distribution is estimated from regionally averaged PWM's. In particular they use GEV/PWM and WAK/PWM for estimating x_T. They specify a heterogeneous region of 20 stations (j=1 ... 20) whose flood populations are GEV distributed with parameters varying over the 20 populations as follows:

μ_j	increases from	μ_1	=	11.6	to μ_{20}	=	19.5
C_{vj}	decreases from	C_{v1}	=	0.60	to C_{v20}	=	0.38
g_j	decreases from	g_1	=	3.0	to g_{20}	=	2.15
while available record length							
n_j	increases from	n_1	=	10	to n_{20}	=	29.

These choices are designed to reflect a transition from small catchments (small j) to large catchments (large j).

A second similar hypothetical region was specified in which g = 3.0 for every station. It is to be expected that the use of a single X distribution for all stations in a heterogeneous region would lead to under-estimation in some cases, such as j = 1 above, and to over-estimation in others, such as j = 20 above. The simulation studies confirm this but nevertheless the regional algorithms give relatively stable estimates for all sites. The change from variable skewness to a fixed skewness of 3.0 had only a small effect, although this would be more remarkable if skewness of the order of 1.14 (EV1 case) had been included in the first data-base.

Lettenmaier et al. (1985), using heterogeneous GEV data bases qualitatively similar to Hosking et al (1985a) compared at-site EV1/PWM and GEV/PWM with a variety of GEV/PWM regional estimating techniques. The clear conclusion from that study is that with record lengths ≤ 30 years, at-site quantile estimates are less reliable than at-site /regional estimates, notwithstanding the regional heterogeneity. At-site GEV/PWM estimates of Q_T have very large se(Q_T) especially if site skewness is large. At-site EV1/PWM estimates have smaller s.e.(Q_T) than GEV/PWM but are severely biased downwards if site skewness $>> 1.14$.

Lettenmaier and Potter (1985), when testing flood frequency estimation methods, use a regional flood generating model in which the logarithms of the first two moments of the annual flood distribution at each site depend on the logarithm of the catchment area. This offers the advantage of a controlled simulation study which recognises the dependence of flood flow distribution on a catchment characteristic and can be used to impose heterogeneity on the flood generating populations. In all they compare eight flood estimating methods (estimators) on data simulated from three different generating populations. Two of the estimators are at-site ones while the other six estimators combine site and regional information. None of them assumes a completely unguaged site; all eight estimators assume the existence of at-site data.

They found that the at-site/regional quantile estimators nearly always had lower root mean square error than the at-site ones, even when the region was moderately heterogenous. An exception to this occurred when data were generated from an EV1 parent with large C_v and/or non-zero C_c (= Coeff. of variation of C_v). However, the estimators used did not include at-site/regional GEV/PWM, which Lettenmaier et al. (1985a) had earlier shown to be superior to at-site EV1/PWM estimates.

Thus in regions of relatively low C_v (< 0.6 say) a small degree of heterogeneity does not negate the benefits of using an at-site/regional estimating method such as GEV/PWM. Increased partitioning of the catchment data set, with a view to achieving increasing internal homogeneity of regions is therefore not necessary. It may be more efficient to have a few larger slightly heterogeneous regions each with many catchments, then many homogeneous regions, each with only a small number of catchments.

Spatial and Temporal Interdependence of Flood Magnitudes

Frequency analysis methods assume that AM flood series form random samples in which all the observations are identically and independently distributed (iid). In a single at-site series this assumption might be violated by serial dependence in certain categories of rivers. Landwehr et al. (1979a) and McMahon and Srikanthan (1982b) have examined the effect of parent population serial correlation on quantile estimates obtained on the idd assumption. They examined EV1 and LP3 populations respectively and found that a small amount of bias and an increase in se($\hat{Q}_T$) occurred under these circumstances. The conclusion is that a small amount of serial correlation in flood series does not greatly affect quantile estimates obtained with the iid assumption providing the appropriate form of parent distribution is used. These studies did not examine robustness against wrong choice of distribution in the serially correlated case.

Spatial interdependence is a greater cause of anxiety to hydrologists who use regional estimation techniques. Intuitively it is felt that contemporary flood series from adjacent river sites must be interdependent in some way, because some flood events in each series occur on the same day and are caused by a single widespread meteorological event. This coincidence is most easily noted for large events, but is less noticeable among small events. Since not all events in such pairs of series have common dates or meteorological cause, the dependency between series is incomplete while not all of those events which have common dates or causes show a strong degree of resemblance.

If a regional estimation method pools all $L = \Sigma_1^M N_j$ standardised flood values $x_{ij} = Q_{ij}/\overline{Q}_j$, ((i = 1,2...$N_j$), j = 1,2...M) together and treats them as a single random sample from an X-population, then the presence of interdependence among the X values would contradict the random sample assumption and lead to bias in estimated quantiles $\hat{x}_T$ and $\hat{Q}_T$. The variance of estimate would also be effected. This single random sample assumption is what is called the "station year" method which essentially says that N observations at each of M sites are statistically equivalent to MN observations at one site .

On the other hand if the regional estimation procedure averages some sample statistics, such as standardised PWMs or dimensionless moments, across or over the samples the effect of inter-site dependence should not of itself introduce bias into quantile estimates obtained by a method which is bias free in the case of independent series. The inter-site dependency reduces the statistical degrees of freedom between sites which leads to an increase in standard error of estimate. Thus there should be no great distortion or bias due to inter-site dependence, only an increase in standard error of estimate.

While hydrologists have voiced concern about regionally estimated flood quantiles obtained from interdependent data, the nature and amount of such interdependence has rarely been articulated in a manner which allows assessment of its effect on flood quantile estimation. However, Hosking and Wallis (1985) have recently modelled the interdependency among some British AM series and their main conclusions, in the context of GEV/PWM regional estimation, are

1. Inter-site dependence affects neither the bias of quantile/growth curve estimators nor the rate at which rmse falls as record lengths increase.
2. Quantile estimate error is due more to $\bar{Q}$ error than to $Q_T/\bar{Q}$ error even in the presence of dependence.
3. rmse of $Q_T/\bar{Q}$ can be seriously affected by inter-site dependence while rmse of $\hat{Q}_T$ is not much affected by such dependence.
4. The effect on rmse is not serious for $T < 10$ and it gets greater as T increases.
5. In independent regions quantile accuracy decreases with heterogeneity of record length but this effect is less noticeable if inter-site dependence exists.

Beneficial aspects of spatial-interdependence. While spatial interdependence of flood peaks is a disadvantage in station-year, and to some extent in index flood, regional estimation methods it forms the very basis of another regional flood estimation technique. This is the record extension technique in which a regression relation is developed between the common values (y_i, x_i, $i = 1,2...n_1$) in neighbouring flood series. If the x_i series has an additional n_2 values, an additional n_2 values of y are obtained from the regression relation, thus giving an extended record of y values. Care has to be taken to incorporate a random residual term in y, otherwise var(y) is underestimated by the augmented sample. The extended record of y values need not be physically calculated if bivariate normal estimation theory (after log transformation of flows Q_x and Q_y to give approximately normally distributed x and y series) is used for estimation of the parameters μ_x, σ_x,μ_y, σ_y, (Moran 1974).

The record extension technique cannot be used indiscriminately because if the dependence between the Y and X populations is too weak there can be no legitimate transfer of information between X and Y samples. Fiering (1962, 1963) showed that the augmented sample provides a more efficient estimate of μ_y only if $\rho^2 > 1/(n_1 - 2)$ where ρ is the population value of correlation between X and Y. This criterion is of no immediate use if only a sample estimate r of the correlation coefficient is available. A further restriction is imposed on ρ if the augmented sample is to provide a more efficient estimate of var(y), (USWRC, 1981).

The record extension technique received much attention in the 1960's and early 1970's and was in those times regarded as a major hydrological tool, (Matalas and Jacobs, 1964; Gilroy 1970). These procedures have been recently reviewed, analysed and improved by Vogel and Stedinger (1985).

Detection and Use of Historical Floods.

A large amount of data about floods in historical times exists for some of the older civilisations, particularly for Chinese rivers (Chen et al. 1975; Hua, 1985; Luo 1985; Pang 1985), for the Nile (Sutcliffe, 1985) and for the Danube (Dyck, 1984) among others. The above records extend over many centuries while in other countries some of the major floods which occurred during the century preceding systematic recording of flows have been noted (NERC, 1975). Usually such records are partial in the sense that only notable floods have been marked, thus forming a Type 1 censored sample. Such marks have been chiselled on natural rocks, on bridges or on buildings. A second source of historical data is provided by the new science of palaeoflood hydrology which uses a combination of morphometric and carbon dating methods (Baker 1985, Xu and Ye 1985). The accuracy of all historical flood flows depends on the hydraulic knowledge available about the site.

Combination of historical and recent records. The traditional manner of dealing with historical floods is a graphical one in which the historic magnitudes are assigned plotting positions appropriate to the largest, second largest events in a record of length $N_1 + N_2$ (Benson (1950)). Such graphical techniques are currently in regular use in China (Chen et al. 1975; Hua, 1985; Luo 1985). The plotting positions used in China are discussed by Zhang (1982). Plotting positions for historical floods are further discussed by Hirsch (1985).

USWRC (1981) recommends combining historical and recent observations in weighted estimates of moments, in log domain, of the assumed LP3 parent population. These expressions are based on the notion of a threshold of perception, Q_H, for historical floods. Historical floods below Q_H are not available. The combined recent plus historical record is complete for peaks exceeding Q_H but incomplete for peaks $< Q_H$. The historically weighted moments adjustment effectively replicates the systematic record of floods $< Q_H$, as many times as necessary, to fill in the empty portions of the historical record. This replication is done by weighting the systematically recorded floods by approximately $W = H/N$ where H = historic record time span and N = length of systematic record.

Censored sample methods. Historical data can be more satisfactorily combined with recent records using censored sample concepts in maximum likelihood estimation (Leese, 1973). If in addition to N_1 years of (recent) systematic AM record at a site, occasional remarkably large peak values may be available over an earlier N_2 year period, giving a total sample size $N = N_1 + N_2$. If there are n values available among the earlier N_2 years there are $(N_2 - n)$ values whose magnitudes are unknown. If it can be assumed that all of these $(N_2 - n)$ values are less than a certain threshold of perception Q_H, then all this information can be incorporated into a likelihood function

$$\mathbf{L}(\mathbf{Q}/\theta) = \{F(Q_H/\theta)^{N_2-n}\} \cdot \Pi_1^{N_1} f(q_i/\theta) \cdot \Pi_1^{n} f(q_j/\theta). \qquad ...(27)$$

whose maximum defines the Maximum Likelihood estimate, θ, of the parameter vector θ.

It is evident that this is an at-site formulation. To extend it to a regional case the observations and Q_H would first have to be standardised by each site's $\bar{Q}$ and the M quantities $\mathbf{L}_j(\mathbf{X}/\theta)$, analogous to that defined in eqn. (27), would then contribute to a single likelihood function for the standardised variate $X = Q/\bar{Q}$.

$$\mathbf{L(X/\theta)} \quad = \quad \Pi_1^M \; L_j(\mathbf{X/\theta}) \qquad \text{...(28)}$$

This formulation of the likelihood function implies the space-time equivalence assumption of the station year method which can result in biassed estimates in the presence of inter-site dependence.

The maximum likelihood formulation can also be adapted to take account of measurement truncation in which all peak magnitudes in a recent or historical greater than a truncation value Q_t were unmeasurable but whose occurrences have been noted.

Hosking and Wallis (1984) have examined the effect of including the single largest historical flood, from the earlier N_2 year period at one site, in a regional estimating scheme based on the station year method using data simulated from GEV populations. They concluded that it offered little improvement, if the model chosen for the standardised variate X has two parameters but that its inclusion contributed a noticeable improvement if a three parameter distribution is assumed for X. However, even in the latter case, the usefulness of including the historical flood was diminished if it itself was subject to any reasonable degree of measurement error.

Stedinger and Cohn (1983, 1985) using data simulated from log normal distributions have concluded that incorporation of historical flood data is valuable and that knowledge of the number of occurrences of large floods exceeding a known threshold during the historical N_2 year period is very valuable, even if their precise magnitudes are unknown. The usefulness of historical data is increased according as the threshold of perception Q_H is reduced.

Hosking et al (1985a) express the contrary view that historical floods may not be worth incorporating into a regional flood frequency estimation scheme because "in general, historic information is likely to be subject to bias and lack of precision and these effects may well be so large that the use of historic information may cause a degradation rather than an improvement in the estimates of quantiles or growth factors." However, notwithstanding these reservations it will take some extensive and extended arguments to dissuade hydrologists from trying to make use of whatever data happen to be available.

CONCLUSIONS

Statistical flood frequency estimation models based on annual maxima have been intensively researched during the past two decades. Peaks over a threshold (POT) models have also been put on a solid footing in that the required probabilistic algebra has been presented but the properties of such models as flood estimators have been less rigorously tested.

A selective historical survey of the development of flood frequency estimation has been presented in tabular form. It is suggested that the development can be viewed in three phases; an early phase of problem definition, a second phase of technical development beginning with the introduction of Extreme Value theory and a final phase in which questions are addressed which could only be attempted with the aid of computers

The search for a suitable flood estimation procedure should take into account both a models's descriptive ability and its predictive ability. To ignore the former might lead to irrelevant estimators while to ignore the latter might lead to useless (large rmse) estimators.

Annual maximum flood data are positively skewed with mean values in the range 0.5 to 3.0 with C_v values in the range 0.1 to 1.0 depending on climate. There is evidence that the AM population distribution in some countries is thick tailed. Traditional distributions may not be able to mimic that behaviour, leaving only Wakeby and TCEV distributions as genuine contenders.

Regional index flood estimation techniques which pool all standardised statistics in a region together in order to estimate the parameters of a standardised distribution,particularly those based on probability weighted moments (PWM),are generally more efficient than at-site estimators which are based on single samples. While traditionally, regional estimators assume spatial homogeneity of standardised flood behaviour, such estimators outperform at-site ones even in the presence of moderate degrees of heterogeneity. Therefore regional estimates of growth factors, $Q_T/\bar{Q}$, should be used at sites for which AM data are available, in conjunction with the at-site estimate of $\bar{Q}$. Ungauged catchment flood estimation should use regional $Q_T/\bar{Q}$ along with $\bar{Q}$ estimated from a regionally calibrated relation between $\bar{Q}$ and catchment characteristics. Where records are short or non-existent the uncertainty associated with the $\bar{Q}$ estimate overwhelms that associated with the growth factor.

Bayesian estimators have been developed which allow combination of regional and at-site data in a manner which enables the results to be used in a decision-theory framework. However, the prior information, based on regional regression relations, used by these models is weak and their quantile estimating abilities do not compare well with index flood methods.

The log Pearson Type 3 distribution, for 20 years recommended for use by U.S. Federal Agencies, has been shown to be less efficient and less robust than other flood estimation methods currently available.

The increased use of regional estimating techniques has given an impetus to the investigation of regional homogeneity of flood statistics and recent developments in finding expressions which define homogeneity have been summarised. Geographically defined regions are not always mutually heterogeneous nor always internally homogeneous. Definition of regions within the catchment characteristic space is possible but sometimes such regions are not internally homogeneous. However, it has been shown that from the point of view of flood quantile estimation, regions do not have to be perfectly homogeneous to be of value.

Flood magnitudes may be spatially correlated, but if regional quantile estimating methods which average flood statistics across regions, rather than those which depend on the station year equivalence assumption, are used then the amount of dependence actually present has been shown to have very little effect on quantile estimates.

A considerable amount of historical flood information is available from the older civilisations while the new science of palaeoflood hydrology provides an alternative source of pre-record information. Historical flood data can be formally and beneficially incorporated into flood estimators. However some doubts have been expressed about the utility of including historical floods in estimating schemes especially if they are inaccurately measured.

Current at-site/regional flood estimates allow more reliable flood estimates to be obtained now than twenty years ago. Further, considerably more is now understood about the robustness of such estimators. Continued efforts must be made to ensure that additional empirical knowledge of flood behaviour is accumulated and studied and that observed flood behaviour be accounted for even better in future models.

The largest single obstacle to progress lies in the ungauged catchment case where the relaion between mean annual (Index) flood and catchment characteristics remains stubbornly imprecise. It is hoped that physically based flood frequency models may be of some assistance in this in the future.

REFERENCES

Acreman, M.C. and Sinclair, C.D.: Classification of Drainage Basins According to their Physical Characteristics; An Application for Flood Frequency Analysis in Scotland. *J. Hydrol., 84, 365 - 380, 1986.*

Adamowski, K.: Nonparametric Kernel Estimation of Flood Frequencies.*Water Resour. Res., 21(11), 1585 - 1590, 1985.*

Arnell, N. and Gabrielle, S.: Regional Flood Frequency Analysis with the Two-Component ExtremeValue Distribution: An Assessment using Computer Simulation Experiments. *Workshop on Combined Efficiency of Direct and Indirect Estimations for Point and Regional Flood Prediction, Perugia, Italy, December 1985.*

Ashkar, F. and Rousselle, J.: Some remarks on the truncation used in partial duration flood series models.*Water Resour. Res., 19(2), 477-480, 1983a.*

Ashkar, F. and Rousselle, J.: The effect of certain restrictions imposed on the inter-arrival times of floods. *Water Resour. Res., 19(2), 481-485, 1983b*

Atkins,G.P.: Regional flood frequency analysis in Papua New Guinea. *Ph.D thesis, Univ. Technology, Lae, Papua, New Guinea, 1980.*

Baker, V.R.: Palaeoflood Hydrology of Extraordinary Flood Events. *Paper pres. at US-China Bilateral Symposium on the analysis of Extraordinary Flood Events, Nanjing, Oct. l985.*

Beable, M.E. and McKerchar, A.I.: Regional Flood Estimation in New Zealand. *New Zealand National Water and Soil Cons. Org., Water and Soil Tech. Publ. no. 20, 132pp, 1982.*

Benson, M.A.: Use of historical data in flood frequency analysis. *Trans. Amer. Geophys. Union, 31(3), 419 - 424, 1950.*

Benson, M.A.: Evolution of Methods for Evaluating the Occurrence of Floods. *U.S. Geol. Surv., Water Supply Paper 1550-A, 1962.*

Benson, M.A.: Factors influencing the occurence of floods in a humid region of diverse terrain. *U.S. Geol. Surv., Water Supply Paper No. 1580 - B., 64p, 1962.*

Benson, M.A.: Uniform flood frequency estimating methods for federal agencies.*Water Resour. Res. 4(5), 891 - 908, 1968.*

Beran, M.A. and Nozdryn-Plotnicki, M.J.: Estimation of low return period floods. *Hydrological Sciences Bull., 22, 275 - 281, 1977.*

Beran, M.A., Hosking, J.R.M. and Arnell, N.: Comment on "Two Component Extreme Value Distribution for Flood Frequency Analysis" by Rossi et al. (1984). *Water Resour. Res., 22(2), 263 - 266, 1986.*

Bernier, J.: Sur le theorie du renouvellement et son application en hydrologie. *Electricite de France, HYD, 67(10), 1967a.*

Bernier, J.: Methodes Bayesiennes en hydrologie statistique. *Proc. Intl. Hydrol. Symp., Fort Collins, Colorado, U.S.A., 459 -470, 1967b.*

Biswas, A.K. and Fleming, G.: Floods in Scotland: Magnitude and Frequency. *Water and Water Engg., 246 - 252, 1966.*

Bobee, B.: Sample error of T Year events computed by fitting a Pearson Type 3 distribution.*Water Resour. Res., 9(5), 1264 - 1270, 1973.*

Borgman, L.E.: Risk Criteria. *Jour. Waterways and Harb. Div., ASCE. 89, 1 - 35, 1963.*

Burges, S.J., Lettenmaier, D.P. and Bates, C.L.: Properties of the three-parameter lognormal distribuation. *Water Resour. Res., 11(2), 229 - 235, 1975.*

Cervantes, J.E., Kavvas, M.L and Delleur, J.W.: Cluster Model for Flood Analysis. *Water Resour. Res., 19(1), 209-224, 1983.*

Chen Jia-Qi, Ye Yong-yi and Tan Wei : The Important Role of Historical Flood Data in the Estimation of Spillway Design Floods. *Scientia Sinica, 18(5), 669 - 680, 1975.*

Chow, V.T.: Discussion of "Annual Floods and the Partial Duration Series" by W.B. Langbein. *Trans. Am. Geophys. Union, 31(6), 939 - 941, 1950.*

Chow, V.T.: A general formula for hydrologic frequency analysis.*Tran. Amer. Geophys. Union, 32, 231 - 237, 1951.*

Chow, V.T.: Frequency analysis of hydrologic data with special application to rainfall intensities. *Univ. Illinois Eng. Expt. St., Bull. 414, 1953.*

Chow. V.T.: The log probability law and its engineering applications. *Proc. ASCE, 80(536), 1 - 25, 1954.*

Cole, G.: An Application of the Regional Analysis of Flood Flows. *Institn. Civil Engineers, Symposium on River Flood Hydrology (1965), 39-57, 1966.*

Condie, R.: The Log Pearson Type 3 Distribution: The T-Year Event and Its Asymptotic Standard Error by Maximium Likelihood Theory. *Water Resources Research, 13(6), 987-991, 1977.*

Condie, R. and Lee, K.A.: Flood frequency analysis with historic information. *J. Hydrol., 58(1/2), 47-62, 1983.*

Costa, J.E.: Holocene Stratigraphy in Flood Frequency Analysis.*Water Resources Res., 14(4), 626 - 632, 1978.*

Costa, J.E.: Interpretation of the Largest Rainfall-runoff Floods Measured by Indirect Methods on Small Drainage Basins in the Coterminus United States. *Paper presented at US-China Bilateral Symposium on the Analysis of Extraordinary Flood Events, Nanjing, Oct. l985.*

Cunnane, C. and Nash, J.E.: Bayesian estimation of frequency of hydrologic events. *Proc. Warsaw Symp. on Math. Models in Hydrology, July 1971. (IAHS Sci.Publ. 100, pp47-55, 1974).*

Cunnane, C.: A particular comparison of annual maximum and partial duration series methods of flood frequency estimation. *J. Hydrol., 18, 257 - 273, 1973, with ERRATA, 19, p377.*

Cunnane, C.: Unbiassed Plotting Positions - a Review. *J. Hydrol., 37(3/4), 205 - 222, 1978.*

Cunnane, C.: A note on the Poisson assumption in partial duration series models. *Water Resour. Res., 15(2), 489 - 494, 1979.*

Cunnane, C.: Condition of separation of skewness of random samples from the General Extreme Value distribution. *Unpubl. Rep., Dept. Eng. Hydrol., Univ. Coll., Galway, 7pp, 1984.*

Dalrymple, T.: Flood freqency methods. *U.S. Geol. Survey, Water Supply Paper 1543 A, pp 11 - 51, Washington, 1960.*

Davis, D.R., Kisiel, C.C. and Duckstein, L.: Bayesian Decision Theory applied to Design in Hydrology. *Water Resources Res., 33 - 41, 1972.*

De Coursey, D.G.: Objective Regionalisation by Peak Flow Rates. *Proc. Sec. Int. Sympos. in Hydrology, Fort Collins, Colorado, pp 385 - 405, 1972.*

Dyck, S.: Historical Data Analysis. *Pres. at Workshop on Frequency Analysis, Univ. of Birmingham, England, Feb. 1984.*

Eychaner, J.H.: Estimation of magnitude and frequency of floods in Pima County, Arizona with comparisons of alternative methods. *Water Resour. Invest., Rep. 84-4142, U.S. Geol. Surv., Tucson, Arizona, 1984.*

Farhan Y.I.: Regionalisation of surface water catchments in east bank of Jordan. *Proc. Sympos. on "Problems in Regional Hydrology", Univ. of Freiburg, Germany, 1984.*

Fiering, M.B.: On the Use of Correlation to Augment Data. *J. Am. Stat. Assoc. 57, 20 - 32, 1962.*

Fiering, M.B.: Use of Correlation to Improve Estimates of the Mean and Variance. *U.S. Geol. Surv. Prof. Pap. 434-C, C1-C9, 1963.*

Fiering, M.B.: Streamflow Synthesis. *Macmillan, London, 1969 (Chapt. 3).*

Fiorentino, M. and Gabriele, S.: A correction for the bias of maximum likelihood estimators of Gumbel parameters. *J. Hydrol. 73 (1/2), 39 - 50, July 1984.*

Foster, H.A.: Theoretical frequency curves and their application to engineering problems. *Trans. ASCE, 87, 142 - 173, 1924.*

Fuller, W.E.: Flood Flows. *Trans. Am. Soc. Civl Engrs., 77, 564 - 617, 1914.*

Gilroy, E.J.: Reliability of a Variance Estimate Obtained from a Sample Augmented by Multivariate Regression. *Water Resour. Res., 6(6), 1595 - 1600, 1970.*

Greenwood, J.A., Landwehr, J.M., Matalas, N.C. and Wallis, J.R.: Probability Weighted Moments: Definition and relation to parameters of distributions expressible in inverse form. *Water Res. Res., 15(5) 1049 -1054, Oct. 1979.*

Greis, N.P.: Flood Frequency Analysis:- A Review of 1979 - 1982. *Rev. Geophysics and Space Phys., 21(3), 699 - 706, 1983.*

Greis, N.P. and Wood, E.F.: Regional flood frequency estimation and network design. *Water Resour. Res., 17(4), 1167 - 1177, 1981.*

Gumbel, E.J.: The return period of flood flows. *Ann. Math. Statist., 12(2), 163 - 190, 1941.*

Gumbel, E.J.: Statistics of Extremes. *Columbia Univ. Press, 375 pp, 1958.*

Gumbel, E.J.: Extreme Value Analysis of Hydrologic Data. *Proc. Hydrol. Sympos. No.5, "Statistical Methods in Hydrology" McGill Univ., Publ. by Queen's Printer, Ottawa, pp147 - 169, 1966.*

Guo, Sheng-Lian : Regional Flood Frequency Analysis with reference to Hubei Province, China. *Thesis subm. in part. fulfilment of M.Sc.,Dept. of Engng.*

Hydrol., Univ. Coll., Galway, 1986

Hall, M.J. and O'Connell, P.E.: Time series analysis of mean daily river flows. *Water and Water Engng., 76, 125-133, 1972.*

Hasselbad, V.: Estimation of Finite Mixtures of Distributions from the Exponential Family. *J.Am. Stat. Assoc., 64, 1459 - 1471, 1969.*

Hazen, A.: Flood flows. *John Wiley, New York, 199 pp, 1932.*

Hebson, C.S. and Cunnane, C.: Assessment of use of At-site and Regional Flood Data for Flood Frequency Estimation. *Paper presented at Int. Sympos. on Flood Frequency and Risk Analyses, Baton Rouge, May 1986.*

Hirsch, R.M.: Probability Plotting Position Formulas for Flood Records with Historical Information. *Paper pres. at US-China Bilateral Symposium on the Analysis of Extraordinary Flood Events, Nanjing, Oct.1985.*

Hoshi, K., and Burges, S.J.: 1981a. Sampling properties of parameter estimates for the log Pearson Type 3 distribution using moments in real space.*J. Hydrol., 53: 305-316.*

Hosking, J.R.M.: Comment on "A correction for the bias of Maximum Likelihood Estimates of Gumbel Parameters", by Fiorentino and Gabrielle. *J. Hydrol., 78(314), 393-396, 1985.*

Hosking, J.R.M. and Wallis, J.R.: Palaeoflood Hydrology and Flood Frequency analysis. *EOS, 65(45), p890, Nov. 6th l984, (Presented AGU Fall Meeting, SanFrancisco, Dec.l984).*

Hosking, J.R.M. and Wallis, J.R.: The Effect of Inter-site Dependence on Regional Flood Frequency Analysis. *EOS, 66(46), p906, Nov. 12th l985, (presented AGU Fall Meeting, San Francisco, Dec.1985).*

Hosking, J.R.M., Wallis, J.R., and Wood, E.F.: An appraisal of the regional flood frequency procedure in the UK Flood Studies Report. *Hydrol. Sci. J., 30(1), 85-109, 1985a.*

Hosking, J.R.M., Wallis, J.R. and Wood, E.F.: Estimation of the Generalised Extreme Value distribution by the method of probability weighted moments. *Technometrics, 27(3), 251-261, Aug. 1895b.*

Houghton, J.C.: Robust estimation of the frequency of extreme events in a flood frequency context. *Ph.D. Dissertation, Div. of Appl. Sci., Harvard Univ., Cambridge, Mass., 1977.*

Houghton, J.C.: Birth of a parent: The Wakeby distribution for modelling flood flows. *Water Resour. Res., 14(6), 1105 - 1109, 1978a.*

Hua, Shi-Qian: A general survey of Flood Frequency Analysis in China. *Paper pres. at US-China Bilateral Symposium on the Analysis of Extraordinary Flood Events, Nanjing, Oct. 1985.*

Jenkinson, A.F.: The frequency distribution of the annual maximum (or minimum) values of meteorological elements. *Quart. J. Roy. Meteor. Soc., 81, 158 - 171, 1955.*

Jenkinson, A.F.: Statistics of extremes. *In: Estimation of maximum floods. WMO Tech. Note no. 98, 183 - 228, 1969.*

Kaczmarek, Z.: Efficiency of the estimation of floods with a given return period. *Proc. Toronto Sympos., IAHS Public. No. 45, 145 - 159, 1957.*

Kimball, B.F.: Sufficient Statistical Estimation Functions for the Parameters of the Distribution of Maximum Values. *Ann. Math. Statist., 17, 299 - , 1946.*

Kirby, W.: Algebraic boundness of sample statistics.*Water Resour. Res., 10(2), 220 - 222, 1974.*

Kirby, W. and Moss, M.E.: U.S. Contributions to the Analysis of Extraordinary Events. *US-China Bilateral Symposium on the Analysis of Extraordinary Flood Events, Nanjing, October 1985.*

Kochel, R.C., Baker, V.R. and Patton, P.C.: Palaeohydrology of Southern Texas. *Water Resources Res., 18(4), 1165 - 1183, 1982.*

Kottegoda, N.T.: Investigation of Outliers in Annual Maximum Flow Series. *J. Hydrol., 72, pp. 105 - 137, 1984.*

Kuczera, G.: Combining site-specific and regional information: An empirical Bayes approach. *Water Resour. Res., 18(2), 306 - 314, 1982 a.*

Kuczera, G.: Robust flood frequency models.*Water Resour. Res., 18(2), 315 - 324, 1982 b.*

Landwehr, J.M., Matalas, N.C. and Wallis, J.R.: Some comparisons of Flood Statistics in Real and Log Space.*Water Res. Res., 14(5), 902 - 920, Oct. 1978.*

Landwehr, J.M., Matalas, N.C. and Wallis, J.R.: Probability weighted moments compared with some traditional techniques in estimating Gumbel parameters and quantiles.*Water Res. Res., 15(5), 1055 - 1064, Oct. 1979a.*

Landwehr, J.M., Matalas, N.C. and Wallis, J.R.: Estimation of parameters and quantiles of Wakeby distribution.*Water Res. Res. 15(6), 1361 - 1379, 1979, b.c.*

Landwehr, J.M., Matalas, N.C. and Wallis J.R.: Quantile estimation with more or less flood-like distributions. *Water Resour. Res., 16(3), 547-555, 1980.*

Langbein, W.B. and others: Topographic Characteristics of Drainage Basins. *U.S. Geol. Surv. Prof. Paper 968-C, pp 125 - 157, 1947.*

Langbein, W.B.: Annual floods and the partial duration flood series. *Trans. Am. Geophys. Union, 30, 879 - 881, 1949.*

Leese, M.N.: Use of censored data in the estimation of Gumbel distribution parameters for annual maximum flood series. *Water Resources Research, 9(6), 1534 - 1542, 1973.*

Lettenmaier, D.P: Regionalisation in Flood Frequency Analysis:- Is it the Answer? *Paper pres. at US-China Bilateral Symposium on the Analysis of Extraordinary Flood Events, Nanjing, Oct. l985.*

Lettenmaier, D.P., and Potter, K.W.: Testing Flood Frequency Estimation Methods Using a Regional Flood Generating Model. *Water Resources Res., 21(12) 1903 ; 1914, 1985.*

Lettenmaier, D.P., Wallis, J.R. and Wood, E.F.: Note on the Comparative Robustness of Estimates of Extreme Flood Quantiles, *Presented at American Geophysical Union, Spring Meeting, Baltimore, May, 1985.*

Lettenmaier, D.P., Wallis, J.R. and Wood, E.F.: Effect of Heterogeneity on Flood Frequency Estimation. *Submitted to Water Resources Res., 1987.*

Lowery, M.D. and Nash, J.E.: A comparison of methods of fitting the double exponential distribution. *J. Hydrol., 10(3), 259 - 275, 1970.*

Luo, Cheng-Zheng: A survey of Historical Flood and its Regionalisation in China. *Paper pres. at US-China Bilateral Symposium on the Analysis of Extraordinary Flood Events, Nanjing, Oct.1985.*

Marin, C.M.: Parameter Estimation in Water Resources Planning and Management: Optimal Actions or Optimal Parameters? *Water Resour. Res. 22(3), 353 - 360, 1986.*

Matalas, N.C. and Jacobs, B.: A Correlation Procedure for Augmentng Hydrologic Data. *U.S. Geol. Surv. Prof. Pap., 434-E, E1-E7, 1964.*

Matalas, N..C. and Wallis, J.R.: An approach to formulating strategies for flood frequency analysis. *Proc. Int. Syump. on Uncertainties in Hydrologic and Water Resource Systems, Tucson, Arizona, 940 - 961, 1972.*

Matalas, N.C. and Wallis, J.R.: Eureka! It fits a Pearson Type 3 distribution. *Water Resour. Res. 9(2), 281 - 289, 1973.*

Matalas, N.C., Wallis J.R and Slack J.R.: Regional Skew in search of a parent. *Water Resour. Res., 11(6), 815 - 826, 1975.*

McMahon, T.A.. and Srikanthan, R.: Log Pearson Type 3 distribution - is it applicable to flood frequency analysis of Australian streams? *J. Hydrol., 52, 139 - 147, 1981 a.*

Moran, M.A.: On Estimators Obtained from a Sample Augmented by Multiple Regression. *Water Resour. Res., 10(1), 81- 85, 1984.*

Mosley, M.P.: Delimitation of New Zealand Hydrological Regions. *J. Hydrol., 49, 173 ; 192, 1981.*

Nash J.E. and Amorocho, J.: The Accuracy of the Prediction of Floods of High Return Periods. *Water Resour. Res., 2(2), 191 - 198, 1966.*

Nash, J.E. and Shaw, B.L.: Flood frequency as a function of catchment characteristics. *Proc. Sympos. on River Flood Hydrology, Inst. Civ. Engnrs., London, 115 - 136, 1966.*

NERC: Flood Studies Report. *Natur. Environ. Res. Council, London, Vols. 1 - 5, 1100 pp, 1975..*

Njenga, M.: Simulation applied to the Inference Problem of the Underlying Distribution of Hydrologic Random Variables. *Thesis subm. in partial fulfillment of M.Sc., Dept. Eng. Hydrol., Univ.Coll., Galway, 150pp, May 1985.*

Nozdryn-Plotnicki, M.J. and Watt, W.E.: Assessment of fitting techniques for the log Pearson type 3 distribution using Monte Carlo simulation. *Water Resour. Res., 15(3), 714-718, 1979.*

O'Donnell, T., Hall, M.J. and O'Connell, P.E.: Some Applications of Stochastic Hydrological Models. *International Symposium on Modelling Techniques in Water ResourcesSystems, (Editor AK Biswas) Vol.1, pp 227 ; 239, Publ. by Environment, Canada, Ottawa, 1972.*

Otten, A. and van Montfort, M.A.J.: Maximum Likelihood Estimation of the General Extreme Value Distribution Parameters.*J. Hydrol., 47, 187 - 192, 1980.*

Pang, K.D.: Extraordinary Floods in Early Chinese History and their Absolute Dates. *Paper pres. at US-China Bilateral Symposium on the Analysis of Extraordinary Flood Events, Nanjing, Oct.1985.*

Phien, H.N.,and Lung-Cheng Hsu: Variance of the T-year event in the log Pearson Type 3 Distribution. *J. Hydrology 77, 141 - 158, 1984.*

Quimpo, R.G.: Stochastic Model of Daily Flow Sequences. *Hydrology Paper No. l8, Colorado State Univ., 1967.*

Reich, B.M.: Lysenkoism in U.S. flood determinations. *Special session on flood frequency methods, Amer. Geophys. Union, San Francisco, 13 pp, Dec. 1977.*

Roeske, R.H.: Methods for estimating magnitude and frequency of floods in Arizona. *Open File Rep., U.S. Geol. Surv., Reston, Virginia, 1978.*

Rosbjerg, D.: Estimation in partial duration series with independent and dependent peak values. *Journal of Hydrology, 76, 183 - 195, Jan. 1985.*

Rossi, F., Fiorentino, M. and Versace, P.: Two component extreme value distribution for flood frequency analysis. *Water Resources Research, 20(7), 847 - 856, 1984.*

Rossi, F., Fiorentino,M. and Versace, P.: Reply to Comment by Beran et al. (1986). *Water Resources Res., 22(2), 267 - 269, 1986.*

Sangal, B.P., and Biswas, A.K.: The three-parameter log normal distribution and its application in hydrology, *Water Resour. Res., 6(2), 505-515, 1970.*

Senaratne, S.: Frequency Analysis by Peaks over a Threshold Method with high Thresholds. *Thesis subm. in Partial Fulfillment of M.Sc., Dept. Eng. Hydrol., Univ. Coll., Galway, May l986.*

Shane, R.M. and Lynn, W.R.: Mathematical model for flood risk evaluation. *J. Hydraul. Div., ASCE, 90, 1 - 20, 1964.*

Slack, J.R., Wallis, J.R. and Matalas, N.C.: On the value of information to flood frequency analysis. *Water Resour. Res., 11(5), 629-647, 1975.*

Slade, J.J.: The Reliability of Statistical Methods in the Determination of Flood Frequencies. *U.S. Geol. Surv., Water Supply Paper 771, pp 421 - 432, 1936.*

Stedinger, J.R.: Fitting log normal distributions to hydrologic data. *Water Resour. Res., 16(3) 481-490, 1980.*

Srikanthan, R andMcMahon, T.A. : Log Pearson Type 3 distribution effect of dependence, distribution parameters and sample size on peak annual flood estimates.*J. Hydrol., 52, 149 - 159, 1981 a.*

Stedinger, J.R. and Cohn, T.A.: The Use of Historical Flood Information in Flood Frequency Analysis. *Paper pres. at US-China Bilateral Symposium on the Analysis of Extraordinary Flood Events, Nanjing, Oct. 1985.*

Sutcliffe, J.V.: The Use of Historical Records in Flood Frequency Analysis. *Paper pres. at US-China Bilateral Symposium on the Analysis of Extraordinary Flood Events, Nanjing, Oct. 1985.*

Tabios, G., Shen, H.W., Obeysekera, J.T.B. and Salas, J.D.: Outlier Detection and Effects of Outliers on Parameter Estimation through Contaminants.*Paper pres. at US-China Bilateral Symposium on the Analysis of Extraordinary Flood Events, Nanjing, Oct. 1985.*

Taesombut, V. and Yevjevich, V.: Use of partial flood series for estimating distribution of maximum annual flood peak. *Hydrol. Paper No. 97, Colorado State University, Fort Collins, 71 pp, 1978.*

Takeuchi, K. : Annual maximum series and partial duration series - evaluation of Langbein's formula and Chow's discussion. *J. Hydrol., (68), 275 - 284, 1984.*

Tasker, G.D. and Moss, M.E.: Analysis of Arizona flood data network for regional information. *Water Resour. Res., 15(6), 1791-1796, 1979.*

Tavares, L.V. and Da Silva, J.E.: Partial duration series method revisited. *J. Hydrol., 64, 1 - 24, 1983.*

Thomas, H.A.: Frequency of minor floods. *Journ. Boston Soc. Civ. Engrs., 35, p. 425, 1949.*

Thomas, D.M. and M.A. Benson, : Generalization of Streamflow Characteristics from Drainage-Basin Characteristics, *U.S.G.S. Water-Supply Paper No. 1975, 1970.*

Todorovic, P.: On some problems involving random number of random variables. *Annals Math. Statist., 41(3), 1059 - 1063, 1970.*

Todorovic, P. and Zelenhasic, E.: A stochastic model for flood analysis. *Water Resour. Res., 6(6), 1641 - 1648, 1970.*

Todorovic, P. and Rousselle, J.: Some problems of flood analysis. *Water Resour. Res., 7(5), 1144 - 1150, 1971.*

U.S. W.R.C.: Guidelines for determining flood flow frequency. *Bulletin 15, Hydrology Committee, Water Resources Council, Washington, 1967. (Also revised versions, Bulletin 17A, 1977 and Bulletin 17B, 1981).*

van Montfort, M.A.J. and Witter, J.V.: Testing Exponentiality against Generalised Pareto Distribution.*J.Hydrol., 78(314), 305 - 316, 1985.*

Vogel, R.M. and Stedinger, J.R.: Minimum Variance Streamflow Record Augmentation Procedures. *Water Resour. Res., 21(5), 715 - 723, 1985.*

Wallis, J.R.: Risk and uncertainties in the evaluation of flood events for the design of hydrologc structures. *Keynote address at "Seminar on Extreme Hydrological Events - Floods and Droughts", Erice, Italy, 33 pp, 1980.*

Wallis, J.R.: Probable and Improbable Rainfall in California. *IBM, Environmental Science Research Report, RC9350, Yorktown, New York , 26pp, 1982a.*

Wallis, J.R.: Hydrological Problems Associated with Oil Shale Development. *In Environmental Systems Analysis and Management. Ed. S. Rinaldi), North Holland Publ. Co. pp185 ; 102, 1982b.*

Wallis J.R. and Wood E.F.: Relative Accuracy of Log Pearson III procedures. *A.S.C.E. Journal of Hydraulic Engineering, 111(7), 1043 - 1056, July 1985.*

Wallis, J.R., Matalas N.C. and Slack J.R.: Just a moment! *Water Resour. Res., 10(2), 211 - 219, 1974.*

Water and Energy Commission (WEC): Hydrological Studies of Nepal, Volumes I and II. *WEC, Ministry of Water Resources, Kathmandu, (c.200 and c.330 pages resp.) 1982.*

Waylen, P. and Woo, Ming-ko: Prediction of annual floods generated by mixed processes. *Water Resour. Res., 18(4), 1283 - 1286, 1982.*

Weiss, G.: Shot Noise Models for Synthetic Generation of Multi-Site Daily Streamflow Data. *IAHS Sympos. on Design of Water Resource Projects with Inadequate Data, Madrid, 1973.*

White, E.L.: Factor Analysis of Drainage Basin Properties: Classification of Flood Behaviour in Terms of Basin Geomorphology. *Water Resources Bulletin, 11(4), 676 - 687, 1975.*

Wiltshire, S.W.: Grouping Basins for Regional Flood Frequency Analysis. *Hydrol. Sci. Journ., 30(1), 151 - 159, 1985.*

Wiltshire, S.W.: Identification of Homogeneous Regions for Flood Frequency Analysis. *J. Hydrol., 84, 287 - 302, 1986a.*

Wiltshire, S.W.: Regional Flood Frequency Analysis I: Homogeneity Statistics. *Hydrol. Sci. Jour., 31(3), 321-333, 1986b.*

Wiltshire, S.W.: Regional Flood Frequency Analysis II: Multivariate Classification of Drainage Basins in Britain. *Hydrol. Sci. Jour., 31(3), 335-346, 1986c.*

Wood, E.F. and Rodriguez-Iturbe, I.: Bayesian inference and decision making for extreme hydrologic events. *Water Resour. Res., 11(4), 533 - 542, 1975.*

World Meteorological Organisation: Estimation of maximum floods. *World Met. Organisation, Tech. Note No. 98, pp 183 -228, 1969.*

Wu, B. and Goodridge, J.D.: Selection of frequency distributions for hydrologic frequency analysis. *Dept. of Water Ressources, State of California, Sacramento, 85 pages, June 1976.*

Xu Runzi and Ye Bangham: Estimating Rare Floods by Geomorphological Methods, *Paper pres. at US-China Bilateral Symposium on the Analysis of Extraordinary Flood Events, Nanjing, Oct. 1985.*

Yevjevich, V. and Taesombut, V.: Information on flood peaks in daily flow series. *Proc. Int. Symp. on Risk and Reliability in Water Resources, Univ. of Waterloo, Ontario, Canada, pp. 451 - 470, 1978.*

SYMBOLS

Symbols having a common meaning throughout are defined here. Other locally used symbols are defined where they occur.

A	Catchment area
AM	Annual maximum
C_v	Coefficient of variation
C_s	Coefficient of skewness (= g)
C_k	Coefficient of kurtosis
EV1	Extreme value Type 1
F	Non-exceedance probability
GEV	General Extreme Value
g	Coefficient of skewness (= C_s)
G	Regional average of g
g_z	g of logarithm values
L	Total years of record in region
L(./.)	Likelihood
LN	Lognormal (LN2 or LN3)
LP3	Log Pearson Type 3
M	Number of sites in region
ML	Maximum Likelihood
MoM	Method of Moments estimation
N, (N_j)	Number of AM values at site, (at site j)
POT	Peaks over a threshold
PWM	Probability weighted moments
Q	Annual Maximum Flood
$\bar{Q}$	Mean of AM floods at a site
$\bar{R}$	Average annual catchment rainfall
S	Catchment overland slope
T	Return period
TCEV	Two component extreme value
WAK	Wakeby distribution
X	Standardised flood (= $Q/\bar{Q}$)
Y	Standardised variate (= (Q - Location Parameter)/Scale Parameter)
Z	Logarithm of Q
$\boldsymbol{\theta}$	Vector of parameter values

Table 1: Outline of Development of Flood Frequency Analysis Methods in Three Phases Under Selected Headings.

THREE PHASES OF DEVELOPMENT	1.
Early Phase	
Recognition that there is no such thing as a single design flood but rather a choice of different return period floods depending on circumstances.	
Middle Phase	
Introduction of Extreme Value theory, algebraic development of alternative parametric model types and estimation schemes.	
Recent Phase	
Development of models which have been stimulated by the outcome of simulation experiments	

ANNUAL MAXIMUM MODEL DEVELOPMENT	2.
Fuller (1914)	Concept of return period
Foster (1924)	Theoretical frequency curves
Hazen (1932)	Large body of empirical knowledge
Gumbel (1941)	Deductive EV knowledge
Kimball (1946)	ML estimation
Langbein (1949)	Linked AM and POT
Thomas (1948) Adamowski (1985)	Distribution-free methods
Chow (1951)	Frequency Factor K_T
Jenkinson (1955, 1969)	GEV Distribution
Kaczmarek (1957)	Study of Standard Errors
Matalas et al. (1975)	Condition of Separation
Houghton (1978)	Wakeby Distribution
Rossi et al (1984)	TCEV Regional Model

POT MODEL DEVELOPMENT	3
Langbein (1949)	Linked AM and POT
Borgman (1963) Shane & Lynn ('64) Bernier (1967a)	POT Model as Renewal Process
Todorovic (1970) Todorovic & Zelenhasic (1970) Todorovic & Rousselle (1971)	Generalised Probability Model
Cervantes (1983) Kavvas (1982 a,b,)	Trigger-type cluster models
Cunnane (1979) Ashkar and Rousselle (1983a,b)	Test of Poisson Rates of occurence

USE OF TIME SERIES MODELS	4
Quimpo (1967) Hall and O'Connell (1972) O'Donnell, Hall and O'Connell (1972)	Seasonal means and AR Residuals
Weiss (1973) NERC (1975, I.2.9)	Shot-noise Models

COMPARISON OF AM AND POT MODELS	5
Langbein (1949) Chow (1950) Takeuchi (1984)	Theoretical comparison of T_{AM} and T_{POT}
Beran and Nozdryn-Plotnicki (1977) Yevjevich and Taesombut (1978) Taesombut and Yevjevich (1978) Takeuchi (1984)	Empirical Comparison of T_{AM} and T_{POT}
Cunnane (1973) Yevjevich and Taesombut (1978) Taesombut and Yevjevich (1978) Tavares and da Silva (1983) Rosjberg (1985)	Comparison of efficiency of AM and POT estimates

REGIONAL FLOOD ESTIMATES	6
Fuller (1914) Hazen (1932)	Station Year Method
Dalrymple (1960)	Index flood and distribution of standardised variate $X = Q/\bar{Q}$, and choice of homogeneous regions
Cole (1965)	Application of above to UK data
Nash and Shaw (1965)	Relation of regional average dimensionless moments to catchment characteristics
USWRC (1977)	Map of Regional Skew
NERC (1975 I.2.6) Beable and McKercher (1982)	Regional average dimensionless order statistics
Wallis (1980)	Regional average standardised PWM's.

AM DISTRIBUTIONS RECOMMENDED FOR USE	7
(a) on basis of regional studies	
Benson (1968) USWRC (1967,1977,1981)	LP3
NERC (1975)	GEV
Houghton (1978)	Wakeby
Wu and Goodridge (1976)	P3/LN
McMahon & Srikanthan (1981a)	LP3
Hua (1985)	P3
(b) on basis of simulation results	
Hosking et al (1985a)	GEV/PWM
Wallis and Wood (1985)	Wakeby/PWM
Rossi et al (1984)	TCEV

PARAMETER ESTIMATION	8
Kimball (1946)	ML for EV1
Chow (1953)	Least Squares
Gumbel (1958)	Geometric mean of two least squares lines
Jenkinson (1969)	ML and Sextiles for GEV
Greenwood et al (1979)	PWM for distributions expressible in inverse form
Wallis (1980)	Regionally averaged standardised PWM's for index flood method
Rossi et al (1984)	Iterative ML procedure for regional TCEV

STANDARD ERRORS	9
See Table 2	

SIMULATION AND BEHAVIOUR ANALYSIS	10
Nash and Amorocho (1966) Lowery and Nash (1970)	EV1 quantile standard error
Matalas & Wallis (1973)	Pearson 3 estimation
Wallis et al. (1974)	Properties of small sample statistics
Matalas et al (1975)	Regional skewness condition of separation
Landwehr et al (1978)	Relationships between real and log space statistics
Rossi et al (1984) Hosking et al (1985a) Wallis and Wood (1985) Lettenmaier et al (1985) Lettenmaier & Potter (1985) Arnell & Gabrielle (1985)	Statistical properties of regional flood estimation procedures. Selection of some regional parameter values.

BAYESIAN ANALYSIS	11
Bernier (1967b)	Non-informative prior distribution
Cunnane and Nash (1971/74)	Informative Regional prior distribution
Davis et al (1972)	Bayesian decision theory for flood alleviation design
Wood and Rodriguez-Iturbe (1975a)	Informative conjugate Regional prior distributions for flood protection design
Kuczera (1982a)	Linear empirical Bayes procedures

DECISION RELATED PROCEDURES	12
Matalas and Wallis (1972) Slack et al (1975)	Study on usefulness to decision-making of prior information about distribution type.
Davis (1972) Wood and Rodriguez-Iturbe (1975) Kuczera (1982 a,b; 1983)	Use of Bayesian decision theory

USE OF HISTORICAL DATA	13
Benson (1950)	Graphical Procedure
Chen et al. (1975) Hua (1985) Luo (1985)	Use of historical floods in China
Pang (1985)	Floods in early Chinese history
Sutcliffe (1985)	Historical floods on River Nile
Baker (1985) Xu and Ye (1985)	Palaeoflood hydrology and geomorphological methods of quantifying historical floods
Leese (1973)	Censored estimation techniques
Condie and Lee (1983)	Censored estimation and Monte Carlo evaluation
Hosking et al. (1985a)	Question the statistical wisdom of using historical data
Hosking and Wallis (1984) Stedinger and Cohn (1985)	Evaluations of usefulness of historical data

REGIONAL HOMOGENEITY	14
Langbein (1947) Dalrymple (1960)	Homogeneity test based on plot of $\hat{T}$ against record length.
Mosley (1981) Wiltshire (1985) Wiltshire (1986 a,b,c) Acreman and Sinclair (1986)	Question the use of geographical regions and seek homogeneous regions in catchment characteristic space.

Table 2: Selected References to Investigations into Sampling Properties of Quantile Estimates

Model Distribution	At-site	Regional	Reference
Extreme Value	*		Kaczmarek (1957)
Type 1	*		Nash and Amorocho (1966)
	*		Lowery and Nash (1970)
	**		Landwehr et al. (1979a)
		§	Greis and Wood (1981)
	*		Fiorentino and Gabrielle (1984)
	**	§§	Lettenmaier et al. (1985)
	**		Lettenmaier and Potter (1985)
Lognormal	*		Kaczmarek (1957)
	*		Sangal and Biswas (1970)
	*		Burges et al. (1975)
	**	§§	Kuczera (1982b)
	**		Stedinger (1980)
	**	§§	Lettenmaier and Potter (1985)
Pearson Type 3	*		Matalas and Wallis (1973)
	*		Bobee (1973)
Log Pearson 3	*		Condie (1977)
	*		Nozdryn-Plotnicki and Watt (1979)
	*		Hoshi and Burges (1981)
	*		Phien and Hsu (1984)
	**	§§	Wallis and Wood (1985)
General Extreme Value (GEV) and Log EV1	*		Jenkinson (1969)
		§§	Hosking et al. (1985a)
	*		Hosking et al. (1985b)
	**	§§	Wallis and Wood (1985)
	**	§	Lettenmaier et al. (1985)
		§§	Arnell and Gabrielle
Wakeby	*		Landwehr et al. (1979b,c)
		§	Wallis (1980)
		§§	Hosking et al. (1985a)
		§§	Wallis and Wood (1985)
		§§	Arnell and Gabrielle (1985)
TCEV		§§	Arnell and Gabrielle (1985)

* Parent and assumed model distributions are the same.
** Also tests model distribution under different parent distribution assumption(s) (Robustness test).
§ and §§ For regional cases correspond to * and ** in at-site estimation

Table 3: Parent and Model distributions used in intercomparisons of regional quantile estimating schemes. (GEV and WAK models have parameter estimation by Regional average PWM. LEB = Linear Empircal Bayes).

Study	Parent Distributions	Model Distributions	Heterogeneous as well as homogeneous regions
Kuczera (1982a,b)	LN x 5	LN/LEB	Yes
	WAK		
Hosking et al. (1985a)	GEV	FSR, WAK	Yes
	WAK	GEV	
	WEIB	Weibull	
Lettenmaier et al. (1985)	GEV	GEV	Yes
	EV1	EV1	
Wallis and Wood (1985)	GEV	GEV	
	WAK	WAK	
	LP3	LP3	
Lettenmaier and Potter (1985)	EV1	EV1	Yes
	LN2	LN2/LEB x 2	
	P3	EV1/LEB x 2	
Arnell and Gabrielle (1985)	TCEV	FSR, WAK,	No
	WAK	GEV	
	GEV	TCEV	

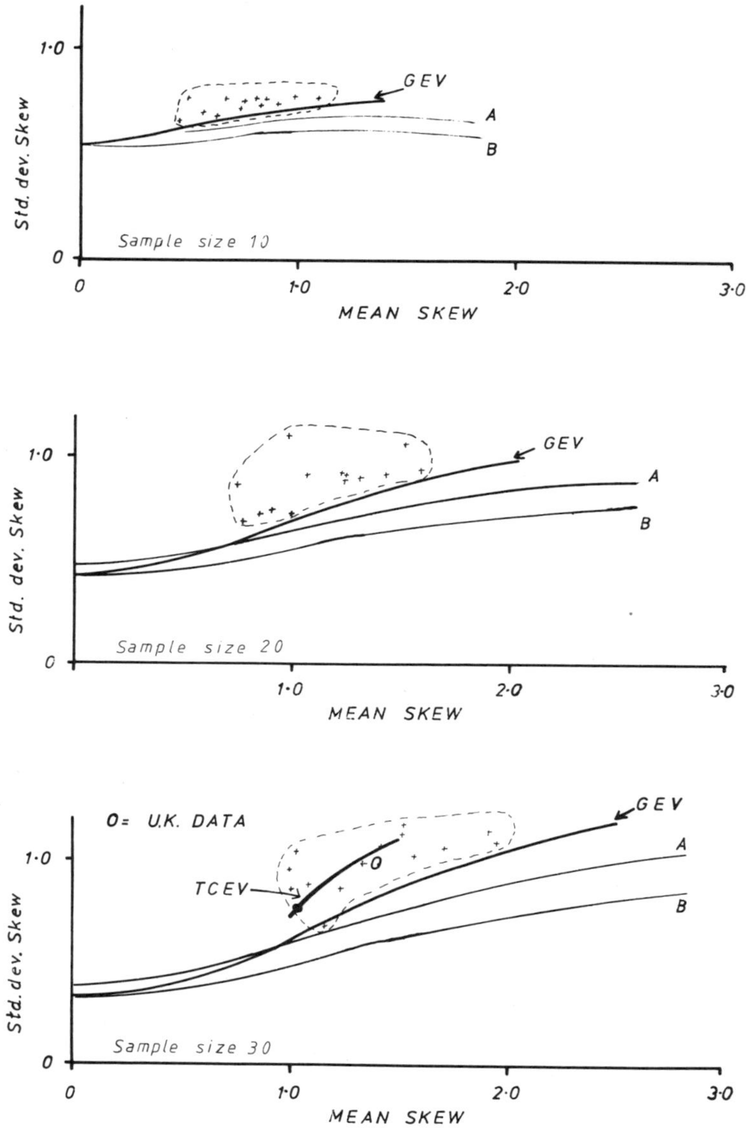

Figure 1: Standard deviation of sample skewness versus mean sample skewness for samples of size 10, 20 and 30 from GEV parents, (from Cunnane,1984). Curves A and B enclose results from all distributions studied by Matalas et al. (1975). TCEV result and UK data points are from Beran et al. (1986).

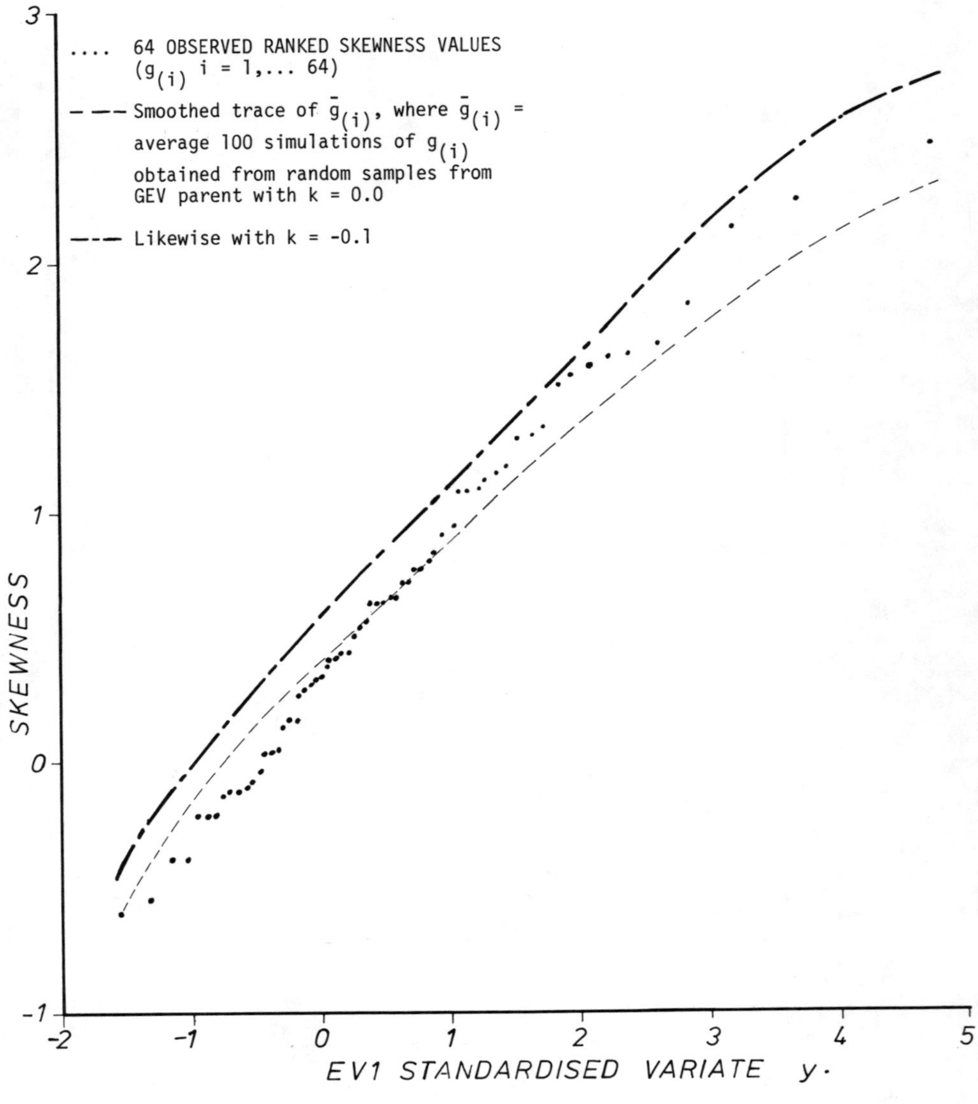

Figure 2: Sampling distribution of observed AM skewness in Ireland (64 stations, 15 years each) compared with trace of expected value of corresponding quantities obtained on assumption of GEV parent populations with k = 0.0 and -0.1. (After Njenga, 1985).

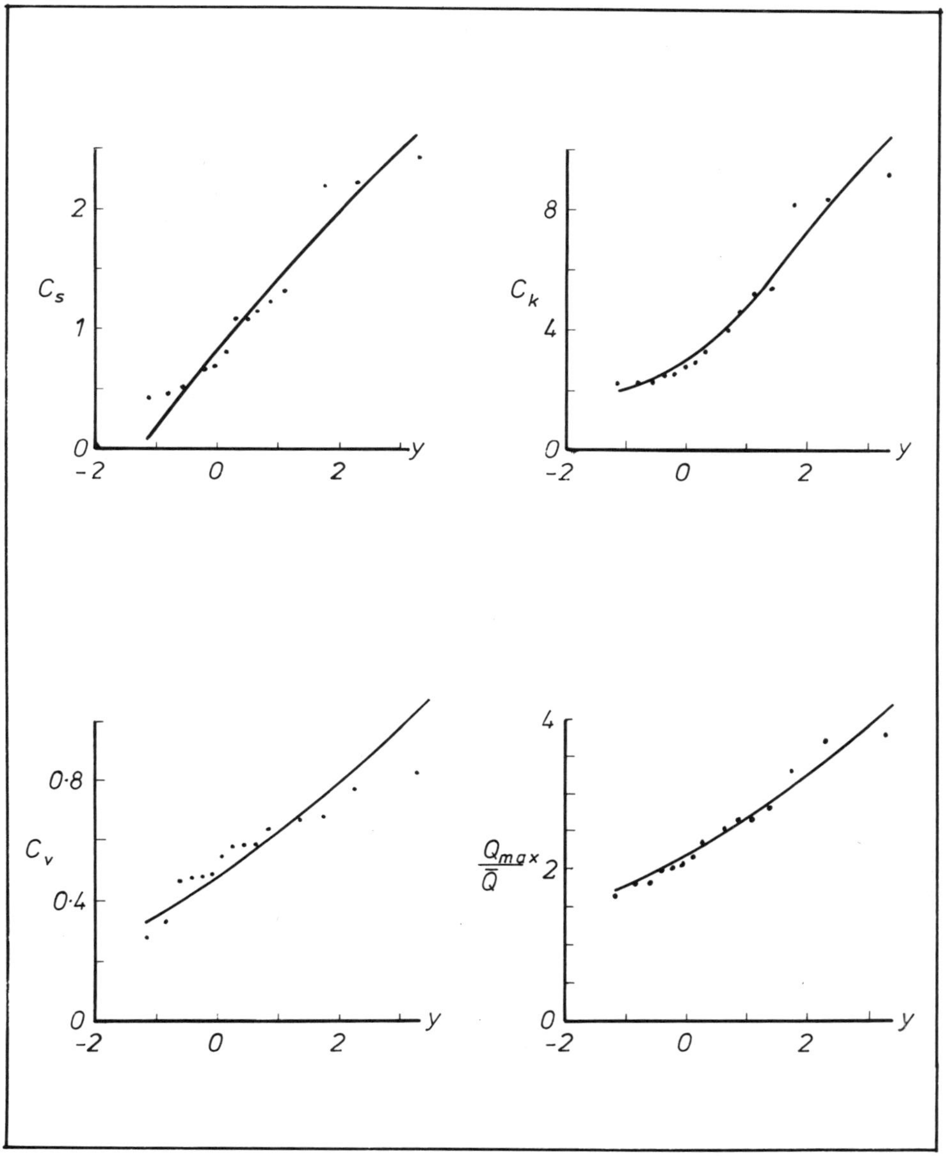

Figure 3: Observed sampling distributions of AM sample statistics for Hubei Province, China. (16 stations 20 years each) compared with trace of expected values of corresponding quantities obtained on assumption of GEV parent population with k = -0.14. (After Guo, 1986).

ESTIMATION OF FLOOD FREQUENCIES BY A NONPARAMETRIC DENSITY PROCEDURE

Kaz Adamowski and Charles Labatiuk
Department of Civil Engineering
University of Ottawa
Ottawa, Canada
K1N 6N5

ABSTRACT. Annual flood risk is usually estimated by fitting an a priori assumed probability distribution function to the observed annual extreme peak series (parametric method). The main shortcomings of such a procedure are: the selection of a distribution, reliability of parameters (especially for skewed data with short record length), inability to analyze multimodal distributions resulting from flooding due to snowmelt versus thunderstorm activity, and treatment of outliers.

A new methodology has been developed based on the nonparametric concepts for estimation of probability density function. The nonparametric method overcomes some of the limitations inherent in the parametric method, and is developed based on a few and very mild assumptions.

Based on the numerical results using real data and Monte Carlo simulation, it was found that the nonparametric flood estimates are accurate and suitable for multimodal densities.

INTRODUCTION

The objective of statistical flood frequency analysis is the inference of form and parameter values of the statistical distributions which characterize the population on the basis of the sample of recorded data. To the hydrologist, flood frequency analysis has evolved as the science and art of fitting distributions to annual flood series (AFS) or partial duration series (PDS) and the associated problems of estimation of parameters, goodness-of-fit testing, treatment of outliers, plotting position, etc.

The histogram is widely used to represent the shape of the probability density function (PDF). To obtain a histogram, balancing of the number of observations against the class width is critical, as are the range, shape, skewness and any tendency for multimodality because of the subjectivity involved in obtaining a histogram. Based on the histogram, a particular parametric PDF family is hypothesized and then the hypothesis is tested.

There are a large number of procedures and distributions available

V. P. Singh (ed.), Hydrologic Frequency Modeling, 97–106.

for the analysis of AFS. The normal distribution has been studied in more detail than any other distribution, so its use, with log-transformation to account for skewness of flood data, was inevitable.

In 1902 Pearson introduced a system of densities as a set of solutions to the differential equations, and thus identified many distributions, including Pearson-Type III which have been extensively applied in hydrology.

In 1941 Gumbel postulated the use of the double exponential Type-I Extreme Value distribution law for evaluating floods, and this was generalized by Jenkinson in 1955 as a General Extreme Value (GEV), and it has been adopted as a base method in the United Kingdom (Hall, 1984).

Most of the common parametric forms, including the ones discussed so far, are unimodal, thus representing homogeneous phenomena. However, floods may be generated by either snowmelt or rainfall, and different types of storms can give rise to different flood characteristics. The parametric procedure presents serious difficulties in the analysis of assymmetrical and multimodal phenomena.

Therefore, various more complex models, either bimodal or multimodal have been proposed. Singh and Nakashima (1983) have concluded that "the mixed distribution method coupled with the detection and modification of outliers and inliers may be the best availble method for flood frequency analysis at present." Rossi et al. (1984) has suggested that the two- component extreme value (TCEV) distribution can be assumed as a parent flood distribution. The TCEV distribution assumes that individual floods arise from a mixture of two exponential components and its four parameters can be estimated by the maximum likelihood method.

Based on the discussion so far, it is evident that there is no particular frequency distribution which is clearly preferable to the others. In view of this, the imposition of a choice, such as LP III by the WRC (1982) is perhaps not such a bad idea. It should be noted that the Australian Institution for Engineers also recommended the use of the LP III distribution, and they too have become bogged down in the LP III debate. These efforts by various countries to achieve some degree of uniformity are very desirable (Thomas, 1985), but are difficult if not impossible to accomplish with the parametric methods (Wallis and Wood, 1985).

Due to recent developments in statistical theory, another possible approach to flood frequency analysis is to estimate the density nonparametrically. This approach is attractive for a number of reasons, the main one being that the problem of a priori distribution choice is bypassed altogether. As the density is not assumed to be of any functional form, nonhomogeneous data can be considered. Also, the parametric approach gives more weight to the central part of the distribution, especially when its parameters are estimated by the method of moments; however, the nonparameteric approach allows a greater weight in the tails.

Adamowski (1985) has suggested that the nonparametric approach may well provide an alternative to making unjustified assumptions concerning the unknown distribution in flood frequency analysis.

NONPARAMETRIC DENSITY ESTIMATION

Many different types of nonparametric density estimators are available, including the histogram, K-nearest neighbor, maximum-likelihood, kernel, orthogonal series estimator, spline estimator and penalized-likelihood estimator. The histogram is probably the oldest probability density estimator, and it is the classic nonparametric statistical tool for the graphical display of data. While the histogram is computationally efficient, it is quite inefficient in terms of integrated mean squared error (IMSE). Given a random sample $x_1,\ldots,x_N$, the IMSE is asymptotically minimized proportional to $N^{-2/3}$ when the histogram class interval width is optimally chosen (Devroye and Gyorfi, 1985).

Histogram construction is such a routine process that we may fail to realize that even when using identical class interval widths, different choices of their locations may result in very different looking histograms for the same data. This fallibility of the classical histogram motivated research into 'running' or 'shiftable' histograms. Rosenblatt in 1956 developed the 'shiftable' histogram in which the intervals are shifted so that every sample point is in the center of an interval. If the sample points are likened to an energy source, the potential at any of these points attains a peak value, and then decreases rapidly at any point away from the sample point. The 'shiftable' histogram has become known as the kernel estimator.

Parzen in 1962 generalized and investigated the consistency properties of the kernel estimator. He popularized it to such an extent that the estimator is also called the Parzen estimator. The kernel estimator is statistically efficient with IMSE rate convergence in the order of $N^{-4/5}$.

During the last twenty-five years a very extensive literature has developed in the area of nonparametric density estimation. Almost all of the literature is in the form of papers, except for a few recent books, i.e., Devroye and Gyorfi (1985).

The purpose of this paper is to review some recent applications of the nonparametric method to determination of flood frequencies.

Kernel Estimator

The kernel estimator can be written as (Adamowski, 1985)

$$f(x) = \frac{1}{Nh} \sum_{i=1}^{N} K\left(\frac{x-X_i}{h}\right) \tag{1}$$

where X_1, X_2, ..., X_N is a random sample of observations, $f(.)$ the unknown density, $K(.)$ the kernel, and h a design parameter commonly called the smoothing factor.

The value of h is critical and in practise not easy. Too large an h implies large bias, an oversmoothed estimate, and consequent loss of information. Too small an h implies large variance and too rough an estimate. As h approaches zero, $K(.)$ approaches a Dirac delta function centered at X_i, and $f(x)$ approaches a superposition of delta functions centered at the samples.

Many investigators including Epanechnikov (1969) have determined that the choice of kernel seems to matter little. Two basic properties of probability densities, and thus also of kernels, are that they be nonnegative, and integrate to one. Some useful kernels are given in Table 1. The rectangular kernel is computationally efficient due to finite support and use of absolute values; however, estimates tend to be rough. The Gaussain kernel is computationally inefficient due to infinite support and use of exponentiation. The quadratic kernel has been shown by Epanechnikow (1969) to be optimal in terms of mean square error rate of convergence for univariate kernal estimation regardless of the sample size or true underlying density.

Since all error measures depend on the unknown density, they generally cannot be used in deriving analytical expressions for selecting the smoothing factor. However, a global optimal h can be obtainted by minimizing the IMSE and substituting f(x) for f. Adamowski (1985) presents this asymptotically optimal choice for the smoothing factor h as follows:

$$h = \alpha(K)\ \beta(f)\ N^{-1/5} \tag{1}$$

where

$$\alpha(K) = [\int K(y)^2 dy]^{1/5}\ [\int K(y) y^2 dt]^{-2/5} \tag{2}$$

and

$$\beta(f) = [\int f''(x)^2 dx]^{-1/5} \tag{3}$$

Thus it can be seen that the smoothing factor h depends on the true density f, the kernel K(.), and the sample size N.

As h is to be computed from data samples, some statisticians have advocated the ad hoc procedure where plots of f are examined, starting with h values which are too large and then decreasing h until density estimates which are too rough are obtained. This method is very useful in data exploration; however it requires an interactive environment, is time consuming and subjective, and is best done by a skilled statistician. All of these factors limit the applicability of ad hoc procedures in flood frequency analysis.

An interesting modification is to examine plots of f" rather than of f itself. The second derivative is much more sensitive to small changes in h then the density function itself. However, the method is still subjective.

In recent years algorithms have been developed that automatically choose a smoothing factor suitable for a given data set. Various data-based algorithms which experimentally compute the h value have been investigated by Labatiuk (1985)who concluded that the method of Adamowski and formula of Flanagan (1980) give best results. The value of h can be estimated either by numerical approximations (i.e., method of Adamowski) or by means of an analytical formula (i.e., method of Flanagan).

The numerical approximation of the smoothing factor h can be determined by minimizing the following mean square error expression (Adamowski, 1985)

$$\sum_{j=1}^{N} [p_j - p(x_j)]^2 \qquad (4)$$

where the unknown probability $p(x_j)$ is estimated by p_j by probability plotting formula.

Flanagan (1980) provides the following formula for the optimal bandwidth when using the quadratic kernel

$$h = \frac{\sum_{i=2}^{N} \sum_{j=1}^{i-1} (X_i - X_j)}{\sqrt{5}\, N(N-10/3)} \qquad (5)$$

SIMULATION STUDY

In order to compare the parametric with the nonparametric procedure, a simulation study was performed. It was decided to select a Log-Pearson III distribution as a parent for data generation. The selection was made essentially on the basis of the widespread use of LP III distribution rather than on physical justification. The values of parameters used are: A = -0.06, B = 25, and C = 7, and are representative of many real conditions (Adamowski, 1985). The statistics of the generated data are: mean = 255, standard deviation = 73, and the coefficient of skew = .444. The simulation data was obtained from Environment Canada (J.P. Pilon, personal communication).

DATA ANALYSIS

For the numerical analysis the Northeast Margaree River (Station WSC01FB001) in Canada was selected. The AFS consisted of maximum daily flows for the period of record from 1917 to 1983 with the arithmetic mean equal to 171 m^3/s, standard deviation equal to 57 m^3/s and the coefficient of skew equal to 1.073.

The data has been analyzed using four different parametric distributions (Condie et al., 1985), that is: log-Normal 3 parameters (LN3), Generalized Extreme Value (GEV), Log-Pearson III (LPIII) and Wakeby, and the nonparametric method with value of h determined by the method of Adamowski and Flanagan. The results are presented in Table 2. For illustration purposes, typical graphical results for both densities and distributions are presented in Fig. 1. It can be observed in Fig. 1 that the density is distinctly bimodal, perhaps reflecting causative factors of spring/snowmelt and fall floods. To further investigate these causative factors, fall floods (Sept-Dec) have been separated from the rest of the year, and are presented in Table 3, and as it can be observed the possible physical explanation for multimodal distribution can be justified.

The simulation results are presented in Table 4. The errors computed were the bias and root mean square error for each event

$$\text{MSE} = (\text{Bias})^2 + \text{Variance} \qquad (6)$$
$$\text{RMS error} = (\text{MSE})^{1/2}$$

TABLE 1

Typical kernels

Kernel, K_i	Interval of Support
$K_1(y) = 1/2$	$y \leqslant 1$
$K_2(y) = \frac{1}{\sqrt{2\pi}} e^{-y^2/2}$	$y \leqslant \infty$
$K_3(y) = \frac{3}{4}(1-y^2)$	$y \leqslant 1$

TABLE 2

Flood Estimates for Northeast Margaree River at Margaree Valley (N = 67, Record from 1917-1983)

	Nonparametric Analysis				Parametric Analysis			
	Untransformed Data		Log Transformed Data		LN3	GEV	LP3	
	AF	F	A	F	ML	ML	ML	Wak.
Event:								
Q 10	247	249	247	251	247	246	246	249
Q 50	345	343	340	339	323	323	322	332
Q100	353	358	358	367	355	357	355	366
Q200	359	367	370	387	387	391	389	399
Parameter:								
h	12.65	14.49	0.0832	0.0837				

TABLE 3

Possible Physical Basis for Bimodal Densities

Estimate of the mean (X), cu. secs.

X from Nonparametric		X from Data
Left Peak	167.26	Fall Floods (Sep-Dec): 14 obser., x = 185.5
Right Peak	199.67	Rest of Year (Jan-Aug): 52 obser., x = 168.6
Overall Mean	= 172.58	Overall mean (66 obser.) = 172.2

TABLE 4

Nonparametric Simulation Results

Data 1-Untransformed, Sample Size = 50, Replications = 100

		Mean		St. Deviation		Bias(%)		RMSE	
Event	Pop.	A	F	A	F	A	F	A	F
Q 10	354	344	357	19.9	19.3	0.2	0.8	19.9	19.5
Q 50	424	421	432	32.5	35.8	-0.7	1.8	32.6	36.5
Q100	450	446	452	41.5	41.2	-0.8	0.4	41.7	41.2
Q200	474	455	465	42.8	43.5	-3.9	-2.0	46.7	44.5
Parameter									
h		21.72	19.49	3.369	1.965				

TABLE 5

Comparison of Parametric and Nonparametric Simulation Results

Sample Size = 50, Replications = 1000, LP3 ML 36 Failed

		Mean Estimate					Bias(%)					Root Mean Square Error				
Event	Pop	LP3 ML	LN3 ML	Wak	GEV ML	A Log	LP3 ML	LN3 ML	Wak	GEV ML	A Log	LP3 ML	LN3 ML	Wak	GEV ML	A Log
Q 50	424	420	429	435	422	425	-.9	1.1	2.6	- .5	.2	32.0	31.1	33.2	27.8	32.2
Q100	450	446	459	467	434	450	-.9	2.0	3.9	-3.6	.1	41.7	40.0	44.6	37.5	40.7

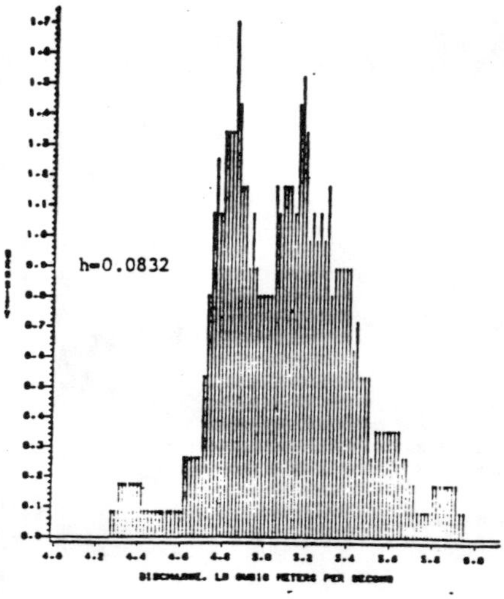

h by Adamowski

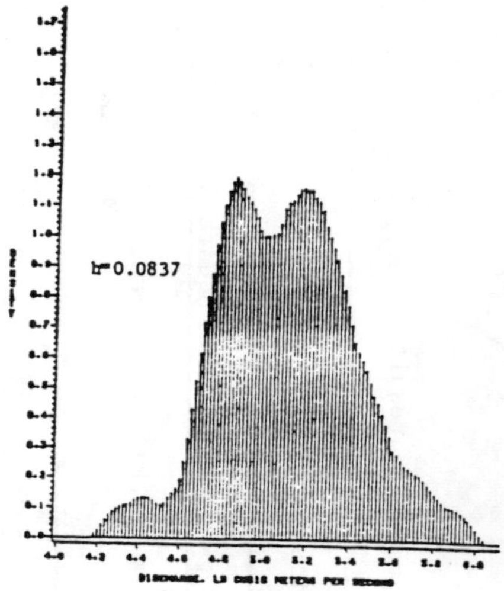

h by Flanagan's Formula

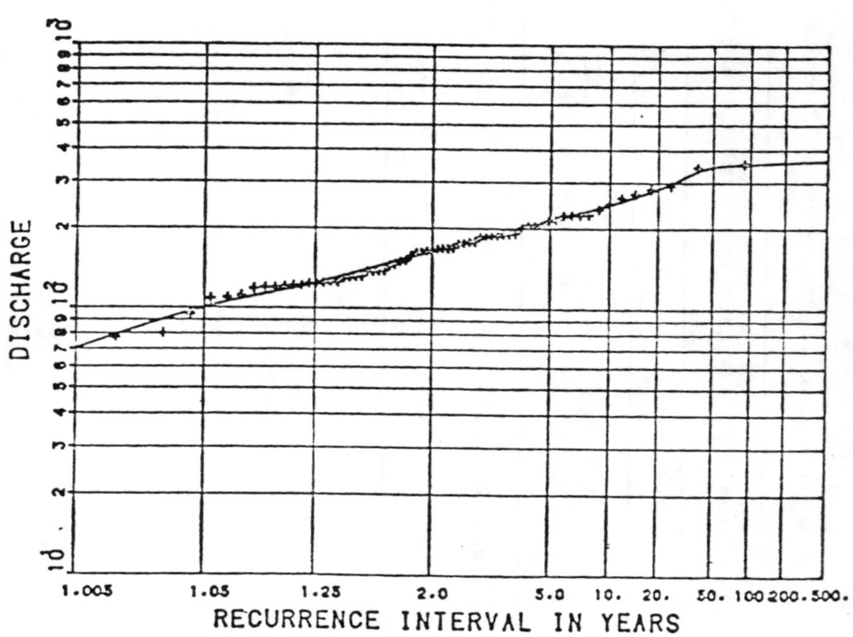

Small sample simulation results for the nonparametric methods are compared with those of parametric methods in Table 5. As can be seen, the RMS error of nonparametric method approaches those of LP3, which is the assumed true value used in data generated. While the information in Table 5 is limited, it clearly indicates the potential of the kernel method in flood frequency analysis.

CONCLUSIONS

The nonparametric kernel method appears to be a viable alternative in flood frequency analysis with the advantage of not requiring a distributional assumption, and providing a uniform procedure.

Nonparametric density estimation indicates the presence of bimodal density, perhaps reflecting causative factors of floods.

REFERENCES

Adamowski, K. (1985) "Nonparametric Kernel Estimation of Flood Frequencies", Water Resour. Res. **21(11)**, 1885-1590.

Condie, R., G.A. Nix and L.G. Boone (1981) Flood Damage Program Flood Frequency Analysis, Engineering Hydrology Section, Environment Canada, (Updated 1985, Unpublished).

Devroye, L. and L. Gyorfi (1985) Nonparametric Density Estimation: The L_1 View, J. Wiley and Sons, 356 pages.

Epanechnikov, V.A. (1969) "Nonparametric estimation of a multivariate probability density", Theory of Probability and its Application, 14, 153-158.

Flanagan, P. (1980) Computer Programs for Kernel Estimates of a Probability Density Function, M.S. Thesis, University of South Carolina.

Hall, M.J. (1984) Urban Hydrology, Elsevier Applied Science Publishers, London, 299 pages.

Houghton, J.C. (1978) "Birth of a Parent: The Wakeby Distribution for Modeling Flood Flows", Water Resour. Res. **14(6)**, 1105-9.

Labatiuk, C. (1985) A Nonparametric Approach to Flood Frequency Analysis, M.A.Sc. Thesis, Univeristy of Ottawa, 184 pages.

Rossi, F., M. Fiorentino and P. Versace (1984) "Two-Component Extreme Value Distribution for Flood Frequency Analysis", Water Resour. Res. **20(7)**, 847-56.

Singh, K.P. and M. Nakashima (1983) A New Methodology for Flood Frequency Analysis with Objective Detection and Modification of

Outliers/Inliers, Frontiers in Hydrology, Water Resources Publications, Littleton, Colorado, p. 187-201.

Thomas, Jr., W.O. (1985) "A Uniform Technique for Flood Frequency Analysis", A.S.C.E., Journal of Water Resources Planning and Management, **Vol. III, No. 3**, p.321-337.

U.S. Geological Survey's Interagency Advisory Committee on Water Data (1982), Guidelines for Determining Flood Flow Frequency, Bull. 17B.

Wallis, J.R., and E.F. Wood (1985) "Relative Accuracy of Log Pearson III Procedures", A.S.C.E., Journal of Hydraulic Engineering, **Vol. III, No.** 7, p.1043-1056.

STATISTICAL MODELS FOR FLOOD FREQUENCY ESTIMATION OF THE MISSISSIPPI AND YAZOO RIVERS

Edward Nissan
Department of Economics
University of Southern Mississippi
Hattiesburg, Mississippi 39406
U.S.A.

ABSTRACT. This paper investigates the application of the theory of extreme values to predit the magnitudes of annual floods for the Mississippi and Yazoo Rivers based on maximum annual levels of the rivers between 1925 and 1975. First, the data were tested to establish an initial probability distribution of the yearly extreme values. It was found by means of goodness of fit tests that both the normal distribution and the log-normal distribution adequately describe the extreme probability distributions for both rivers. Then by using techniques developed by Cramér, the expectations and variances of the mth extremes were calculated. From the estimates of these two parameters, probability predications of the level of floods for any return period may be obtained. The paper presents predicted values of floods for return periods of 100, 200, and 300 years.

1. INTRODUCTION

The purpose of this paper is to develop a statistical treatment of maximum annual levels for the Mississippi River at the Vicksburg station and the Yazoo River at the Greenwood station; both stations are located in the state of Mississippi. The maximum annual level is a term used to denote the largest record of a river during a year measured in feet. Therefore, during a particular year, there is only one measure which describes such a maximum even though there may be many days of the same record.

An important statistical tool for the study of the magnitudes of annual floods is the application of the theory of extreme values which is a special case of order statistics. Order statistics is based on distribution—free theory. The objective is a prediction of the number of cases which are larger than a given extreme observed in a sample of size n in the next set of trials on the assumption that these trials are independent and that they are taken from the same continuous population from which the sample is drawn. Specifically, one desires to calculate the probability that the mth largest among n past observations will be exceeded x times in N future trials. For instance,

V. P. Singh (ed.), Hydrologic Frequency Modeling, 107–115.

let m = 1 signifies the largest observation in n = 20 past observations. Then a question may be asked regarding the probability that in N = 5 future trials, this first extreme will be exceeded by x = 2 observations. A similar question may be asked for the second extreme where in this case m = 2. Thus the distribution-free theory by means of combinatorial mathematics allows the calculations of the frequency of extremes.

Gumbel (1958, p. 59) has shown that a fundamental formula for the distribution of exceedances based on distribution-free theory is

$$w(n,m,N,x)=\frac{\binom{n}{m}m\binom{n}{x}}{(N+n)\binom{N+n-1}{x+m-1}}\ ;\ \sum_{x=o}^{N}w(n,m,N,x)=1;$$

where n,m,N and x are as defined above.

If the initial distribution of a variate is shown to follow a theoretical distribution, then the exact distribution of the extremes can be derived along with their asymptotic distributions. For instance, the initial distribution may be hypothesized to be normally distributed, exponentially distributed, log-normally distributed and so on. Specific tests are required to ascertain the initial probability function, the "Goodness of Fit" tests. If the hypothesis is judged acceptable, then the derivations to obtain the distributions of extremes may be attainable. The results presented here depend on the assumption of an initial parametric probability function.

The next section will discuss the nature of the data, and the choice of the appropriate initial probability function. The third section will take up the distribution of the extremes, the fourth section will discuss the results, and the final section will provide conclusions and summaries.

2. DATA DESCRIPTION

Table I presents maximum annual data x_i, arranged in ascending order, for the Mississippi and Yazoo Rivers for 1925-75. The data were supplied by the Corps of Engineers Office in Vicksburg. Floods are levels above 43 feet for the Mississippi River and 35 feet for the Yazoo River, one foot = 0.3048 m. Therefore, there were 19 years out of the 51 years where the Mississippi River exceeded the flooding level while there were 18 such years for the Yazoo River.

TABLE I: Historical Data in Ascending Order for the Mississippi and Yazoo Rivers, 1925-1975.

Mississippi River		Yazoo River	
Year	Value x_i (feet)	Year	Value x_i (feet)
1954	22.07	1941	23.30
1941	29.20	1963	23.95

TABLE I - Continue

1959	29.39	1954	24.16
1956	32.33	1967	25.21
1953	33.52	1940	25.70
1960	35.08	1934	26.00
1940	35.10	1936	27.10
1934	35.20	1959	29.00
1925	35.20	1952	29.33
1967	35.60	1960	29.49
1966	35.80	1956	29.74
1951	36.11	1971	29.79
1942	36.30	1942	29.80
1964	36.40	1972	29.93
1931	36.84	1925	30.00
1955	37.36	1969	30.00
1968	37.95	1943	30.00
1958	38.44	1965	30.53
1952	38.49	1970	30.65
1972	38.85	1966	31.04
1971	39.04	1964	31.65
1947	39.10	1929	32.20
1963	39.90	1955	32.31
1957	40.10	1928	32.50
1965	40.30	1938	32.70
1969	40.55	1957	32.98
1926	40.80	1968	33.16
1946	40.80	1961	33.46
1948	41.40	1953	33.71
1938	41.70	1962	33.73
1970	42.14	1975	33.97
1949	42.40	1958	34.15
1936	44.00	1930	34.20
1962	44.20	1935	35.60
1939	44.40	1947	35.72
1950	44.96	1939	35.90
1974	45.56	1931	36.00
1944	45.60	1974	36.35
1930	45.70	1937	36.40
1943	45.80	1951	36.92
1961	47.30	1926	37.00
1935	48.00	1945	37.20
1933	48.92	1933	38.10
1928	49.30	1973	38.37
1945	49.80	1944	38.40
1975	49.81	1927	38.40
1932	51.90	1950	38.67
1973	53.10	1949	38.84
1929	55.20	1946	39.80
1937	55.50	1948	40.00
1927	58.40	1932	40.10

A goodness of fit tests were conducted to compare the sampling frequency functions of the 51 values of the Mississippi and Yazoo Rivers with a selected set of theoretical probability functions. Among these were the exponential, the gamma, the normal, and log-normal distributions. The Kolmogrov-Smirnov test is employed for this purposes as suggested by Rohatgi (1984). The exponential distribution was rejected, the gamma distribution converged to the normal distribution and both the normal and the log-normal distributions could not be rejected.

Based on these results, the forthcoming section of this paper will undertake the analysis primarily under the assumptions that the initial sample distributions of the extreme values of the Mississippi and Yazoo Rivers follow the normal and the log-normal theoretical distributions.

3. THE DISTRIBUTION OF THE mth EXTREMES

The theory regarding the distribution of the mth extreme is primarily based on order statistics. That is, in a sample of size n observations the range is finite, and it is the difference between the largest and the smallest observations. It should be pointed out that the probability that a given value x is the largest observation and a given value y is the smallest observation decreases when the sample size increases. As an example, assume that in the last 20 years the coldest temperature registered in a city is -6°F. There is no assurance that this minimum will not be broken in a future date. Therefore, the mth extreme has a sampling distribution which implies that the common characteristics associated with probability distributions such as expectation, variance, range, and mode may be obtained.

Assume that a population is characterized by a continuous distribution function $F(x) = P(X \leq x)$ and a frequency function $f(x) = F'(x)$. Assume further that X_m denotes the mth largest value obtained from a sample of size n from this population, after the n values are arranged in an increasing order of magnitude. Thus m = 1 is the largest observation, m = 2 is the second largest, and so on. In order to derive an expression for the probability element of the mth value $g_m(X_m)dX_m$ from the sampling distribution, it can be observed that this probability element is equivalent to the statement that in each trial, (n-m) values are less than X_m with a probability $F(x_m)^{n-m}$, (m-1) values greater than (x_m+dx_m) with a probability $\{1-F(x_m)\}^{m-1}$ and the remaining mth value itself with a probability $f(x_m)dx_m$. Because of the assumption of independence, these probabilities are multiplied. Taking into account all possible combinations of the values in each trial given by $\binom{n-1}{m-1}$ and since there are n trials (sample size), the probability element of the mth value as shown by Fisz (1963, p. 376) may be expressed in the form

$$g_m(X_m)dx_m = n\binom{n-1}{m-1}\{F(x_m)\}^{n-m}\{1-F(x_m)\}^{m-1}f(X_m)dx_m. \quad (1)$$

This is a general form of the distribution of the mth largest observation from a sample of size n. Gumbel (1958) has given an extensive systematic treatment of this distribution and has also developed some of its asymtotic theory. Cramèr (1957, pp. 370-371) has constructed an asymptotoc distribution to (1) by introducing a new variable ξ_m defined by

$$\xi_m = n\{1-F(x_m)\} \tag{2}$$

where $0 \leq \xi_m \leq n$. By proper substitutions, the probability frequency function for the new variable ξ_m is

$$h_m(\xi_m) = \binom{n-1}{m-1}\left(\frac{\xi_m}{n}\right)^{m-1}\left(1-\frac{\xi_m}{n}\right)^{n-m}. \tag{3}$$

The asymptotic limiting distribution when $n \to \infty$ in (3) converges for fixed m to the gamma distribution

$$h_m(\xi_m) = \frac{\xi_m^{\ m-1}\exp(-\xi_m)}{\Gamma(m)}, \quad \xi_m \geq 0, \ \Gamma(m) = (m-1)!. \tag{4}$$

So far in the derivation of the frequency function $h_m(\xi_m)$ no initial theoretical distribution was assumed for the distribution function F(x). It was only necessary to specify the probability in terms of the ranks of the sample observations by assigning $F(x_m) = q$ for the probability of a value less than x_m and a probability $1-F(x_m) = p$, for a value greater than x_m+dx_m where $0 \leq p+q < 1$, and a probability equals to $f(x_m)dx_m$ for the mth extreme itself.

However, if the initial distribution F(x) is known, then ξ_m, given in (2) can be expressed in terms of the known distribution. Cramèr (1957, pp. 372-378) gives the solutions for the rectangular, triangular, Cauchy, Laplace, and the normal distributions. Of particular interest to this paper are the results obtained for the normal distribution.

Assume that the n sample observations are obtained from an initial distribution F(x) which is normal with mean μ and standard deviation σ. Let X_m be defined as the mth largest value, then according to Cramèr, the expected value of X_m is

$$E\{X_m\} = \mu + \sigma\left((2\ \ln x)^{\frac{1}{2}} - \frac{\ln\ln n + \ln 4\Pi + 2\ (S_1-C)}{2(2\ \ln n)^{\frac{1}{2}}} + \right.$$
$$\left. + O\left(\frac{1}{\ln n}\right)\right) \tag{5}$$

and the variance of x_m is

$$\sigma^2_{x_m} = \frac{\sigma^2}{2\ln n}\left(\frac{\Pi^2}{6} - S_2\right) + O\left(\frac{1}{\ln^2 n}\right) \tag{6}$$

where

$$S_1 = \frac{1}{1} + \frac{1}{2} + \cdots + \frac{1}{m-1}, \quad S_2 = \frac{1}{1^2} + \frac{1}{2^2} + \cdots + \frac{1}{(m-1)^2},$$

C = .5772...is known as Euler's constant, and the symbols $O(1/\ln n)$ and $O(1/\ln^2 n)$ mean that the remainder is at most of the order $(1/\ln n)$ and $(1/\ln^2 n)$ respectively, which are usually neglected in computations.

By replacing the population parameters μ and σ in equations (5) and (6) by their sample estimates $\bar{x}$ and s^2, estimates of the mean and variance of x_m can be obtained. The results for both the Mississippi River and the Yazoo River are discussed in the following section.

4. RESULTS

As outlined above, the initial distributions of the extreme levels of the Mississippi and Yazoo Rivers may be assumed to follow the normal or log-normal distributions. The normal distribution is given by

$$f(x;\mu,\sigma) = (\sqrt{2\pi}\sigma)^{-1} \exp(-\frac{1}{2}(\frac{x-\mu}{\sigma})^2), \quad -\infty<x<\infty,$$

where μ and σ are the population mean and standard deviation respectively. If on the other hand lnx is normally distributed with mean α and standard deviation β, then x has the log-normal distribution given by

$$f(x;\alpha,\beta) = (x\beta\sqrt{2\pi})^{-1} \exp(-\frac{1}{2}(\frac{\ln x-\alpha}{\beta})^2), \quad x>o.$$

The mean μ and variance σ^2 of x, the log-normally distributed variable are shown by Hawkins and Weber (1980, p. 141) to be

$$\mu = \exp(\alpha + \frac{1}{2}\beta^2), \tag{7}$$

and

$$\sigma^2 = \exp(2\alpha + \beta^2)\{\exp \beta^2 - 1\}. \tag{8}$$

In the following subsections, the above models are used to generate the predicted values of X_m which are then compared to the observed values for both the Mississippi and Yazoo Rivers.

4.1. The Mississippi River

Using the data of Table I (Mississippi River data) the estimates for μ and σ for the normal distribution are $\bar{x} = 41.390$ and $s = 7.17849$ respectively; for the log-normal distribution the estimates of α and β are 3.70771 and 0.17972, respectively, and from (7) and (8), the estimates for μ and σ are $\bar{x} = 41.42397$ and $s = 7.50424$.

Table II shows the means and standard deviations of x_m, $m = 1, 2,\ldots,6$ as computed by formulas (5) and (6) along with the actual corresponding extreme values for the sample period 1925-75.

As can be ovserved from Table II, the absolute difference between the actual observations and their means are small and may be attributed to random fluctuations or the smallness of sample since a sample of size 51 is comparatively small for a study of this nature.

If the probability distribution is assumed to remain unchanged for a long period of time, then projections may be made by using formulas (5) and (6). Table II also shows the estimated projections for the expected values of the highest six extremes x_m, $m = 1, 2,\ldots,6$ as well

Table II: Expected Water Levels of the Mississippi River - 1925-2225.

m	x_m actual	normal								log-normal							
		n = 51		n = 100		2 = 200		n = 300		n = 50		n = 100		n = 200		n = 300	
		$E(x_m)$	S_{x_m}	$E(x_m)$	S_{x_m}	$E(x_m)$	S_{x_m}	$E(x_m)$	S_{x_m}	$E(x_m)$	S_{x_m}	$E(x_m)$	S_{x_m}	$E(x_m)$	S_{x_m}	$E(x_m)$	S_{x_m}
1	58.40	58.01	3.28	59.74	3.03	61.40	2.83	62.32	2.73	58.80	3.43	60.61	3.17	62.35	2.96	63.31	2.85
2	55.50	55.45	2.06	57.38	1.90	59.20	1.77	60.20	1.71	56.12	2.15	58.14	1.99	60.04	1.85	61.09	1.78
3	55.20	54.17	1.61	56.19	1.49	58.09	1.39	59.13	1.34	54.78	1.68	56.90	2.55	58.89	1.45	59.98	1.40
4	53.10	53.31	1.36	55.41	1.26	57.36	1.17	58.43	1.13	53.89	1.43	56.08	1.32	58.12	1.23	59.24	1.18
5	51.90	52.67	1.20	54.81	1.11	56.81	1.04	57.89	1.00	53.22	1.26	55.46	1.16	57.54	1.08	58.68	1.05
6	49.81	52.16	1.09	54.34	1.01	56.37	0.94	57.47	0.91	52.69	1.14	54.96	1.05	57.08	0.98	58.24	0.95

Table III: Expected Water Levels of the Yazoo River, 1925-2225.

m	x_m	normal								log-normal							
		n = 51		n = 100		n = 200		n = 300		n = 51		n = 100		n = 200		n = 300	
	actual	$E(x_m)$	S_{x_m}	$E(x_m)$	S_{x_m}	$E(x_m)$	S_{x_m}	$E(x_m)$	S_{x_m}	$E(x_m)$	S_{x_m}	$E(x_m)$	S_{x_m}	$E(x_m)$	S_{x_m}	$E(x_m)$	S_{x_m}
1	40.10	43.22	2.06	44.31	1.90	45.35	1.77	45.92	1.71	43.68	2.15	44.82	1.98	45.90	1.85	46.50	1.78
2	40.00	41.62	1.29	42.83	1.19	43.97	1.11	44.59	1.07	42.01	1.34	43.27	1.24	44.46	1.16	45.11	1.12
3	39.80	40.81	1.01	42.08	0.93	43.27	0.87	43.93	0.84	41.17	1.05	42.50	0.97	43.74	0.91	44.42	0.87
4	38.84	40.28	0.85	41.59	0.79	42.81	0.74	43.48	0.71	40.61	0.89	41.98	0.82	43.26	0.77	43.96	0.74
5	38.67	39.88	0.75	41.22	0.70	42.47	0.65	43.15	0.63	40.20	0.79	41.60	0.73	42.90	0.68	43.61	0.65
6	38.40	39.29	0.68	40.92	0.63	42.19	0.59	42.88	0.57	39.86	0.71	41.29	0.66	42.61	0.61	43.33	0.59

as the standard deviations for n = 100, 200 and 300 years for both the normal and log-normal distributions.

From this table some valuable information may be obtained. For instance, in 200 years it is highly improbable that a flood would exceed $E(x_1)$ by more than (3) standard deviation, or 69.89 feet (61.40 + 3(2.83)). The table may be used in a similar manner for the other extremes and time periods.

4.2. The Yazoo River

Using data of the Yazoo River (Table I), $\bar{x}$ = 32.808 and s = 4.49795 for the estimates of μ and σ for the normal distribution and $\bar{x}$ = 32.82298 and s = 4.69079 for the log-normal distribution where the estimates of α and β from equation (7) and (8) are 3.48102 and 0.14218, respectively.

Table III shows the means and standard deivations of x_m, m = 1, 2,...,6 along with the actual observed values. Also in the same table the estimated projections for the levels of the Yazoo River obtained from equations (5) and (6) are shown. It includes a period which spans 300 years for the six largest extremes.

5. SUMMARY AND CONCLUDING REMARKS

This paper has attempted to portray a profile of the flooding experience of the Mississippi and Yazoo Rivers. It was shown that both the normal and log-normal distributions may be used to advantage to fit a sample of size n = 51 years of the maximum yearly observations. It has further introduced a technique suggested by Cramèr to estimate the expectations of exceedances as well as to generate projections of future floods.

The methods utilized gave exceedingly good results, even though a small sample size of 51 observations was used. Therefore, it can be said that better results could be obtianed if the initial sample size was considerably larger.

Furthermore, the statistical methods suggested are easy to follow and calculations are easy to carry out. Also, they are amenable to a variety of uses. For instance, the procedure may be used to analyze data for a particular month or a particular week within a year.

REFERENCES

Cramèr, H. 1957. Mathematical Methods of Statistics, Princeton University Press, Princeton.

Fisz, M. 1963. Probability Theory and Mathematical Statistics, John Wiley & Sons, Inc., New York.

Gumbel, E. J. 1958. Statistics of Extremes, Columbia University Press, New York.

Hawkins, C. A. and J. E. Weber. 1980. Statistical Analysis, Harper & Row, New York.

Rohatgi, V. K. 1984. Statistical Inference, John Wiley & Sons, Inc., New York.

PARTIAL DURATION SERIES WITH LOG-NORMAL DISTRIBUTED PEAK VALUES

Dan Rosbjerg
Institute of Hydrodynamics and Hydraulic Engineering
Technical University of Denmark
DK-2800 Lyngby

ABSTRACT. The T-year estimate in a partial duration (PD) series with log-normal distributed peak exceedances is analysed in case of parameter estimation by means of the maximum likelihood method and the method of moments, respectively. Approximate expressions for bias and variance of the T-year estimate are developed, and it is hereby found that the sample variance of the log-normal distribution parameters is the main reason for the uncertainty of the T-year estimate, whereas the sample variance of the intensity of the process is of minor importance. Providing smaller bias (in absolute terms) and smaller variance the maximum likelihood method is found to be superior to the method of moments. The two methods produce bias of different signs, positive bias in case of maximum likelihood estimation and negative in case of estimation by method of moments. The variance expression in case of maximum likelihood estimation is to some extent verified by Monte Carlo simulation obtained from a study of significant wave heights. The purpose of the investigation was to assess the 50-year significant wave height on the basis of 17 years of hindcasted wave data together with the standard deviation of the estimate.

INTRODUCTION

By imposing a certain threshold q_0 in a hydrograph series Q(t) and taking only peaks above q_0 into account it is often found that the peak exceedances are exponentially distributed. Consequently, by assuming Poissonian occurrence times of the peak values, the maximum peak exceedance during a given time period is found to be double-exponentially distributed, bounded to the left (Todorovic, 1978). However, in some cases it is found necessary to employ a very high threshold in order to get the assumption of exponentially distributed peak exceedances verified. The high threshold may imply that only very few peak exceedances are available for the statistical analysis resulting in very unreliable estimates of extreme events. If a somewhat lower threshold is chosen, a sufficient number of peak exceedances can be obtained, but this can be at the expense of the exponential assumption. Too low a threshold may cause problems like, for instance, dependency between peak exceedances

V. P. Singh (ed.), Hydrologic Frequency Modeling, 117–129.

and inclusion of peak values which do not belong to the same population as those extreme events the prediction of which is the purpose of the analysis.

In the following, a model for prediction of extreme events based on Poissonian occurrence times and log-normal distributed peak exceedances will be examined with maintenance of the assumption of independent peak exceedances. In particular, the bias and the variance of the T-year estimate due to the sampling variance of the estimated model parameters will be analysed. The implications of alternative methods for estimation of the log-normal parameters will be evaluated as a consequence of the dominating effect which these distributions parameters have on the resulting uncertainty of the T-year estimate.

The analysis will be illustrated by an example in which a 50-year significant wave height was estimated on the basis of hindcasted wave data. Some problems in connection with the application of the model will be briefly commented, and the analytical results will be compared to results obtained by Monte Carlo simulation.

T-YEAR EVENT

Assume that the independent events X in a PD-series follow the log-normal distribution with parameters μ_y and $\sigma_y{}^2$. The cumulative distribution function is then given by

$$F(x) = \Phi\left(\frac{\ln x - \mu_y}{\sigma_y}\right) \tag{1}$$

where $\Phi(\cdot)$ denotes the standard normal distribution function. The relations between the mean value μ_x and variance $\sigma_x{}^2$ of X and the distribution parameters are

$$\mu_y = \ln\frac{\mu_x}{\left(1+\sigma_x^2/\mu_x^2\right)^{\frac{1}{2}}} \quad ; \quad \mu_x = \exp\left(\mu_y + \sigma_y^2/2\right)$$

$$\sigma_y^2 = \ln\left(1+\sigma_x^2/\mu_x^2\right) \quad ; \quad \sigma_x^2 = \exp\left(2\mu_y+\sigma_y^2\right)\left[\exp\left(\sigma_y^2\right)-1\right] \tag{2}$$

Assume further that the occurrence of the events corresponds to a time homogeneous Poisson process with intensity λ. The distribution function for the maximum event in time t, $X_t{}'$, is then given by

$$P\left\{X_t' \leqslant x\right\} = \exp\left(-\lambda_x t\right) \tag{3}$$

where

$$\lambda_x = \lambda\left[1-F(x)\right] = \lambda\left[1-\Phi\left(\frac{\ln x - \mu_y}{\sigma_y}\right)\right] \tag{4}$$

The T-year event, x_T, is defined as the level which on the average will be exceeded once per T years. Accordingly x_T can be found from Eq. (4) by requesting $\lambda_x = T^{-1}$,

$$x_T = F^{-1}\left(1-\frac{1}{\lambda T}\right) = \exp\left(\mu_y + \sigma_y\, \Phi^{-1}\left(1-\frac{1}{\lambda T}\right)\right)$$

$$= \exp\left(\mu_y + y_T\sigma_y\right) \tag{5}$$

where y_T has been introduced as the $\left(1-\frac{1}{\lambda T}\right)$-quantile in the standard normal distribution.

Alternatively, x_T can be expressed by means of a frequency factor K_T as

$$x_T = \mu_x + K_T\sigma_x = \mu_x\left(1 + K_T z\right) \tag{6}$$

where $z = \sigma_x/\mu_x$ denotes the coefficient of variation, and K_T is given by

$$K_T = \frac{\exp\left[y_T\left(\ln(1+z^2)\right)^{\frac{1}{2}} - \frac{1}{2}\ln(1+z^2)\right] - 1}{z} \tag{7}$$

In the case $\lambda = 2$, $\mu_y = 1$ and $\sigma_y = 0.5$ is obtained $\mu_x = 3.0802$, $\sigma_x = 1.6416$ and $z = 0.5329$, and for $T = 50$ years is found $y_T = 2.3268$, $K_T = 3.4238$ and $x_T = 8.7006$. This will be used as a reference example in the following.

T-YEAR ESTIMATE

An estimate of the T-year event can be obtained by inserting estimates of μ_y, σ_y and y_T in Eq. (5)

$$\hat{x}_T = \exp\left(\hat{\mu}_y + \hat{y}_T\,\hat{\sigma}_y\right) \tag{8}$$

or alternatively estimates of μ_x, σ_x and K_T in Eq. (6)

$$\hat{x}_T = \hat{\mu}_x + \hat{K}_T\,\hat{\sigma}_x \tag{9}$$

In the first case maximum likelihood estimates of μ_y and σ_y are used, while in the last case the method of moments is used to estimate μ_x and σ_x. Due to the different qualities of the estimation principles the T-year estimate will possess different characteristics depending on which method has been used. This will be examined in detail with respect to bias and variance which will be calculated approximately in the two cases. Also the effect of the sampling uncertainty of the intensity λ on the T-year estimate will be examined. The analysis will be initiated with an approximate calculation of the bias and variance of the quantile estimate, $\hat{y}_T$.

Estimation of Quantile in Standard Normal Distribution

The intensity λ in the Poisson process is estimated as the average number of events per time unit

$$\hat{\lambda} = \frac{N_1}{t_1} \tag{10}$$

where N_1, the number of events during the time period t_1, is Poisson distributed with λt_1 as parameter. Recognizing that the threshold always will be chosen so that at least one event will occur during the time period t_1, the mean and variance of $\hat{\lambda}$ become

$$E\{\hat{\lambda}\} = \frac{E\{N_1 | N_1 > 0\}}{t_1} = \frac{\lambda}{1 - \exp(-\lambda t_1)} \tag{11}$$

$$Var\{\hat{\lambda}\} = \frac{Var\{N_1 | N_1 > 0\}}{t_1^2} = \frac{\lambda}{t_1(1-\exp(-\lambda t_1))} \tag{12}$$

Estimation of the $(1-\frac{1}{\lambda T})$-quantile in the standard normal distribution is performed by insertion of the estimate of the intensity in the quantile expression

$$\hat{y}_T = \Phi^{-1} \left(1-\frac{1}{\hat{\lambda}T}\right) = g(\hat{\lambda} ; T) \tag{13}$$

Mean and variance of the estimator $\hat{y}_T$ can be found approximately by Taylor series expansion

$$E\{\hat{y}_T\} \simeq (g(\hat{\lambda} ; T))_m + \tfrac{1}{2} \left(\frac{\partial^2 g}{\partial \hat{\lambda}^2}\right)_m Var\{\hat{\lambda}\} \tag{14}$$

$$Var\{\hat{y}_T\} \simeq \left(\frac{\partial g}{\partial \hat{\lambda}}\right)_m^2 Var\{\hat{\lambda}\} \tag{15}$$

where index m denotes that the expressions are to be evaluated at the mean value of $\hat{\lambda}$. This gives

$$(g(\hat{\lambda}))_m = \Phi^{-1}\left(1 - \frac{1-\exp(-\lambda t_1)}{\lambda T}\right) = y_{Tc} \tag{16}$$

$$\left(\frac{\partial g}{\partial \hat{\lambda}}\right)_m = \frac{1}{\phi(y_{Tc})} \frac{(1-\exp(-\lambda t_1))^2}{\lambda^2 T} \tag{17}$$

$$\left(\frac{\partial g^2}{\partial \hat{\lambda}^2}\right)_m = - \frac{2(1-\exp(-\lambda t_1))^3}{\phi(y_{Tc})\lambda^3 T} + \frac{y_{Tc}(1-\exp(-\lambda t_1))^4}{\lambda^4 T^2 (\phi(y_{Tc}))^2} \tag{18}$$

Hence the mean and variance of $\hat{y}_T$ become approximately

$$E\{\hat{y}_T\} = y_{Tc} - \frac{(1-\exp(-\lambda t_1))^2}{(\lambda t_1)(\lambda T)} \left[1 - \frac{y_{Tc}(1-\exp(-\lambda t_1))}{\lambda T\ \phi(y_{Tc})}\right] \tag{19}$$

$$\mathrm{Var}\{\hat{y}_T\} = \frac{(1-\exp(-\lambda t_1))^3}{\lambda t_1(\lambda T\ \phi(y_{Tc}))^2} \tag{20}$$

In case of $\lambda = 2$ and $T = 50$ years the second order term in Eq. (17) is negligible even for small values of t_1. Hence the mean value of $\hat{y}_T$ is here taken to be

$$E\{\hat{y}_T\} = y_{Tc} \tag{21}$$

where y_{Tc} has been defined through Eq. (16).

Estimation by the Maximum Likelihood Method

The properties of $\hat{x}_T$ given by Eq. (8) with the maximum likelihood estimators

$$\hat{\mu}_y = \frac{1}{N_1}\sum_{i=1}^{N_1} \ln x_i$$

$$\hat{\sigma}_y^2 = \frac{1}{N_1}\sum_{i=1}^{N_1} (\ln x_i - \hat{\mu}_y)^2 \tag{22}$$

and $\hat{\lambda}$ given by Eq. (10) inserted are now examined with respect to bias and variance. Considering $\hat{x}_T$ as a function of $\hat{\mu}_y$, $\hat{\sigma}_y$ and $\hat{y}_T$

$$\hat{x}_T = \exp(\hat{\mu}_y + \hat{y}_T\hat{\sigma}_y) = h(\hat{\mu}_y, \hat{\sigma}_y, \hat{y}_T\ ; T) \tag{23}$$

the expectation and variance become approximately

$$E\{\hat{x}_T\} = (h)_m + \tfrac{1}{2}\left(\frac{\partial^2 h}{\partial\hat{\mu}_y^2}\right)_m \mathrm{Var}\{\hat{\mu}_y\}$$

$$+ \tfrac{1}{2}\left(\frac{\partial^2 h}{\partial\hat{\sigma}_y^2}\right)_m \mathrm{Var}\{\hat{\sigma}_y\} + \tfrac{1}{2}\left(\frac{\partial^2 h}{\partial\hat{y}_T^2}\right)_m \mathrm{Var}\{\hat{y}_T\} \tag{24}$$

$$\mathrm{Var}\{\hat{x}_T\} = \left(\frac{\partial h}{\partial\hat{\mu}_y}\right)_m^2 \mathrm{Var}\{\hat{\mu}_y\} + \left(\frac{\partial h}{\partial\hat{\sigma}_y}\right)_m^2 \mathrm{Var}\{\hat{\sigma}_y\}$$

$$+ \left(\frac{\partial h}{\partial\hat{y}_T}\right)_m^2 \mathrm{Var}\{\hat{y}_T\} \tag{25}$$

where the stochastic independence between $\hat{\mu}_y$, $\hat{\sigma}_y$ and $\hat{y}_T$ has been utilized.

The mean and variance of $\hat{\mu}_y$ can be obtained as

$$E\{\hat{\mu}_y\} = E\{E\{\hat{\mu}_y | N_1, N_1 > 0\}\} = \mu_y \tag{26}$$

$$\begin{aligned} Var\{\hat{\mu}_y\} &= E\{Var\{\hat{\mu}_y | N_1, N_1 > 0\}\} + Var\{E\{\hat{\mu}_y | N_1, N_1 > 0\}\} \\ &= E\{\frac{\sigma_y^2}{N_1} | N_1 > 0\} + Var\{\mu_y | N_1 > 0\} \\ &= k\,\frac{\sigma_y^2}{\lambda t_1} \end{aligned} \tag{27}$$

where the correction factor k is given by Rosbjerg (1985).

Correspondingly, mean and variance of $\hat{\sigma}_y$ are approximately found to be

$$E\{\hat{\sigma}_y\} = \sigma_y \tag{28}$$

$$Var\{\hat{\sigma}_y\} = k\,\frac{\sigma_y{}^2}{2\lambda t_1} \tag{29}$$

Mean and variance of $\hat{y}_T$ are given by Eqs. (21) and (20).

Development of the partial derivatives in Eqs. (24) and (25), which is straightforward, and insertion of Eqs. (20)-(21) and (26)-(29) herein result in

$$E\{\hat{x}_T\} = x_{Tc}\left[1 + \frac{\sigma_y^2}{2\lambda t_1}\left(k\left(1+\frac{y_{Tc}{}^2}{2}\right) + \frac{(1-\exp(-\lambda t_1))^3}{(\lambda T\ \phi(y_{Tc}))^2}\right)\right] \tag{30}$$

$$Var\{\hat{x}_T\} = x_{Tc}{}^2\,\frac{\sigma_y^2}{\lambda t_1}\left[k\left(1+\frac{y_{Tc}{}^2}{2}\right) + \frac{(1-\exp(-\lambda t_1))^3}{(\lambda T\ \phi(y_{Tc}))^2}\right] \tag{31}$$

where x_{Tc} is given by

$$x_{Tc} = \exp(\mu_y + y_{Tc}\sigma_y) \tag{32}$$

These results are illustrated in Figs. 1 and 2 where the bias defined as $B\{\hat{x}_T\} = E\{\hat{x}_T\} - x_T$ and the standard deviation $D\{\hat{x}_T\} = (Var\{\hat{x}_T\})^{\frac{1}{2}}$ have been depicted as functions of t_1 in the case $\mu_y = 1$, $\sigma_y = 0.5$, $\lambda = 2$ and $T = 50$ years.

It should be noted that the sample variance of the intensity of the process only to a very small extent contributes to the bias and variance of the T-year estimate. The uncertainty of the T-year estimate

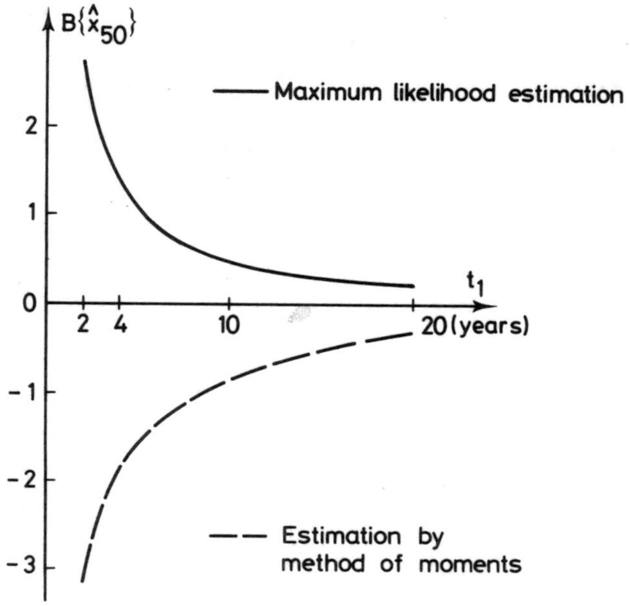

Fig. 1 Approximate bias of the 50-year estimate in case of $\lambda = 2$, $\mu_y = 1$ and $\sigma_y = 0.5$.

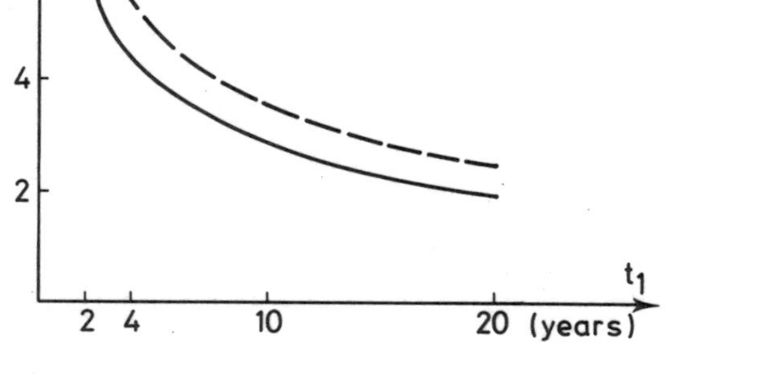

Fig. 2 Approximate standard deviation of 50-year estimate in case of $\lambda = 2$, $\mu_y = 1$ and $\sigma_y = 0.5$.

is found to be dominated by the sample variance of the parameters in the log normal distribution. Therefore the estimation of these parameters is of crucial importance. In order to throw more light on this problem, the results given above will be compared to results obtained on the basis of estimation by means of the method of moments.

Estimation by the Method of Moments

The starting point in this case is Eq. (9) where μ_x and $\sigma_x{}^2$ are estimated by the sample mean and sample variance, respectively

$$\hat{\mu}_x = \frac{1}{N_1}\sum_{i=1}^{N_1} x_i$$
$$\hat{\sigma}_x^2 = \frac{1}{N_1}\sum_{i=1}^{N_1} (x_i - \hat{\mu}_x)^2 \tag{33}$$

$\hat{K}_T$ is defined through Eq. (7) with y_T and z replaced by $\hat{y}_T$ and and $\hat{z} = \hat{\sigma}_x/\hat{\mu}_x$, respectively. Hence $\hat{x}_T$ in Eq. (9) can be regarded as a function of $\hat{\mu}_x$, $\hat{\sigma}_x{}^2$ and $\hat{y}_T$

$$\hat{x}_T = \hat{\mu}_x + \hat{K}_T\sqrt{\hat{\sigma}_x^2} = h(\hat{\mu}_x,\ \hat{\sigma}_x^2,\ \hat{y}_T\ ;\ T) \tag{34}$$

The expectation and variance of $\hat{x}_T$ become approximately

$$E\{\hat{x}_T\} = (h)_m + \tfrac{1}{2}\left(\frac{\partial^2 h}{\partial\hat{\mu}_x^2}\right)_m \mathrm{Var}\{\hat{\mu}_x\} + \tfrac{1}{2}\left(\frac{\partial^2 h}{\partial(\hat{\sigma}_x^2)^2}\right)_m \mathrm{Var}\{\hat{\sigma}_x^2\}$$
$$+ \left(\frac{\partial^2 h}{\partial\hat{\mu}_x\partial(\hat{\sigma}_x^2)}\right)_m \mathrm{Cov}\{\hat{\mu}_x,\hat{\sigma}_x^2\} + \tfrac{1}{2}\left(\frac{\partial^2 h}{\partial\hat{y}_T^2}\right)_m \mathrm{Var}\{\hat{y}_T\} \tag{35}$$

$$\mathrm{Var}\{\hat{x}_T\} = \left(\frac{\partial h}{\partial\hat{\mu}_x}\right)_m^2 \mathrm{Var}\{\hat{\mu}_x\} + \left(\frac{\partial h}{\partial(\hat{\sigma}_x^2)}\right)_m^2 \mathrm{Var}\{\hat{\sigma}_x^2\}$$
$$+ 2\left(\frac{\partial h}{\partial\hat{\mu}_x}\right)_m \left(\frac{\partial h}{\partial(\hat{\sigma}_x^2)}\right)_m \mathrm{Cov}\{\hat{\mu}_x,\hat{\sigma}_x^2\}$$
$$+ \left(\frac{\partial h}{\partial\hat{y}_T}\right)_m^2 \mathrm{Var}\{\hat{y}_T\} \tag{36}$$

With $\gamma_1 = E\{(X-\mu_x)^3\}/\sigma_x^3$ and $\gamma_2 = E\{(X-\mu_x)^4\}/\sigma_x^4$ the following formulae apply, at least approximately

$$\mathrm{Var}\{\hat{\mu}_x\} = k \frac{\sigma_x^2}{\lambda t_1}$$

$$\mathrm{Var}\{\hat{\sigma}_x^2\} = k \frac{\sigma_x^4}{\lambda t_1} (\gamma_2 - 1) \tag{37}$$

$$\mathrm{Cov}\{\hat{\mu}_x, \hat{\sigma}_x^2\} = k \frac{\sigma_x^3}{\lambda t_1} \gamma_1$$

The variance of $\hat{y}_T$ is given by Eq. 20. The partial derivatives in Eqs. 35 and 36 are evaluated for $E\{\hat{\mu}_x\} = \mu_x$, $E\{\hat{\sigma}_x^2\} \simeq \sigma_x^2$ and $E\{\hat{y}_T\} \simeq y_{TC}$. The following results are obtained

$$\left(\frac{\partial h}{\partial \hat{\mu}_x}\right)_m = 1 - z^2 K_{Tc}'$$

$$\left(\frac{\partial^2 h}{\partial \hat{\mu}_x^2}\right)_m = \frac{1}{\sigma_x} (2\, z^3\, K_{Tc}' + z^4\, K_{Tc}'')$$

$$\left(\frac{\partial h}{\partial (\hat{\sigma}_x^2)}\right)_m = \frac{1}{2\sigma_x} (K_{Tc} + z\, K_{Tc}')$$

$$\left(\frac{\partial^2 h}{\partial (\hat{\sigma}_x^2)^2}\right)_m = -\frac{1}{4\sigma_x^3} (K_{Tc} - z\, K_{Tc}' - z^2 K_{Tc}'') \tag{38}$$

$$\left(\frac{\partial^2 h}{\partial \hat{\mu}_x\, \partial (\hat{\sigma}_x^2)}\right)_m = -\frac{1}{2\sigma_x^2} (2z^2\, K_{Tc}' + z^3\, K_{Tc}'')$$

$$\left(\frac{\partial h}{\partial \hat{y}_T}\right)_m = \sigma_x \left(K_{TC} + \frac{1}{z}\right) \left(\ln(1 + z^2)\right)^{\frac{1}{2}}$$

$$\left(\frac{\partial^2 h}{\partial \hat{y}_T^2}\right)_m = \sigma_x \left(K_{Tc} + \frac{1}{z}\right) \ln(1 + z^2)$$

where

$$K_{Tc} = \frac{1}{z}\{\exp[y_{Tc}(\ln(1 + z^2))^{1/2} - 1/2 \ln(1 + z^2)] - 1\}$$

$$K_{Tc}' = \left(\frac{\partial \hat{K}_T}{\partial \hat{z}}\right)_m = \frac{1 + zK_{Tc}}{1 + z^2}\,[y_{Tc}(\ln(1+z^2))^{-1/2}-1] - \frac{K_{Tc}}{z}$$

$$K_{Tc}'' = \left(\frac{\partial^2 \hat{K}_T}{\partial \hat{z}^2}\right)_m = \frac{K_{Tc} + z\,K_{Tc}'}{1 + z^2}\,[y_{Tc}\,(\ln(1 + z^2))^{-1/2} - 3]$$

$$- \frac{z(1 + z\,K_{Tc})}{(1 + z^2)^2}\,y_{Tc}(\ln(1 + z^2))^{-3/2} + \frac{K_{Tc} - zK_{Tc}'}{z^2} \tag{39}$$

By insertion of Eqs. (37) - (39) in Eqs. (35) and (36) the mean and variance of $\hat{x}_T$ are found to be approximately

$$E\{\hat{x}_T\} = \mu_x + K_{Tc}\sigma_x + \frac{\sigma_x}{2\lambda t_1}\,\{k(2z^3K_{Tc}' + z^4\,K_{Tc}'')$$

$$- k(K_{Tc}-zK_{Tc}'-z^2K_{Tc}'')\frac{\gamma_2-1}{4} - k(2z^2K_{Tc}'+z^3K_{Tc}'')\gamma_1$$

$$+ \exp[y_{Tc}(\ln(1+z^2))^{1/2} - 1/2\ln(1+z^2)]$$

$$\frac{\ln(1+z^2)}{z}\,\frac{(1-\exp(-\lambda t_1))^3}{(\lambda T\ \phi(y_{Tc}))^2}\} \tag{40}$$

$$Var\{\hat{x}_T\} = \frac{\sigma_x^2}{\lambda t_1}\,\{k(1-z^2\,K_{Tc}')^2 + k(K_{Tc} + zK_{Tc}')^2\,\frac{\gamma_2-1}{4}$$

$$+ k(1 - z^2\,K_{Tc}')\,(K_{Tc}+ z\,K_{Tc}')\gamma_1$$

$$+ \exp[2y_{Tc}(\ln(1+z^2))^{1/2} - \ln(1+z^2)]$$

$$\frac{\ln(1+z^2)}{z^2}\,\frac{(1-\exp(-\lambda t_1))^3}{(\lambda T\phi(y_{Tc}))^2}\} \tag{41}$$

where

$$\gamma_1 = z^3 + 3z \tag{42}$$

$$\gamma_2 = z^8 + 6z^6 + 15z^4 + 16z^2 + 3 \tag{43}$$

Also in the case of estimation by method of moments the uncertainty in the estimation of $\hat{x}_T$ is completely dominated by the sample variances of the log-normal distribution parameters. The last terms in the brackets in Eqs. (40) and (41) are relatively insignificant, like the last terms in Eqs. (30) and (31).

The bias and standard deviation obtained on the basis of Eqs. (40) and (41) have been calculated in the case $\mu_x = 3.08022$, $\sigma_x = 1.64157$ (corresponding to $\mu_y = 1$, $\sigma_y = 0.5$), $\lambda = 2$ and $T = 50$ years and plotted in Figs. 1 and 2 for comparison with the results obtained on the basis of maximum likelihood estimation.

The comparison shows that the best results are obtained by means of the maximum likelihood method: smaller standard deviation and numerically smaller bias. It is further seen that the two estimation methods cause bias with opposite sign, the maximum likelihood method giving the positive and the method of moments the negative bias.

APPLICATION EXAMPLE

The application example concerns the estimation of a 50-year significant wave height for a specific location in the North Sea. The estimation was based on hindcasted wave data during 17 years, where the hindcasted values have been obtained on the basis of a wind-wave-model. The significant wave height was calculated every 3 hours as the integral of the wave spectral density (the variance of the wave process). Storm periods were then selected as periods with significant wave heights above a threshold of 5.0 m. On inspection it was found that this threshold implied the hereby defined storm events to be independent of each other. Otherwise, further criteria should be imposed in order to get the fundamental assumption of independence accepted, because an interdependence between successive peak events will significantly widen the confidence interval of the T-year event (Rosbjerg, 1985).

The peak value in connection with each excursion above the threshold of the process of significant wave heights was then selected for further analysis. A number of 34 peak values corresponding to 34 independent storms during the 17 years ($\hat{\lambda} = 2$ storms per year) was found. The histogram of the peak exceedance values suggested that the log-normal distribution could be an appropriate choice of distribution function for the peak exceedances, but nevertheless this hypothesis could neither pass a χ^2-test nor a Kolmogorov-Smirnov test. Alternatively, it was tried to use the peak values instead of the peak exceedances with the result that the distribution assumption no longer was rejected by the tests. The maximum likelihood estimates of the parameters were calculated to be $\hat{\mu}_y = 1.818$ and $\hat{\sigma}_y = 0.124$. Fig. 3 shows the histogram of the selected peak values together with the estimated log-normal probability density function.

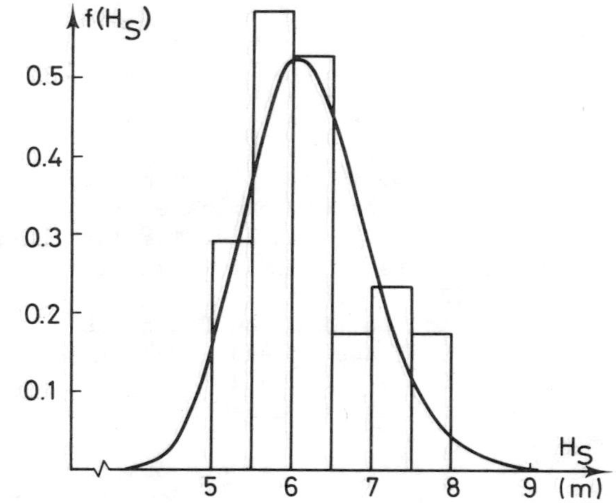

Fig. 3 Histogram for peak values of significant wave heights compared to estimated log-normal pdf.

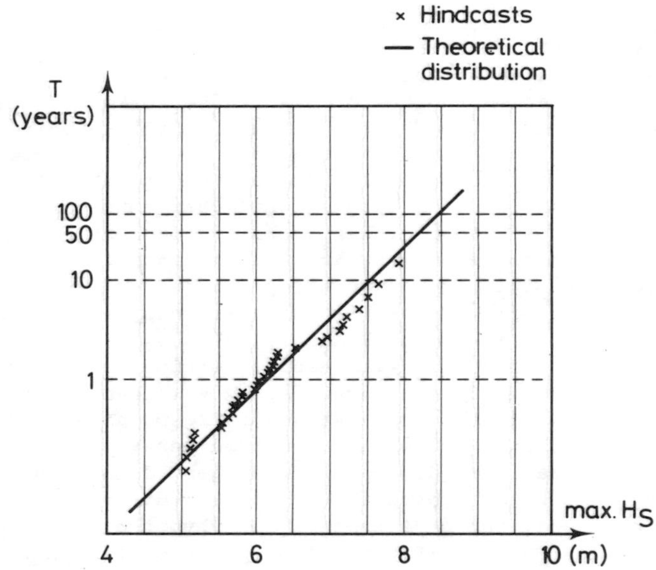

Fig. 4 Relation between peak value of significant wave height and return period with hindcasted values plotted for comparison.

Table 1 T-year estimates with analytically and Monte Carlo based standard deviations.

T Return period (years)	$\hat{x}_T$ T-year estimate (m)	Standard deviation of estimate Analytical (m)	Monto Carlo (m)
10	7.55	0.26	0.28
50	8.22	0.35	0.38
100	8.48	0.39	0.43

On the basis of the estimated parameters and by means of Eqs. (13) and (8) the relation between T and $\hat{x}_T$ is easily established. The relation is shown in Fig. 4 together with the hindcasted peak values. By means of Eq. (31) (where $\hat{\lambda} t_1 = 34 \Rightarrow k = 1.03$) the standard deviations of the T-year estimates are easily obtained too. Values corresponding to T = 10, 50 and 100 years are shown in Table 1.

For comparison standard deviations obtained by means of Monte Carlo simulation are given too. However, a discrepancy should be expected because of the inclusion of a feature in the Monte Carlo simulation not accounted for in the theory. The simulation procedure went on as follows. First the number of storms, N_j, during the jth 34-year period was generated on the basis of $\hat{\lambda} = 2$. Then N_j log-normal distributed events based on $\hat{\mu}_y = 1.818$ and $\hat{\sigma}_y = 0.124$ were generated. To account for the hindcast uncertainty, each of the generated events were superimposed by a small normal-distributed error term, which was obtained from a regression of hindcasted wave heights on measured values. On the basis of the hereby obtained N_j events a recalibration of the parameters was performed and the T-year estimate $\hat{y}_{T,j}$ was calculated. By repetition of the procedure for j = 1, 2, ..., 500, a number of 500 T-year estimates were obtained from which the standard deviation was calculated. Due to the additional hindcast uncertainty the standard deviation obtained by means of the Monto Carlo simulation was expected to be greater than the theoretical values. However, the additional error term has only been of minor importance for the uncertainty of the T-year estimate, as seen from Table 1.

REFERENCES

Rosbjerg, D. 1985. 'Estimation in partial duration series with independent and dependent peak values'. Journal of Hydrology, Vol. **76**, pp. 183-195.

Todorovic, P. 1978. 'Stochastic models of floods'. Water Resources Research, Vol. **14**, No. 2, pp. 345-356.

ESTIMATION OF A PRIOR DISTRIBUTION FOR THE BAYESIAN ESTIMATION OF PEARSON III SKEWS

Upmanu Lall
Assistant Professor
Department of Civil Engineering
University of Utah
Salt Lake City UT 84112

ABSTRACT. The Pearson III distribution is often used as a model for the annual maximum streamflow process. Estimates of flood quantiles and the probabilities of exceedance of flood magnitudes, using this distribution and the small samples usually available, are rather biased. A number of investigators have shown that the sampling distributions of, and the bias in the estimates of the Pearson III parameters, and consequently the flood quantiles and probabilities, are dependent only on the magnitude of the skew γ of the distribution and the sample size n. The importance of unbiased estimates of skew is thus recognized. The development of such unbiased estimators, using regional and/or site information, has met with limited success. This has contributed to the controversy surrounding the use of the Pearson III distribution to model the flood frequency process.

The usefulness of Bayesian estimation techniques in combining regional and site specific information has been recognized and stressed in the literature. One problem faced by the Bayesian estimator is the specification of an appropriate prior distribution to describe skews in a region of interest. The lack of this development has inhibited the use of Bayesian techniques. This paper presents a general framework for the estimation of a prior distribution for skews for a region where some data on annual maximum streamflows is available.

Sampling distributions $\phi_{1n}(g|\gamma)$, for sample skew g estimated from a sample of size n, where the parent population skew γ is known, are available in the literature for a number of values of n and γ. Sampling distributions $\phi_{2n}(g)$, for sample skew g estimated from a sample size n, using flood flow data from regional stations, where the parent population skew γ is unknown, can be derived for a number of sample sizes n, through resampling and subsampling techniques. The objective is then the determination of a prior distribution $\pi(\gamma)$, that yields the observed sampling distributions $\phi_{2n}(g)$, upon convolution with the sampling distributions $\phi_{1n}(g|\gamma)$, over some admissible set for γ, and for all sample sizes n. A discrete set of population skews γ_k (e.g. -6(0.5)6), is considered. The prior distribution $\pi(\gamma)$ is then approximated as a set of probabilities p_k associated with each of the admissible γ_k. The strategy for the minimum error estimation of $\pi(\gamma)$ then entails the determination of $\pi(\gamma)$ through the solution of a constrained least squares problem, where the objective is to minimize the sum of square of errors in predicting the quantiles of the sampling distributions $\phi_{2n}(g)$ through the convolution of $\pi(\gamma)$ and quantiles of the sampling distributions $\phi_{1n}(g|\gamma)$, over all sample sizes n for which data is available.

The prior probability distribution derived for regional skew can be used to develop unbiased Bayesian estimators of skew using site data. A framework is thus provided for the use of regional information in developing at site skew estimators. The framework can be extended to the estimation of flood quantiles and probabilities with uncertain skew. The techniques developed are applied to data from 314 U.S. gaging stations, that have no diversion or regulation of flow, and have at least 60 years of annual maximum data.

V. P. Singh (ed.), Hydrologic Frequency Modeling, 131–147.

1. INTRODUCTION

The Pearson III distribution is often used as a model for the annual maximum streamflow process. Estimates of flood quantiles and the probabilities of exceedance of flood magnitudes, using this distribution and the small samples usually available, are rather biased. The bias in the estimates of the Pearson III parameters, is dependent only on the magnitude of the skew γ of the distribution and on the sample size n. The importance of unbiased estimates of skew is thus recognized. The development of such unbiased estimators, using regional and/or site information, has met with limited success. This has contributed to the controversy surrounding the use of the Pearson III distribution to model the flood frequency process.

Bayesian estimation techniques can be used for combining regional and site specific information to develop estimates of skew. A problem faced by the Bayesian estimator is the specification of an appropriate prior distribution to describe the skews in a region of interest. This paper presents a general framework for the estimation of a prior distribution for skews for a region where some data on annual maximum streamflows is available. Information on the sampling distributions of skew computed from annual maximum flow data and synthetically generated data from Pearson III populations with specified skew, is used for the determination of a regional prior distribution for skew. The prior distribution estimated is then combined with sampling distributions from Pearson III populations with known skew, to obtain posterior probabilities for skew as a function of sample size n and the computed value of skew at the site. A Bayesian estimator for skew is then developed as a function of sample size n and the estimate of skew at the site.

1.1. Review

The Pearson type III distribution has often been proposed as a suitable probability model for annual maximum streamflow data. It is described functionally as:

$$f(x; \alpha,\beta,\lambda) = (\alpha\, \Gamma(\beta))^{-1}\ ((x-\lambda)/\alpha)^{\beta-1} \exp(-(x-\lambda)/\alpha) \tag{1}$$

where

β	a shape parameter
α	a scale parameter
λ	a location parameter
$\Gamma(\,)$	the Gamma function
x	the Pearson III random variate

The relationships between the sample moments and the above parameters are expressed for a sample of size n as:

Mean

$$m = \Sigma_i\, x_i / n = \alpha'\beta' + \lambda' \tag{2}$$

Variance

$$s^2 = (\Sigma_i\, x_i^2/n - m^2) = \alpha'^2\beta' \tag{3}$$

Coefficient of Skewness

$$g = (\Sigma_i\, x_i^3/n - 3\,ms^2 - m^3)/s^3 = 2/\beta'^{1/2} \tag{4}$$

Fisher (1921) shows that (1) the method of moments estimates for the Pearson type III parameters are highly inefficient and (2) the relative efficiency of estimation of all three parameters is a function only of β, the shape parameter (skew). As skew increases, the efficiency of estimation decreases rapidly. The need for unbiased and efficient estimates of skew is thus highlighted.

1.2. Previous Results

Fisher (1929) specified the factor $(n(n-1))^{1/2}/(n-2)$ to remove the bias in the estimation of skew, if the parent population is Gaussian Normal. Hazen (1930) specified an additional factor of (1+8.5/n), where the Pearson type III distribution is used. No analytical solutions to the general problem of bias removal in the estimation of skew are available for this distribution. Further, Kirby (1974) shows that there is a bound SB_n on the absolute value of sample skew g, given as

$$SB_n = (n-2)/(n-1)^{1/2} \tag{5}$$

Wallis et al (1974) used a Monte Carlo simulation to derive bias factors and sampling distributions of skew in samples of size n, ranging from 10 to 90, for various values of parent population skew. Their results showed only an approximate agreement with Hazen's factor (recommended for use by the U.S. Water Resources Council (1977)) for the range of 0.5 to 2.0 for parent population skew γ. Bobee and Robitaille (1975) used Wallis et al's (1974) results to formalize a bias correction factor for average sample skew of the form

$$G_s = g_s [(1 + 6.51/n + 20.2/n^2) + (1.48/n + 6.77/n^2)g_s^2] \tag{6}$$

where G_s is the unbiased estimate of parent population skew and g_s is average sample skew.

The Monte Carlo Simulation of Wallis et al (1974) and the relationship derived by Bobee and Robitaille (1975) have a limited utility, since at best, they can be used to determine an unbiased estimate of parent population skew, where the average sample skew for samples of the same size n, is known for a number of regional annual maximum streamflow gages, and one assumes that the data from all those gages comes from the same parent population with a unique skew. Essentially, the relationships developed by these investigators are based on $E[g|\gamma]$, rather than on $E[\gamma|g]$. The latter is the statistic of interest since a sample skew g is usually the statistic available from a given annual maximum streamflow sample, and γ is not known.

Lall and Beard (1982) addressed the estimation of $E[\gamma|g]$ for the Pearson type III distribution using a Bayesian estimation scheme and Monte Carlo simulation. A non-informative prior distribution $\pi(\gamma)$ was specified for parent population skew γ, with γ ranging from -12 to 12. The use of other prior distributions was discussed and estimates for $E[\gamma|g]$ derived in terms of sample size n, sample skew g, and the bound on sample skew SB_n. This relationship resulting from the use of the non-informative prior is expressed as

$$E[\gamma|g] = 6.55\, g^{1.21} SB_n^{-0.75} \tag{7}$$

Lall and Beard (1982) point out that the use of a non-informative prior defined over the above range leads to a maximum adjustment to the value of g, in the average case. The "true" adjustment for bias possibly lies within the range indicated by their results and those of Wallis et al (1974). Wallis et al's (1974) results may be viewed as a situation where the prior distribution $\pi(\gamma)$ has a probability of 1 for a specific value of γ and is 0 elsewhere. A procedure to determine an appropriate specification of the prior distribution $\pi(\gamma)$ is thus needed. A desirable feature of this procedure would be to make use of any or all of the available regional information on annual maximum streamflows.

1.3. Regional Skew

A number of investigators have also considered the use of regional or generalized skew to aid the estimation of skew for the Pearson III distribution. Hardison (1974) presents isopleths of regional skew (skews estimated frim long-record stations) on a map of the United States. The Hydrology Committee of the U.S. Water Resources Council (1977), also recommended the use of such procedures. The intent was to use the regional skew from the map in combination with the at site skew estimate in order to (presumably) obtain a better estimate of skew at the site. Various weighting schemes to combine the regional and site estimates of skew have been recommended by the U.S. Water Resources Council (1977), McCuen (1979), Matalas et al (1975), Tasker (1978), and Tung and Mays (1980). Of these the procedure of Tung and Mays (1980) is perhaps the most attractive. They use the Jacknife (Miller, ((1974)) and the Bootstrap (Efron, (1977)), to determine the at-site skew estimator and its variance. A weighted average of the regional and at-site estimators of skew is then used to compute an estimate of skew. The weights are selected in proportion to the relative variance of the regional and site skew estimators.This estimator has the minimum variance of all weighted combinations of the regional and site skew estimates, but does not guarantee that an unbiased estimate of the population skew is obtained. The authors also suggest the use of regression relationships between skew and drainage basin parameters to compute regional estimators of skew and their variance. While this is an attractive approach where regional skew is used, its primary shortcoming is the reliance on estimates of skew computed from a number of gaging stations with records that are not equal in length. Where regressions are employed to determine regional skew, an additional element of uncertainty (that of the validity of the established relationship) is also introduced. Further, where map skew is used, the rationale for the delineation of geographic isopleths of skew is somewhat questionable.

The Bayesian approach to constructing a regional prior probability distribution for skew and then using it to determine an estimate of skew appears to be more attractive and theoretically justifiable than approaches based on determining a weighted estimator.

2. GENERAL FRAMEWORK

This section presents the general framework developed for the estimation of prior and posterior probability distributions of skew, given sampling distributions for sample skew for regional annual maximum flow data and for samples generated from populations with known skew. Basic concepts are defined, followed by subsections on procedures for the evaluation of the sampling and prior distributions.

The sample space for skew g, estimated from a sample of size n, can be discretized into NS categories ranging from $-SB_n$ to SB_n. The regional sampling distribution of g estimated for a sample of size n, from annual maximum streamflow data can be approximated by a probability content f_{jn} in each of the j categories. The probability content f_{jn} can be estimated as the ratio (m_{jn}/MS_n) of the number of samples m_{jn} that yield a value of sample skew g in the category j, to the total number of samples MS_n of sample size n. As MS_n increases the computed f_{jn} become good estimators of the probability content associated with the average sample skew in that skew category. The set of values f_{jn} are then an approximation of the sampling distribution $\phi_{2n}(g)$ of skew for regional data for each sample of size n.

Next, consider an admissible (bounded) set $[\gamma_1,\gamma_2]$ for parent population skew γ. Discretize this set into NG categories with an increment $\Delta\gamma$. This yields a discrete set of parent population skews γ defined as $\gamma_1,(\Delta\gamma),\gamma_2$. If we consider the generation of MG samples each of size n, from each of NG parent population skews γ_k, then we can approximate sampling distributions $\phi_{1n}(g|\gamma)$ for each parent population skew γ_k, and for each sample size n, through probability contents h_{jkn} for each category j, with h_{jkn} defined as the ratio (l_{jkn}/MG), where l_{jkn} is the number of samples from parent population skew γ_k that yield a sample skew in the category j, for a sample of size n.

For each sample of size n, the regional sampling distributions $\phi_{2n}(g)$ are related to the sampling distributions $\phi_{1n}(g|\gamma)$ from parent population skew γ, through a prior probability distribution function $\pi(\gamma)$ as

$$\phi_{2n}(g) \quad = \quad \phi_{1n}(g|\gamma)\,\pi(\gamma) \tag{8}$$

Given, the discrete set $\gamma_1,(\Delta\gamma),\gamma_2$ used for representing the domain of γ, the prior probability distribution $\pi(\gamma)$ is approximated through a set of probability contents p_k associated with each value γ_k. Then the previous relationship can be stated for each sample skew category j, for each sample size n, in terms of discrete variables as

$$f_{jn} \quad = \quad \Sigma_k\, h_{jkn}\, p_k \tag{9}$$

Presuming that f_{jn} and h_{jkn} can be evaluated, the problem of determining an appropriate prior distribution $\pi(\gamma)$ is then reduced to the determination of NG values p_k such that the total error in satisfying all equations of the form (9) is minimized. The total number of these equations is NS*NN, where NN is the total number of different sample sizes n considered. Typically NS*NN is quite large, implying that the selection of NG as NS*NN is likely to be computationally prohibitive. A constrained least squares approach used for the determination of the values p_k, is described in a later subsection.

Once the prior distribution $\pi(\gamma)$ is available (through the approximation p_k), Bayes rule can be used to evaluate, for each sample size n, the posterior probabilites asssociated with each population skew in terms of sample skew as:

$$\psi(\gamma|g) \quad = \quad \phi_{1n}(g|\gamma)\,\pi(\gamma) \,/\, \int \phi_{1n}(g|\gamma)\,\pi(\gamma)d\gamma \tag{10}$$

In the discrete framework, the posterior probability t_{jkn} associated with a parent population skew γ_k, given a skew g estimated from a sample of size n, that falls in a sample skew category j, can be evaluated as:

$$t_{jkn} \quad = \quad h_{jkn}\,p_k \,/\, (\, \Sigma_k\, h_{jkn}\,p_k\,) \tag{11}$$

A Bayesian estimator G_b of skew, given a sample estimate g (skew category j), from a sample of size n, is then computed as the expectation of γ over the posterior distribution $\psi(\gamma|g)$ as

$$G_b \quad = \quad E[\,\gamma|g] \; = \; \int \gamma\; \psi(\gamma|g)\, d\gamma \quad = \quad \Sigma_k\, \gamma_k\, t_{jkn} \tag{12}$$

Its variance $V(G_b)$ is computed as

$$V(G_b) \quad = \int (\gamma - G_b)^2\, \psi(\gamma|g)\, d\gamma \; = \; \Sigma_k\, t_{jkn}\,(\gamma_k \; - \; G_b)^2 \tag{13}$$

2.1. Evaluation of Sampling Distributions

The determination of regional sampling distributions $\phi_{2n}(g)$, through the evaluation of the probability contents f_{jn} for each of NN sample sizes n, is considered first. It is assumed that annual maximum streamflow data are available at a number of gaging stations in the region. The length of record at each gaging station need not be the same. The data at each location are, however, assumed to be independent and identically distributed. The gaging stations selected should have records that are free of diversion, regulation or other flow modification.

The first step then is to design a scheme for subsampling each of these records (with or without replacement) to construct MS_n independent and unique samples for each sample of size n. The choice of NN is dictated by computational resources, the number of gages available and the distribution of their length of record, as well as by a desire to keep MS_n as large as possible for each sample of size n. Once a set of samples MS_n for each sample size n, has been generated, the probability contents f_{jn} for each of the NN sample sizes n, can be readily evaluated by computing the sample skew g (equation (4)), for each sample, incrementing the frequency count m_{jn} for the sample skew category j to which it belongs, and finally computing the ratio m_{jn}/MS_n.

The determination of sampling distributions $\phi_{1n}(g|\gamma)$ is a similar process, with the exception that instead of using streamflow data, random samples with mean zero, variance 1, and skew γ_k are generated for each sample size n, and each parent population γ_k. Both Wallis et al (1974) and Lall and Beard (1982), outline procedures for the evaluation of these sampling distributions. The experiment design of Lall and Beard (1982) is appropriate for the current discussion.

2.2. Estimation of Prior Probability Distribution

The focus of this subsection is on the development of a procedure for the determination of the prior probabilities p_k (i.e. the distribution $\pi(\gamma)$) through a minimum error solution of the set of equations defined by equation (9). The problem can be stated as a constrained least squares problem with the objective of minimizing the total sum of squares of errors in replicating the regional sampling distributions of skew at each sample skew category and for each sample size. This objective is stated as

$$\text{Minimize} \quad \Sigma_n \Sigma_j \, (f_{jn} - \Sigma_k \, h_{jkn} \, p_k)^2 \tag{14}$$

The constraints associated with this problem are (1) each p_k is bounded between 0 and 1, and (2) the sum of the NG values p_k should equal 1. These are formally stated as

$$0 \leq p_k \geq 1 \qquad k = 1.......NG \tag{15}$$

$$\Sigma_k \; p_k = 1 \tag{16}$$

This yields a quadratic programming problem in NG variables and NG + 1 constraints. A number of algorithms are available for the solution of such a problem. In addition to the above constraints, it was felt that a unimodal prior probability density function may be desirable. Assuming that γ_{k*} represents the mode of the prior probability distribution $\pi(\gamma)$, the following additional (NG-1) constraints are considered (if unimodality is desired).

$$p_k \geq p_{k+1} \qquad k = k^*.......NG\text{-}1 \tag{17}$$

$$p_k \leq p_{k+1} \qquad k = 1........k^*\text{-}1 \tag{18}$$

Clearly, the mode γ_{k*} is not known a priori, and some trial and error is necessary in its specification. A solution of the problem without the unimodality constraints can be used to provide a good indication for k*.

Other constraints to limit the error for each sample size or for each sample skew category can also be imposed.

3. EXAMPLE APPLICATION

Sampling distributions of skew for data from 314 U.S. annual maximum flow records were evaluated by Lall (1981). The gaging stations selected had at least 60 years of record and were unaffected by diversion or regulation. All data was first logarithmically transformed. Five (NN) sample sizes of size (n) 5, 10, 15, 20 and 30 respectively were selected. Six hundred and twenty eight (MS_n) samples were constructed for each sample size, by first splitting each record of 60 into 2 halves of length 30 each, using alternate values in the original record for each half. One sample of size n was then selected from each half of the original record, by taking the first n values of the half-record.

Twenty (NS) categories were used to discretize the sample skew space (from $-SB_n$ to SB_n) for each sample size n. Twentyfive (NG) values of parent population skew γ_k were used ranging as -6,(0.5),6.0. One thousand (MG) samples of each size n were then generated for each γ_k, following the procedure of Lall and Beard (1982).

The sampling distributions $\phi_{1n}(g|\gamma)$ and $\phi_{2n}(g)$ were evaluated for each sample size n. Table I presents the values of f_{jn}, representing the distributions $\phi_{2n}(g)$. Table II presents values of h_{jkn}, representing the distributions $\phi_{1n}(g|\gamma)$. It should be noted that only results for skews 0 through 6 are presented. The distributions for skews -0.5 to -6.0 are the same as for 0.5 to 6.0 if one reverses the order on the skew categories (i.e. for a parent skew of -0.5, the h_{jkn} values are obtained as the values $h_{j'kn}$ for parent skew 0.5, where j' = NS+1-j). This follows from the observation that if the random variable x is transformed to a random variable y, such that y=-x, then the distribution of y has a skew $-\gamma$ if the distribution of x has skew γ.

The quadratic programming problem stated by equations (14) to (18) for the least square error determination of prior probabilities p_k was solved using the Successive Linear Programming algorithm of Zhang et al (1985). Initial solutions were attempted without the unimodality constraints, and indicated a mode at $\gamma = 0$. Subsequent solutions were attempted with the mode located at $\gamma = -0.5$, 0, and 0.5 respectively. The best solution in terms of total square error was obtained with the mode $\gamma_{k*} = 0$. This solution is presented as Table III. As can be seen from Table III, most of the prior probability is centered close to a skew of 0, with a diffuse probability distribution at larger (absolute value) skews. The optimal solution indicated that 97.47% (1 - ratio of sum of squares of errors to sum of squares for f_{jn}) of the variability in the values of f_{jn} was explained by the solution for p_k.

The sampling distributions $\phi_{1n}(g|\gamma)$ were re-evaluated for sample sizes n ranging from 10 to 100 and for 50 (NS) categories for sample skew. The prior probabilities p_k for γ_k were then used to determine posterior probabilities t_{jkn} and the Bayesian estimator G_b of equation (12), and its variance $V(G_b)$. These results for G_b and $V(G_b)$ are presented in Tables IV and V respectively in terms of a quantity ∂_j, defined as the ratio of average sample skew in category j to the bound on sample skew for that sample size. Following Lall and Beard (1982) a relationship between G_b, SB_n, g, and n could also be established from Table IV values using parametric or non-parametric regression.

As an illustration of the use of Tables IV and V, consider a sample skew of 0.25 calculated from a sample of size 20. For this sample size the bound on sample skew is 4.13, implying that ∂_j is .06, and that the Bayes estimate G_b of skew is 0.1051. The variance $V(G_b)$ of this estimate is 0.5623. The experience of Lall (1981) and Lall and Beard (1982) suggests that with a 1000 samples generated from each parent population, the results of the Monte Carlo simulation have adequate resolution for up to a value of ∂_j of approximately 0.38. For the example application reported here the sampling distributions are based on 628 and 1000 samples. Caution should thus be exercised in using Table IV values beyond an absolute value of ∂_j of approximately 0.3. The threshold of ∂_j acceptability levels for Table V is perhaps lower. A more exhaustive application using higher values of MS_n and MG would circumvent these limitations.

4. DISCUSSION

Bayes estimation provides a useful vehicle for integrating regional and at-site information to obtain unbiased estimates of a parameter of interest. The problem of estimating the skew, where the random variable (annual maximum floods) is considered to follow the Pearson III distribution was addressed, with particular focus on incorporating regional information into an appropriate prior probability distribution for skew. A general framework was developed and exemplified with some data from the U.S.A.. The example application was somewhat restricted by computational resources, in that (1) it did not completely exploit the regional data set (by creating as many random samples of each sample size as possible), (2) limited the number of samples (MG) generated for each sample size n, for each parent skew, and (3) considered only a limited discretization in terms of the number of samples of different sizes considered and the number of parent population skews. The level of discretization in the application may however be adequate for the estimation of expected values of skew. The concepts presented appear promising in light of the results of the application.

Another issue of concern in determining the prior probability distribution, is a possible need to attach appropriate weights to the sampling distributions from samples of different sizes n. A much higher degree of variability in skew is observed in shorter records, and possibly more weight should be attached to matching the frequency distributions from longer records. If the samples from annual maximum flow data are randomly constructed with replacement, and the maximum number of samples MS_n are constructed for each sample size n, then this problem may be partially resolved. Since MS_n decreases as n increases, using a greater number of samples for the smaller sample sizes should offset their higher variability per sample. More formal solutions to this problem could also be investigated.

In summary, the techniques developed (1) provide a formal basis for the Bayesian estimation of skew for annual maximum streamflow data, (2) provide a mechanism for incorporating regional information that is more appealing than the techniques currently in use, and (3) offer a framework where once prior probabilities have been estimated for a region , the Bayesian estimation of skew could be condensed into a single bias correction equation or table-lookup procedure, that is attractive for practitioners. The framework presented could be extended to the Bayesian estimation of flood frequencies and quantiles.

TABLE I

Sampling distributions $\phi_{2n}(g)$ of Skew for U.S. Flood Data
Values of f_{jn}

Category j	Sample size n: 5	10	15	20	30
1	0.000000	0.000000	0.000000	0.000000	0.000000
2	0.000000	0.000000	0.000000	0.000000	0.000000
3	0.046178	0.009554	0.003185	0.003185	0.000000
4	0.041401	0.036624	0.003185	0.000000	0.022293
5	0.090764	0.042994	0.058917	0.035032	0.019108
6	0.047771	0.042994	0.028662	0.023885	0.009554
7	0.065287	0.063694	0.047771	0.057325	0.011146
8	0.081210	0.081210	0.076433	0.068471	0.057325
9	0.062102	0.105096	0.141720	0.143312	0.170382
10	0.062102	0.133758	0.189490	0.232484	0.250000
11	0.078025	0.132166	0.165605	0.195860	0.261146
12	0.090764	0.113057	0.117834	0.114650	0.122611
13	0.079618	0.078025	0.076433	0.065287	0.035032
14	0.062102	0.055732	0.030255	0.028662	0.022293
15	0.047771	0.033439	0.028662	0.015924	0.009554
16	0.070064	0.047771	0.019108	0.009554	0.007962
17	0.055732	0.019108	0.012739	0.006369	0.001592
18	0.019108	0.004777	0.000000	0.000000	0.000000
19	0.000000	0.000000	0.000000	0.000000	0.000000
20	0.000000	0.000000	0.000000	0.000000	0.000000
Av. Skew	-.03	-.02	-.05	-.06	-.05

MS - # of samples for each sample size = 628

TABLE II

Sampling Distribution of Skew $\phi_{1n}(g|\gamma)$ - Values of h_{jkn} (x1000)

Sample Size n = 5

Category j	Parent Population Skew γ_k 0.	0.5	1.0	1.5	2.0	2.5	3.0	3.5	4.0	4.5	5.0	5.5	6.0
1	10	2	1	1	4	2	1	0	0	0	0	0	0
2	16	8	5	3	4	1	1	0	1	0	0	0	0
3	24	25	11	6	2	1	1	1	0	1	0	3	0
4	33	27	21	10	6	4	1	2	1	0	0	0	0
5	49	28	15	13	8	5	4	4	2	0	0	0	0
6	57	28	33	20	9	5	6	4	0	2	0	1	0
7	62	45	35	21	21	11	9	9	4	2	3	1	0
8	83	71	64	54	35	24	26	16	15	9	7	6	4
9	86	75	63	57	40	32	22	19	10	10	4	7	4
10	99	81	87	67	58	44	32	23	14	6	13	9	8
11	85	82	84	73	68	43	34	33	32	23	14	5	14
12	77	98	102	103	87	86	54	39	37	28	25	17	16
13	87	112	115	127	98	101	89	78	62	75	62	50	48
14	65	71	82	72	94	105	94	72	79	77	58	65	55
15	50	69	57	81	77	72	73	70	66	68	71	49	49
16	40	44	61	66	85	66	83	68	71	53	60	42	51
17	35	49	47	57	72	75	84	82	65	57	58	44	45
18	19	33	62	71	59	94	83	83	76	78	74	74	72
19	14	34	35	65	80	98	102	115	117	106	97	105	92
20	9	18	20	33	93	131	201	282	348	405	454	522	542

Sample Size n = 10

j	0.	0.5	1.0	1.5	2.0	2.5	3.0	3.5	4.0	4.5	5.0	5.5	6.0
1	0	0	0	0	0	0	0	0	0	0	0	0	0
2	0	0	0	0	0	0	0	0	0	0	0	0	0
3	2	0	0	0	0	0	0	0	0	0	0	0	0
4	3	1	1	0	0	0	0	0	0	0	0	0	0
5	10	1	0	0	0	0	0	0	0	0	0	0	0
6	27	2	4	0	0	0	0	0	0	0	0	0	0
7	54	23	5	5	0	0	0	0	0	0	0	0	0
8	97	44	21	7	4	1	1	0	0	0	0	0	0
9	127	83	58	28	4	5	3	0	1	0	0	0	0
10	190	149	124	61	29	13	7	5	5	0	0	1	1
11	171	186	157	129	77	55	19	15	9	3	2	2	0
12	164	190	196	165	124	87	64	25	32	16	10	4	8
13	77	131	169	181	183	130	98	79	57	40	37	28	20
14	35	97	116	149	156	168	144	138	103	84	77	43	46
15	26	43	61	114	159	162	152	129	140	124	99	94	87
16	13	30	44	72	96	155	172	158	162	180	176	156	144
17	3	12	27	41	78	98	112	133	126	109	137	115	119
18	1	6	11	29	47	57	99	120.	125	115	105	118	116
19	0	2	6	15	34	50	82	105	116	146	135	143	143
20	0	0	0	4	9	19	47	93	124	183	222	296	316

TABLE II (continued)

Sampling Distribution of Skew $\phi_{1n}(g|\gamma)$ - Values of h_{jkn} (x1000)

Sample Size n = 15

Category j	Parent Population Skew γ_k 0.	0.5	1.0	1.5	2.0	2.5	3.0	3.5	4.0	4.5	5.0	5.5	6.0
1	0	0	0	0	0	0	0	0	0	0	0	0	0
2	0	0	0	0	0	0	0	0	0	0	0	0	0
3	0	0	0	0	0	0	0	0	0	0	0	0	0
4	1	0	0	0	0	0	0	0	0	0	0	0	0
5	2	0	0	0	0	0	0	0	0	0	0	0	0
6	3	1	0	0	0	0	0	0	0	0	0	0	0
7	19	2	0	0	0	0	0	0	0	0	0	0	0
8	72	16	4	1	0	0	0	0	0	0	0	0	0
9	163	60	27	3	0	1	0	0	0	0	0	0	0
10	258	171	90	37	6	3	1	0	0	0	0	0	0
11	248	284	243	132	53	25	9	6	3	0	0	1	1
12	153	247	252	233	169	95	71	18	19	12	5	2	1
13	57	127	185	224	255	182	119	85	55	41	26	19	4
14	17	61	114	167	181	231	182	176	120	103	90	59	47
15	6	20	43	120	150	188	193	184	200	180	146	131	123
16	1.	6	29	46	95	128	164	182	188	189	178	170	161
17	0	5	9	17	47	66	104	122	133	138	164	178	170
18	0	0	3	16	37	56	91	111	114	125	117	133	134
19	0	0	1	4	7	19	51	79	99	116	146	133	148
20	0	0	0	0	0	6	15	37	69	96	128	174	211

Sample Size n = 20

Category j	0.	0.5	1.0	1.5	2.0	2.5	3.0	3.5	4.0	4.5	5.0	5.5	6.0
1	0	0	0	0	0	0	0	0	0	0	0	0	0
2	0	0	0	0	0	0	0	0	0	0	0	0	0
3	0	0	0	0	0	0	0	0	0	0	0	0	0
4	0	0	0	0	0	0	0	0	0	0	0	0	0
5	0	0	0	0	0	0	0	0	0	0	0	0	0
6	2	0	0	0	0	0	0	0	0	0	0	0	0
7	7	1	0	0	0	0	0	0	0	0	0	0	0
8	38	2	0	0	0	0	0	0	0	0	0	0	0
9	155	46	11	0	0	0	0	0	0	0	0	0	0
10	321	183	73	13	2	1	1	0	0	0	0	0	0
11	297	342	252	137	46	7	6	2	0	0	0	0	0
12	145	260	310	275	188	121	54	18	14	5	2	4	1
13	26	125	221	257	294	209	154	97	54	42	25	16	4
14	8	34	85	178	218	269	218	187	159	129	104	74	56
15	1	5	30	81	146	190	225	233	234	203	199	161	134
16	0	2	14	43	55	106	157	209	183	209	196	198	181
17	0	0	3	11	35	47	86	101	139	146	168	203	198
18	0	0	0	4	15	35	62	83	115	125	117	133	149
19	0	0	1	1	1	14	29	49	68	96	113	117	136
20	0	0	0	0	0	1	8	21	34	45	76	94	141

TABLE II (continued)

Sampling Distribution of Skew $\phi_{1n}(g|\gamma)$ - Values of h_{jkn} (x1000)

Sample Size n = 30

Category j	Parent Population Skew γ_k 0.	0.5	1.0	1.5	2.0	2.5	3.0	3.5	4.0	4.5	5.0	5.5	6.0
1	0	0	0	0	0	0	0	0	0	0	0	0	0
2	0	0	0	0	0	0	0	0	0	0	0	0	0
3	0	0	0	0	0	0	0	0	0	0	0	0	0
4	0	0	0	0	0	0	0	0	0	0	0	0	0
5	0	0	0	0	0	0	0	0	0	0	0	0	0
6	1	0	0	0	0	0	0	0	0	0	0	0	0
7	0	0	0	0	0	0	0	0	0	0	0	0	0
8	10	0	0	0	0	0	0	0	0	0	0	0	0
9	99	13	0	0	0	0	0	0	0	0	0	0	0
10	413	161	32	1	0	0	0	0	0	0	0	0	0
11	394	479	293	114	29	5	2	0	0	0	0	0	0
12	79	272	444	364	236	106	46	8	8	4	1	0	0
13	4	66	167	329	358	319	221	132	86	63	36	20	8
14	0	8	52	131	228	303	293	290	240	183	136	102	83
15	0	1	11	39	93	150	223	270	255	254	253	233	182
16	0	0	1	17	40	78	130	146	173	199	227	225	239
17	0	0	0	5	12	30	48	79	131	139	164	195	188
18	0	0	0	0	3	6	23	45	64	86	98	99	132
19	0	0	0	0	1	3	12	28	33	56	56	73	104
20	0	0	0	0	0	0	2	2	10	16	29	53	64

TABLE III

Estimated Prior Probability Distribution $\pi(\gamma)$

k	Skew γ_k	Probability p_k
1	-6.0	0.0000000
2	-5.5	0.0000000
3	-5.0	0.0094974
4	-4.5	0.0095310
5	-4.0	0.0095371
6	-3.5	0.0095371
7	-3.0	0.0095371
8	-2.5	0.0095371
9	-2.0	0.0095371
10	-1.5	0.0526740
11	-1.0	0.1422000
12	-0.5	0.1422000
13	0.0	0.2417600
14	0.5	0.2417600
15	1.0	0.0215480
16	1.5	0.0215480
17	2.0	0.0215480
18	2.5	0.0215480
19	3.0	0.0215480
20	3.5	0.0049544
21	4.0	0.0000000
22	4.5	0.0000000
23	5.0	0.0000000
24	5.5	0.0000000
25	6.0	0.0000000

Sum of squares of errors	=	0.0150358
Sum of squares of f_{jn}	=	0.5962665
Percent of variance explained	=	97.478
NS - # of sample skew categories	=	20
NG - # of parent skew values	=	50
Number of constraints/bounds	=	100

Table IV

Expected Value of Skew G_b

∂_j	Sample Size 10	15	20	30	40	50	60	80	100
-.98	-4.4104	-4.5597	-4.6405	-4.5376	-4.6993	-5.0000			
-.94	-4.1708	-4.3003	-4.4406	-4.5825	-4.7496	-4.8119	-4.4990		
-.90	-3.8240	-4.1647	-4.2997	-4.4536	-4.5528	-4.6996	-5.0000	-5.0000	
-.86	-3.2189	-4.0477	-4.2988	-4.3517	-4.3615	-4.6661	-4.7496	-4.4990	-4.0000
-.82	-2.8596	-3.5038	-4.1062	-4.3797	-4.4045	-4.5596	-4.7494	-4.7496	-4.5990
-.78	-2.6376	-3.5805	-3.8095	-4.1333	-4.2489	-4.3039	-4.5642	-4.7216	-5.0000
-.74	-2.6062	-3.4043	-3.8432	-4.2094	-4.1184	-4.1288	-4.2087	-4.4106	-4.5822
-.70	-2.6911	-3.4192	-3.7638	-4.0717	-4.3484	-4.3303	-4.1925	-4.5508	-4.5326
-.66	-1.9625	-2.8532	-3.4258	-4.0116	-4.1791	-4.2587	-4.4046	-4.4446	-4.5312
-.62	-2.2908	-3.2213	-3.5206	-3.9523	-4.0894	-4.2173	-4.2309	-4.3674	-4.5557
-.58	-2.1451	-2.4937	-3.0794	-3.9084	-4.0759	-4.1525	-4.2971	-4.3888	-4.5182
-.54	-1.9336	-2.5466	-2.8150	-3.8141	-3.9075	-4.0464	-4.2778	-4.3724	-4.4386
-.50	-1.6161	-2.2494	-2.5893	-3.4380	-3.8391	-3.8760	-3.9830	-4.2553	-4.3688
-.46	-1.4013	-1.8926	-2.4918	-3.0688	-3.7624	-3.9602	-4.0888	-4.2280	-4.3691
-.42	-1.2559	-2.0590	-2.3569	-2.6387	-3.4557	-3.7303	-3.8543	-4.0931	-4.3083
-.38	-1.1049	-1.6455	-1.9760	-2.4561	-2.9818	-3.5373	-3.7867	-3.9903	-4.1248
-.34	-.9531	-1.2982	-1.6196	-2.1999	-2.4353	-3.0807	-3.5049	-3.7766	-4.0227
-.30	-.7240	-1.1092	-1.4844	-1.8669	-2.2254	-2.4976	-2.9157	-3.4248	-3.7673
-.26	-.6434	-.9255	-1.1691	-1.5019	-1.7426	-1.9087	-2.2518	-3.0366	-3.2570
-.22	-.5786	-.7526	-.9016	-1.1722	-1.4124	-1.5805	-1.7362	-2.1044	-2.6318
-.18	-.4379	-.6280	-.7691	-.9956	-1.1306	-1.2657	-1.3286	-1.5424	-1.7599
-.14	-.3335	-.4558	-.6275	-.7960	-.9255	-1.0132	-1.1104	-1.2060	-1.3185
-.10	-.3291	-.3616	-.3864	-.5548	-.6667	-.7694	-.8510	-.9433	-1.0300
-.06	-.2231	-.1724	-.2060	-.3029	-.3802	-.4372	-.4711	-.5963	-.6757
-.02	-.1066	-.1212	-.0932	-.0882	-.0726	-.1034	-.1231	-.1370	-.1697

NOTE: 1. A missing value indicates no samples were generated for category.
2. ∂_j = Sample Skew g/Bound on Sample Skew SB_n.

Table IV(continued)

Expected Value of Skew G_b

Sample Size

∂_j	10	15	20	30	40	50	60	80	100
.02	-.0358	.0081	.0584	.1122	.1350	.1673	.1813	.1998	.2278
.06	.1008	.1051	.1529	.2578	.3185	.3666	.3994	.4641	.4969
.10	.1179	.2679	.3084	.4171	.4876	.5490	.5993	.7182	.8770
.14	.2621	.3509	.4567	.6124	.7568	.8634	1.0778	1.3190	1.6527
.18	.3320	.4742	.6462	.9153	1.1251	1.4024	1.5879	1.9952	2.1936
.22	.4646	.6556	.8196	1.2403	1.6388	1.9421	2.1894	2.3901	2.5458
.26	.5139	.8426	1.1941	1.6923	2.0552	2.2115	2.4066	2.6292	2.7016
.30	.5837	1.1000	1.6232	2.0276	2.3787	2.4892	2.5701	2.7037	2.7793
.34	.8435	1.2213	1.6352	2.1843	2.3742	2.5882	2.7288	2.7458	2.8899
.38	.9860	1.6648	1.6829	2.3433	2.5152	2.7107	2.7799	2.8408	2.9448
.42	1.0472	1.8500	2.2469	2.4925	2.6866	2.7208	2.7447	2.9043	2.8605
.46	1.2619	1.5935	2.3718	2.5538	2.7010	2.7774	2.8138	2.8924	2.8845
.50	1.4358	2.0364	2.1466	2.5253	2.7212	2.8058	2.8990	2.8497	2.9122
.54	1.5182	2.2284	2.1310	2.7227	2.7330	2.8560	2.9497	2.9745	3.0282
.58	1.5768	2.2697	2.6264	2.7851	2.9166	2.8823	2.9871	2.7620	3.2899
.62	1.4445	2.4717	2.4447	2.8187	2.8357	2.9325	2.7068	2.9791	3.0935
.66	1.7357	2.4415	2.6418	2.7895	2.8902	2.7773	2.9992	3.0735	3.5000
.70	2.4012	2.6320	2.7268	2.8295	2.8696	2.9580	2.7943	3.5000	3.0000
.74	2.1856	2.3346	2.7250	2.8902	2.7152	2.7931	3.0606	3.0516	3.0000
.78	2.1041	2.3788	2.7217	2.8651	2.8756	2.8902	3.1575		
.82	2.1047	2.6296	2.9311	2.9743	2.8243	3.2041			
.86	2.4135	2.8289	2.9423	2.8429	2.9918				
.90	2.6993	2.8179	2.9394	3.0145	3.5000				
.94	2.9289	2.9947	3.1058	3.1575					
.98	2.9864	2.9580	2.6869	3.5000					

NOTE: 1. A missing value indicates no samples were generated for category.
2. ∂_j = Sample Skew g/Bound on Sample Skew SB_n.

Table V

Variance of Skew $V(G_b)$

Sample Size

∂_j	10	15	20	30	40	50	60	80	100
-.98	.6278	.5745	.5181	.4986	.4002				
-.94	.7804	.7120	.5829	.5180	.2500	.3484	.5000		
-.90	1.0108	.7947	.7207	.6418	.4695	.2449			
-.86	1.4545	.8554	.7355	.7480	.6892	.3118	.2500	.5000	
-.82	1.5623	1.3465	.7502	.6397	.7340	.5745	.3287	.2500	.5835
-.78	1.5070	1.3368	1.2169	.7865	.7608	.7147	.5957	.3426	
-.74	1.5061	1.4196	.9859	.7081	.9849	.8650	.7045	.6912	.7320
-.70	1.3636	1.2542	1.0736	.9231	.7465	.7149	.7744	.5145	.5314
-.66	1.5916	1.4201	1.2865	.9633	.8452	.8986	.6538	.6363	.5813
-.62	1.4838	1.4018	1.3107	1.0975	.8886	.7191	.9097	.6525	.5292
-.58	1.6661	1.4725	1.4349	.9314	.8373	.9043	.6669	.7472	.5717
-.54	1.4902	1.4703	1.4966	1.0036	1.0804	1.0127	.7180	.6511	.6106
-.50	1.4163	1.4532	1.5220	1.2785	1.0036	1.0882	1.0652	.7891	.6600
-.46	1.2864	1.3556	1.4007	1.3473	1.0394	.8615	.8567	.7338	.6663
-.42	1.1313	1.3879	1.3410	1.4304	1.1545	1.0300	.9902	.9712	.7281
-.38	1.2265	1.1733	1.3234	1.3882	1.3211	1.0921	.9555	.8632	.8874
-.34	1.0851	1.1020	1.0848	1.2275	1.3519	1.2299	1.0142	.9525	.8047
-.30	1.0274	.9478	.9752	1.0988	1.1528	1.2391	1.2225	1.0533	.9206
-.26	.9114	.8610	.8475	.8739	.9696	1.0393	1.0877	1.0235	1.0145
-.22	.8412	.7727	.6736	.6809	.6979	.7728	.8716	1.0025	.9872
-.18	.7844	.7050	.6651	.5576	.5305	.5336	.5414	.6343	.7698
-.14	.7363	.6950	.5954	.5146	.4547	.4016	.4058	.3938	.4133
-.10	.7602	.6293	.5881	.4910	.4400	.4032	.3586	.3399	.3151
-.06	.6985	.6039	.5613	.5144	.4486	.4230	.4035	.3579	.3314
-.02	.7502	.6121	.5317	.4644	.4181	.3752	.3476	.3127	.2913

NOTE: 1. A missing value indicates no samples were generated for category.
2. ∂_j = Sample Skew g/Bound on Sample Skew SB_n.

Table V(continued)

Variance of Skew $V(G_b)$

Sample Size

∂_j	10	15	20	30	40	50	60	80	100
.02	.6886	.5875	.5013	.3957	.3623	.3268	.3083	.2819	.2685
.06	.6797	.5623	.4781	.3966	.3247	.2943	.2718	.2278	.2129
.10	.7148	.5832	.5087	.4036	.3619	.3646	.3597	.4144	.4877
.14	.7446	.6798	.5800	.5437	.5756	.5666	.6462	.6649	.5961
.18	.8090	.7170	.7407	.7458	.7423	.7878	.7614	.6591	.5865
.22	.8427	.7994	.7810	.8699	.8772	.7645	.6661	.5751	.5279
.26	.9021	.8924	.9539	.9330	.7956	.7528	.6457	.5181	.4945
.30	.9190	.9608	.9667	.8369	.6774	.6644	.5646	.5155	.4995
.34	.9945	1.0367	.9638	.8515	.7570	.5639	.5017	.4968	.4020
.38	1.1404	.9907	1.0653	.7813	.6913	.5285	.4670	.4245	.4240
.42	1.0571	.9918	.8406	.7333	.5394	.5221	.5007	.4691	.4263
.46	1.1020	1.0612	.7562	.6127	.4970	.4774	.4739	.4299	.3932
.50	1.1291	1.0007	.9318	.7534	.5368	.4731	.4757	.4670	.4719
.54	1.1760	.8853	.9967	.5215	.5184	.4733	.3764	.5192	.5444
.58	1.1980	.9177	.6232	.4544	.3894	.5058	.4310	.4893	.2468
.62	1.1419	.7940	.8390	.5594	.5276	.3977	.5539	.6163	.1949
.66	1.1009	.6676	.5799	.5399	.4585	.5020	.5346	.1771	
.70	.6611	.5634	.5293	.4700	.3418	.3974	.4557		
.74	.9228	.9799	.5539	.4066	.5474	.4804	.1632	.1521	
.78	.9770	.9561	.6284	.3338	.4272	.4878	.2323		
.82	1.0004	.6257	.3605	.4285	.4681	.2458			
.86	.9034	.4633	.4644	.4747	.3122				
.90	.5353	.4741	.2796	.3227					
.94	.3911	.2515	.2042	.2323					
.98	.3605	.3357	.3899						

NOTE: 1. A missing value indicates no samples were generated for category.
2. ∂_j = Sample Skew g/Bound on Sample Skew SB_n.

REFERENCES

1. Bobee, B., and Robitaille, R. 1975. Correction of Bias in the Estimation of the Coefficient of Skewnewss. Water Resources Research, Vol. **11**, No. 6, pp. 851-854.

2. Efron, B. 1977. Bootstrap Methods: Another Look at the Jacknife. Technical Report No. **32**, Division of Biostatistics, Stanford University, Palo Alto, California.

3. Fisher, R.A. 1921. On the Mathematical Foundations of Theoretical Statistics. Philosophical Transcations of the Royal Society, Series A, Volume **222**, pp. 309.

4. Fisher, R.A. 1929. Moments and Product Moments of Sampling Distributions. Proceedings of the London Mathematical Society, Volume **2**, pp.30.

5. Hardison, C.H. 1974. Generalized Skew Coefficients of Annual Floods in the United States and Their Application. Water Resources Research, Vol. **10**, No. 4, pp. 745-752.

6. Hazen, A. 1930. Flood Flows. John Wiley, New York.

7. Kirby, W. 1974. Algebraic Boundedness of Sample Statistics. Water Resources Research, Volume **10**, No. 2, pp. 220-222.

8. Lall, U. 1981. The Value of Data in Relation to Uncertainty and Risk. PhD. Dissertation. University of Texas at Austin.

9. Lall, U. and Beard, L.R. 1982. Estimation of Pearson III Moments. Water Resources Research, Vol. **18**, No. 5, pp. 1563-1569.

10. Matalas, N.C., Slack, J.R., and Wallis, J.R. 1975. Regional Skew in Search of a Parent. Water Resources Research, Vol. **11**, No. 6, pp. 815-826.

11. McCuen, R.H. 1979. Map Skew. Journal of Water Resources Planning and Management Division, American Society of Civil Engineers, Vol. **105**, WR2, pp. 269-278.

12. Miller, R.G. 1974. The Jacknife - A Review. Biometrika, Vol. **6**, No. 1, pp. 1-17.

13. Tasker, F.D. 1978. Flood Frequency Analysis with a Generalized Skew Coefficient. Water Resources Research, Vol. **14**, No. 2, pp. 373-376.

14. Tung, Y.K. and Mays, L.W. 1980. Optimal Risk Based Design of Water Resource Engineering Projects. Center for Research in Water Resources Report **171**. University of Texas at Austin.

15. U.S. Water Resources Council 1977. Guidelines for Determining Flood Flow Frequency. Bulletin **17A**. Washington D.C..

16. Wallis, J.R., Matalas, N.C., and Slack, J.R. 1974. Just a Moment. Water Resources Research, Vol. **10**, No. 2, pp. 211-219.

17. Zhang, J., Kim, N.H., and Lasdon, L.S. 1985, An Improved Successive Linear Programming Algorithm. Management Science, Vol. **31**, No. 10, pp. 1312-1331.

COMPARISON OF FLOOD-FREQUENCY ESTIMATES BASED ON OBSERVED AND MODEL-GENERATED PEAK FLOWS

Wilbert O. Thomas, Jr.
U.S. Geological Survey
Water Resources Division
Mail Stop 415, National Center
Reston, Virginia 22092
USA

ABSTRACT. Comparisons were made of flood-frequency estimates from a rainfall/runoff model and from observed annual peak data to determine if these estimates exhibited the same statistical variability. The analysis was based on data at 173 stations in 10 States where there were 20 or more years of observed annual peak data and flood-frequency estimates from a rainfall/runoff model. Paired t-tests and analysis-of-variance techniques indicated that there were statistically significant differences between the two sets of flood-frequency estimates. These differences existed for different recurrence interval floods (2-, 10-, 50-, and 100-years), different watershed sizes, and different States. These differences appeared to be primarily related to three infiltration parameters in the rainfall/runoff model and to a lesser extent the observed record length at the station. Furthermore, the ratio of the model-generated 100-year flood discharge to the 2-year flood discharge was 3.98 whereas the same ratio based on observed peak flows was 5.72. Additional research is needed to identify what factors cause flood estimates from the rainfall/runoff model appear to exhibit less variability than flood estimates based on observed peak flow records.

INTRODUCTION

During the last 20 years, the U.S. Geological Survey has undertaken and completed many rainfall/runoff modeling studies of small rural watersheds usually in cooperation with the Federal Highway Administration and various State highway departments. In these studies a rainfall/runoff model was first calibrated using short-term (approximately 10 years) concurrent rainfall, evaporation, and runoff data. Using long-term (60-70 years) rainfall and evaporation data and the calibrated rainfall/runoff model, a long series of peak flows were simulated. The logarithms of the annual maximum peak flows were fit to a Pearson Type III distribution to define a synthetic flood-frequency curve. The rainfall/runoff model utilized in these studies was a lumped, unit-hydrograph model documented by Dawdy et al. (1972). Most of the small

V. P. Singh (ed.), Hydrologic Frequency Modeling, 149–161.

stream studies began in the early to mid 1960's and terminated in the mid to late 1970's. Annual peak flow data are still being collected at many sites so that 20 or more years of observed peak flow data are now available at these sites. The objectives of this paper are to (1) compare flood-frequency estimates based on model-generated and observed record, (2) quantify the differences and determine if these differences are statistically significant, and (3) identify reasons for these differences, if they exist, (4) stress the need for additional research relative to the accuracy and variability of flood estimates from a rainfall/runoff model.

DATA BASE

A review of the U.S. Geological Survey's small streams program revealed that 10 States--Alabama, Georgia, Illinois, Mississippi, Missouri, Tennessee, Delaware, Maryland, Kansas, and Oklahoma--had watersheds with flood estimates based on a rainfall/runoff model (Dawdy et al., 1972) and 20 or more years of observed annual peak data. The criteria of 20 or more years of record was generally considered a minimum for computing flood estimates with recurrence intervals of 50 to 100 years. A total of 173 gaging stations meeting the criteria of this study were identified. The number of stations in each State, their average record length, and watershed size are summarized in Table I.

TABLE I. Summary of number of stations, average record length, and watershed size by State for the stations used in the study

State	Number of stations	Average record length, in years	Average watershed size, in mi^2	(km^2)
Alabama	3	26.3	6.84	(17.72)
Georgia	22	21.8	3.68	(9.53)
Illinois	22	24.4	1.50	(3.89)
Mississippi	34	23.1	0.75	(1.94)
Missouri	36	23.9	2.00	(5.18)
Tennessee	16	21.1	10.30	(26.68)
Delaware	4	34.0	5.32	(13.78)
Maryland	3	25.3	6.41	(16.60)
Kansas	13	25.9	3.98	(10.31)
Oklahoma	20	20.4	10.64	(27.56)
Total	173	23.3	4.06	(10.52)

It would be desirable to compare the synthetic (model-generated) flood estimates to observed estimates with record lengths much greater than 20 years. However, Delaware, with four watersheds averaging 34 years in length, has the longest watershed records where synthetic flood estimates from a particular rainfall/runoff model (Dawdy et al., 1972) are also available. There is considerable variation in flood estimates based on 20-25 years of record, particularly for the larger recurrence interval floods. Even considering this variability in the observed flood estimates, the comparisons given in this report are considered indicative of the bias of flood estimates computed using the indicated rainfall/runoff model. The assumption is made that the observed flood records yield unbiased estimates of the 2-, 10-, 50-, and 100-year floods. Most of the 173 station records were collected during the period 1955-85 which is considered to be a representative sample in time. An analysis of the annual 2-day maximum rainfall at 13 long-term rain gages in the study area indicated that the period 1955-85 was representative of a longer 60-70 year period.

The synthetic flood estimates for recurrence intervals of 2-, 10-, 50-, and 100-years were compiled from the following reports: Alabama (Olin and Bingham, 1977), Georgia (Golden and Price, 1976), Illinois (Curtis, 1977), Mississippi (Colson and Hudson, 1976), Missouri (Hauth, 1974), Tennessee (Wibben, 1976), Delaware (Simmons and Carpenter, 1978), Maryland (Carpenter, 1980), Kansas (Clement, 1984), and Oklahoma (Thomas and Corley, 1977). For Illinois the synthetic flood estimates were available for only the 2- and 50-year recurrence intervals. The observed flood estimates for the study were determined by fitting the logarithms of the observed annual peaks to a Pearson Type III distribution as described in Bulletin 17B (Hydrology Subcommittee, 1982). Data were generally available up to 1985 except for a few States where the most recent data available were in 1984.

DATA ANALYSIS

The analysis technique included the following steps: (1) quantify differences in the synthetic and observed flood estimates for recurrence intervals of 2-, 10-, 50-, and 100-years, (2) use paired t-tests and analysis of variance techniques (ANOVA) to identify if the differences are statistically significant, and (3) try to relate the differences to watershed size, length of record of observed annual peak data, and rainfall/runoff model parameters.

Computation of Bias

The differences between the synthetic and observed flood estimates, for lack of a better word, are referred to as biases. These biases were computed as follows:

$$BIAS_i = (Q_iS - Q_i)/Q_i \qquad (1)$$

where $BIAS_i$ = difference in the two estimates divided by the observed flood estimate for recurrence interval i,

Q_iS = synthetic flood estimate for recurrence interval i in ft^3/s (m^3/s), and
Q_i = observed flood estimate for recurrence interval i, in ft^3/s (m^3/s).

In equation 1, the difference is divided by the observed estimate to remove the effect of watershed size so that large differences for the larger watersheds are given appropriate weight. If the bias from equation 1 is multiplied by 100, it is simply the percent difference in the two estimates.

The biases from equation 1, converted to percentages, are shown in Table II for each State and recurrence interval and an average value is provided for all States. The positive percentages in Table II indicate the synthetic flood estimates are higher than the observed flood estimates and conversely the negative percentages indicate the observed flood estimates are higher. The general tendency, as shown by the average values, is for the 2-year synthetic flood estimates to be 13 percent larger and for the observed flood estimates to be 12 percent larger at the 100-year recurrence interval. Similar results have previously been reported by Lichty and Liscum (1978) and Thomas (1982). The present data base includes more stations with longer records than the previous two studies. The overestimation-underestimation trend illustrated in Table II implies that the slopes of the synthetic frequency curves are, in general, flatter than the observed frequency curves and therefore exhibit less variability. This "loss of variance"

TABLE II. Summary of BIAS in percent for each State and recurrence interval

State	BIAS (in percent)				Standard error, in percent, of observed 100-year flood
	2-year	10-year	50-year	100-year	
Alabama	11.3	-9.3	-22.9	-27.6	30.8
Georgia	15.9	5.3	0.4	-1.2	27.6
Illinois	0.6	*	5.1	*	25.6
Mississippi	-6.2	-12.5	-16.4	-17.7	20.2
Missouri	8.6	-8.6	-17.4	-20.0	28.3
Tennessee	-2.5	-16.2	-26.2	-29.3	32.0
Delaware	39.4	59.4	76.0	85.3	25.7
Maryland	-12.0	-3.7	-1.0	-0.1	38.5
Kansas	37.0	-5.4	-20.7	-24.8	34.4
Oklahoma	62.5	19.1	5.0	1.1	38.1
Average	13.3	-2.4	-8.1	-12.3	

*Not available.

is common to many types of deterministic and statistical models as discussed by Kirby (1975). Although there is considerable variability in the bias values from State to State, this tendency for flatter synthetic frequency curves is prevalent except for Delaware and Maryland. Delaware and Maryland have only four and three stations, respectively, so the results may not be statistically significant. However, Delaware has the longest observed record of any State and the synthetic flood estimates are consistently higher for all four watersheds for all four recurrence intervals. This may imply that the rainfall/runoff model is not appropriate for the types of watersheds on the coastal plains of Delaware. These watersheds are characterized by a large number of lakes and swamps and extremely slow response time to direct runoff.

To put the biases in Table II in proper perspective, the standard error (time-sampling error) of the observed 100-year flood estimate is also given in Table II. The differences in the two flood estimates (biases) for the 100-year recurrence interval are generally within the computed standard error in percent. Delaware is the only State where the difference in 100-year flood estimates is greater than the standard error. Only the standard error for the 100-year flood is given because it is the largest value.

To further illustrate the loss of variance effect for the synthetic flood estimates, the ratios of the 50- and 100-year flood discharges to the 2-year discharge were computed. These ratios were computed for both the synthetic and observed flood estimates and are shown in Table III. On the average, the ratio of the synthetic 100-year flood discharge to the 2-year discharge is 3.98 whereas the same ratio for the observed flood estimates is 5.72. The observed flood-frequency curves are consistently steeper with the exception of these frequency curves in Delaware and Maryland.

TABLE III. Summary of flood-frequency ratios

State	$\frac{Q_{100}S}{Q_2S}$	$\frac{Q_{100}}{Q_2}$	$\frac{Q_{50}S}{Q_2S}$	$\frac{Q_{50}}{Q_2}$	Number of watersheds
Alabama	3.82	6.18	3.24	4.86	3
Georgia	3.93	4.93	3.37	4.09	22
Illinois	*	*	4.06	4.09	22
Mississippi	2.78	3.52	2.47	3.00	34
Missouri	4.07	6.18	3.49	5.00	36
Tennessee	4.21	6.38	3.59	5.07	16
Delaware	6.49	5.52	5.06	4.34	4
Maryland	8.69	8.62	6.42	6.34	3
Kansas	3.97	7.61	3.41	6.09	13
Oklahoma	4.55	8.09	3.88	6.39	20
Total	3.98	5.72	3.48	4.63	173

Tests of Significance

Paired t-tests and analysis of variance (ANOVA) techniques were used to determine if the biases computed by equation 1 were significantly different from zero. In other words, are the synthetic flood estimates, on the average, significantly different from the observed flood estimates at a given level of significance.

A simple way to determine if the BIAS values in Table II are significantly different from zero is to use a paired t-test. The t-statistic is the difference in the two flood-frequency estimates (i.e., BIAS values) divided by the standard error of this difference. The paired t-test was applied to all States where there was a sufficient number of stations for the test to be significant. The results for the six States having 13 or more stations are summarized in Table IV. For four of the six States, the 50- and 100-year bias values are significantly different from zero at the 5 percent level of significance. This implies there is a significant difference between the synthetic and observed flood estimates at those recurrence intervals. For all six States, the BIAS values are significantly different from zero for at least one recurrence interval.

TABLE IV. Tests of significance to determine if BIAS is significantly different from zero (t-tests by State)

State	Number of stations	BIAS for indicated recurrence interval			
		2-year	10-year	50-year	100-year
Georgia	22	yes	no	no	no
Mississippi	34	no	yes	yes	yes
Missouri	36	no	no	yes	yes
Tennessee	16	no	yes	yes	yes
Kansas	13	no	no	yes	yes
Oklahoma	20	yes	no	no	no

The analysis of variance technique is useful in identifying factors affecting a criterion variable such as bias. The analysis is based on 151 stations (excluding Illinois) to make the experimental design as balanced as possible. Furthermore, slightly different techniques were used to compute Q_2S and $Q_{50}S$ for Illinois and so this further source of uncertainty was avoided. The ANOVA experimental design was set up assuming that the following three factors are likely to affect the value of bias—the effect of watershed size, the effect of recurrence interval, and the effect of different analysts in different States making the synthetic flood estimates. Given these three main effects, there are also three interactions of these main effects.

The interactions represent the failure of one main effect to produce a constant effect for various values of another main effect. The ANOVA model that describes the relationship between bias and the three factors above is as follows:

$$BIAS = SIZE + RI + STATE + SIZE * RI + SIZE * STATE + RI * STATE + \text{an error term} \tag{2}$$

where BIAS = the value of BIAS from equation 1 for a given watershed and recurrence interval

SIZE = main effect of watershed size
= 1 if watershed size ≤ 1.0 mi^2 (≤ 2.59 km^2)
= 2 if watershed size ≥ 1 to 5.0 mi^2 (>2.59 and ≤ 12.95 km^2)
= 3 if watershed size >5.0 to 10 mi^2 (>12.95 and ≤ 25.9 km^2)
= 4 if watershed size >10 mi^2 (>25.9 km^2)

RI = main effect of recurrence interval
= 1 if recurrence interval equals 2 years
= 2 if recurrence interval equals 10 years
= 3 if recurrence interval equals 50 years
= 4 if recurrence interval equals 100 years

STATE = main effect of the State in which analysis was done
= 1 for Alabama
= 2 for Georgia
= 3 for Illinois (Illinois not included in analysis)
= 4 for Mississippi
= 5 for Missouri
= 6 for Tennessee
= 7 for Delaware
= 8 for Maryland
= 9 for Kansas
= 10 for Oklahoma

SIZE * RI = the interaction of watershed size and recurrence interval
SIZE * STATE = the interaction of watershed size and the State
RI * STATE = the interaction of recurrence interval and the State

The General Linear Models (GLM) Procedure in the Statistical Analysis System (SAS Institute, Inc., 1982) was used to analyze the data. All three main effects (SIZE, RI, STATE) and the interaction effect SIZE * STATE were significant at the 1 percent level of significance. This implies there is a difference in bias due to watershed size, recurrence interval, and the State in which the analysis was completed and that bias varies for different watershed sizes across the various States. Duncan's multiple range test (Hines and Montgomery, 1980, or SAS Institute, Inc., 1982) can be used to decide which watershed sizes or recurrence intervals or States are significantly different. Table V summarizes the results of the Duncan's multiple range test for the main effect SIZE. Table V provides the mean BIAS for the various watershed sizes (1, 2, 3, 4). The number of observations, N, includes bias values for all four recurrence intervals. Therefore, the number of observations (N) must be divided by four to obtain the number of watersheds in each watershed size grouping. There are 52 watersheds

TABLE V. Results for Duncan's multiple range test for the main effect SIZE

Duncan Grouping	Mean BIAS	N	SIZE
A	0.19022	68	4
B	.04281	216	2
B	-.04274	112	3
C	-.15388	208	1

less than 1 square mile (2.59 km^2), 54 watersheds in the grouping 1 to 5 square miles (2.59 to 12.95 km^2), 28 watersheds in the grouping 5 to 10 square miles (12.95 to 25.9 km^2), and 17 watersheds greater than 10 square miles (25.9 km^2). These groupings were chosen somewhat arbitrary but should be sufficient for demonstrating the variability of BIAS across various watershed sizes. The letter under the Duncan Grouping heading indicates those watershed sizes for which the mean BIAS values are or are not significantly different at the 5 percent level of significance. The values of BIAS with the same letter are not significantly different while those with different letters are significantly different. For example, in Table V the Duncan's multiple range test indicates that (1) the mean BIAS for watershed sizes 2 and 3 are not significantly different from each other but watershed sizes 1 and 4 are significantly different from the other watershed sizes, and (2) the mean BIAS for watershed sizes 1 and 4 are significantly different from each other. The results in Table V indicate that there is a tendency for the rainfall/runoff model to underestimate flood peaks for small watersheds and to overestimate for larger watersheds. This result is obviously related to the size of watersheds gaged in each State. For example, Mississippi and Missouri have the smallest watersheds, on the average, and there is a tendency in these States for the rainfall/runoff model to underestimate the flood peaks. On the other hand, Oklahoma has the largest watersheds and there is a tendency for overestimation in this State. This explains why the SIZE * STATE interaction was significant in the analysis of variance.

The mean BIAS by State is shown in Table VI which summarizes the results for Duncan's multiple range test for the main effect STATE. The mean BIAS values are averaged across all four recurrence intervals and therefore differ from the BIAS values in Table II which are summarized by recurrence interval. The format and definition of variables is the same as for Table V. The two-letter abbreviation is shown for each State to help the reader identify the results. The results in Table VI indicate the following: (1) the mean BIAS for watersheds in Delaware is significantly different from the other States (the applicability of the rainfall/runoff model for the types of watersheds in Delaware was questioned earlier), (2) the mean BIAS for watersheds in Oklahoma and Georgia are not significantly different from each other, Oklahoma is significantly different from the other States, while

TABLE VI. Results for Duncan's multiple range test for the main effect STATE

Duncan Grouping				Mean BIAS	N	STATE
			A	0.6503	16	7 (DE)
		B		.2194	80	10 (OK)
C		B		.0509	88	2 (GA)
C	D			-.0347	52	9 (KS)
C	D			-.0420	12	8 (MD)
C	D			-.0937	144	5 (MO)
C	D			-.1212	12	1 (AL)
C	D			-.1320	136	4 (MS)
	D			-.1850	64	6 (TN)

Georgia is different from Delaware and Tennessee, (3) the mean BIAS for watersheds in Georgia, Kansas, Maryland, Missouri, Alabama, and Mississippi are not significantly different from each other but are significantly different from Delaware, Oklahoma, and Tennessee, and finally, (4) the mean BIAS for watersheds in Kansas, Maryland, Missouri, Alabama, Mississippi, and Tennessee are not significantly different from each other but are different from Oklahoma, Delaware, and Georgia. The obvious question is why is there so much difference in mean BIAS from State to State. Is it due to different types of hydrology or differences in how the rainfall/runoff model was calibrated? Further investigation is needed to answer these questions.

The results for Duncan's multiple range test for the main effect RI are summarized in Table VII. The format and definition of variables is the same as for Table V. The results in Table VII indicate, for example, that the mean BIAS for a recurrence interval of 2 years is significantly different from the other three recurrence intervals, the mean BIAS for the 10- and 50-year recurrence intervals are not significantly different from each other, and the mean BIAS for the 50- and 100-year recurrence interval are not significantly different. Again these results indicate that the synthetic flood-frequency curves have

TABLE VII. Results for Duncan's multiple range test for the main effect RI

Duncan Grouping			Mean BIAS	N	RI
		A	0.15200	151	1 (2 years)
	B		-.02445	151	2 (10 years)
C	B		-.10114	151	3 (50 years)
C			-.12318	151	4 (100 years)

flatter slopes, on average, than the observed flood-frequency curves and therefore exhibit less variability.

The paired t-tests, analysis of variance techniques, and Duncan's multiple range tests indicated that there are differences in mean BIAS across the States for various watershed sizes and recurrence intervals and identified which ones were different. The next section of the paper attempts to relate these differences to causative factors.

Relating Bias to Causative Factors

The previous analyses have identified that the rainfall/runoff model (Dawdy et al., 1972) overestimates or underestimates flood peaks for various recurrence intervals and watershed sizes. This section of the paper attempts to identify reasons or factors affecting this bias.

The lumped unit-hydrograph model documented by Dawdy et al. (1972) has nine parameters that are optimized in the calibration of the model. These parameters are defined in Table VIII. The first seven parameters defined in Table VIII determine the rate of infiltration during the storm runoff and the antecedent moisture conditions. The model uses the Green-Ampt infiltration equation as documented by Philip (1954). The last two parameters, KSW and TC, are routing parameters that determine how fast the rainfall excess leaves the watershed. Model parameters were compiled for 138 of the 151 watersheds used in the earlier ANOVA analysis. In the calibration of the rainfall/runoff model, EVC and RR were constrained to nearly constant values so these two parameters are not included in the subsequent analysis.

Multiple regression analyses were undertaken relating bias for each of the recurrence intervals (four separate analyses) to the model parameters in Table VIII (except RR and EVC), the drainage area of the watershed, and the length of record (L, in years) for the observed annual peak data. The results of these analyses indicated that the parameters most related to the bias were PSP, RGF, and DRN. For the 2-year bias values, PSP was the only significant parameter. For the 10-, 50-, and 100-year bias values, all three parameters were highly significant. It should be noted that these infiltration parameters only explained about 20 percent of the variation in the bias values. All relationships were inverse for all recurrence intervals meaning, for example, as PSP and RGF increased, the bias decreased. Increasing the values of DRN, RGF, and PSP would tend to decrease the lower recurrence interval floods more than the high ones. This adjustment would tend to steepen the slopes of the synthetic flood-frequency curves more in correspondence with the observed curves. However, these results are very preliminary and further investigations are needed to fully understand their implications. Lichty and Liscum (1978) identified inadequacies in determining antecedent soil-moisture and infiltration computations as one reason for the apparent bias in the model-generated flood estimates. This present analysis tends to confirm this hypothesis. For the 50- and 100-year recurrence intervals, the record length of the observed annual peak data was mildly significant in explaining the variation in bias. This suggests that the biases may be related to time-sampling errors in the observed flood estimates.

TABLE VII. The nine model parameters and their application in the modeling process

Parameter identifier	Units	Application
PSP	inches (centimeters)	The combined effect of initial moisture content and soil suction at the wetted front at field capacity - function of soil type
RGF		Ratio of the suction at the wetted front for soil moisture at wilting point to that at field capacity
KSAT	inches (centimeters) per hour	The minimum (saturated) value of hydraulic conductivity used to determine infiltration soil rates - function of soil type
BMSM	inches (centimeters)	Soil moisture storage volume at field capacity - function of average depth of soil zone
EVC		Coefficient to convert pan evaporation to potential evapotranspiration values
DRN	inches (centimeters) per hour	A constant drainage rate for redistribution of soil moisture - function of soil type
RR		Proportion of daily rainfall that infiltrates the soil - function of hydrologic conditions in basin
KSW	hours	Time storage coefficient for linear reservoir routing - function of basin characteristics
TC	minutes	Base time of unit translation hydrograph which is the time of concentration - function of basin characteristics

SUMMARY AND CONCLUSIONS

Rainfall/runoff models are currently being used in conjunction with long-term climatic data to simulate a long-time series of annual peak flows for flood-frequency analysis. This is particularly true for urban watersheds where changing land-use conditions preclude obtaining a long observed peak flow record. Additional research is needed to determine if flood estimates from a rainfall/runoff model exhibit the same variability as those estimates based on observed records. Results in the present study indicate that synthetic flood estimates from a particular rainfall/runoff model do not exhibit the same variability. In particular, there is a tendency to overestimate lower recurrence interval floods and underestimate higher recurrence interval floods. If linear regression theory can be extended to the rainfall/runoff model, then the "loss of variance" correction suggested by Kirby (1975) can be utilized. However, the applicability of this adjustment procedure needs further evaluation. It should be noted that the results in this study pertain to using long-term (historical) climatic data and a calibrated model to simulate a long series of annual peaks and not the use of a design rainfall storm for flood-frequency analysis.

Furthermore, the results in this study should be evaluated with the understanding that the model-generated flood estimates were compared to observed flood estimates based on 20-25 years of observed data. The assumption was made earlier that the period 1955-85, on which the observed flood estimates were based, was a representative sample in time. Based on an analysis of annual 2-day maximum rainfalls at 13 long-term rain gages, this assumption seems reasonable. However, a more detailed analysis should be undertaken to determine if flood-frequency estimates based on data collected during 1955-85 are representative of a longer time period. There was some indication in the study that the bias of the 50- and 100-year synthetic flood estimates decreased as the record length of the observed peak flows increased.

Finally, the bias appeared to be related to three infiltration parameters (RGF, DRN, PSP) of the model. The analysis indicated that the bias values for all recurrence intervals decreased as values of RGF, DRN, and PSP increased. This implies that the determination of the infiltration parameters in the modeling process may have a significant effect on the accuracy of simulated flood-frequency estimates.

The results presented in this paper are considered indicative of other unit-hydrograph type models using infiltration algorithms similar to the Green-Ampt/Philip equation used in the model documented by Dawdy et al. (1972). Since these types of models are likely to be used in future flood-frequency analyses, additional research is needed to better estimate the appropriate values of these infiltration parameters.

ACKNOWLEDGMENTS

The author is indebted to his wife Neva who provided valuable assistance in data entry and editing and to Nancy Owens who accurately typed the manuscript under severe time constraints.

REFERENCES

Carpenter, D.H. 1980. 'Techniques for Estimating Magnitude and Frequency of Floods in Maryland.' U.S. Geological Survey Water-Resources Investigations Open-File Report 80-1016, 79 pp.

Clement, R.W. 1984. 'Improvement of Flood-Frequency Estimates for Selected Small Watersheds in Eastern Kansas Using a Rainfall-Runoff Model.' U.S. Geological Survey Water-Resources Investigations Report 83-4110.

Colson, B.E. and Hudson, J.W. 1976. 'Flood Frequency of Mississippi Streams.' Mississippi State Highway Department Publication, 34 pp.

Curtis, G.W. 1977. 'Frequency Analysis of Illinois Floods Using Observed and Synthetic Streamflow Records.' U.S. Geological Survey Water-Resources Investigations 77-104, 32 pp.

Dawdy, D.R., Lichty, R.W. and Bergmann, J.M. 1972. 'A Rainfall-Runoff Simulation Model for Estimation of Flood Peaks for Small Drainage Basins.' U.S. Geological Survey Professional Paper 506-B, pp. B1-B28.

Golden, H.G. and Price, M. 1976. 'Flood-Frequency Analysis for Small Natural Streams in Georgia.' U.S. Geological Survey Open-File Report 76-511, 75 pp.

Hauth, L.D. 1974. 'Model Synthesis in Frequency Analysis of Missouri Floods.' U.S. Geological Survey Circular 708, 16 pp.

Hines, W.W. and Montgomery, D.C. 1980. Probability and Statistics in Engineering and Management Science, Second Edition, John Wiley and Sons, Inc., New York, 634 pp.

Hydrology Subcommittee of the Interagency Advisory Commmittee on Water Data. 1982. Bulletin 17B, Guidelines for Determining Flood Flow Frequency. U.S. Geological Survey, Office of Water Data Coordination, Reston, Virginia.

Kirby, W.H. 1975. 'Model Smoothing Effect Diminishes Simulated Flood Peak Variances (Abstract).' American Geophysical Union Transactions, Vol. 56, No. 6, p. 361.

Lichty, R. W. and Liscum, F. 1978. 'A Rainfall-Runoff Modeling Procedure for Improving Estimates of T-year (annual) Floods for Small Drainage Basins.' U.S. Geological Survey Water-Resources Investigations 78-7, 44 pp.

Olin, D.A. and Bingham, R.H. 1977. 'Flood Frequency of Small Streams in Alabama.' Alabama Highway Department HPR Report No. 83, 44 pp.

Philip, J.R. 1954. 'An Infiltration Equation with Physical Significance.' Soil Science, Vol. 77, No. 2, pp. 153-157.

SAS Institute, Inc. 1982. 'SAS User's Guide: Statistics, 1982 Edition.' SAS Institute, Inc., Cary, North Carolina, 584 pp.

Simmons, R.H. and Carpenter, D.H. 1978. 'Technique for Estimating the Magnitude and Frequency of Floods in Delaware.' U.S. Geological Survey Water-Resources Investigations Open-File Report 78-93, 69 pp.

Thomas, W.O., Jr. 1982. 'An Evaluation of Flood Frequency Estimates Based on Rainfall/Runoff Modeling.' Water Resources Bulletin, Vol. 18, No. 2, pp. 221-230.

Thomas, W.O., Jr. and Corley, R.K. 1977. 'Techniques for Estimating Flood Discharges for Oklahoma Streams.' U.S. Geological Survey Water-Resources Investigations 77-54, 170 pp.

Wibben, H.C. 1976. 'Application of the U.S. Geological Survey Rainfall-Runoff Simulation Model to Improve Flood Frequency Estimates on Small Tennessee Streams.' U.S. Geological Survey Water-Resources Investigations 76-120, 53 pp.

FLOOD FREQUENCY ANALYSIS USING BOX-COX TRANSFORMATION BASED GUMBEL EV-I DISTRIBUTION

M.Perumal
Scientist 'B'
National Institute of Hydrology
Roorkee-247667
India

R.D.Singh
Scientist 'B'
National Institute of Hydrology
Roorkee - 247667
India

S.M.Seth
Scientist 'F'
National Institute of Hydrology
Roorkee - 247667
India

ABSTRACT. Box-Cox transformation is one of the powerful procedure for transforming the data series to near normalization which has been employed in flood frequency analysis. In this study, an attempt has been made to develop methodology for transforming the annual peak flood series to the form of Gumbel EV-I distribution using Box-Cox transformation. The exponent λ of the Box-Cox transformation has been estimated by trial and error using the method of maximum likelihood (MML) and method of probability weighted moments (MPWM), so as to obtain nearly the same estimates of log likelihood functions by both the methods. This methodology has been applied to 1000 samples of various sample sizes of randomly generated synthetic "flood"series which follow the Pearson type-III distribution. The statistical estimates of the reduced variates of the Gumbel EV-I distributed transformed series, viz.mean and standard deviation, have been found nearer to 0.5772 and 1.2825 as required from theoretical considerations andthus verifying the applicability of the proposed methodology for transforming to Gumbel EV-I distribution.

1. INTRODUCTION

The objective of flood frequency analysis is to estimate the design flood from the limited sample data. An assumption is generally

V. P. Singh (ed.), Hydrologic Frequency Modeling, 163–172.

made of a theoretical frequency distribution which fits the population events and the parameters of the assumed distribution are computed from the sample data. However, data arising from various situations form their own distributions. As such instead of assuming that the data follows a particular distribution, it is more appropriate to transform the data to a particular distribution of known characteristics for the purpose of frequency analysis. Since the properties of a normal distribution are completely defined, it is a common procedure to transform the given data series to a normally distributed series. Power transformation or Box-Cox transformation (Box and Cox, 1964) is one such procedure of normalization. This transformation has been found suitable or flood frequency analysis (Chander et al.,1978). Recently Kuczera (1983) has used this transformation method for regional flood frequency analysis.

In this study, an attempt has been made to use the Box-Cox transformation for transforming the synthetic "flood series to Gumbel EV-I distribution. Since Gumbel EV-I distribution can be more easily handled by theoretical analysis, the transformation of data to this distribution would be useful for flood frequency analysis.

2. METHODOLOGY

A given data series x_i is transformed to series z_i of the Gumbel EV-I distribution using Box-Cox transformation as follows:

$$z_i = \frac{x_i^{\lambda} - 1}{\lambda} \qquad \text{in which } \lambda \neq 0$$
$$z_i = \log x_i \text{ in which } \quad \lambda \to 0 \tag{1}$$

where,

λ = an exponent used for transformation and its value is found by trial and error, such that the series z_i is approximately Gumbel EV-I distributed.

The series z_i is approximately Gumbel EV-I distributed, as the moments of the transformed series higher than second order will not be strictly verified for their closeness with the theoretical values in the proposed methodology. It can also be seen that when $\lambda \to 0$, the transformed series follows the log Gumbel distribution.

The probability density function of the Gumbel EV-I distribution is given as:

$$p(z_i) = (1/\alpha) e^{-(z_i - u)/\alpha - e^{-(z_i - u)/\alpha}} \tag{2}$$

in which,

$(z_i - u)/\alpha$ is the reduced variate y_i, and α and u are the parameters of the distribution.

The exponent λ of equation (1) can not be determined in closed form. It is determined by trial and error in such a manner that the values of the parameters u and α of the Gumbel EV-I distribution estimated using method of maximum likelihood (MML) on the transformed data, when used in method of probability weighted moments (MPWM) (Landwehr et al,1979) yield the same parameter estimates for u and α . For this condition, the ratio of log likelihood functions of the MML and MPWM is nearer to unity and also the coefficient of variation of the reduced variate series y_i nearer to 2.222 as required from theoretical considerations.

2.1 Linking MML and MPWM

Landwehr et al.(1979) have found that the estimates of $\hat{u}$ and $\hat{\alpha}$ obtained using MPWM are unbiased. However, they have shown that the estimates of MML has lesser variance than that of MPWM. In the proposed method, the estimates $\hat{u}_1$ and $\hat{\alpha}_1$ have been obtained first by using MML on the transformed series z_i for a given λ , and using these estimates the probability of non-exceedence F_i is determined corresponding to each z_i, by the following expression:

$$F_i = \exp\left[-\exp\left(\frac{z_i - \hat{u}_1}{\hat{\alpha}_1} \right) \right] \tag{3}$$

This expression for F_i is used to estimate the probability weighted moments $\hat{M}_o$ and $\hat{M}_1$ required for the estimation of the Gumbel EV-I distribution parameter estimates $\hat{u}_2$ and $\hat{\alpha}_2$. The log likelihood functions of MPWM is calculted using these estimates, $\hat{u}_2$ and $\hat{\alpha}_2$. The ratio R of log likelihood functions of MML and MPWM is computed as:

$$R = \frac{\ln\,[\,p\,(z_1)\,p\,(z_2)\,\ldots.p(z_n]_{MML}}{\ln\,[\,p\,(z_1)\,p\,(z_2)\,\ldots.p(z_n)]_{MPWM}} \tag{4}$$

The ratio R would be nearer to unity corresponding to that trial value of λ which yield the values of $\hat{u}_1$ and $\hat{\alpha}_1$ nearly the same as that of $\hat{u}_2$ and $\hat{\alpha}_2$ respectively. Therefore, the iteration procedure for estimating λ could be repeated until the value of R is nearer to unity. For this condition, the coefficient of variation of reduced variate $y_i = (z_i - u)/\alpha$ would also be nearer to 2.222 as required from theoretical considerations.

3. MONTE CARLO EXPERIMENTS

In order to judge the applicability of this methodology, the Monte Carlo experiments were used. For this purpose the skewed distribution random numbers which follow the Pearson type III distribution were generated using the following approximate relationship (Clarke, 1973) between the reduced variate of Pearson type III distribution and the coefficient of skewness:

$$y \approx p \left[1- \frac{1}{9p} + \frac{t}{3\sqrt{p}} \right]^3 \qquad (5)$$

where,

$$p = \frac{4}{g^2} \qquad (6)$$

g being the given coefficient of skewness and t, the standard normal deviate.

The frequency factor, K of Pearson type-III distribution is written as (Kite, 1977):

$$K = - \frac{2}{g} + \frac{gy}{2} \qquad (7)$$

The Pearson type-III distributed synthetic "floods" were generated using the following equation (Chow, 1964):

$$x = \mu + \sigma K \qquad (8)$$

where,

μ and σ respectively the given population mean and standard deviation.

Two sets of 100,000 numbers which follow the Peason type-III distribution approximately with coefficient of skewness 1.0 and 0.5 were generated using the above described procedure. For both the sets, population mean and standard deviation were supplied as 500 and 200 respectively, although any other mean and standard deviation could have been used.

Since, the relationship between the reduced variate and the coefficient of skewness given by equation (5) is only approximate the coefficient of skewness of 100,000 generated numbers need not be nearly eual to the supplied population coefficient of skewness. Therefore, these generated two sets of 100,000 numbers were considered as representative of population and their respective computed mean, standard deviation and coefficient of skewness were treated as population parameters. The mean, standard deviation and coefficient of skewness thus computed for the two sets of synthetic "flood" series are respectively given as 528, 289 and 1.336 for case-I series and 520, 275 and 0.750 for case-II series

Population estimates of "floods" for return periods of 20,50, 100, 200, 500 and 1000 years were obtained using the population mean, standard deviation and coefficient of skewness as given above. Using the 100,000 generated number of each of the two series are considered, 1000 samples for each of the sample sizes N = 10, 20, 30, 40, 50, 60, 70, 80, 90, and 100 were studied using the proposed methodology. For each of the sample size considered, 1000 sample estimates for return periods of 20,50, 100, 200,500 and 1000 years were computed. The estimates in the original domain were estimated from those in the transformed domain using the following inverse transformation relationship:

Perform

$$x_T = (z_T \lambda + 1)^{1/\lambda} \tag{9}$$

These estimates were subjected to the performance tests adopting the criteria discussed below:

3.1 Performance Criteria

The performance indices used for judging the accuracy of the suggested methodology are bias, coefficient of variation (CV) and root mean square error (RMSE) as defined below. These indices were normalize by the population values for the purpose of comparison (Lettenmaier and Burges, 1982).

3.1.1 Bias. It is the tendency to overestimate or underestimate a given event level corresponding to the population estimate. A positive bias indicates the over estimation and a negative bias indicates the underestimation. It is measured as:

$$\text{bias } \delta = \frac{E(\hat{x}_T) - x_T}{x_T} \tag{10}$$

where

$E(\hat{x}_T)$ = mean of the estimates of $\hat{x}_T$ for a given sample size.

$\hat{x}_T$ = the estimate of flood corresponding to T-year recurrence internal.

3.1.2 Coefficient of Variation (CV). The coefficient of variation is a measure of the precision of estimation or scatter of the estimates derived from many samples of the same sample size. It is measured as:

$$CV = \frac{[E|\hat{x}_T - E(\hat{x}_T)|^2]^{1/2}}{x_T} \tag{11}$$

3.1.3 Root mean square error (RMSE). RMSE is a common statistical measure which combines the effects of bias and variability. It measures the accuracy of the suggested methodology in fitting the population estimate. It is measured as:

$$RMSE = (\delta^2 + CV^2)^{1/2} = \frac{[| E (\hat{x}_T - x_T) |^2]^{1/2}}{x_T} \tag{12}$$

4. ANALYSIS AND DISCUSSION ON MONTE CARLO TESTS

Table I shows the expected value of mean, standard deviation, and coefficient of skewness of the reduced variate series of the Gumbel EV-I distribution obtained by using the suggested method of transformation on 1000 samples of various sample sizes derived from case-I and case-II series. It can be seen that the expected values of mean computed over the reduced variates of 1000 samples for all the sample sizes of both the series considered yield a value of 0.5772 which corresponds to the theoretical value of Gumbel EV-I distributed reduced variate series. Similarly the expected values of the standard deviations computed over the reduced variates of 1000 samples for all the samples sizes of both the series considered yield a value close to 1.2825 which is the theoretical value for the Gumbel EV-I distributed reduced variate series. However, the expected value of coefficient of skewness of the reduced, variates of case-II series are nearer to 1.139, the theoretical value of Gumbel EV-I distribution, indicating that case-II series could be transformed more close to Gumbel EV-I distribution as compared to case-I series.

Figures 1 and 2 illustrate the bias and RMSE, in percentages of the estimates of different return period "floods" obtained by using the proposed methodology on different sample sizes with 1000 samples each considered from case-I and case-II series. Since the valuesof RMSE and CV do not differ much, only RMSE has been considered for analysis.

From figure 1 of the case-I series, it can be inferred that the bias in the estimates are less than 5% except in the estimates corresponding to the return periods of 500 and 1000 years which are computed as 16% and 28% respectively for sample size equal to 10 (not shown in the figures). It can be seen from the variation of bias values that the lower return period "floods" are under estimated and the higher return period floods are over estimated. The RMSE of the estimates reduces as the sample size increases for a given return period and it increases with return periods. This has resulted in very high values of the RMSE as 115% and 185% respectively for the estimates of 500 and 1000 year return periods obtained using sample size 10. The reduction in RMSE Of the estimates of all return periods is quite significant when the sample size increases from 10 to 20 indicating the importance of sample size in the frequency analysis using the proposed methodology.

It can be seen from figure 2 that the variation of bias and RMSE of the estimates obtained using case-II series also indicate somewhat similar behaviour as for case-I series, but with smaller

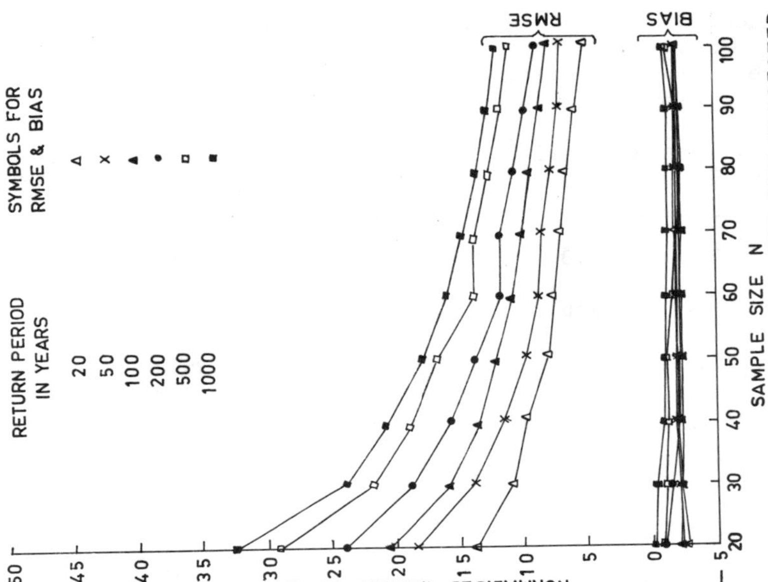

FIG. 2 PERFORMANCE TESTS FOR THE SUGGESTED TRANSFORMATION PROCEDURE USING CASE – II SERIES

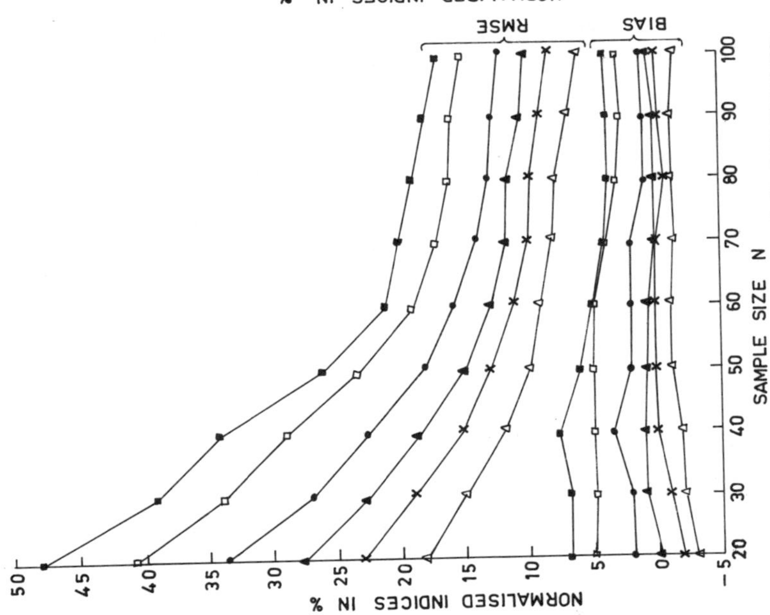

FIG. 1 PERFORMANCE TESTS FOR THE SUGGESTED TRANSFORMATION PROCEDURE USING CASE – I SERIES

Table I. Expected value of mean, standard deviation and coefficient of skewness of the reduced variate series obtained using the suggested transformation procedure on 1000 samples.

Sample Size	Case-I series $\mu_y=E(\bar{y})$	Case-I series $\sigma_y=E(s_y)$	Case-I series $\gamma_y=E(g_y)$	Case-II series $\mu_y=E(\bar{y})$	Case-II series $\sigma_y=E(s_y)$	Case-II series $\gamma_y=E(g_y)$
10	0.5772	1.2813	0.9108	0.5772	1.2816	0.9004
20	0.5772	1.2810	0.9893	0.5772	1.2817	1.0150
30	0.5772	1.2809	0.9971	0.5772	1.2812	1.0580
40	0.5772	1.2805	1.0084	0.5772	1.2813	1.0867
50	0.5772	1.2807	1.0352	0.5772	1.2816	1.0996
60	0.5772	1.2807	1.0489	0.5772	1.2815	1.1109
70	0.5772	1.2808	1.0617	0.5772	1.2815	1.1137
80	0.5772	1.2809	1.0767	0.5772	1.2820	1.1207
90	0.5772	1.2809	1.0784	0.5772	1.2820	1.1186
100	0.5772	1.2808	1.0818	0.5772	1.2816	1.1209

variations. It is observed that the "floods" are generally under estimated. Also the magnitudes of bias and RMSE of the estimates are generally less than that of the estimates arrived using case-I series. This difference in bias and RMSE obtained using case-I and case-II series may be attributed to the difference in skewness of the generated population series.

5. CONCLUSIONS

The present study carried out with Monte Carlo experiments involving two different cases of generated series, though of somewhat limited scope, has shown that Box-Cox transformation could be used for transforming the data series to Gumbel EV-I distribution. The methodology suggested in the study can be used satisfactorily for this purpose, inspite of the effect of coefficient of skewness of the original series on the estimation of "floods", as indicated in the study.

ACKNOWLEDGEMENTS

Thanks are due to Dr Satish Chandra, Director, National Institute of Hydrology for granting permission to publish this paper.

REFERENCES

Box, G.E.P., and Cox, D.R.,1964.'An Analysis of Transformation'. Journal of the Royal Statistical Society, Vol.**B26**, pp.211-252.

Chander, S.,Spolia, S.K., and Arun Kumar.1978'Flood Frequency Analysis by Power Transformation'. Journal of the Hydraulics Division, Proc.A.S.C.E.,Vol.**104**, No.HY11, Nov.,pp.1495-1504.

Chow, V.T.1964,'Statistical and Probability Analysis of Hydrologic Data: Part I.Frequency Analysis'. in :V.T.Chow(Editor), Handbook of Applied Hydrology, McGraw-Hill Book Co.,New York.

Clarke, R.T.1973.Mathematical Models in Hydrology. Food and Agriculture Organisation of the United Nations, Rome.

Greenwood, J.A.,Landwehr, J.M., and Matalas, N.C., and Wallis, J.R.1979.'Probability Weighted Moments' Water Resources Research, Vol.**15**, No.5, October.

Kite, G.W.1978.Frequency and Risk Analysis in Hydrology. Water Resources Publications, Fort Collins, Colorado.

Kuczera, G.1983.'A Bayesian Surrogate for Regional Skew in Flood Frequency Analysis' Water Resources Research, Vol.**19**, No.3, June, pp.821-832.

Landwehr, J.M., and Matalas, N.C., and Wallis, J.R.1979.'Probability Weighted Moments Compared with Some Traditional Techniques in Estimating Gumbel Parameters and Quantiles'. Water Resources Research, Vol.**15**, No.5, October, pp.1055-1064.

Lettenmaier, D.P., and Burges, S.J.1982.'Gumbel's Extreme Value I Distribution: A New Look 'Journal of Hydraulics Division, Proc. A.S.C.E.,Vol.**108**, No.HY4, April, pp.502-514.

FLOOD-FREQUENCY ANALYSIS WITH HISTORICAL DATA IN CHINA

Vernon B. Sauer
U.S. Geological Survey
Water Resources Division
75 Spring Street, S.W., Suite 772
Atlanta, Georgia 30303

ABSTRACT. The U.S. Geological Survey has participated with the People's Republic of China since 1981 in a scientific and technical exchange of hydrologic data and analytical techniques. The Chinese have been documenting historical flood elevations at thousands of sites in China since about 1950. Information has been found at a few sites to establish flood elevations for extraordinary events occurring as much as 2,000 years ago. Peak discharges for many of the historical flood events have been quantified by using stage-discharge relations, hydraulic equations such as by Manning and Chezy, and by hydraulic models. The Chinese have used the historical data extensively in their analyses of flood frequency for design purposes. Frequency curves are developed by a combination of analytical and graphical methods, with the Pearson Type III distribution and Weibull plotting positions. The upper part of the curves are strongly influenced by the historical data.

INTRODUCTION

The U.S. Geological Survey and the Ministry of Water Resources and Electric Power of the People's Republic of China have cooperated since 1981 in a project of hydrologic extremes, which is part of a protocol for scientific and technical exchange of information in surface-water hydrology. This project is subdivided into two subprojects, one on floods and one on droughts. It is the floods subproject which is the principal topic of this paper. Geological Survey personnel have visited the mainland of China on several occasions, and reciprocal visits of the Chinese scientists to the United States have been made. All of the visits have been for the purpose of observing first hand the methods, techniques, and data sources of the two countries. In October 1985, a bilateral U.S./China symposium was held in Nanjing, China, and a post-symposium study tour was taken to several sites in eastern China by many of the delegates. As a result of these exchanges and the symposium, much has been learned about the state of the art of flood analysis in China.

V. P. Singh (ed.), Hydrologic Frequency Modeling, 173–182.

A significant accomplishment of the Chinese has been the recovery and documentation of thousands of historical floodmarks. In the United States, this kind of information is usually not available for more than 100 to 200 years in the past, but in countries with long cultural histories such as China, historical flood data are available for hundreds of years in the past, and at some sites as far back as 1,000 to 2,000 years ago.

It is the purpose of this paper to describe China's efforts in documenting historical floods and to give a brief, general description of their techniques for analyzing flood magnitude and frequency. Much of the information in this paper is taken from Chen Jia-Qi et al (1975) and Cong Shuzheng et al (1979).

HISTORICAL FLOODS IN CHINA

Soon after the founding of the People's Republic of China in 1949, water conservancy districts were formed, river basin planning was started, and plans were being made for the building of dams and other structures on rivers. The Chinese recognized the shortage of hydrologic data (many gaging stations had just been established) and began the investigation of historical floods in the early 1950's to supplement their data base and improve estimates of flood magnitude and frequency. Extraordinary floods in the years of 1956 and 1960 in north and northeast China, respectively, resulted in significant changes to design flood values and further emphasized the need to search out records of historical floods. To date, the Chinese have investigated thousands of historic floodmarks which are used in frequency analysis with observed data to improve magnitude and frequency estimates. They are also used as supplemental proof of design flood data that are computed by other methods, such as rainfall-runoff methods. And in some instances, where no other data exists, they are used directly as design floods.

The Chinese hydrologists and engineers have explored many avenues in their search for historical flood data. There are many populated areas along and immediately adjacent to the rivers and streams. Generation to generation of the same family can frequently be found living in the same area, and knowledge about past floods of extraordinary size are passed from generation to generation. Old inhabitants can recall from their own memory floods which have occurred in their lifetime. By interviewing these people, a considerable amount of historical flood information can be obtained. To verify this kind of information, they try to find more than one independent source regarding the same flood in a given reach of the river. Residents can frequently point out the height of the historical flood by identifying the location of the highest point reached by the flood on a building or other object.

It is interesting to note that sometimes flood information can even be found in things like old folk songs. For instance, in the area of Yunyang Xian in the upper reaches of the Changjiang (Yangtze) River, a folk song has the following phrase:

> In the fifty-third year of Qianlong's reign, the river flat at Longmentan was inundated; in the ninth year of Tongzhi's reign, the flood had reached Sheqianzu.

This verse of folk song indicates there were large floods in the years of A.D. 1788 and 1870. And in reality, this has been confirmed by other sources of information. The 1870 flood is known to be the largest flood to have occurred along certain reaches of the Changjiang River.

Another extremely valuable source of flood information is the many stones and monuments that have floodmarks and related descriptive information carved on them. The Chinese people will frequently mark the height of a flood on an existing boulder or stone structure that was along the bank of the river. Or in some cases, they may erect a special stone or monument upon which they carve the highwater mark and other information relative to that flood. It has been found that such carvings have been done for extraordinary floods which have occurred hundreds and sometimes thousands of years ago. Along the upper reaches of the Changjiang River, many stone carvings and monuments have been found for floods in 1153, 1227, 1560, 1788, 1796, 1860, and 1870. In a reach of 330 km, from Chongqing to Wan Xian, stone carvings and monuments have been found at 13 different places for the 1788 flood and at 39 places for the 1870 flood. Obviously, with such abundant information it is readily apparent that fairly reliable flood profiles can be constructed.

Documents of various kinds have been researched and found to contain information about some outstanding floods. Geographical and historical books dating back over successive historical periods have written accounts of major disasters such as floods and droughts. Some of these may be dated back to some 2,000 years ago. Earlier descriptions may be somewhat sketchy with little or no detail, but accounts during the Ming and Qing Dynasties (the past 300 to 600 years) are fairly complete. Some books are available which deal specifically with waterways and rivers. For instance, the book "Xing Shui Jin Jian" (The Golden Guide to River Regulation) is a collection of some 480 volumes written over a period of several centuries. Other written sources, such as old private notebooks, diaries, and accounting books, have been found in villages and towns along rivers. Much of the written information is qualitative in nature, with only general descriptions of floods and possible relative comparisons of two or more floods. In some cases though, specific information on flood height or depth is given. For example, in the book "Shui Jing Zhu" (The Commentary on the Waterways Classic, written in A.D. 527), it is recorded that an abnormal flood occurred in A.D. 223 on the Yi River at Longmenzhen. It stated that the water rose to a height of 45 "chih," which is equivalent to 10.9 meters. Later investigators believe that this was the greatest flood since 182 B.C.

A particularly good source of flood information has been the many temples and religious sites located along rivers. Both written documents and inscriptions on plaques and temple walls have been found that describe historical floods. An excellent example of this is found at Huangling Temple along the Changjiang River in the Three Gorges area. This temple, which was built in the Tang Dynasty (800-900 A.D.), has a wall plaque describing the 1860 and 1870 floods. The 1870 flood is the greatest flood known at this site and the actual floodmark (mud line) is still plainly visible on several of the interior structural members of the temple. The 1860 flood was lower than the 1870 flood and can also be seen very faintly about 2 meters below the 1870 mark.

Sediment deposits have been located at some sites which can be related to specific floods. Although these are not precise indicators of flood height, they do give considerable information regarding certain floods. A good example is along the Huang (Yellow) River at Xiaolangdi gaging station and proposed dam site. At this site, local residents have confirmed that a clearly visible sediment layer was deposited during the flood of 1843. Although it is not known how deep the water was over the sediment, it is known that this flood represents the greatest flood in about 1,000 years and is the largest flood known in this reach of the river.

Historical floodmarks obtained in reaches of rivers that are fairly stable, such as bedrock channels, can usually be used with fair confidence that they relate to current conditions. On the other hand, rivers that tend to meander considerably, or that have alluvial channels that scour and fill frequently, may be much more difficult to analyze. The Chinese have researched some sites and found information that can be helpful in defining changes to the river. Sometimes descriptions are found that describe the river bed before and after a historical flood. Some of these may relate to something as simple as where women washed clothes in the river. The location of the footings of old bridge piers and piles and how they relate to corresponding plans and descriptions are sometimes used. Old maps can provide information regarding channel locations and, in some cases, may even show objects at a location of the present channel. An especially interesting example was found for a site along the Huang River. At this site, a series of old drawings and more recent photographs show a specific culvert that can be identified in the years 1400, 1855, 1874, 1931, 1935, 1976, and 1982. Although the oldest (1400) of these is not very clear, the others show a progression of sediment buildup in the vicinity of the culvert that can be assumed to relate to the general buildup of the river channel.

ANALYTICAL METHODS USED IN CHINA

Historical flood elevations, in order to be used in frequency analysis, must be converted to flood discharges. Several methods are used in China and are generally in conformance with those used in other countries. These methods and a general description of the Chinese techniques of flood-frequency analysis are described in the following paragraphs.

Quantifying Peak Discharge

Determining the peak discharge of a historical flood is a difficult problem which the Chinese have approached in several different ways. The simplest is by extrapolation of a stage-discharge relation, provided there is sufficient data available. They use both arithmetic and logarithmic plotting procedures for stage-discharge curves, and make attempts to verify that channel conditions are essentially the same for the historic flood and the applicable stage-discharge relation.

A second method used for estimating a peak discharge is by hydraulic formula, such as the Manning or Chezy equations. This may be the most widely used method. Water-surface slope is an essential data requirement for this method. It is usually determined from water-surface profiles established from highwater marks of the historical flood in question, or from highwater marks of other floods, either historical or recent. Frequently the data available for establishing a water-surface profile may be only a few highwater marks many kilometers apart. Such a profile may be reasonably accurate for depicting the general slope of the flood, but may be considerably in error for a localized reach of channel where the historic flood of interest is located. This could be a significant source of error, especially in rivers where large expansions and contractions of the cross section occur. The channel characteristics, roughness, area, and hydraulic radius are also factors which must be determined for the site of the historic flood. In river reaches where stable channels exist, such as the bedrock channels of the Three Gorges of the Changjiang River, there is usually good reason to use the measured channel characteristics as they presently exist. For other streams, like the Huang River where there is much sediment transport, scour and fill, and shifting of the channel both vertically and laterally over a period of time, it becomes extremely difficult to ascertain the size, shape, and roughness of the channel at the time of the historic flood. For those cases, information is sought that will help to define changes that may have occurred over the period of years since the flood. Scour-fill characteristics of recent years are sometimes used to estimate conditions of prior years. Roughness coefficients are usually estimated from existing channel conditions, although if there is information available such as old drawings or descriptions, it may be used to adjust the roughness coefficient. There has been some work done to relate roughness to water-surface slope and flood depth.

A third method of computing peak discharge is through the use of hydraulic models. Little information is available on the use of this method.

Peak discharges computed for historical floods are usually compared for consistency with peaks computed for the same flood at other sites on the same stream or on adjacent streams. These comparisons may consist of simply plotting the flood peaks and corresponding dates on a map of the region of interest. By visual examination, anomalies can be spotted and set aside for further study. Likewise, plots of peak discharge versus basin characteristics, such as drainage area, can be used for a specified region. These plots also provide a method to identify possible errors.

There are obviously many uncertainties, some of which have already been described, when calculating the peak discharge of a historical flood. Scour and fill of the channel, as well as other unknowns regarding the size, shape, roughness, and location of the channel, have been discussed above. Water-surface profiles may be difficult to estimate due to local abnormalities in the bed and banks, or to unknown conditions which existed during the flood such as channel blockages from debris or ice. Uncertainty of the historical high-water elevation itself could cause considerable error in the computed discharge. Other factors such as seasonal vegetation and temperature may cause uncertainty in the roughness coefficients. Even though all of these sources of error are considered and evaluated, the passage of time makes them all the more uncertain and it is probable that in some cases the error in peak discharge exceeds 50 percent, possibly even 100 percent.

Frequency Analysis

The most widely used method of flood-frequency analysis in China is the Pearson Type III distribution. Estimating the statistical parameters of this distribution, as the Chinese do it, involves a combination of graphical and mathematical procedures. First, historical data are examined to establish the period of record to which they belong. The largest known flood, for instance, may be the largest in a period of time that dates back even prior to the earliest known flood. If this can be established or inferred from other information, then it is assigned a rank of 1 ($m = 1$) for a length of record, n, corresponding to this period of time. In a similar way, other lower floods are ranked in accordance with the period of time to which they belong.

The second step is to plot flood magnitude versus exceedance probability. Plotting positions are computed by the Weibull formula,

$$P = m/(n+1)$$

and points are plotted on arithmetic probability paper. Following this, values are assumed or calculated for the Pearson III parameters, Cv (coefficient of variation), and Cs (coefficient of skewness). Some Chinese investigators calculate Cv by the method of moments, but Cong Shuzheng et al (1979) indicates this results in a biased estimate and recommends using an assumed value. Adequacy of the assumed Cv is based on the subsequent fitting of the frequency curve to the data. The parameter, Cs, is considered a function of Cv and usually ranges from 2 to 4 times the value of Cv, although it can be outside these limits. The mean of the peak discharges, Qo, is computed by the method of moments using both the available systematic record and historic data. Using these parameters, sufficient points are calculated and plotted to draw a graphical curve. The measure of goodness of fit is usually based on a visual interpretation. However, in some cases the fit may be based on minimization of an objective function. Chinese research recommends that the objective function be computed as the sum of the absolute differences in discharges between the curve and the data points. These procedures, visual or objective function, are referred to as "the curve fitting method," and are usually done by trial and error. If the first trial values of the parameters do not produce an acceptable fit, new values are assumed, and a new curve is calculated.

Example Design Flood Analysis

The following example of a Chinese design flood analysis is taken from a paper by Chen Jia-Qi et al (1975). Information regarding the length of time assigned to the various historical floods is not complete; so, it is not possible to duplicate the plotting positions for some of the floods, but the example does serve as a good illustration of the effect of historical floods on design floods.

The example is for the Hutuo River at the Huangbizhuang gaging station, located in north China. In 1954, it was desired to compute a design flood of 1,000 years recurrence interval (probability = 0.1%). At that time, there were 20 years of observed data, even though the station was started in 1918. The largest peak discharge among the observed records was 3,700 cms. Historical information was also available for the following floods:

Year	Peak Discharge, in cms
1794	20,000 to 25,000
1853	16,000 to 18,000
1917	13,500
1939	8,300

Prevailing opinion in 1954 was that the peak discharges calculated for these floods were not reliable, and that there was not enough information about the variation characteristics of annual peak discharges in this region. Consequently, the first analysis was made with only the 20 years of observed data. This curve is shown as curve I in figure 1 (figure 1 is reprinted directly from the Chinese paper). The Pearson III statistical parameters for curve I are, Qo = 1,360 cms, Cv = 0.80, and Cs = 3Cv. Using these values, the design flood of 1,000-year recurrence interval was calculated as 8,400 cms.

Two years later in 1956, a very large flood with a peak discharge of 13,100 cms occurred at the gaging station. Then in 1963, another large flood with a peak discharge of 12,000 cms occurred. Both of these floods were larger than the estimated 1,000-year flood record of 8,400 cms. It became evident that the 20-year observed record was not representative of the flood population at this site. Likewise, it was evident that the historical records should have been used in the 1954 analysis. If the four historical peaks of 1794, 1853, 1917, and 1939 had been used, with the 1794 peak taken as the largest in a period of 160 to 350 years, quite different results would have been obtained.

The statistical parameters then would have been Qo = 1,800 cms, Cv = 1.7, and Cs = 3Cv. The 1,000-year design peak discharge would have been 30,000 cms. Curve II in figure 1 represents such an analysis. Furthermore, if the records through 1963 were used, as well as the four historical peaks, another frequency curve would result and is shown as curve III in figure 1. The statistical parameters for this curve are Qo = 2,200 cms, Cv = 1.5, and Cs = 3Cv. The 1,000-year design flood becomes 31,000 cms, which is in good agreement with what would have been computed in 1954 if the four historical floods had been used. This example shows the dramatic difference made by the inclusion of historic data. This and other similar examples are used by the Chinese to support their position that historical data are important.

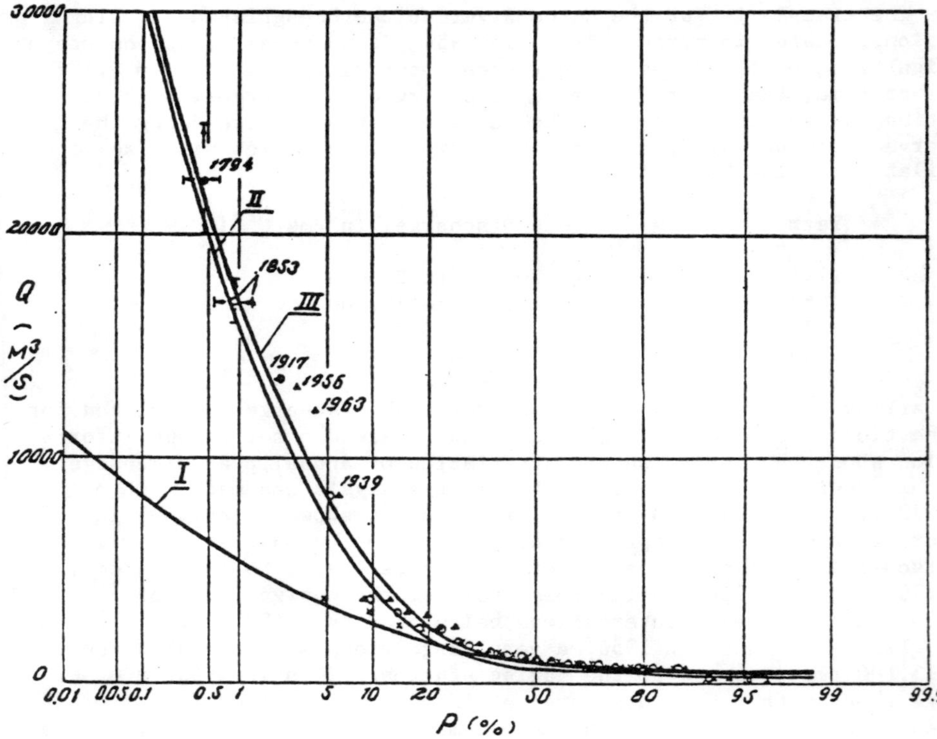

Figure 1. Frequency curves of the annual maximum discharges at Huangbizhuang station on the Hutuo River. (Figure printed directly from Chen Jia-Qi et al, 1975).

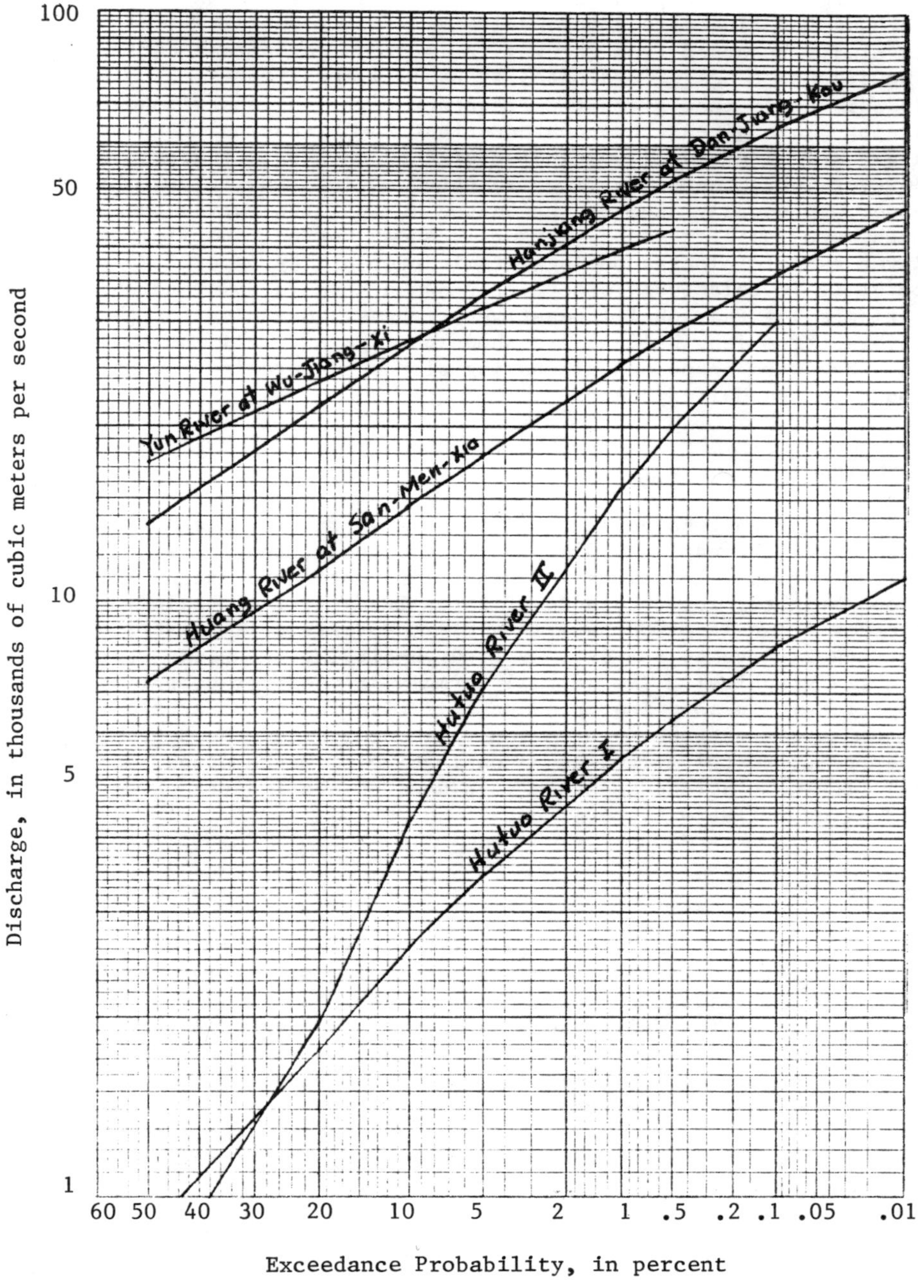

Figure 2. Examples of Chinese flood-frequency curves.

Figure 1 is typical of the examples seen in China of plotting methods for flood-frequency analysis. The probability scale (abcissa) is always reversed from the way it is shown on United States graphs. The discharge scale (ordinate) is usually an arithmetic scale, and it is this scale that gives the frequency curve the sharp concave upward shape. This gives the appearance that the historical data are causing a dramatic increase in peak discharge as probability decreases. However, when the curves are replotted on logarithmic probability paper, as usually used in the United States, the curves take on the much more familiar shape. Figure 2 shows Curve I and II of the Hutuo River example plotted on logarithmic probability paper. Even though curve II is relatively steep, it shows a concave downward shape, indicating a log Pearson III negative skewness. Examples for three other rivers (from Teng Weifen) are shown on figure 2, all indicating concave downward shape and all having slopes comparable to what would be found in the United States.

SUMMARY

China has been involved in an intensive search for historical flood data since the early 1950's; and consequently, they have found reliable flood elevations and other information regarding relative flood heights, for thousands of sites in many river basins, mostly in the eastern half of China. They use the Pearson III distribution for fitting a curve to the data by a curve-fitting technique which is usually graphical. The Weibull formula is used for computing plotting positions. In some instances an objective function is used to minimize the differences between the fitted curve and the data points. They refer to the method as a combination of mathematical and graphical fitting. Strong emphasis is placed on the importance of finding and using reliable historic data.

REFERENCES

Chen Jia-Qi, Ye Yong-yi and Tan Wei-Yan. 1975. 'The Important Role Historical Flood Data in the Estimating of Spillway Design Floods,' Scientia SINICA, **Vol. 18, No. 5.**

Cong Shuzheng, Tan Weiyan, Huang Shouxin, Xu Yingbo, Zhang Weiran and Chen Yuanze. 1979. 'Statistical Testing Research on the Methods of Parameter Estimation in Hydrological Computation' (publication reference not available).

Ten Weifen. 'A Brief Introduction of Flood Estimates for Three Projects in China' (date and publication reference not available).

DEVELOPMENT OF A VERSATILE FLOOD FREQUENCY METHODOLOGY AND ITS APPLICATION TO FLOOD SERIES FROM DIFFERENT COUNTRIES

Krishan P. Singh
Illinois State Water Survey
Champaign, IL 61820
USA

ABSTRACT. Many observed annual flood series exhibit reverse curvatures when plotted on lognormal probability paper. The occurrence of reverse curvature may be attributed to factors such as dominance of within-the-channel or floodplain flow, seasonal variation in flood-producing storm types, variability in antecedent soil moisture and cover conditions, and a mixture of rainfall and snowmelt floods. A transform has been developed which considers the probability distribution of observed flood series as a mixture of the probabilities of two lognormal distributions. Distribution parameters can be significantly biased if the observed flood series has outliers and inliers. The distributions of outliers and inliers for various size samples and at various probabilities or significance levels have been developed from extensive Monte Carlo experiments. The objective detection and modification of any outliers and inliers in an observed flood series is an integral part of the versatile flood frequency methodology. Results presented for flood series from 7 basins from different countries show the versatility and superiority of the proposed methodology.

INTRODUCTION

Many observed annual flood series exhibit reverse curvatures when plotted on lognormal probability paper, instead of no curvature (as for a lognormal distribution) or a concave or convex curvature (as for a log-Pearson type III distribution). The occurrence of these reverse curvatures may be attributed to 1) seasonal variation in flood-producing storm types, 2) dominance of within-the-channel or floodplain flow, and 3) variability in antecedent basin soil moisture and cover conditions. Thus, many flood series have flood events which do not strictly belong to the same population. A mixed distribution model is needed to investigate and analyze such flood series.

Distribution parameters can be significantly biased if the observed flood series has outliers and inliers. The presence of outliers and inliers at both the high and low ends of the flood spectrum needs to be detected and objectively modified so that design

V. P. Singh (ed.), Hydrologic Frequency Modeling, 183–197.

flood estimates with the least bias can be computed. A detection and modification methodology needs to be developed. Analyses of storms producing high floods perceived as outliers and droughts containing low floods perceived as outliers provide a physical basis for their being outliers.

A versatile flood frequency methodology has been developed and computerized. Outlier/inlier detection and modification at various significance levels is an integral part of this methodology. A mixed distribution model is used to simulate the observed flood series, with any detection and modification of outliers/inliers if needed. The methodology has been tested on hundreds of flood series from various parts of the world.

EXISTENCE OF MIXED DISTRIBUTIONS

The lognormal probability plots of many observed annual flood series at various stream gaging stations in the United States and other countries in the world do not follow straight lines or curves with concave or convex curvature, but exhibit reverse curvatures of different shapes. Careful analyses and considerations strongly suggest that these shapes are caused by the heterogeneity or "mixedness" of the population of floods (Singh, 1968).

Moran (1959) calls the expression $a_1\ p_1\{x\} + \ldots + a_k\ p_k\{x\}$ a mixture of the probability distributions $p_1\{x\}, \ldots, p_k\{x\}$, if the $a_1, \ldots, a_k$ are non-negative and $a_1 + \ldots + a_k = 1$. Hald (1952) designates a population formed by combining two populations in a given proportion as a heterogeneous population. Though many hydrologic phenomena possess mixed populations, only the observed annual flood series are considered in this paper. Magnitude of a flood peak depends largely on storm and basin characteristics.

Storm characteristics of interest are the type of storm and intensity and duration of storm. Hiser (1956) gives six type distributions of precipitation for four raingage stations in Illinois. He defines these as cold front, warm front, stationary front, squall line, warm air mass, or cold air mass type. Essenwanger (1960) refers to storms caused by hurricanes, extratropical cyclones, and frontal or air-mass rains. These different types of storms elicit a different response from the basin. High intensity and short duration storms usually cause much higher peak flow than low intensity and long duration storms with the same total precipitation.

With regard to basin characteristics, three main factors are the relative dominance of the within-the-channel or floodplain flow, the antecedent soil moisture condition, and the vegetal cover. Bankfull discharge in a river corresponds to about a 2-year or median flood. With an increase in flow, the water spreads over the floodplain. For low depths of inundation, the mean velocity of the composite flow section is much lower than at the bankfull discharge. It slowly increases with an increase in the depth of flooding and may reach and surpass the bankfull flow velocity as the flow steadily increases. This leads to the introduction of a new storage element and possible flattening and later rise of the flood probability curve at a different slope than for the floods within the channel.

The difference between precipitation and runoff is largely affected by the antecedent soil moisture condition and vegetal cover which vary seasonally and cause the runoff factor to fluctuate from a high value of say 0.6 to a low value of 0.1 or even less. The annual flood peaks, therefore, depend on a number of significant factors which vary from season to season and within a season. The interactions of the distributions of pertinent factors may lead to flood series which may resemble conventional distributions such as lognormal, log-Pearson, or extreme value, or which may exhibit shapes with marked reverse curvatures to be dealt with by the mixed-distribution concept. Mixed distributions have been considered in terms of rainfall and snowmelt floods (Waylen and Woo, 1982) and floods caused by hurricanes and other storms (Canterford and Pierrehumbert, 1977). According to Potter (1958), the dog-leg in the distribution curves of annual flood series occurs because the samples of annual maxima are drawn from two different populations.

MIXED DISTRIBUTION MODEL

The mixed distribution model considers the observed floods (or their logarithms) to belong to two populations with means μ_1 and μ_2, variances σ_1^2 and σ_2^2, and relative weights a and $1-a$. The model developed from various studies (Singh, 1968; Singh and Sinclair, 1972) is based on the following equations:

$$p\{x\} = a\ p_1\{x\} + (1-a)\ p_2\{x\}$$

$$p\{x\} = \frac{1}{\sigma_1 \sqrt{2\pi}} \int_{-\infty}^{x} \exp\left[-\frac{(x'-\mu_1)^2}{2\sigma_1^2}\right] dx'$$

$$p_2\{x\} = \frac{1}{\sigma_2 \sqrt{2\pi}} \int_{-\infty}^{x} \exp\left[\frac{(x'-\mu_2)^2}{2\sigma_2}\right] dx'$$

in which p is the probability of being equal to or less than x, and $x = \log Q$ where Q is the annual flood. The five parameters (μ_1, μ_2, σ_1, σ_2, and a) are linked to statistics: mean, standard deviation, and skew of observed annual flood series (Cohen, 1967). The parameter values can be obtained with a nonlinear programming algorithm (Singh and Nakashima, 1981). Various frequency floods are computed by a reiterative process.

The mixed distribution model considers an observed flood distribution as essentially composed of two component distributions. If such a mixed distribution does not fit the observed distribution satisfactorily, recourse may have to be made to three or more component distributions. However, this can increase the complexity of the problem and make the problem intractable.

OUTLIER/INLIER DETECTION AND MODIFICATION

An outlier in a set of data is defined as an observation or a subset of observations that appears to be inconsistent with the remainder of that set of data (Barnett and Lewis, 1978). When the values for the highest observed floods of an annual flood series are much higher or much lower than expected, these values are designated as outliers and inliers, respectively. When the values for the lowest observed floods are much higher or lower than expected, these are designated as inliers and outliers, respectively. Analyses of storms producing high floods perceived as outliers and droughts containing low floods perceived as outliers provide a physical basis for their being outliers. However, these outliers and inliers need to be detected and properly modified to derive unbiased design flood estimates.

A literature search did not show the existence of statistical tests for checking outliers at higher than the 0.1 level or inliers at any level. Tests have been developed for detection of any outliers and inliers at various levels (Singh and Nakashima, 1981) with experiments on millions of normally distributed numbers generated with the polar method of Box, Jenkins, and Marsaglia (Knuth, 1969). The derived statistics for the high outliers at the 0.01 and 0.05 levels are the same as given by Barnett and Lewis (1978). The developed test statistic is termed a departure and is given by

$$\text{departure, } \Delta i = z_i - z_{si}$$

in which z is the theoretical standard normal deviate and z_s is the sample standardized deviate corresponding to the plotting position, p

$$p = \frac{m - \sigma}{n + 1 - 2\alpha}$$

in which p is the probability of nonexceedance, m is the rank order for the flood series ranked low to high, and α is 0.38. Test values of outlier and inlier departures are given in Table I in which SL is the significance level and there are 6 windows corresponding to SL thresholds of 0.01, 0.05, 0.1, 0.2, 0.3, and 0.4 (Singh, 1983).

The observed flood series needs to be transformed to resemble a series distributed as $N(u,\sigma^2)$. The testing of various transformations resulted in the selection of the power transformation (Box and Cox, 1964) as the best technique. A flood Q is transformed to y

$$y_i = (Q_i^{\lambda} - 1)/\lambda \ ; \ \lambda \neq 0$$

and

$$y_i = \log Q_i \ ; \ \lambda = 0$$

TABLE I. Test Values of Outlier and Inlier Departures

*	SL	Outlier Inlier	Low 1 15-100	Low 2 20-100	Low 3 25-100	Low 4 30-100	Low 5 40-100
1	<0.01	Inlier	<-0.689	<-0.495	<-0.412	<-0.363	<-0.327
	>0.99	Outlier	>1.029	>0.643	>0.498	>0.418	>0.368
2	<0.05	Inlier	<-0.532	<-0.369	<-0.303	<-0.264	<-0.237
	>0.95	Outlier	>0.681	>0.421	>0.337	>0.285	>0.253
3	<0.10	Inlier	<-0.441	<-0.299	<-0.243	<-0.211	<-0.188
	>0.90	Outlier	>0.503	>0.321	>0.254	>0.217	>0.193
4	<0.20	Inlier	<-0.318	<-0.209	<-0.167	<-0.143	<-0.127
	>0.80	Outlier	>0.297	>0.197	>0.159	>0.137	>0.123
5	<0.30	Inlier	<-0.221	<-0.141	<-0.110	<-0.093	<-0.082
	>0.70	Outlier	>0.161	>0.112	>0.092	>0.081	>0.073
6	<0.40	Inlier	<-0.132	<-0.080	<-0.060	<-0.050	<-0.043
	>0.60	Outlier	>0.052	>0.043	>0.037	>0.034	>0.032
			High 1 15-100	High 2 20-100	High 3 25-100	High 4 30-100	High 5 40-100
1	<0.01	Outlier	<-1.054	<-0.654	<-0.511	<-0.429	<-0.377
	>0.99	Inlier	>0.679	>0.488	>0.407	>0.358	>0.323
2	<0.05	Outlier	<-0.683	<-0.433	<-0.341	<-0.290	<-0.256
	>0.95	Inlier	>0.529	>0.369	>0.300	>0.263	>0.235
3	<0.10	Outlier	<-0.500	<-0.322	<-0.258	<-0.221	<-0.195
	>0.90	Inlier	>0.438	>0.299	>0.241	>0.209	>0.186
4	<0.20	Outlier	<-0.295	<-0.197	<-0.161	<-0.139	<-0.124
	>0.80	Inlier	>0.317	>0.209	>0.166	>0.143	>0.126
5	<0.30	Outlier	<-0.159	<-0.112	<-0.094	<-0.082	<-0.074
	>0.70	Inlier	>0.221	>0.140	>0.110	>0.093	>0.082
6	<0.40	Outlier	<-0.051	<-0.043	<-0.039	<-0.035	<-0.032
	>0.60	Inlier	>0.132	>0.079	>0.060	>0.050	<0.043

Notes: 15-100,..., and 40-100 denote the range of sample size *n* in years; * denotes window; SL = significance level

The constant λ can be obtained with the maximum log-likelihood method (Singh, 1980). The power-transformed series, y, has a skew very close to zero but the kurtosis may be significantly different from that for a normal distribution. If the frequency distribution is symmetric (e.g., the 5th moment is close to zero), low values of kurtosis denote platykurtic distribution with rather flatter tops, and higher values of kurtosis denote leptokurtic distribution with rather sharper tops than with normal distribution. The adjustment factors for various values of kurtosis in the case of symmetric distributions are given elsewhere (Singh and Nakashima, 1981). If the transformed series exhibits asymmetry, these adjustment factors may not yield the best results.

The detection and modification procedure starts from window 1. If some outliers and/or inliers are found in window 1, the same are modified to values at level 1, the resulting series is analyzed to test the departures, and the procedure is repeated, if necessary, to ensure that there are no outliers and/or inliers in window 1 or at the 0.01 significance level. The procedure is followed sequentially from one window to the next, computing desired distribution statistics and floods before moving on to the next window. If no outliers and/or inliers are detected in a particular window, no modifications are done in that window.

VERSATILE FLOOD FREQUENCY METHODOLOGY

A computer program package has been developed (Singh and Nakashima, 1981) for objectively detecting and modifying any outliers/inliers sequentially from levels 1 to 6 and for computing various frequency floods with three methods: power transformation method, log-Pearson type III method, and mixed distribution method. The U.S. Water Resources Council (1981) recommends log-Pearson type III for general use.

In the power transformation method, the T-yr flood is computed from

$$Q_T = (\lambda\, y_T + 1)^{1/\lambda} \; ; \; y_T = \overline{y} + Z_T\, s_y$$

in which $\overline{y}$ and s_y are the mean and standard deviation of the power transformed series and Z_T is the standard deviate corresponding to the nonexceedance probability, 1 - 1/T. The Z_T is obtained for both kurtosis = 3 (i.e., normal distribution) and sample kurtosis (i.e., symmetric but not normal distribution). With log-Pearson type III, the T-yr flood is obtained from

$$Q_T = (\overline{x} + k_T\, s)^{10}$$

in which x and s are the mean and standard deviation of the log-transformed flood ($\overline{x}$ = log Q) and k_T is the frequency factor which depends on skew and recurrence interval, T. With the mixed distribution, the x for a desired value of p is obtained by finding p for a trial value of x from the equation

$$p(x) = a\ p_1(x) + (1-a)\ p_2(x)$$

and converging to the desired value of x through a fast reiterative process. The program package prints the various frequency floods, modifications of any detected outliers/inliers at various levels of significance, and distribution statistics for the three methods. These results help in deciding the best-fit distribution and in selecting the storms and droughts causing floods which are perceived as high and low outliers.

EXAMPLES

Seven annual flood series from rivers in Japan, the U.S.S.R., Poland, Czechoslovakia, Italy, and the U.S.A. (UNESCO, 1976) were selected to evaluate the results obtained with the proposed versatile flood frequency methodology. The relevant information; years of record (n) in years; and drainage area, DA, in sq km; are given below.

Highest and lowest floods (without modification SL=0 and with modification SL=0.3), fitted log-Pearson type III or LP3 and mixed distribution or MD statistics, and 100-year and 1000-year floods with LP3 and MD are given in Table II.

Basin No.	River and Gaging Station Location	n, years	DA, sq km
1	Tone River, Yattajima, Japan	34	5110
2	Fuji River, Shimizubata, Japan	51	2120
3	Dep River, Rychkovo, U.S.S.R.	30	8440
4	Dunajec River, Nowy Sacz, Poland	50	4340
5	Zdechovka River, Zdechov, Czechoslovakia	45	4.08
6	Magra River, Calamazza, Italy	40	939
7	Beetree Creek, Swannanoa, U.S.A.	45	14

1. Tone River, Yattajima, Japan. All 34 observed annual floods were caused by rainfall and 32 out of a total of 34 floods occurred during June to October, with the 3 highest floods occurring in September. Only the second highest flood of 9730 m^3/s was detected as an inlier at the 0.2 significance level and modified to 10,555 m^3/s and to 11,269 m^3/s at the 0.3 significance level. The fitted MD curve together with the two component distributions as well as the flood series and modification of the inlier are shown in Figure 1.

2. Fuji River, Shimizubata, Japan. All 51 observed annual floods were caused by rainfall and 50 out of a total of 51 floods occurred during June to October, with the three highest floods occurring in August-September and the lowest flood in April. No high floods were detected as outliers/inliers up to the 0.3 significance level. The second

TABLE II. Highest and Lowest Floods and Fitted Distribution Statistics, and 100-Year and 1000-Year Floods

Item	SL	Basin 1	Basin 2	Basin 3	Basin 4	Basin 5	Basin 6	Basin 7
H1	0	16900	7300	3970	3300	14.7	3480	38.8
	.3	16900	7300	2968	3300	14.7	3609	36.1
H2	0	9730	5600	2000	3300	13.0	3100	23.5
	.3	11269	5600	2000	3064	11.03	3117	23.2
H3	0	9680	5000	1620	2680	8.18	3100	17.1
	.3	9680	5000	1620	2600	7.74	3096	17.1
L3	0	745	400	376	186	0.73	450	2.9
	.3	745	372	353	186	0.73	520	3.2
L2	0	555	400	291	138	0.60	420	2.3
	.3	555	312	291	138	0.60	435	3.0
L1	0	422	210	208	120	0.60	410	1.9
	.3	422	210	221	120	0.57	402	2.6
LP3: μ	0	3.447	3.139	2.872	2.855	0.269	3.128	0.830
	.3	3.449	3.136	2.867	2.854	0.266	3.129	0.835
σ	0	0.403	0.356	0.266	0.345	0.330	0.244	0.247
	.3	0.406	0.361	0.254	0.342	0.325	0.241	0.233
g	0	-0.193	-0.059	0.449	-0.203	0.900	-0.337	0.658
	.3	-0.174	-0.096	0.227	-0.224	0.831	-0.294	0.912
MD: a	0	0.331	0.488	0.857	0.192	0.689	0.356	0.539
	.3	0.285	0.563	0.984	0.303	0.665	0.430	0.721
μ_1	0	3.024	2.869	2.842	2.420	0.121	2.902	0.761
	.3	2.984	2.907	2.856	2.517	0.117	2.946	0.757
μ_2	0	3.656	3.396	3.108	2.958	0.596	3.253	0.911
	.3	3.634	3.431	3.572	3.001	0.559	3.268	1.038
σ_1	0	0.253	0.245	0.232	0.229	0.191	0.192	0.122
	.3	0.237	0.270	0.240	0.258	0.192	0.199	0.151
σ_2	0	0.281	0.236	0.376	0.281	0.338	0.169	0.321
	.3	0.295	0.223	0.051	0.261	0.334	0.164	0.282
Q_{100} LP3	0	21204	8964	3776	4021	17.60	4315	33.24
	.3	21884	8917	3168	3925	16.24	4327	33.71
MD	0	18432	7623	4190	3875	16.66	4133	36.30
	.3	19202	7530	3696	3732	15.42	4126	35.19
Q_{1000} LP3	0	38173	16238	7352	6614	52.23	5854	67.53
	.3	40036	15937	5442	6371	45.48	5930	73.09
MD	0	30860	11903	9984	6422	32.92	5638	67.18
	.3	32889	11557	4598	6011	30.05	5605	62.68

Notes: H and L are the highest and lowest annual floods in m^3/sec
SL = significance level
LP3 = log-Pearson type III; and MD = mixed distribution

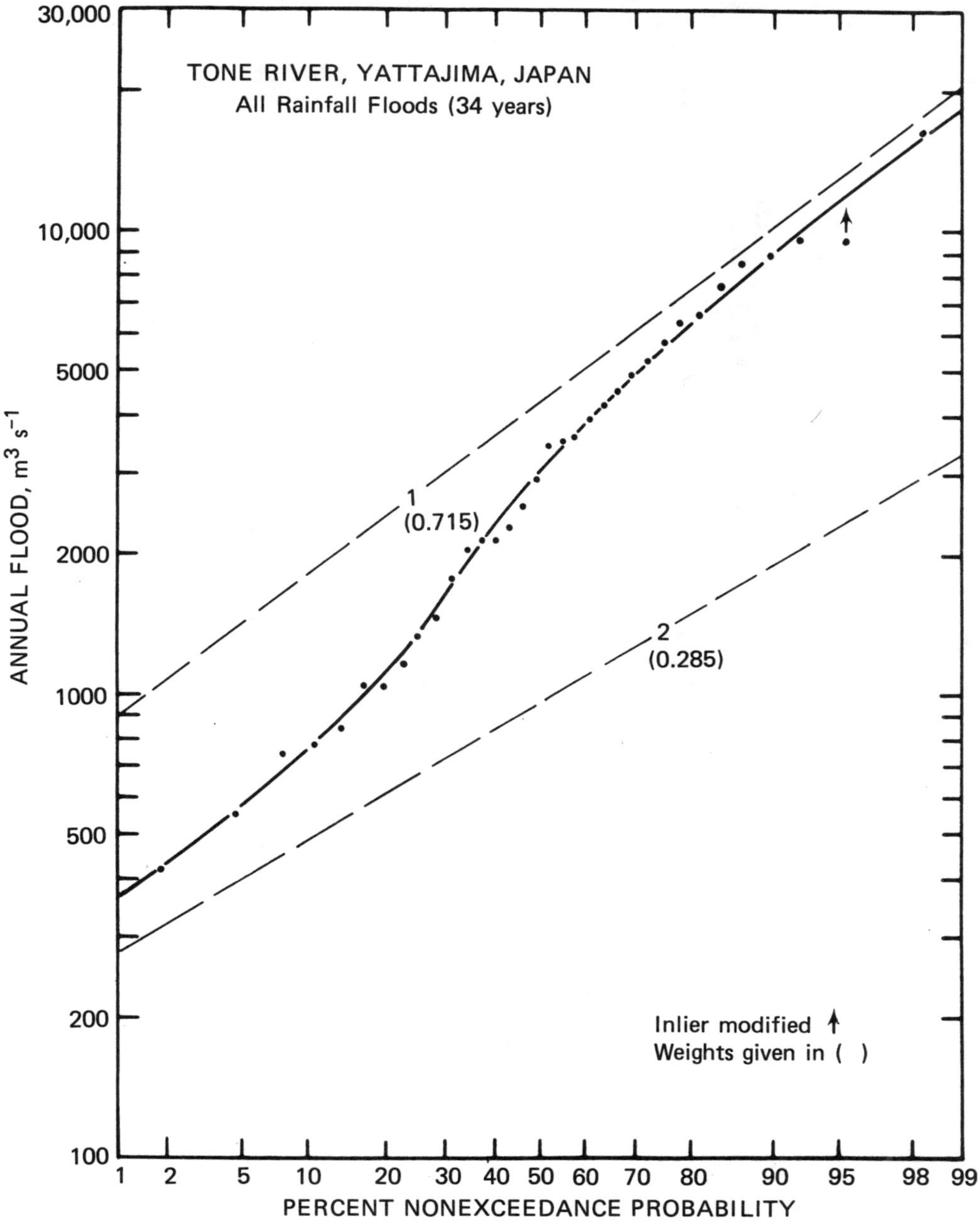

Figure 1. Flood frequency analysis: Tone River, Japan

lowest flood of 400 m^3/s was detected as an inlier at the 0.05 level and finally modified from 400 to 312 m^3/s in the 5th window or the 0.3 significance level. The third lowest flood of 400 m^3/s was detected as an inlier at the 0.2 level and modified to 372 m^3/s at the 0.3 level. The fitted MD curve together with the two component distributions as well as the flood series and modification of the inliers are shown in Figure 2.

3. Dep River, Rychkovo, U.S.S.R. Out of the 30 annual floods, 28 resulted from rainfall and 2 from snowmelt. The rainfall floods occurred in June through September and the snowmelt floods occurred in May. The highest 3 floods occurred in July. The highest flood of 3970 m^3/s (the second highest is 2000 m^3/s) was detected as an outlier at the 0.2 significance level and modified to 3635 m^3/s. It was further modified to 2968 m^3/s at the 0.3 significance level. The fitted MD curve together with the two component distributions, the observed annual floods, and modifications are shown in Figure 2. The second component distribution has a very small weight but it increases the flood magnitude significantly beyond the 50-year recurrence interval because of a large difference between μ_2 and μ_1 values. Modification of the highest flood significantly decreases flood estimates for 100-year and higher recurrence intervals.

4. Dunajec River, Nowy Sacz, Poland. Out of the 50 annual floods, 34 resulted from rainfall and 16 from snowmelt. The rainfall floods occurred in May through October with about one-half occurring in July and August, and the snowmelt floods occurred in February - April with one-half in April. The highest 4 floods occurred in June and July. The second highest flood of 3300 m^3/s was detected as an outlier at the 0.3 significance level and modified to 3064 m^3/s. The third highest flood of 2680 m^3/s was modified to 2600 m^3/s. The fitted MD curve together with the two component distributions, the observed annual floods, and modifications are shown in Figure 3. For this river, the weights are practically proportional to the number of snowmelt and rainfall floods.

5. Zdechovka River, Zdechov, Czechoslovakia. Out of the 45 annual floods, 28 resulted from rainfall and 17 from snowmelt. The rainfall floods occurred in May through November but 19 of them occurred in June and July. The snowmelt floods occurred in December through April. The nine highest floods were all rainfall floods. The second, third, and fourth highest floods were detected as outliers at the 0.3 significance level and slightly modified with the exception of the second highest flood. The fitted MD curve together with the two component distributions, the observed annual floods, and outlier modifications are shown in Figure 3.

6. Magra River, Calamazza, Italy. All 40 annual floods were caused by rainfall. The first highest flood of 3480 m^3/s was detected as an inlier at the 0.3 significance level and modified to 3609 m^3/s. The fourth highest flood was detected as an outlier at the 0.1 significance

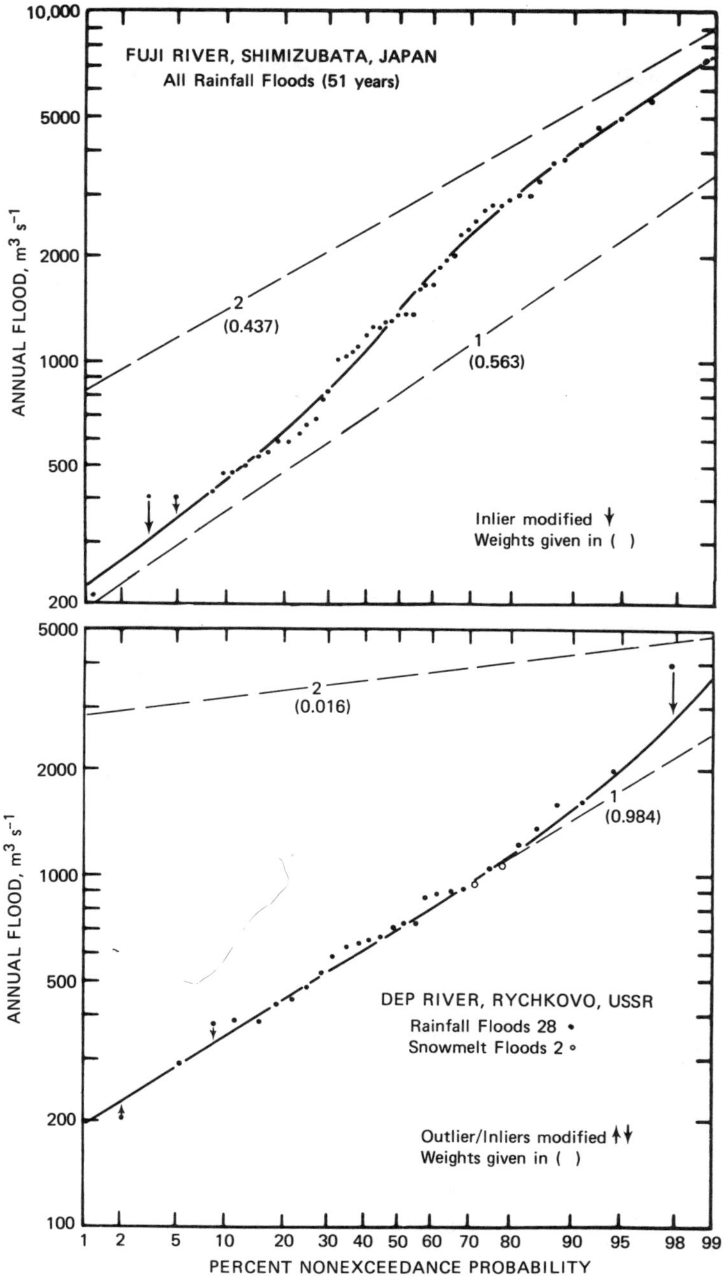

Figure 2. Flood frequency analyses: Fuji River, Japan, and Dep River, U.S.S.R.

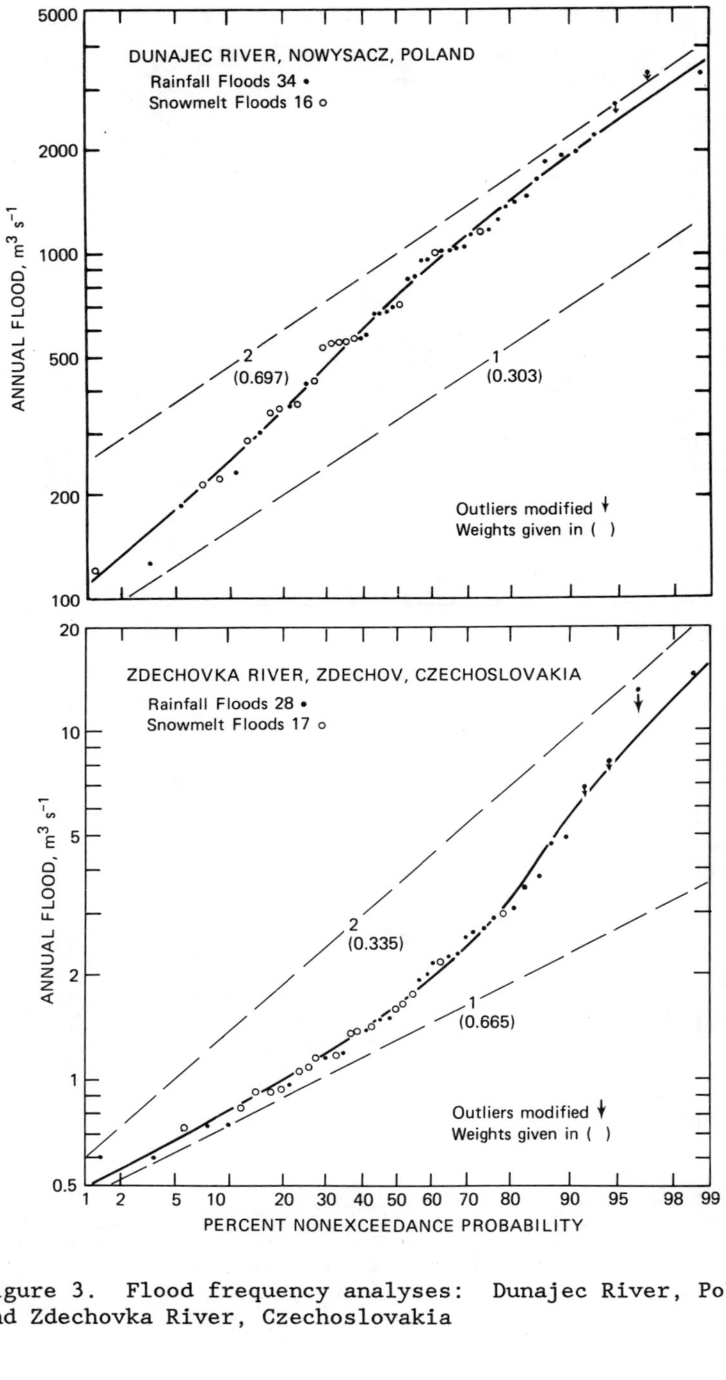

Figure 3. Flood frequency analyses: Dunajec River, Poland, and Zdechovka River, Czechoslovakia

level and modified from 3070 m^3/s to 2847 m^3/s in the 5th window. There were very minor modifications in the second and third highest floods. Out of the four lowest floods, only the third lowest was detected as an outlier and modified from 450 to 520 m^3/s; others were only very slightly modified. The fitted MD curve together with the two component distributions, the observed annual floods, and modifications are shown in Figure 4.

7. Beetree Creek, Swannanoa, U.S.A. All 45 annual floods were caused by rainfall. They occurred in all months, with maximum numbers of 9, 7, and 5 in March, December, and September, respectively. The highest flood of 38.8 m^3/s was detected as an outlier at the 0.2 significance level and later modified to 36.09 m^3/s at the 0.3 level. Second and third highest floods were very slightly modified in the 5th window. The three lowest floods of 1.90, 2.30, and 2.90 m^3/s were detected as outliers at the 0.2 significance level and later modified to 2.58, 2.98, and 3.24 m^3/s at the 0.3 level. The fitted MD curve together with the two component distributions, the observed annual floods, and modifications are shown in Figure 4.

CONCLUSIONS

1. Mixed populations and hence mixed distributions can occur because of a host of factors such as different types of flood-producing storms, dominance of within-the-channel or floodplain flow, antecedent basin soil moisture and vegetal cover conditions, and rainfall and snowmelt floods.
2. A method has been developed and computerized to consider a mixed distribution as consisting of two normal or lognormal distributions (other distributions such as extreme value distribution can be used).
3. An objective method for detection and modification of any outliers and inliers at various significance levels in a flood series has been developed and computerized. Its application to an annual flood series is achieved by power transformation.
4. A versatile flood frequency methodology has been developed and computerized to detect and modify any outliers/inliers at various significance levels and to compute desired-frequency floods with normal distribution (using power-transformed series), log-Pearson type III distribution, and mixed distribution.
5. The results of application to 7 flood series from basins in Japan, the U.S.S.R., Poland, Czechoslovakia, Italy, and the U.S.A. show the versatility and superiority of the mixed distribution method in objectively detecting and modifying any outliers/inliers and yielding reliable flood estimates.

REFERENCES

Barnett, V. and Lewis, T. 1978. Outliers in Statistical Data. John Wiley and Sons, New York, 356 p.

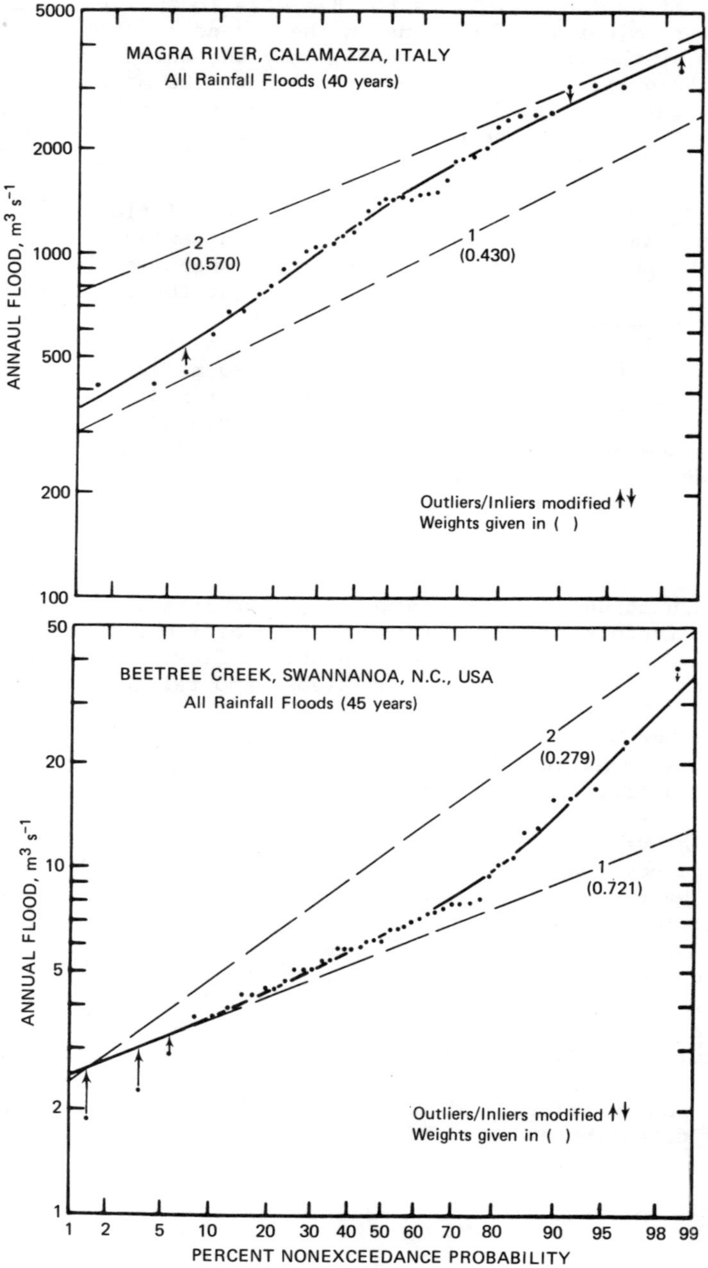

Figure 4. Flood frequency analyses: Magra River, Italy, and Beetree Creek, U.S.A.

Box, G.E.P. and Cox, D.R. 1964. 'An Analysis of Transformation'. Journal of Royal Statistical Society, Vol. **26**, pp. 211-252.

Canterford, R.P. and Pierrehumbert, C.L. 1977. 'Frequency Distributions for Heavy Rainfalls in Tropical Australia'. In: Hydrology Symposium, Institute of Civil Engineers, Brisbane, Australia.

Cohen, A.C. 1967. 'Estimation in Mixtures of Normal Distributions'. Technometrics, Vol. **9**, No. 1, pp. 15-28.

Essenwanger, O. 1960. 'Frequency Distributions of Precipitation'. In: Physics of Precipitation, American Geophysical Union Publication No. 746, pp. 271-278.

Hald, A. 1952. Statistical Theory with Engineering Applications. John Wiley and Sons, New York, pp. 152-154.

Hiser, H.H. 1956. Type Distributions of Precipitation at Selected Stations in Illinois, Transactions, American Geophysical Union, Vol. **37**, pp. 421-424.

Knuth, D.E. 1969. The Art of Computer Programming, Vol. **2**. Addison-Wesley, Reading, Massachusetts, 624 p.

Moran, P.A.P. 1959. The Theory of Storage. John Wiley and Sons, New York, p. 14.

Potter, W.D. 1958. 'Upper and Lower Frequency Curves for Peak Rates of Runoff'. Transactions, American Geophysical Union, Vol. **39**, pp. 100-105.

Singh, K.P. 1968. 'Hydrologic Distributions Resulting from Mixed Populations and Their Computer Simulation'. In: Hydrology Symposium, International Association of Scientific Hydrology, pp. 671-681.

Singh, K.P. and Sinclair, R.A. 1972. 'Two-Distribution Method for Flood Frequency Analysis'. Journal of Hydraulics Division, American Society of Civil Engineers, Vol. **98**, No. HY1, pp. 29-44.

Singh, K.P. 1980. Discussion on "Flood Frequency Analysis by Power Transformation." Journal of Hydraulics Division, American Society of Civil Engineers, Vol. **106**, No. HY3, pp. 462-465.

Singh K.P. and Nakashima, M. 1981. 'A New Methodology for Flood Frequency Analysis with Objective Detection and Modification of Outliers/Inliers'. Illinois State Water Survey Contract Report 272, Champaign, Illinois, 145p.

Singh, K.P. 1983. 'Software Design for a Versatile Flood Frequency Analysis'. Advances in Engineering Software, Vol. **5**, No. 2, pp. 107-112.

UNESCO. 1976. World Catalogue of Very Large Floods. The Unesco Press, Paris, 424 p.

U.S. Water Resources Council. 1981. Guidelines for Determining Flood Frequency. Bulletin 17B of the Hydrology Committee, U.S. Government Printing Office, Washington, D.C.

Waylen, P. and Woo, M.K. 1982. 'Prediction of Annual Floods Generated by Mixed Processes'. Water Resources Research, Vol. **18**, No. 4, pp. 1283-1286.

HYDROCLIMATICALLY-DEFINED MIXED DISTRIBUTIONS IN PARTIAL DURATION FLOOD SERIES

Katherine K. Hirschboeck
Department of Geography and Anthropology
Louisiana State University
Baton Rouge, Louisiana 70803

ABSTRACT. The possibility of mixed distributions in flood series is widely recognized and attempts to separate records into homogeneous subsets have been encouraged. Climate is often proposed as a source of mixed distributions, but the separation of flood records into climatic subgroups has tended to focus on seasonal divisions or a partitioning of the flood series into rainfall and snowmelt-generated events. In this study, a detailed hydroclimatic analysis is used to categorize floods on the basis of the synoptic weather patterns which produced them. This procedure represents the first attempt to classify events in a partial duration flood series according to the specific climatic flood-generating mechanism responsible for each event. The technique identifies mixed distributions by using physically-based information that is independent of the runoff series, rather than by defining subpopulations on the basis of the shape of the parent distribution itself.

INTRODUCTION

One of the most frequently cited areas of potential research in flood frequency analysis is the problem of mixed distributions or multiple populations in hydrologic time series. Homogeneity in the flood series is a basic underlying assumption for the probabilistic determination of flood magnitudes and frequencies. Wherever this assumption appears in the flood frequency literature, however, it is commonly followed by a disclaimer which states that due to differences in the climatic processes involved in the generation of floods, multiple populations or mixed distributions may be present in the data:

> At some locations flooding is created by different types of events. For example, flooding in some watersheds is created by snowmelt, rainstorms, or by combinations of both snowmelt and rainstorms. Such a record may not be homogeneous and may require special treatment. (U.S. Water Resources Council, 1981, p. 7)

V. P. Singh (ed.), Hydrologic Frequency Modeling, 199–212.

Despite the almost universal recognition that some observed flood samples may not be drawn from a single, climatically homogeneous population, only a handful of researchers have seriously devoted their efforts to the analysis of this problem. Potter (1958) was one of the first to discuss the evidence for two or more distinct populations of peak runoff (as seen in "dogleg" flood frequency curves), and he proposed possible climatic causes for the multiple populations. Singh (1968, 1974) presented a methodology for mathematically simulating mixed distributions in hydrologic samples, but although he referred to climate as a probable cause of multiple populations, his approach was to objectively search the streamflow data alone to define a mixture of distributions, rather than to decompose the data on the basis of additional climatic information. Other studies have attempted to identify mixed distributions in streamflow series by separating the flood record into seasonal subpopulations (Guillot 1973; Browzin et al. 1973). To date, the division of a record into seasons is probably the method most commonly used in attempts to identify the effect of climate on flood series homogeneity.

Jarrett and Costa (1982), Elliott et al. (1982), and Waylen and Woo (1982) moved beyond the simple seasonal division of a flood series and looked at the differences between rainfall and snowmelt-generated floods to examine the problem of mixed distributions in hydrologic data. By detailed examination of both streamflow and weather records, these researchers were able to subdivide flood series into rainfall and snowmelt "populations" so that separate flood frequency curves could be developed from each subset of data.

The separation of events in a flood series according to climate can be taken a step farther by identifying the various synoptic atmospheric circulation mechanisms and patterns that generate each flood event in a series. This is especially appropriate in climatically sensitive regions or in climatic transition zones where floods evolve from a variety of different processes that may be exhibited in complex frequency functions. This new hydroclimatic approach to defining mixed distributions in a flood series holds the promise of both enhancing our understanding of the flooding process and potentially improving flood frequency estimates.

THEORY OF MIXED DISTRIBUTIONS

The concept of a mixed distribution is based on the idea that a set of hydrologic observations, when sampled blindly from a population assumed to represent a single process or phenomenon, can also be interpreted as a composite sample representing several different processes or phenomena. In other words, the overall "parent" population may, in actuality, be composed of two or more subpopulations, each with its own distinct distribution. An example given frequently is that of annual floods arising from two sources, snowmelt and rainfall, each process producing floods having their own characteristic frequency distribution.

Figure 1 depicts a theoretical representation of the mixed distribution model. Events from two separate normal populations with different means and standard deviations may, when combined, form the top distribution in the figure, which is the distribution that is "seen" or represented by the observed mean, standard deviation, and skewness of the data sample. In this example, the combination of the two non-skewed subpopulations results in a slightly skewed "mixed" population with a mean and standard deviation different from either of its two component parts.

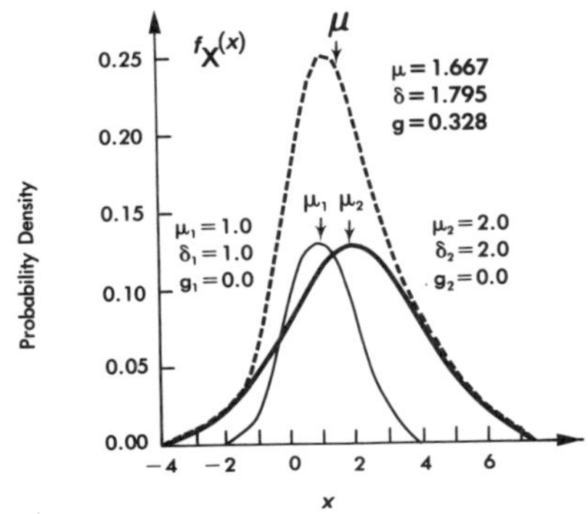

Figure 1. The mixed distribution model.

Different means (μ), standard deviations (σ), and skewness (g) are depicted for two component distributions, and the mixed distribution which they form (dashed line). (Modified from Hawkins, 1974)

Often, complicated frequency functions, such as those with multiple peaks or unusual skew, are more easily interpreted by thinking of the parent distribution as a mixed distribution which can be decomposed into several simpler frequency functions (Figure 2). In the simplest mixed distribution case, depicted in Figure 2a, each of the component distributions is considered to be normal with identical variances, hence the mixed distribution evolves only because of differences in the means of identically distributed frequency functions. An alternative model (Figure 2b) depicts the component distributions as having different means and variances, although still exhibiting normality. Frequency distributions of real-world phenomena need not necessarily decompose into simple normal frequency functions because nature does not always follow a Gaussian law precisely (Anderson 1967). A mixture of skewed and non-skewed distribution functions with various means and variances (Figure 2c) may be the model that best depicts the overall frequency distribution of a process arising from a multiplicity of causes, such as flooding.

The decomposition of hydrologic data into component distributions has significance for simulation studies, for extending frequency curves to predict rare events, and for increasing our understanding of the basic underlying phenomena in the flooding process (Hawkins, 1974). Studies which have concentrated on decomposing a flood series into component parts on the basis of the shape of the parent distribution alone have not always been able to address the underlying physical mechanisms responsible for each component in the distribution. An alternative approach to decomposition of the flood series is presented here. By classifying the events in a flood series according to the different types of climatic mechanisms that produce each flood, hydroclimatically-defined frequency components of the parent distribution can be isolated and examined. This separation of a flood frequency sample into a physically-based mixed distribution sample is a unique way to evaluate the importance of climate as a source of nonhomogeneity in a flood series.

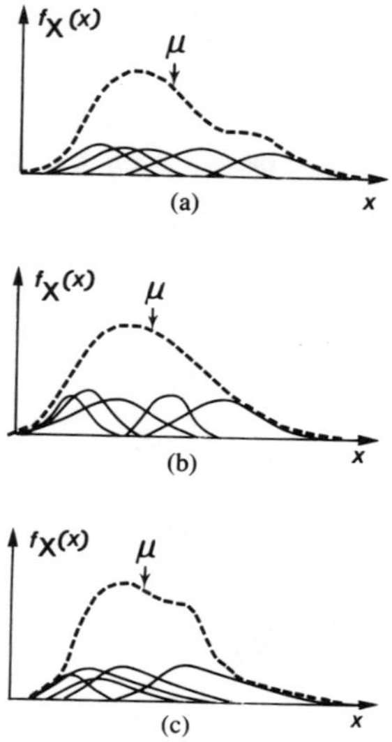

Figure 2. Complicated frequency functions decomposed into simpler components.

(a) different means only
(b) different means and variances
(c) different means, variances, and skewness.

FLOOD HYDROCLIMATOLOGY IN THE GILA RIVER BASIN

In a hydroclimatic approach to flood analysis, hydrologic events are interpreted as occurring within the spatial and temporal framework of regional and global patterns of varying combinations of meteorological elements such as precipitation, storm tracks, fronts, air masses, and atmospheric circulation patterns. Flood hydroclimatology focuses specifically on the climatic genesis of different flood events.

The Gila River basin of central and southern Arizona (Figure 3) is an excellent study area for examining hydroclimatic aspects of the mixed distribution problem because the location, physiography, and climate of this area allow a variety of physical processes to produce flooding in different parts of the basin and at different times during the year. Most of the western part of the basin lies within the Sonoran Desert and is characterized by extremely arid conditions, sparsely vegetated flat-floored alluvial valleys, and ephemeral stream systems which flow largely in response to local convectional showers. The Gulf of California and the Pacific Ocean are close enough, however, to provide a source for large influxes of moisture which are occasionally steered into the region either by tropical storms or other kinds of atmospheric circulation patterns. In the range and basin topography of the central and southeastern portions of the basin, slightly higher elevations and frequent rainfall due to orographic effects result in increased intermittent streamflow. At times, exceptionally large flows and flash floods are possible when excess moisture enters the region from either the Gulf of Mexico, humid regions of Mexico, the Gulf of California, or the Pacific Ocean. In the northern and northeastern parts of the basin, many streams draining the high elevations of the Mogollon Rim are able to maintain perennial flow due to an enhanced orographic effect and the accumulation of sufficient snowfall during winter storms to contribute to spring snowmelt flooding. Local convectional thunderstorms are a common source of flash flooding throughout the basin, especially during the humid summer months, and large frontal storms frequently sweep through the basin in winter.

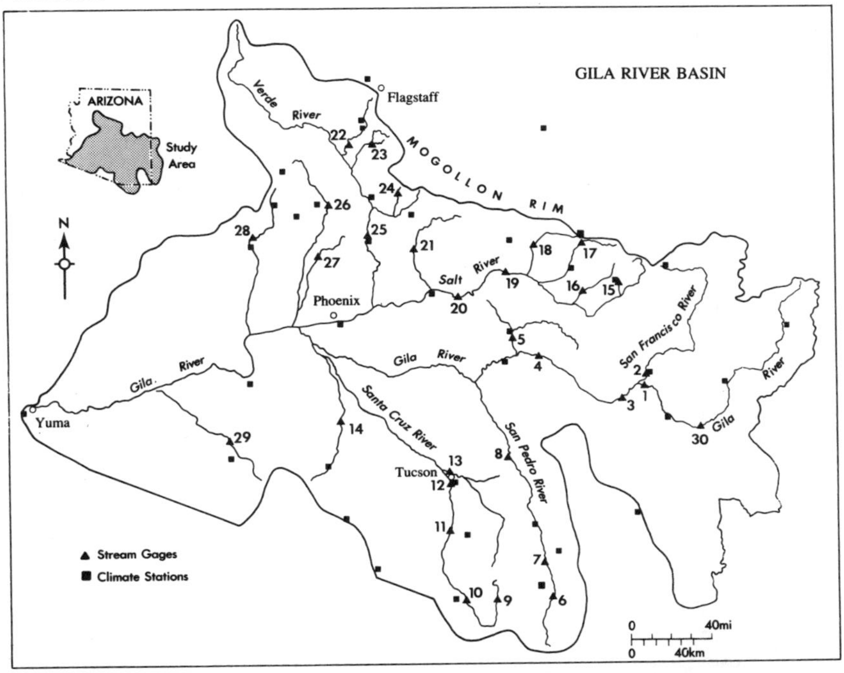

Figure 3. The Gila River basin. Numbered symbols show locations of gaging stations. (A complete description of the stations is given in Hirschboeck, 1985.)

An example of an actual hydroclimatic sequence that resulted in widespread flooding throughout the Gila River basin is depicted in Figure 4. During September 4-6, 1970, several major synoptic features operated together and introduced large amounts of precipitable water vapor into the region, as well as a triggering mechanism to release this moisture. In the upper air circulation, a deep trough of low pressure steered moisture from the Pacific Ocean into southwestern United States, and on September 5th and 6th, the trough began to form a cutoff low circulation, causing the moisture and storminess to persist in the region instead of being moved along rapidly to the east in the upper air westerly flow. At the same time, Tropical Storm Norma tracked very close to the Baja California coast and introduced additional large fluxes of moisture into Arizona at both surface and above-surface levels in the atmosphere. Also concurrently, a surface cold front associated with the upper air trough, moved rapidly across the Gila River basin and acted as a lifting mechanism for the large amounts of moist, unstable air in the region.

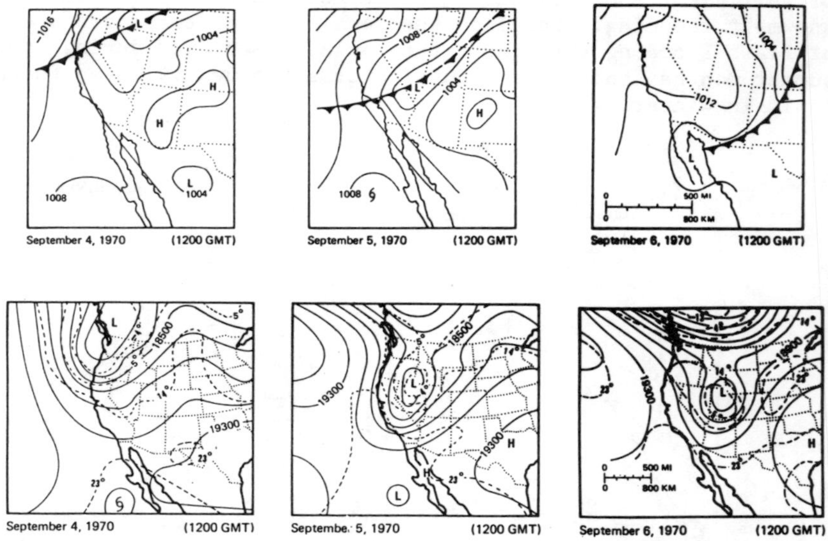

Figure 4. Hydroclimatic sequence which produced widespread flooding in the Gila River basin. (Top) surface weather maps for September 4-6, 1970. (Bottom) Corresponding 500 millibar charts. (Source: Hansen and Schwarz, 1981)

The above combination of synoptic-scale events generated flooding in the basin in a different way than the flooding that typically arises from local summer convectional storms, isolated winter frontal passages, or spring snowmelt. The relevant question for the mixed distribution problem is whether or not these differing flood-generating mechanisms each produce their own unique frequency distribution of flood magnitudes. The arid and semiarid streams of the Gila River basin are especially well-suited to explore this question because sensitivity to climatic variability is high in ephemeral and intermittent streams.

Runoff hydrographs of streamflow in the Gila River basin can be strikingly different at different times of the year, due, in part, to the kinds of atmospheric generating mechanisms which produce the flows. For example, a flow event occurring in response to an intense summer convectional thunderstorm might exhibit a hydrograph similar to Figure 5a, while a flow event produced by a winter frontal storm would show a longer rise time, less of a peak, and a longer recession time (Figure 5b). Floods produced by rare extreme events, such as tropical storms, cutoff lows, or a combination of processes, exhibit complex hydrographs (Figure 5c) which reflect the great magnitude of moisture delivered to the watershed, the complex nature of the storm itself, and the complicated nature of the catchment's response to a synoptic-scale delivery system.

These differences in the flood discharge hydrographs generated by different types of atmospheric processes in the Gila River basin suggest that the frequency distributions of the discharges may also vary under different types of flood-generating atmospheric processes and hence be a major cause of mixed distributions in observed flood series.

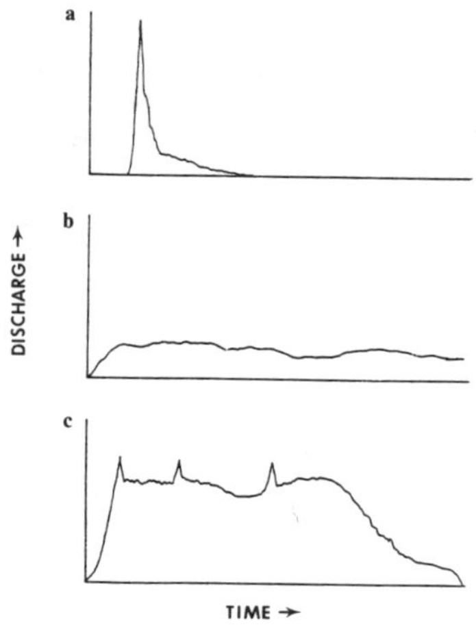

Figure 5. Idealized Gila River basin hydrographs resulting from different atmospheric mechanisms.

(a) summer convectional shower
(b) winter frontal storm
(c) tropical storm/cutoff low combination.
(Modified from Keith, 1981)

DATA AND METHODS

In order to address the question of whether different hydroclimatic flood-generating mechanisms produce their own unique frequency distributions, flood series from thirty U.S. Geological Survey gaging stations in the Gila River basin were analyzed (Figure 3). The stations were selected to reflect a wide range of conditions in the basin and to provide an adequate sample with which to examine multiple causes of flooding. To avoid the influence of nonclimatic sources of nonhomogeneity, care was taken to select watersheds with as few diversions and upstream impoundments as possible, although nearly all streams but the very smallest drainages along the Mogollon Rim were used as sources of irrigation or municipal water supplies to varying degrees. No major reservoirs were located upstream from any of the selected stations.

The flood events studied were those of the U. S. Geological Survey's "peaks-above-base" or partial duration series data for the period 1950 to 1980. Partial duration series data were used so that a large spectrum of floods could be analyzed to better define the shapes of the distributions of various flood-generating mechanisms. In order to compare the relative magnitudes of peak flows from station to station, the flood values were transformed to base ten logarithms and standardized to dimensionless z-scores. After assigning a hydroclimatic classification to each flood event, histograms of the hydroclimatic subgroups were plotted to approximate sample distributions of each type of flood-generating category.

Hydroclimatic Classification of Flood Events

Each flood event was classified into one of eight categories representing the major flood-generating mechanisms in the Gila River basin. The hydroclimatic categories were defined after reviewing daily surface and upper air weather maps and maps of the basin-wide precipitation patterns for each flood episode during 1950 to 1980. In order to be as consistent and objective as possible when assigning flood events to a category, a detailed decision tree or flow chart was designed. Figure 6 shows a greatly simplified version of this chart. Decisions were made at branching points in the chart on the basis of the precipitation pattern, daily synoptic weather maps, and antecedent conditions in the basin. The general pathway through the decision tree consisted of the following.

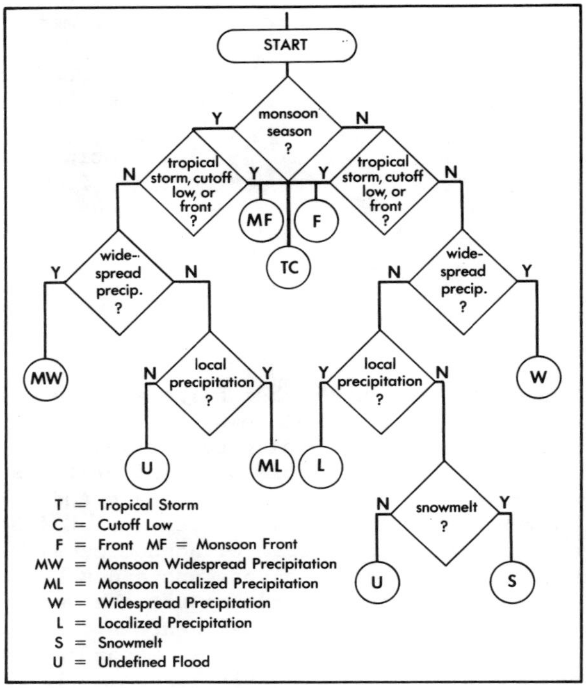

Figure 6. Decision tree for assigning a hydroclimatic classification to each flood event.

After first determining whether or not the humid summer circulation pattern, called the "summer monsoon" (M), had been established, weather maps were examined to see if major synoptic features such as tropical storms (T), cutoff lows (C), or fronts (F) were affecting the precipitation pattern of a developing flood. If no such features were observed, a determination was made of the "widespread synoptic" (W) vs. "local convectional" (L) nature of the precipitation over the basin on the basis of daily precipitation totals. In the widespread synoptic situation, although no major synoptic feature appeared on the weather map, precipitation occurred at stations throughout the basin, indicating that some large scale circulation control was in operation introducing moist and unstable air into the area. In the local convectional situation, precipitation was recorded at only a few stations in the basin, indicating that a very localized weather pattern produced the flood. During the non-monsoon season, evidence for snowmelt as a source of flooding was also evaluated. Combinations of classifications were possible, but an attempt was made to determine the dominating flood-generating feature. Because tropical storms and cutoff lows frequently occurred together and are synoptically related, a combined tropical storm/cutoff low class (TC) was formed. If no classification or probable classification could be identified, the flood event was designated as "undefined" (U).

Results of the Classification

Monthly frequencies of floods produced by each hydroclimatic category are shown in Table 1, along with percentages reflecting the relative importance of each type of flood-generating mechanism in the study area. The most important categories for the Gila River basin as a whole are, in descending order, monsoon widespread (MW), frontal (F), monsoon local (ML), and tropical storm/cutoff low (TC). Table 1 also shows the clear dominance of some categories in certain months, such as the tropical storm/cutoff low type in September and October, and the frontal type in December through March.

Table 1 Monthly Frequencies of Floods in Hydroclimatic Categories

	Oct	Nov	Dec	Jan	Feb	Mar	Apr	May	Jun	Jul	Aug	Sep	Total	%
TC	106	33	5	11		7	11	2	7	14	59	100	355	13
F	20	27	175	157	123	129	19	10	1			20	681	24
L	11	1	1		3	10	49	17	8			6	106	4
W	7	2	13	18	37	18	6	5	4			17	127	4
MF									1	24	17	5	47	2
ML									4	186	310	71	571	20
MW									1	274	472	55	802	28
S			1	4	18	60	48						131	5
Total	144	63	195	190	181	224	133	34	26	498	858	274	2820	100

Table 2 reveals the relative importance of each category at selected individual stations in the Gila River basin. It is immediately apparent that different parts of the study area are affected to varying degrees by the eight categories of flood-generating mechanisms. The more northern stations are strongly dominated by frontal floods, while the more southern stations -- including the southeastern and southwestern stations -- are dominated by monsoon widespread floods. Tropical storms and cutoff lows are an important source of flooding at stations throughout the basin, but snowmelt as a flood-generating mechanism is limited to the northern, high elevation catchments that drain the Mogollon Rim.

Table 2 Classification Frequencies for Selected Stations, 1950-1980

		TC	F	L	W	MF	ML	MW	S
Eastern stations:									
1	Gila near Clifton	14	18	2	2	2	24	27	1
2	San Francisco at Clifton	23	28	2	4	1	21	34	5
5	San Carlos near Peridot	16	29	2	7		28	31	3
Southern stations:									
7	San Pedro at Charleston	12	4		1	3	35	48	
12	Santa Cruz at Tucson	18	11	1	2	4	40	64	
Northern stations:									
20	Salt near Roosevelt	15	50	3	6		7	17	12
21	Tonto near Roosevelt	16	58	1	10		11	30	2
22	Oak Creek near Cornville	10	41		9	2	8	16	13
25	Verde above Horseshoe Dam	12	47		7	1	4	13	14
Western stations:									
26	Agua Fria near Mayer	13	20	4	4	2	30	40	
28	Hassayampa near Wickenburg	20	40		8	2	16	31	

HYDROCLIMATICALLY-DEFINED MIXED DISTRIBUTIONS

Theoretically, if each hydroclimatic category reflects a homogeneous climatic process, the subset of floods generated at a given station by a specific hydroclimatic mechanism should represent a homogeneous sample drawn from the population of all possible floods produced by that particular mechanism. Furthermore, the separation of a flood record into each of its hydroclimatic flood-generating components simulates the decomposition of a single-process frequency distribution into a mixed distribution having a physical explanation for each of the component parts.

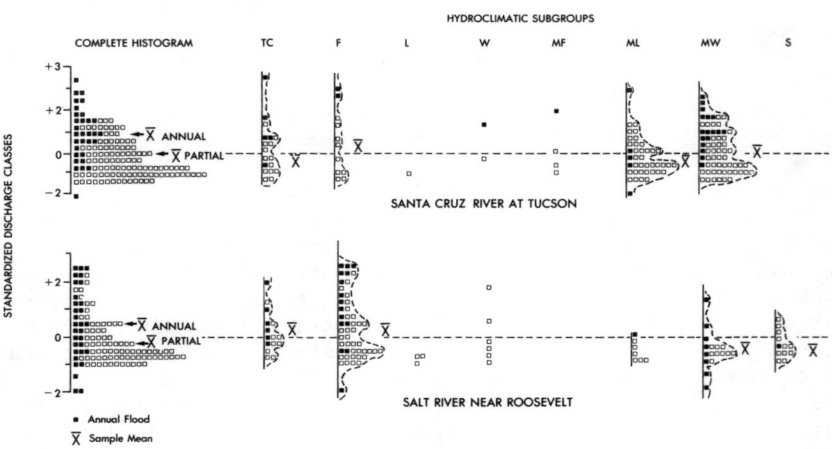

Figure 7. Decomposition of two flood series into hydroclimatic subgroups. Each square represents one flood event.

Histograms of the standardized flood discharges generated by each mechanism serve as an approximation of the decomposition of a flood series into its hydroclimatically-defined component parts. Figure 7 displays this decomposition for two of the Gila basin stations: the Santa Cruz River at Tucson in the southern part of the region with a drainage area of 5755 sq km (Station 12), and the Salt River near Roosevelt in the northern part of the basin with a drainage area of 11,153 sq km (Station 20). The subgroup histograms at each station effectively reveal the nature of the flooding process in these two different parts of the basin. In the southern Santa Cruz drainage, summer monsoon widespread synoptic precipitation (MW) is the major cause of both partial and annual series flooding, and the dominance of this mechanism is reflected in the similarity of shape between the sample frequency distribution of MW floods and the combined histogram of the series. However, some of the largest floods in the record were produced by the tropical storm/cutoff low and frontal precipitation mechanisms. These distributions, (to the degree that they are defined by the sample), are highly positively skewed with large variances. In October 1983, a TC type event associated with Tropical Storm Octave resulted in the largest flood on record for the Santa Cruz River. The inclusion of this event in Figure 7 would skew the TC distribution even more. The monsoon local precipitation mechanism (ML) is also a major generator of floods in the Santa Cruz, but the discharges in this subset tend to be small and few annual floods are produced, despite the high frequency of this type. Snowmelt is not important as a flood-generating mechanism in the Santa Cruz watershed.

The Salt River to the north exhibits a different sensitivity to the eight hydroclimatic categories. The frontal precipitation mechanism generates the most floods, as well as the largest floods. The tropical storm/cutoff low category also contributes some large floods, although unlike the Santa Cruz, the TC mechanism is not responsible for the largest flood on record. Monsoon local precipitation is relatively ineffective as a flood-generator in the large Salt River basin, but the monsoon widespread synoptic precipitation mechanism is responsible for several of the annual floods occurring in summer. Snowmelt produces some flooding in the basin, but discharges are relatively small.

DISCUSSION

The decomposition of a flood series into hydroclimatic subgroups is an effective method of exploring the underlying physical elements of the flooding process and how they combine to produce the flood frequency distribution that is "seen" by a sample of floods recorded at a gaging station. The method is also very useful for identifying which streams will respond in like manner to a given climatic input so that spatially homogeneous stations can be grouped together for regional studies. Due to the occasional occurrence of several atmospheric mechanisms operating together to produce floods, the eight hydroclimatic subgroups may not represent completely homogeneous components of the mixed distribution. In fact, the histograms in Figure 7 suggest that some hydroclimatic categories have multiple-peaked distributions of their own that may reflect two or more modes of operation within a given flood-generating process. In addition, nonclimatic sources of nonhomogeneity, such as land use changes or channel modifications, may obscure the uniqueness of each hydroclimatic subgroup.

The technique needs to be further refined in order to address the statistical significance of frequency distributions derived from different hydroclimatic mechanisms. Due to the small sample sizes of some of the subgroups, special modeling or simulation will be required to obtain parameter estimates of the hydroclimatic component distributions. The results of this study suggest that the means, variances, and shapes of the flood frequency distributions of different hydroclimatic subgroups may have a physical basis that can be linked to the inherent nature of the flood-generating process. For example, tropical storm/cutoff low floods at many of the stations tended to have highly positively skewed distributions with extremely large variances. This can be physically explained by the erratic nature of eastern North Pacific tropical storm tracks and the rare but extremely large discharges that can occur when a storm, on occasion, does move directly into the Gila River basin.

If reliable parameter estimates of the component distributions can be made, hydroclimatically-defined mixed distributions may be of greatest use in extending frequency curves to predict rare events. In the Santa Cruz basin, tropical storms have been responsible for the largest floods on record; while in the Salt River system, frontal storms produced the highest recorded discharges. A better knowledge of the frequency distribution of floods produced by each of these mechanisms would greatly improve our ability to predict the extreme events occurring in the tails of the composite distribution.

SUMMARY AND CONCLUSIONS

Climate is often proposed as a potential source of mixed distributions in flood series, but the separation of flood records into climatic subgroups is usually based on a simple seasonal division that does not take into consideration the variety of atmospheric circulation mechanisms which generate floods at different times of the year in various parts of a large drainage basin. By classifying events in a flood series on the basis of the climatic mechanism which generates each flood, hydroclimatic subgroups of the series can be defined. Theoretically, these subgroups can be interpreted as samples representing the component frequency distributions of a composite, mixed distribution of floods.

The technique presented here defined mixed distributions in peak flow data for thirty gaging stations in the climatically sensitive Gila River basin for the period 1950-1980. The flood events were hydroclimatically classified into eight main categories using surface and upper air weather maps and precipitation data. The analysis demonstrated that floods in the basin originate from a variety of atmospheric processes and suggested that the parameters of the flood frequency distributions of different hydroclimatic subgroups may have a physical basis.

This study is believed to be the first to attempt to define the mixed components of a distribution by classifying each event in a flood series according to the synoptic weather patterns responsible for the event. Refinement of the technique and larger sample sizes could result in improved parameter estimates of the component distributions, as well as a better understanding of the physical basis for rare flood events.

REFERENCES

Anderson, D.V. 1967. 'Review of basic statistical concepts in hydrology,' in: Statistical Methods in Hydrology, proceedings of Hydrology Symposium No. 5, held at McGill Univ., Montreal, Canada, 24-25 Feb, 1966. National Research Council of Canada and Dept. of Energy and Resources, Inland Waters Branch, pp. 3-35.

Browzin, B.S., Baumbusch, C.A. and Pavlides, M.G. 1973. 'Significance of the genesis of floods on probability analysis,' in: Schulz, E.F., Koelzer, V.A., and Mahmood, Khalid, (eds.), Floods and Droughts, proceedings of the Second International Symposium in Hydrology, September 11-13, 1972, Fort Collins. Water Resources Publications, pp. 450-461.

Elliott, J.G., Jarrett, R.D., and Ebling, J.L. 1982. 'Annual snowmelt and rainfall peak-flow data on selected foothills region streams, South Platte River, Arkansas River, and Colorado River Basins, Colorado.' U.S. Geological Survey Open File Report 82-426, 88 p.

Guillot, Pierre. 1973. 'Application of the method of gradex,' in: Schulz, E.F., Koelzer, V.A., and Mahmood, Khalid, (eds.), Floods and Droughts, proceedings of the Second International Symposium in Hydrology, September 11-13, 1972, Fort Collins. Water Resources Publications, pp. 44-49.

Hansen, E.M. and Schwarz, F.K. 1981. 'Meteorology of important rainstorms in the Colorado River and Great Basin drainages.' Hydrometeorological Report No. 50, NOAA, U.S. Dept. of Commerce, 167 p.

Hawkins, R.H. 1974. 'A note on mixed distributions in hydrology,' in: Proceedings of a Symposium on Statistical Hydrology. U.S. Dept. of Agriculture, Agric. Res. Service, Misc. Publ. No. 1275, pp. 336-344.

Hirschboeck, K.K. 1985. Hydroclimatology of flow events in the Gila River basin, central and southern Arizona. Ph.D. dissertation, unpublished, Dept. of Geosciences, University of Arizona, Tucson, 335 pp.

Jarrett, R.D. and Costa, J. E. 1982. 'Multidisciplinary approach to the flood hydrology of foothills streams in Colorado,' in: International Symposium on Hydrometeorology, American Water Resources Association, pp. 565-569.

Keith, S.J. 1981. Stream channel recharge in the Tucson basin and its implications for ground-water management. M.S. thesis, unpublished, Dept. of Hydrology and Water Resources Administration, University of Arizona, Tucson.

Potter, W.D. 1958. 'Upper and lower frequency curves for peak rates of runoff.' Transactions of the American Geophysical Union, Vol. **39**, pp. 100-105.

Singh, K.P. 1968. 'Hydrologic distributions resulting from mixed populations and their computer simulation,' in: The use of analog and digital computers in hydrology, Vol. **2**. Symposium organized by the International Assoc. of Scientific Hydrologists, Tucson, Ariz., 1968, Publ. No. 81, pp. 671-681.

_________. 1974. 'A two-distribution method for fitting mixed distributions in hydrology,' in: Proceedings of a Symposium on Statistical Hydrology. U.S. Dept. of Agriculture, Agric. Res. Service, Misc. Publ. No. 1275, pp. 371-382.

U.S. Water Resources Council. 1981. Guidelines for determining flood flow frequency. Bulletin #17B, U.S. Water Resources Council, Washington D.C.

Waylen, Peter and Woo, Ming-ko. 1982. 'Prediction of annual floods generated by mixed processes.' Water Resources Research, Vol. **18**, pp. 1283-1286.

MIXED FLOOD DISTRIBUTIONS IN WISCONSIN

Tim Diehl
Research Assistant
Department of Civil and Environmental Engineering
University of Wisconsin

Kenneth W. Potter
Associate Professor
Department of Civil and Environmental Engineering
University of Wisconsin

ABSTRACT

Traditional flood frequency analysis is predicated on the assumption that the annual flood series can be considered to be a sample from a single population. In Wisconsin this is not a valid assumption. Wisconsin floods are of two types, which are hydrologically and statistically distinct.

An analysis of the time of occurrence of Wisconsin floods indicates two dominant flood seasons: spring and summer. Late May is the boundary between these two seasons.

Using May 20 as the date of separation, we constructed parallel spring and summer flood series for all Wisconsin partial flood series with ten or more years of record (29 series). (A seasonal series is the series of largest spring floods from each year of record.) Based on the two-sample Kolmogorov-Smirnov test the hypothesis that spring and summer floods are identically distributed was rejected for 18 of the 29 pairs of flood series at the 5 percent significance level and for 23 pairs of flood series at the 10 percent significance level. We interpret this as convincing evidence that spring and summer floods in Wisconsin generally have distinct statistical distributions.

An analysis of runoff data for eight of the 29 partial series gages demonstrates that spring and summer floods in Wisconsin are also hydrologically distinct. Spring floods show relatively high runoff-precipitation ratios, presumably due to the presence of frozen ground and/or snowmelt. Summer floods show relatively low runoff-precipitation ratios.

Future work is needed, both to better understand the hydrology of spring and summer floods, and to determine if and how this understanding can improve quantile estimation.

INTRODUCTION

Traditional flood frequency analysis is predicated on the

V. P. Singh (ed.), Hydrologic Frequency Modeling, 213–226.

assumption that the annual flood series can be considered to be a sample from a single population. Several recent studies have convincingly documented that this assumption is not always valid, that in some regions the annual flood series is a sample of a mixture of two or more populations. Following a review of previous work, this paper demonstrates that Wisconsin is also such a region.

PREVIOUS WORK

A number of investigators, including Potter (1958), Singh and Sinclair (1972), Singh (1974), Hawkins (1974), Canfield (1978), and Rossi, Fiorentino and Versace (1984) have treated floods as coming from a mixed distribution. For the most part these investigators approached the problem from a statistical perspective. Other investigators have focused on the physical processes contributing to mixed flood distributions.

Waylen and Woo (1982, 1983) found that floods in British Columbia could be divided into rain-generated and snowmelt-generated events. Snowfall and intense rain are limited to winter, while snowmelt occurs in spring. This seasonal separation of processes allowed them to separate the flood data into two categories based on the time of occurrence. For each data set they then constructed parallel seasonal series of rain floods and snowmelt floods, fit the extreme value type one distribution to each series, and used the product of the two fitted cumulative distributions as an estimate of the cumulative distribution function of the annual flood series.

Jarrett and Costa (1982) analyzed floods in the Colorado Rockies and foothills. Like Waylen and Woo they distinguished between rain-generated and snowmelt-generated floods; however, they had more difficulty in classifying the data. While snowmelt floods in their study area generally occur earlier in the year than rain floods, the periods of activity of the two processes overlap at many gages. Hence it was necessary to analyze the hydrograph of each flood in order to identify its generating process. Once they had classified the floods at each gage, they constructed parallel seasonal series of rain floods and snowmelt floods. They then fit the log Pearson type 3 distribution to each series, and used the product of the fitted cumulative distribution functions to estimate the annual flood distribution.

Hirschboeck (1985) analyzed floods and their associated weather patterns in New Mexico. She distinguished six major meteorological processes capable of producing floods, and was able to separate gaged basins into regional groups based on their characteristic response to the various types of meteorological processes.

MIXTURES IN WISCONSIN

We have observed that floods in Wisconsin are of two distinct types. The separation can be based on time of occurrence, leading to flood types which differ both hydrologically and statistically.

Wisconsin Floods Fall into Two Seasonal Groups

In analyzing the seasonality of Wisconsin floods, we considered two sets of data. One set includes the maximum observed discharge from each of 42 gages selected for record length of 30 or more years and minimal influence by regulation or urbanization. The second set includes all of the 29 available partial series with ten or more years of record.

Figure 1 is a plot of the thirty day moving average of the number of maximum floods observed through the calendar year. It shows that a little less than half of these floods occurred in a short flood season in March and April. The remainder occurred in a longer flood season from June through September.

Figure 2 is a plot of the fifteen day moving average of the number of partial floods observed through the calendar year. Most of these floods occurred in the spring season. The break between the spring and summer seasons is less distinct than in the case of the maximum floods, but it is clearly present. Based on this plot we chose May 20 as the date separating the two flood seasons.

Spring and Summer Floods Have Different Distributions

For the 29 gages with partial duration series we constructed parallel spring and summer flood series. In each year of record the largest partial flood in the period from January 1 through May 20 was designated the spring flood; the largest partial flood in the period from May 21 through December 31 was designated the summer flood. When no partial flood existed in a given period (that is, when no flood exceeded the threshold), the value of the seasonal flood for that period of that year was set equal to zero.

For each of the 29 gages we constructed lognormal probability plots of the parallel seasonal series. In general the plots show distinct differences in distribution between the parallel series; these differences vary from gage to gage. Figures 3, 4, 5, and 6 are typical examples. The most common pattern is one in which most of the partial floods occurred in spring, even though very large floods occur in about equal numbers in the two seasons.

To verify that the apparent differences in distribution are statistically significant, we applied the two-sample Kolgomorov-Smirnov test (Benjamin and Cornell, 1970). Truncation of the flood series at the partial duration series threshold makes the test conservative; the addition of information about sub-threshold floods could only increase the significance level of a given test.

Application of the Kolmogorov-Smirnov test resulted in rejection of the null hypothesis (that spring and summer floods are identically distributed) at 18 of 29 gages at the 5 percent significance level, and at 23 of 29 gages at the 10 percent significance level.

From a statistical standpoint the existence of a mixed distribution is adequately demonstrated if an objective criterion such as time of year can be used to separate floods into two groups with different statistical distributions. This would be unsatisfactory from a hydrological point of view, however, if it were not also true that the two groups are hydrologically distinct.

Spring and Summer Floods are Hydrologically Distinct

Data given in Dalrymple (1965) can be used to demonstrate that spring and summer floods in Wisconsin are hydrologically distinct. Dalrymple compiled hydrologic data for annual floods at 797 selected gages in the United States for the period of record through 1960. Depth of runoff was calculated by hydrograph separation. Depth of associated precipitation was based on a Thiessen weighting of measurements at nearby rain gages.

Dalrymple analyzed eight Wisconsin gages, with a total of 142 annual floods. At each of these gages floods before May 20 show relatively high ratios of runoff to associated precipitation. (Figure 7 illustrates a typical example.) Many of these floods have runoff-precipitation ratios greater than one, indicating contributions from snowmelt. In some cases there is no associated precipitation.

For each of these eight gages a threshold value for the runoff-precipitation ratio can be chosen that is less than the runoff-precipitation ratio of nearly all the spring floods, and greater than the runoff-precipitation ratio of nearly all the summer floods. Of the 142 floods for which data are given by Dalrymple, only four spring floods have runoff-precipitation ratios less than the threshold value; only one summer flood has a runoff-precipitation ratio greater than the threshold.

This difference in the runoff-precipitation ratio of spring and summer floods is clearly linked to soil-climate relationships. In the spring flood season the soil is likely to be either frozen, or wet due to snowmelt (Figures 8, 9). In either case, infiltration capacities tend to be low. However, rainfalls in the spring are unlikely to be very intense. Large spring floods occur when moderate rainfalls occur in conjunction with snowmelt or frozen ground, hence the high runoff-precipitation ratios.

By late May snow and frost have disappeared throughout Wisconsin, and infiltration capacities begin to rise as the ground dries. At the same time, the potential for very intense rainfall increases. Most summer floods are caused by intense rain of which less than half runs off.

CONCLUSIONS AND FUTURE WORK

Floods in Wisconsin are of two types, which are hydrologically and statistically distinct. Spring floods, which occur before the end of May, are characterized by high runoff-precipitation ratios. The critical factor in their occurrence is low soil infiltration capacity due to frozen ground and/or snowmelt. Summer floods, which occur after late May, are characterized by low runoff-precipitation ratios. Critical to summer flooding is the occurrence of very intense rainfalls. For many gaged streams in Wisconsin, spring and summer have statistically distinct distributions.

The question remains as to whether quantile estimation will benefit from the recognition that annual floods in Wisconsin come from mixed distributions. (The analogous question has not been answered in any region where mixtures have been identified.) Monte Carlo studies are needed to demonstrate if and when it is statistically advantageous to use quantile estimation procedures which exploit the presence of mixtures.

Questions also remain about the hydrology of Wisconsin floods. For example, is it possible to identify physical factors which explain the relative importance of spring and summer floods? In some basins the largest annual floods occur in summer, even though most of the annual floods occur in spring. Might this be a function of the soils in the basin? In basins with permeable soils, most annual floods will be associated with reduced infiltration capacities due to frozen ground and/or snowmelt. However, very large floods might be possible

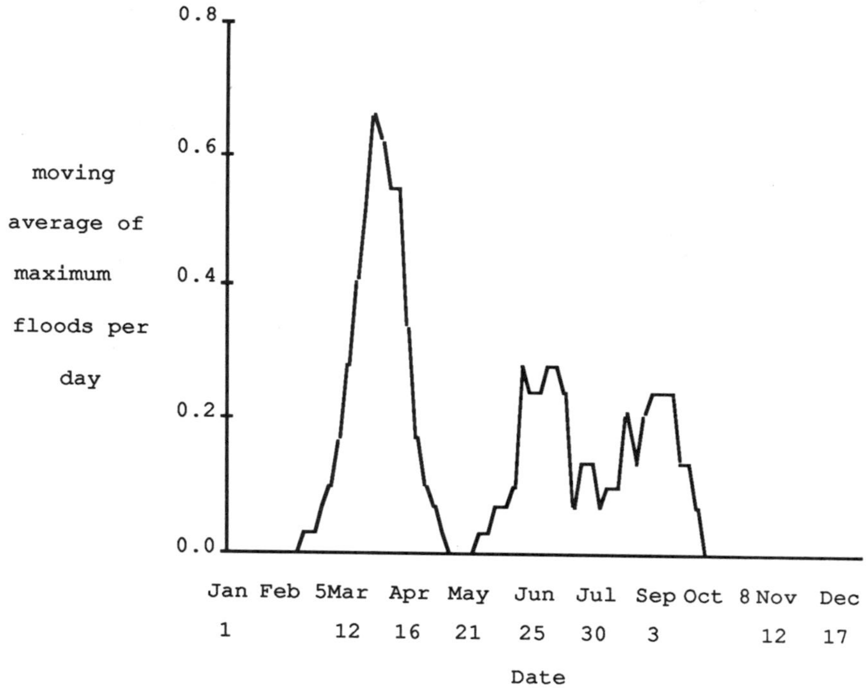

Figure 1: Thirty-day moving average of the occurence of maximum observed floods at 29 Wisconsin gages.

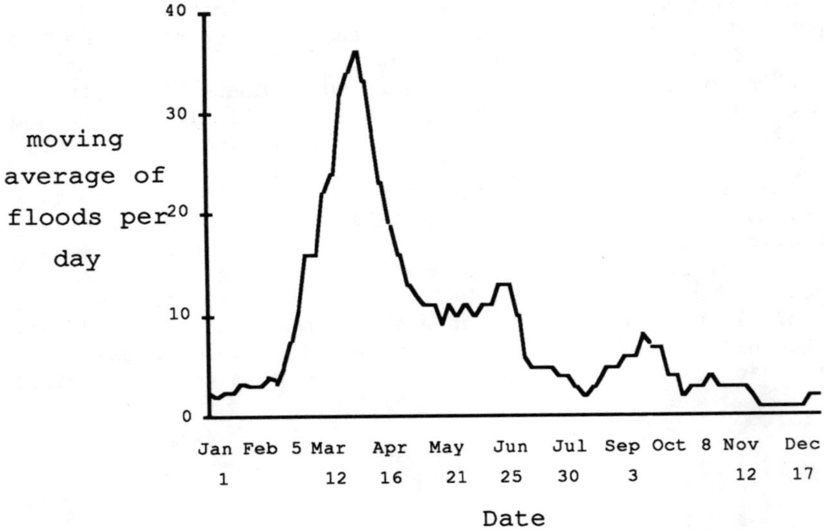

Figure 2: Fifteen-day moving average of the occurence of partial duration series floods at 29 Wisconsin gages..

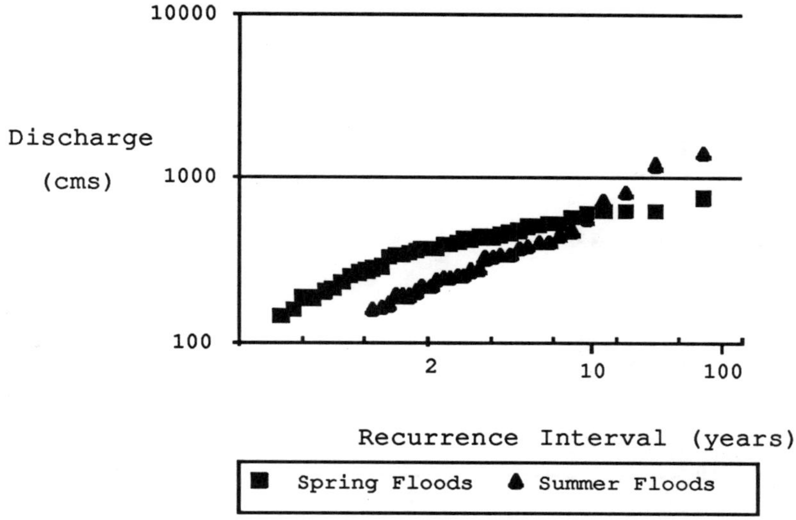

Figure 3: Parallel seasonal flood series for the Black River at Niellsville (gage no. 538100).

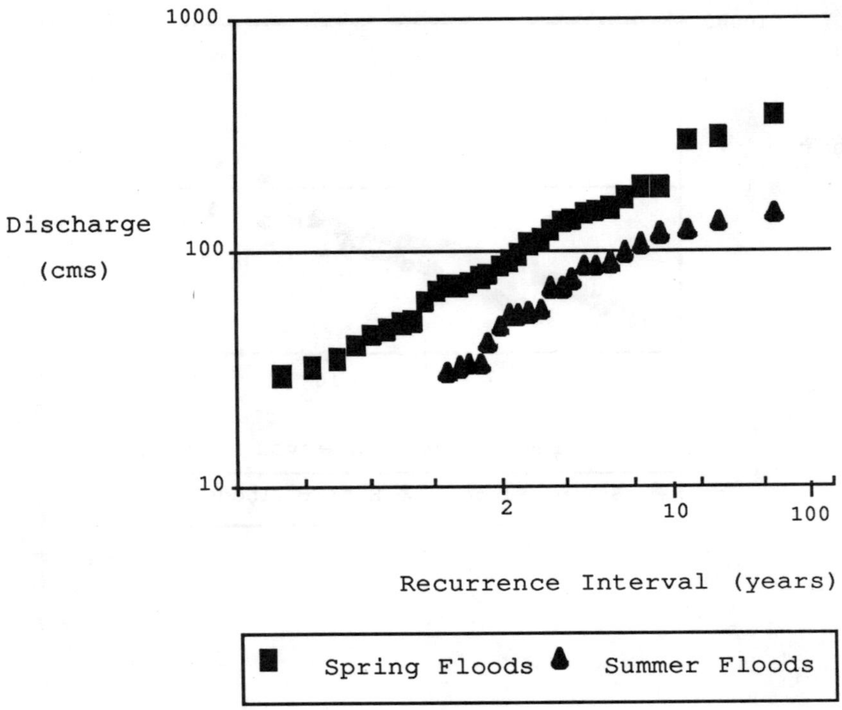

Figure 4: Parallel seasonal flood series for the Hay River at Wheeler (gage no. 536800).

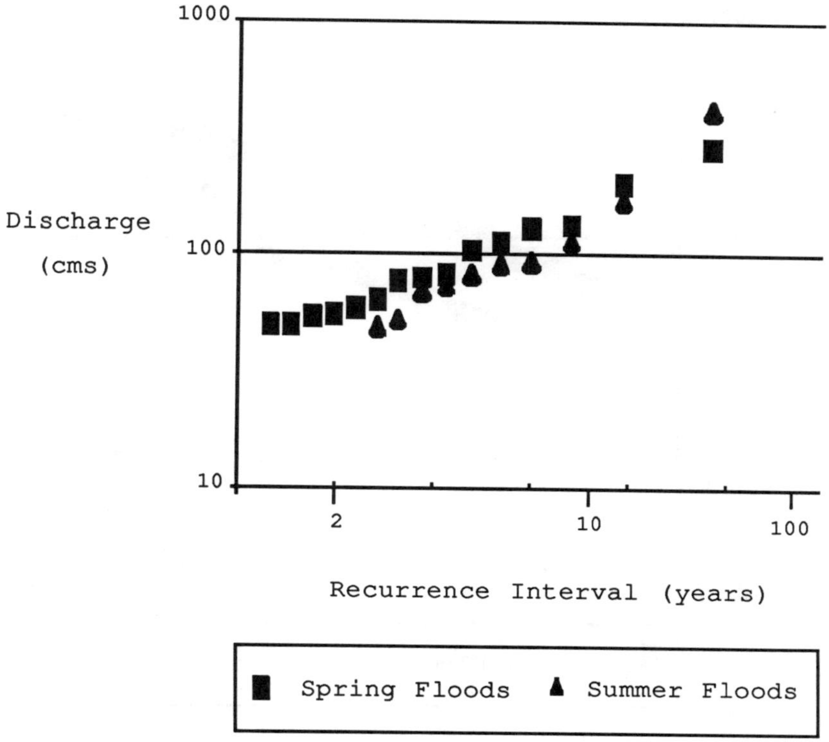

Figure 5: Parallel seasonal flood series for the Kickapoo River at La Farge (gage no. 540800).

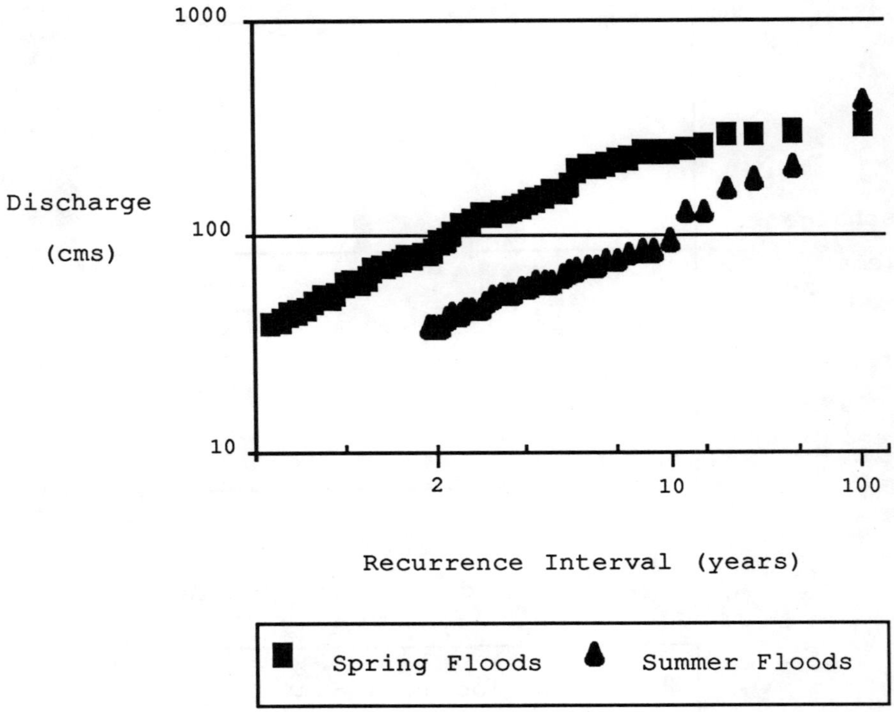

Figure 6: Parallel seasonal flood series for the Sugar River near Brodhead
(gage no. 543650)

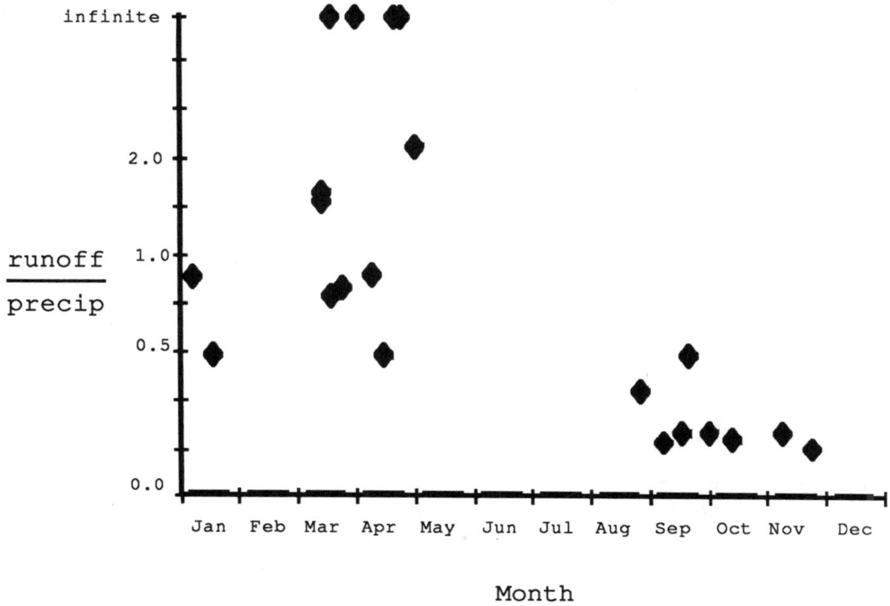

Figure 7: Ratio of runoff depth to depth of associated precipitation for floods on the Galena River near Buncombe, plotted against date of occurence. Data are from Dalrymple (1965). (Note: vertical scale was transformed by taking arctangent.)

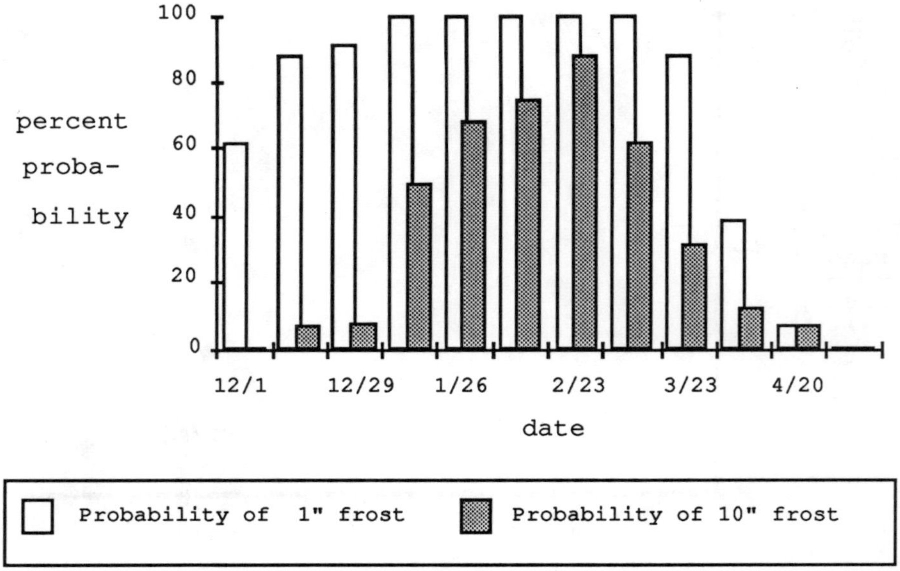

Figure 8: Biweekly probability of one inch and ten inches of frozen ground in south central Wisconsin. Data are from Wisconsin Agricultural Reporting Service (1979).

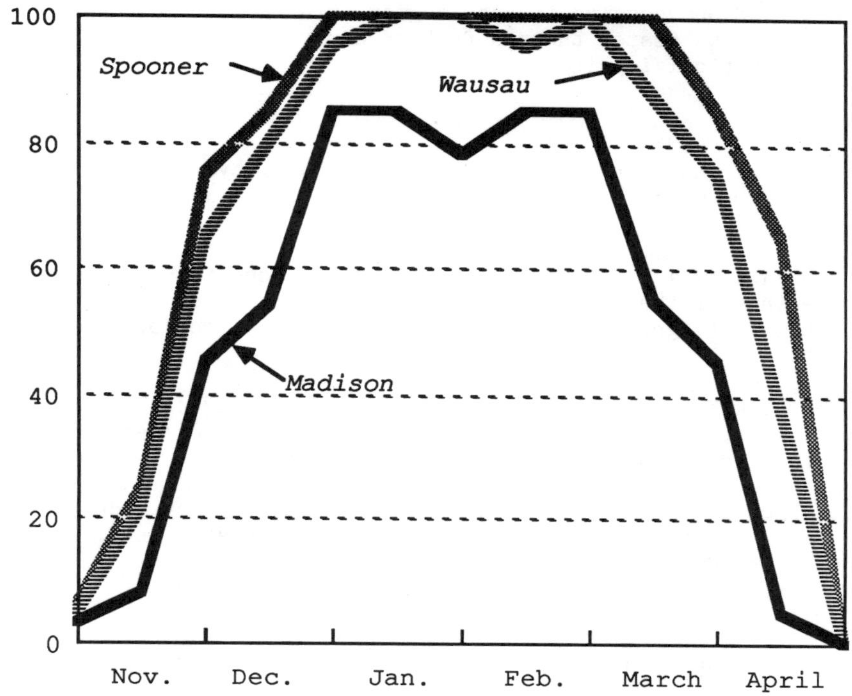

Figure 9: Percentage frequncy of snow on the ground at Madison, Wausau, and Spooner, Wisconsin, 1961-1977. From Wisconsin Agricultural Reporting Service (1978).

as a result of succcessive intense summer rainstorms. Such an hypothesis can be tested through further analysis, and may have important practical implications.

REFERENCES

Benjamin, J.R. and Cornell, C.A. 1970. Probability, Statistics, and Decision for Civil Engineers, McGraw-Hill, Inc., New York, pp. 509-510.

Hawkins, R.A. 1974. A note on mixed distributions in hydrology. Proceedings of the Symposium on Statistical Hydrology, U.S. Department of Agriculture Miscellaneous Publication No. 1275, pp. 336-349.

Hirschboeck, K.K. 1985. Hydroclimatology of flow events in the Gila River basin, central and southern Arizona. Ph.D. Dissertation, Department of Geosciences, University of Arizona.

Jarrett, R.D. and Costa, J.E. 1982. Multi-disciplinary approach to the flood hydrology of foothill streams in Colorado, in A. Ivan Johnson and Robert A. Clark (Editors), pp. 565-569, International Symposium on Hydrometeorology, American Water Resources Association, Bethesda, Maryland.

Potter, W.D. 1958. Upper and lower frequency curves for peak rates of runoff. Transactions, American Geophysical Union, Vol. 39, pp. 100-105.

Rossi, F., Fiorentino, M. and Versace, P. 1984. Two-component extreme value distribution for flood frequency analysis. Water Resources Research, Vol. 20, pp. 847-856.

Singh, K.P. 1974. Two-distribution method for fitting mixed distributions in hydrology. Proceedings of the Symposium on Statistical Hydrology, U.S. Department of Agriculture Miscellaneous Publication No. 1275, pp. 371-382.

Singh, K.P. and Sinclair, R.A. 1972. Two-distribution method for flood frequency analysis. Journal of the Hydraulics Division, American Society of Civil Engineera, Vol. 98, pp. 29-44.

Waylen, P. and Woo, M.K. 1982. Prediction of annual floods generated by mixed processes. Water Resources Research, Vol. 18, pp. 1283-1286.

Waylen, P. and Woo, M.K. 1983. Annual floods in southwestern British Columbia, Canada. Journal of Hydrology, Vol. 62, pp. 95-105.

Wisconsin Agricultural Reporting Service. 1978. Snow and Frost in Wisconsin 1961-1977. Madison, Wisconsin.

FLOOD DATA, UNDERLYING DISTRIBUTION, ANALYSIS, AND REFINEMENT

Krishan P. Singh
Illinois State Water Survey
Champaign, Illinois 61820
U.S.A.

ABSTRACT. Annual flood peak data must conform to certain standards of quality control, sample size, and homogeneity for use in flood-frequency analysis. Sometimes the values for the highest observed floods of an annual flood series are much higher or much lower than expected; these values are designated as outliers and inliers, respectively. When the values for the lowest observed floods are much higher or lower than expected, these are designated inliers and outliers, respectively. Analyses of storms producing high floods perceived as outliers and droughts containing low floods perceived as outliers provide a physical basis for their being outliers. However, these outliers and inliers need to be detected and properly modified to derive unbiased design flood estimates. Test statistics have been developed for objective detection of any outliers/inliers in a given flood series after converting it to a normally distributed series with the power transformation. A design flood estimation methodology is presented.

INTRODUCTION

The annual flood series available at a gaging station is used for developing estimates of various frequency floods with an underlying distribution such as Gumbel, lognormal, or log-Pearson type III. The distribution parameters are fitted using the methods of moments, maximum likelihood, or optimization. The annual flood peak data must conform to certain standards of quality control, sample size, and homogeneity for use in flood-frequency analyses. A continuous annual flood record is desirable. Most of the high flood peaks are estimated from extended or extrapolated stage-discharge or rating curves. These extrapolations need to be critically reviewed and validated by simulation of flood flows with physical information on the stream section and the floodplain. Usually a minimum of 25 years of record is needed but a larger record becomes necessary if the drainage area comprises 2 or more distinct regions with different hydrological and/or meteorological regimes. A flood series used in flood frequency

V. P. Singh (ed.), Hydrologic Frequency Modeling, 227–241.

analysis belongs to a population of floods with characteristics practically invariant with time. Any change in conditions will require adjustment of the previous floods to the new condition. There are methods available for considering snowmelt and rainfall floods as two components of a flood series or as two component series constituting the flood series.

Graphical and/or analytical methods are used in selecting an underlying distribution that best fits the observed annual flood series. The graphical methods rely on plotting the observed annual floods on various probability papers (normal, lognormal. extreme-value, etc.) to determine which of the conventional distributions yields a best fit to the plotted flood values. The analytical methods rely on computing statistical parameters of the flood series: mean, standard deviation, and skew coefficient. The interrelationships between these parameters define the feasibility regions for various distributions. Sometimes, two or more candidate distributions are fitted to the flood series and either their fit with the observed floods or a numerical measure of deviation of the observed floods from the fitted distribution curve is used in selecting the underlying distribution.

The most commonly used distributions in flood-frequency analyses are the Pearson and log-Pearson, normal and lognormal, and Gumbel (extreme value) and log-Gumbel. There are standard methods available for evaluating the parameters of these distributions. None of these distributions, however, fit an observed flood series with reverse curvature. The reverse curvature can occur because of great variation in channel and floodplain geometry, seasonal characteristics of flood-producing storms, and rainfall and snowmelt floods. A versatile mixed distribution transform has been developed (Singh, 1983).

Sometimes the values for the highest observed floods of an annual flood series are much higher or lower than expected; these values are designated as outliers and inliers, respectively. When the values for the lowest observed floods are much higher or lower than expected, these are designated inliers and outliers, respectively. These outliers and inliers need to be detected and properly treated to derive unbiased flood estimates.

FLOOD MEASUREMENT ERROR

In flood frequency analysis it is generally assumed that peak flood discharges are measured without significant error. The U.S. Geological Survey (USGS) maintains gaging stations or various creeks, streams, and rivers in the United States. River stage is continuously recorded at a gaging station. The stage is converted to discharge with a rating curve (or stage versus discharge curve) developed from 10 to 20 flow measurements made each year and kept current to minimize errors. It is the large flood discharges, which exceed the upper limits of the established rating curve, that are susceptible to significant errors because of the extrapolation involved. Potter and Walker (1985) define gaging ratio as the ratio of the largest directly measured discharge to the largest observed discharge. Low values of gaging ratio, found mostly for small drainage areas, cause higher values of standard deviation and skew for some flood series with high outliers which may be caused by errors in extrapolation. Characteristics of the storm causing such an outlier can be analyzed to determine if the perceived outlier is a genuine outlier or is caused by extrapolation. Knowledge of the gaging ratio can be used in assigning weights to flood estimates from various gaging stations to develop regional flood frequency curves.

FLOW SECTION CHARACTERISTICS

Bankfull discharge in a river is considered to correspond to about a 2-year or median flood. At higher flows, the water gradually covers the floodplain; the depth and width of this flooding depend on the floodplain geometry. The interaction between the fast-moving channel flow and the slow-moving floodplain flow at small depths of flooding leads to the transport of the longitudinal momentum from the main channel to the floodplain (Rajaratnam and Ahmadi, 1979). This reduces the flow velocity in the channel as well as the mean velocity for the composite section in which the discharge increases, with a disproportionate increase in the width and area of the flow section. This is equivalent to the introduction of an additional storage element into the system.

The curves of mean velocity V, water width W, and area of flow section A_f with respect to the discharge Q observed in the Salt Creek, Greenview, Illinois, are drawn in Figure 1 from flow measurement data spread over a period of 35 years. Maximum flow of record, 41,200 cfs (1167 cms) occurred on May 19, 1943. Maximum velocity of about 3.2 ft/sec (0.98 m/sec) occurs at a bankfull discharge of 7500 cfs (212 cms) which corresponds to a recurrence interval of 1.4 years. The mean velocity declines to a minimum of about 2 ft/sec (0.61 m/sec) at a discharge of 10,500 cfs (297 cms) with a recurrence interval of 1.8 years. A_{f_o} denotes extra flow section added by the floodplain. A plot of observed annual floods on lognormal probability paper indicates a change in curvature in the range of p = 0.3 to 0.6, where p is the probability of being equal to or less than a given flood. This change in curvature may be attributed to a change from a predominantly channel flow to a predominantly floodplain flow. The relative magnitude of floodplain and channel flow sections during the transition from mainly channel to mainly floodplain flow can affect the observed flood distribution shape.

PLOTTING POSITION

The plotting position for observed floods has been a matter of considerable controversy. A general formula for computing the plotting position (Harter, 1971) is:

$$p = (m-a)/(n-a-b+1) \tag{1}$$

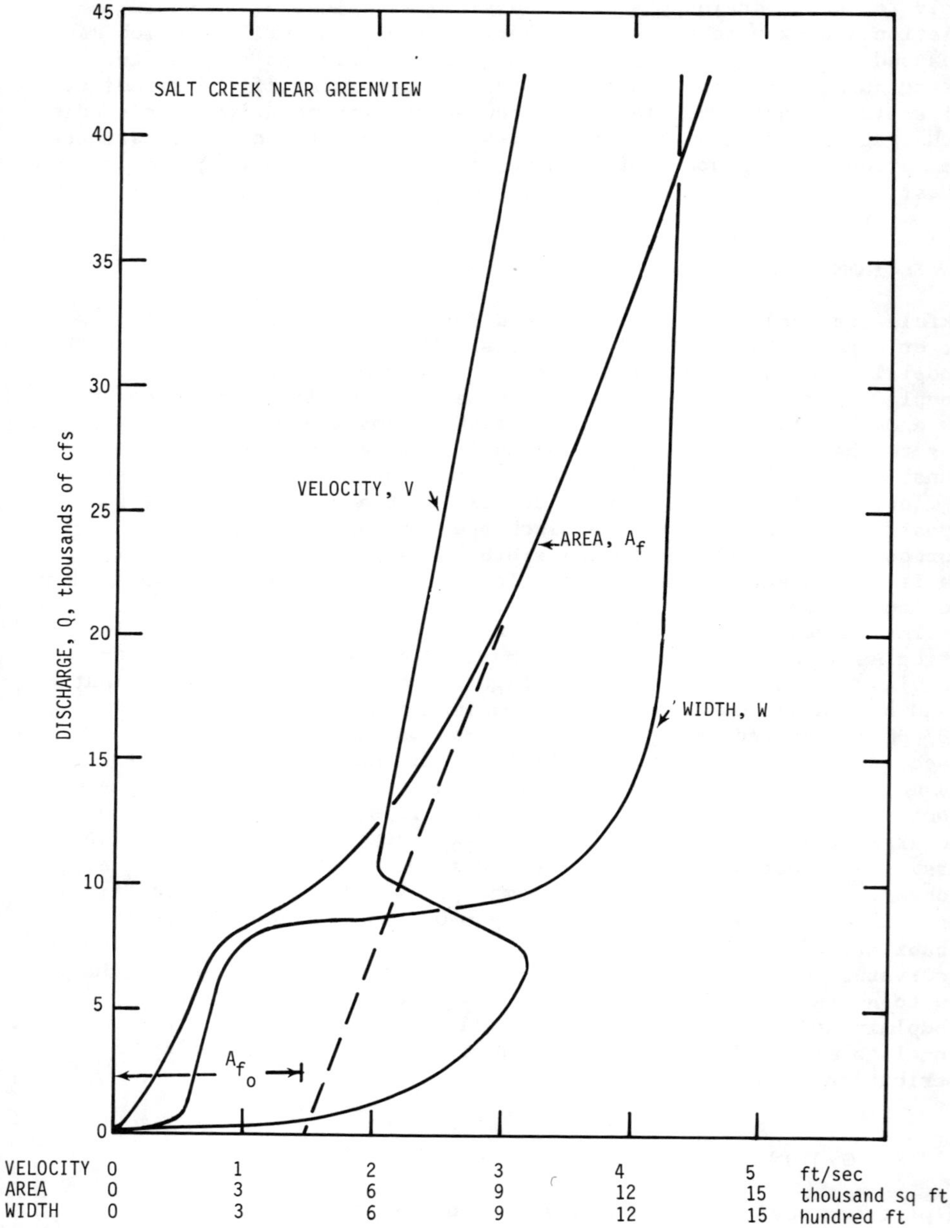

Figure 1. Discharge, velocity, area of flow section, and width relationships for the Salt Creek near Greenview, Illinois (1 ft = 0.3048m; 1 sq ft = 0.0929 sq m: 1 cfs = 0.02832 cms)

in which m is the rank order of flood values arranged in an ascending order of magnitude, p is the cumulative probability of non-exceedance, and a and b depend on the distribution. For a symmetrical distribution, $\alpha = a = b$, and therefore:

$$p = (m-\alpha)/(n-2\alpha+1) \tag{2}$$

The commonly used Weibull plotting position is obtained by putting α equal to zero. However, an α value of 0.38 (Cunnane, 1978; Blom, 1958) is the best for normal distribution. Cunnane states that the Weibull plotting formula is exact when the distribution is uniform and that the Gringorten formula with $\alpha = 0.44$ is satisfactory for exponential distribution. Use of $\alpha = 0.38$ assigns a recurrence interval of 63.3 years instead of 40 years with $\alpha = 0$ for the highest flood in a 39-year series. The difference between the recurrence intervals decreases for the lower flood; for example, the fifth highest flood will have a recurrence interval of 8.5 years instead of 8 years.

FLOOD-PRODUCING STORMS

Information on the duration and magnitude of a flood-producing storm and on the antecedent soil moisture condition can be used to check on any perceived high outlier in a flood series to be analyzed. The flood peak is mainly a function of storm precipitation and runoff factor for similar duration storms. The runoff factor varies with storm magnitude, antecedent soil moisture condition, AMC, and season (Singh, 1982). Day, month, and year of a high flood observed at a gaging station indicate the approximate time of the storm event; the storm usually precedes the flood peak by less than an hour to several hours for small- to medium-size basins and by several hours to a few days for medium- to large-size drainage areas. Hourly and/or daily catchment precipitations are developed from the data at the relevant raingage stations. Daily precipitations for the 3-week period preceding the storm are obtained for assessing the soil moisture conditions. The catchment precipitation is converted to the point estimate with the area-depth curves (Hershfield, 1961). The point estimate and storm duration are used in obtaining the storm frequency from tables containing depth of precipitation for various recurrence intervals and duration. Two examples of the use of storm analyses in investigating perceived outliers are given below. The ratio Q/P is defined as the ratio of flood peak in inches per hour (or mm per hour) to the average storm rainfall intensity P in inches per hour (or mm per hour).

1. Sangamon River near Oakford, Illinois; 5093 sq mi (13,191 sq km); record = 1910 - 1983

5 highest flood peaks:	123,000	68,700	55,900	46,300	45,800	cfs
	3,483	1,946	1,583	1,311	1,297	cms

Perceived high outlier: 123,000 cfs or 3,483 cms

Ratio Q/P: 0.47 for max flood; 0.21 average for other 4 floods
Storms: 50-yr (max flood); 10- to 50-yr (other 4 floods)
AMC: very wet (max flood); average (others)
Very wet antecedent condition and extensive areal coverage by the storm caused an extremely high flood (1.79 times the next highest flood), perceived as an outlier.

2. Skillet Fork near Wayne City, Illinois; 464 sq mi (1202 sq km); record = 1929 - 1983

5 highest flood peaks:	51,000	22,800	20,000	18,500	18,000	cfs
	1,444	646	566	524	510	cms

Perceived high outlier: 51,000 cfs or 1,444 cms
Ratio Q/P: 0.64 average for the 5 floods
Storms: 300- to 500-yr (max flood); 2- to 15-yr (other 4 floods)
AMC: wet (max flood); wet, frozen ground (other 4 floods)
A 300- to 500-yr storm caused the maximum flood (2.24 times the next highest flood), perceived as an outlier. It also shows that the other 4 floods would have been higher with 10- to 30-yr storms.

FLOODS DURING DROUGHTS

Existence of a low outlier(s) in a flood series can significantly increase the standard deviation and greatly distort the skew values. The skew parameter in probability distributions such as log-Pearson type III is greatly affected by the presence of low outliers. The intent of flood frequency analysis is to develop satisfactory estimates of floods expected once in 10, 25, 50, 100 or more years. Floods below the 2-year flood are of practically no interest except in delineation of the flood distribution over the entire range of the cumulative probabilities. The U.S. Water Resources Council (1981) recommends deleting of a low outlier if the following criteria are met:

$$Q_1 \leqq Q_L \quad \text{and} \quad \log Q_L = \overline{x} - k_n S \tag{3}$$

in which Q_1 is the lowest observed flood in the series; Q_L denotes the low outlier threshold at the 0.1 significance level; k_n is the corresponding frequency factor for a sample size of n; and $\overline{x}$ and S are the mean and standard deviation of the log-transformed floods.

The low outliers usually occur during extended drought periods if precipitation from a large storm does not occur during such a period. This is exemplified from flood records at two gaging stations in central Illinois where a prolonged drought occurred during 1953-1955 with the worst dry period in 1954. This drought is considered to be a 1 in 80 years drought.

1. Lake Fork near Cornland, Illinois; 214 sq mi (554 sq km); record = 1948 - 1983

5 lowest flood peaks:	152	548	680	786	1000	cfs
	4.3	15.5	19.3	22.3	28.3	cms

Perceived low outlier: 152 cfs or 4.3 cms

Annual flow: mean = 154 Max = 336 min = 8.9 cfs
= 4.4 = 9.5 = 0.25 cms
The minimum annual flow occurred in 1954, the year in which the lowest flood (perceived as an outlier) occurred.

2. North Fork Embarras River near Oblong, Illinois; 318 sq mi (824 sq km); record = 1924 - 1983
5 lowest flood peaks: 228 1110 1880 1900 2520 cfs
6.5 31.4 53.2 53.8 71.4 cms
Perceived low outlier: 228 cfs or 6.5 cms
Annual flow: mean = 256 max = 644 min = 3.9 cfs
= 7.2 = 18.2 = 0.11 cms
The minimum annual flow occurred in 1954, the year in which the lowest flood (perceived as an outlier) occurred.

OUTLIERS AND INLIERS

An outlier in a set of data is defined as an observation or subset of observations which appears to be inconsistent with the remainder of that set of data (Barnett and Lewis, 1978). The inconsistency can be interpreted as the observation being either significantly higher or lower at the high end than the values indicated by the rest of the data, in which case the observation is termed an outlier or an inlier, respectively. If the observation is significantly higher or lower than expected at the low end, it will be termed an inlier or outlier, respectively. In conventional flood frequency analyses, it has been a matter of subjective judgment on the part of the analyst whether or not he picks up some observation for scrutiny. A literature search did not show the existence of statistical tests for checking outliers at higher than the 10% significance level or inliers at any level. Because the outliers and inliers, if present in a flood series, seriously distort the distribution parameters, the development of suitable statistical tests for their detection is necessary to improve the reliability of flood frequency analyses.

Test Statistics

Prerequisites for deriving satisfactory estimates of design floods are the objective detection and modification of any outliers/inliers at desired levels of significance and a versatile technique of flood frequency analysis. Test statistics for objective detection of outliers/inliers in a given flood series were developed from extensive Monte Carlo experiments in terms of departures by generating millions of standard normal deviates. The departure is defined as the theoretical standard normal deviate corresponding to a given p minus the generated standard normal deviate at the desired significance or probability level. The best values of α for the 5 highest and lowest ranks for sample size n varying from 10 to 100 are given on the next page.

n	α values for the highests and lowest ranks 1	2	3	4	5
10	0.425	0.474	0.492	0.506	0.511
20	0.408	0.455	0.478	0.491	0.501
30	0.404	0.443	0.467	0.481	0.491
40	0.403	0.440	0.459	0.473	0.482
50	0.403	0.440	0.454	0.467	0.475
60	0.403	0.440	0.451	0.462	0.469
75	0.403	0.440	0.450	0.458	0.468
100	0.403	0.440	0.450	0.456	0.460

Considering at the most three outliers/inliers at the high and/or low end of the ranked flood series, the α values are 0.40, 0.44, and 0.45 for n = 30 to 100. The distribution of departures for both the low and high ends is shown in Figure 2. Percent exceedance probability corresponds to percent significance level. For the same level of significance, the outliers have higher departures than the inliers. The outlier can depart considerably from the assumed underlying distribution, but the inlier departs by a lesser amount because the next observation can replace an inlier. For use of the developed departure values, a given flood series has to be transformed to a normally distributed series in order to detect any outliers/inliers.

METHODS OF TRANSFORMATION

Three methods of transforming an observed flood series to a series distributed as N (μ,σ^2), where μ is the mean and σ^2 is the variance, were tested extensively on 50 flood series in Illinois. These methods are: power transformation, Wilson-Hilferty transformation, and 3-parameter lognormal transformation. The power transformation was found to be the best of the tested transformations for converting an observed flood series so that it approximates a normally distributed series. Power transformed series were found to be more stable and consistent (Singh and Nakashima, 1981) even when some outliers/inliers were present.

Box and Cox (1964) suggested the power transformation for normalizing a data set:

$$y_i = (Q_i^{\lambda} - 1) / \lambda, \quad \lambda \neq 0 \tag{4}$$

and

$$y_i = \log Q_i, \ \lambda = 0 \tag{5}$$

in which Q_i is the annual flood from a sample of size n, λ is a constant of transformation, and i = 1,2,......,n. It is a general power transformation and the logarithmic, reciprocal, and square-root transformations can be considered as its special cases. The maximum

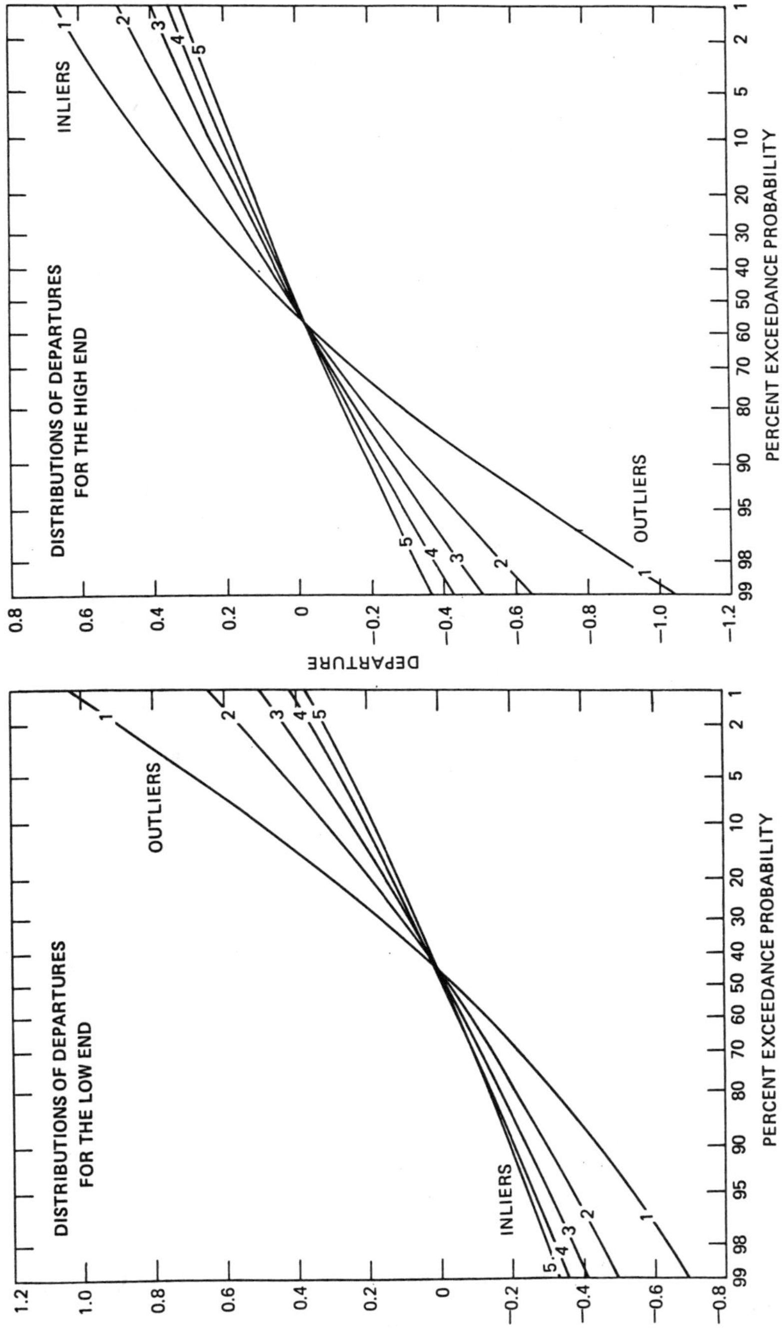

Figure 2. Distribution of departures for both the high and low end

log-likelihood estimator of λ, when $\lambda \neq 0$, can be obtained from Singh (1980):

$$L_{max}(\lambda) = -1/2\ n \log \hat{\sigma}_y^2 + \log J(\lambda;Q) \tag{6}$$

$$\log J(\lambda;Q) = (\lambda-1) \sum_{i=1}^{n} \log Q_i \tag{7}$$

A computer algorithm can easily be developed for determining $L_{max}(\lambda)$.

OUTLIER/INLIER DETECTION AND MODIFICATION

The detection and modification are based on the concept of levels and windows as shown in Figure 3 in which departures Δ for the first outlier/inlier at the high end are plotted on normal probability paper. The six levels 1,2,3,4,5, and 6 correspond to probability pairs (0.01, 0.99), (0.05, 0.95), (0.1, 0.9), (0.2, 0.8), (0.3, 0.7), and (0.4, 0.6). For the outliers, window 1 contains Δ values $\leqq -1.054$, window 2 contains Δ values such that $-1.054 < \Delta \leqq -0.683$, and so on for windows 3 through 6. The levels and windows for the first inlier/outlier at the low end are also shown in the figure.

The detection and modification procedure starts from window 1. If some outliers and/or inliers are found in window 1, they are modified to respective values at level 1, the resulting series is analyzed to test the departures, and the procedure is repeated, if necessary, to ensure that there are no outliers and/or inliers in window 1 or at the 0.01 significance level. The procedure is followed sequentially from one window to the other, with desired distribution statistics and floods computed before moving on to the next window. If no outliers and/or inliers are detected in a particular window, no modifications are done in that window.

UNDERLYING DISTRIBUTION

The primary purpose of flood-frequency analysis is to define a relation or distribution that conforms well to the data and represents an orderly variation of probability rather than the chance erratic variations usually found in a flood series (Singh and Sinclair, 1972). There is no physical rationale for assuming a single, underlying distribution for a flood series. Observed flood data do not always conform well to the conventional distributions in use, such as log-Pearson type III, lognormal, and log-Gumbel. Various hydrologists, engineers, and researchers have found that many observed flood series, when plotted on lognormal probability paper, exhibit reverse curvatures, doglegs, and S-curve shapes in addition to convex or concave curvature (Armentrout and Bissell, 1970; Potter, 1958).

There are no valid reasons to expect that an observed flood series would conform to a chosen statistical distribution. Variations in

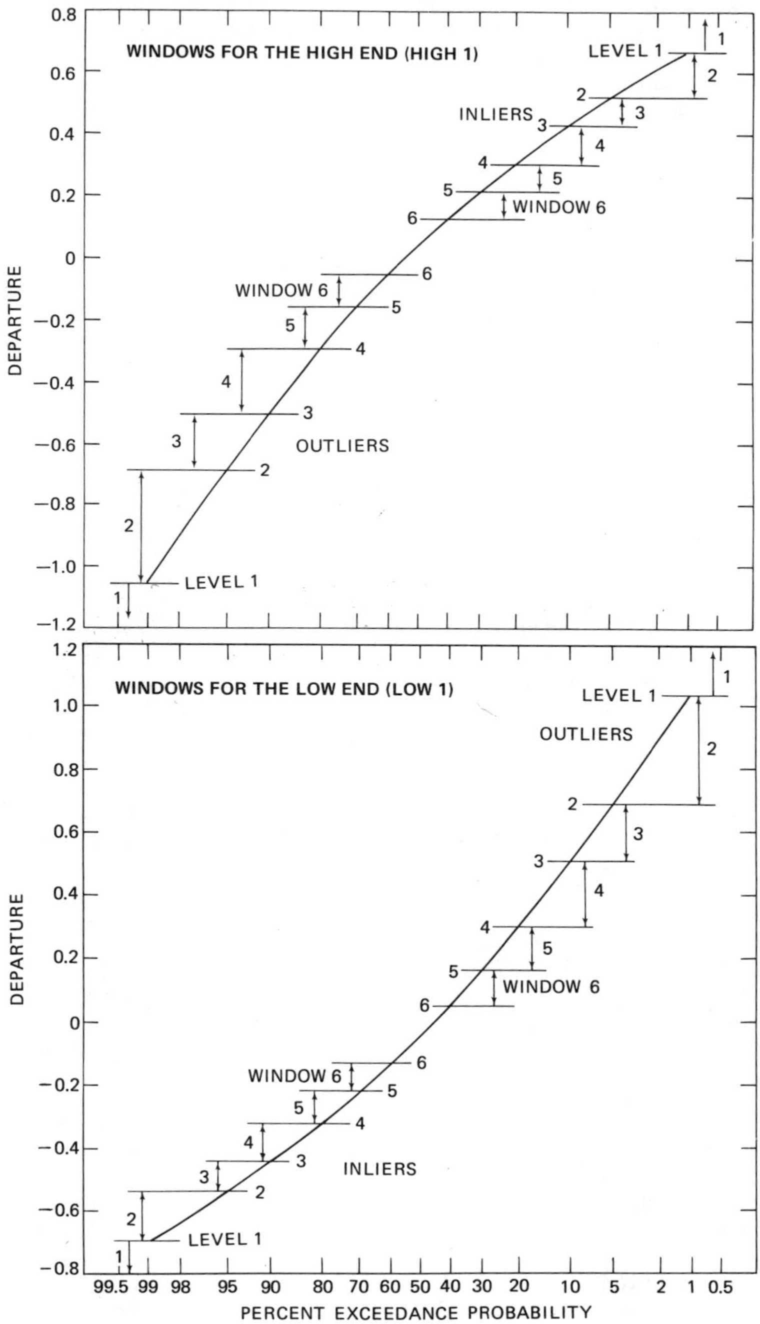

Figure 3. Levels and windows for the outliers and inliers

storm types, antecedent soil moisture conditions, and channel and floodplain storage cause variations in the basin response and flood peaks. Therefore, the flood series can hardly be expected to follow any particular theoretical distribution. According to Anderson (1967), complicated distributions can be decomposed into a few simpler distributions. Thus, an observed probability distribution P(x) may be expressed by:

$$P(x) = a_1 P_1(x) + a_2 P_2(x) + \ldots + a_k P_k(x) \quad (8)$$

in which $a_1, a_2, \ldots, a_k$ are nonegative and $a_1 + a_2 + \ldots + a_k = 1$. Such a probability distribution is designated as a mixed distribution. Singh (1968) developed the concept of a mixed distribution with two component distributions. The component distributions can be lognormal, exponential, or some other distribution.

DESIGN FLOOD ESTIMATION

A computer program package has been developed for objectively detecting and modifying any outliers/inliers sequentially from levels 1 to 6 and for computing floods at 2- to 1000-year recurrence intervals with three methods: 1) normal distribution applied to power-transformed series, with kurtosis (kt) equal to 3 and with sample kurtosis; 2) log-Pearson type III distribution, with sample skew and with weighted skew; and 3) mixed distribution having two component lognormal distributions. Extensive analyses of about 600 flood series in Indiana, Illinois, other states in the USA, and other countries in the world indicate that the maximum number of outliers/inliers at either end of the flood spectrum do not exceed the integer values of [n/10] and are not more than 5. These analyses also indicate that design flood estimates obtained in window 5 are satisfactory even when there are significant outliers and inliers. They also show that the flood estimates with the mixed distribution are very stable and lead to better regional estimates. Some examples are given below to address flood series with outliers and inliers. Level 0 or window 0 corresponds to the condition of no detection and modification of outliers/inliers.

Example 1. North Fork Embarras River near Oblong, Illinois. It has a drainage area of 318 sq mi (824 sq km) and a 43-years (1941-1983) record. The lowest flood (Figure 4) is detected as an outlier in window 4 and modified. It is followed by modifications in the succeeding two windows. The second low flood is modified only slightly in window 5. The highest flood remains practically the same but the next 3 floods are somewhat reduced. Sample skew and standard deviation for the log-Pearson type III or LP3 change from -1.39 to -0.73 and 0.377 to 0.329, respectively, from window 0 to window 5.

Example 2. Skillet Fork near Wayne City, Illinois. It has a drainage area of 464 sq mi (1202 sq km) and a 55-year (1929-1953) record. The highest flood (Figure 4) is detected as an outlier in window 3 and modified. It is followed by modification in the succeeding 3 windows. Four other high floods are picked as minor inliers in

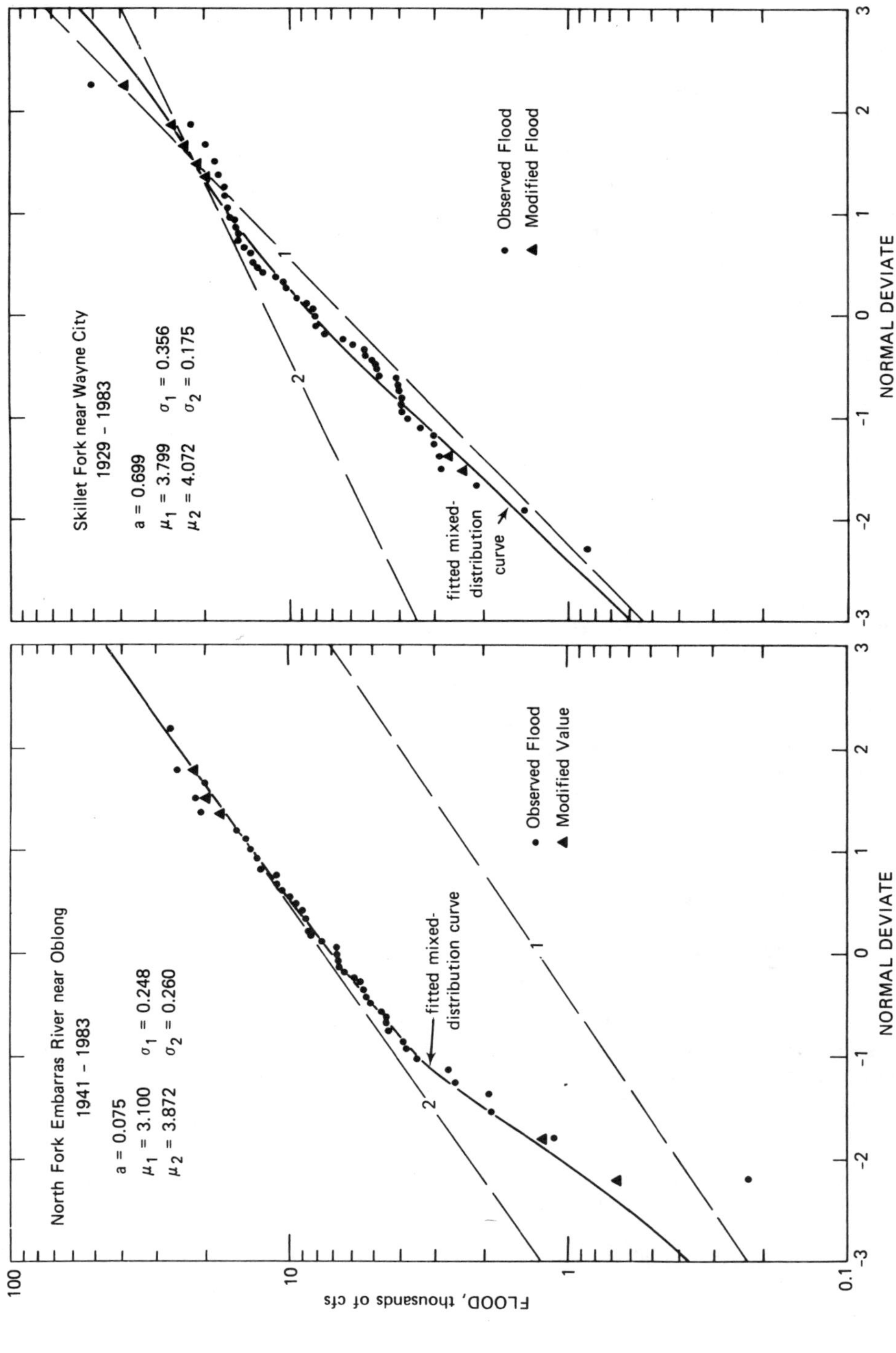

Figure 4. Observed and modified floods and the fitted mixed distribution curve (windows 5)

window 2. The low floods are modified only slightly. Sample skew and standard deviation for the log-Pearson type III change from -0.35 to -0.39 and 0.332 to 0.337, respectively, from window 0 to window 5.

CONCLUSIONS

1. The observed annual flood series at gaging stations with low gaging ratio (ratio of largest discharge measured to the largest discharge in a flood series) may have a few high floods (extrapolated from the established rating curves) which may either be much higher or lower than expected. Storm analyses and regional analyses can be used to modify such flood estimates.

2. Generally, a median or a 2-year flood is contained within the main channel. For low levels of inundation, the floodplain provides more storage than conveyance for flood flow. The relative magnitude of floodplain and channel flow sections during the transition from mainly channel to mainly floodplain flow can affect the observed flood distribution shapes.

3. A statistically best plotting position is obtained with $\alpha = 0.38$ in the formula: $p = (m-\alpha)/(m-2\alpha + 1)$.

4. Analysis of a few high flood-producing storms and associated antecedent soil moisture conditions is useful in checking the potential of a perceived outlier to be a true outlier. An extremely low flood, perceived as an outlier at the low end of the flood spectrum, can also be checked from the analysis of long-term droughts.

5. Objective detection and modification of any outliers/inliers in a flood series is necessary for deriving satisfactory estimates of design floods. An inlier at the high end is the maximum flood which appears to be much lower than expected from the rest of the data. Test statistics have been developed for detection of outliers/inliers at significance levels of 0.01, 0.05, 0.1, 0.2, 0.3, and 0.4.

6. Power transformation is considered the best for converting an observed flood series to a normally distributed series. The transformation allows the use of test statistics for detection of any outliers/inliers in a given flood series.

7. Objective modification of any detected outliers/inliers has been formalized through the concept of six levels and windows, moving sequentially from the first to the sixth.

8. There are no valid reasons to assign *a priori* a theoretical distribution as the underlying distribution for a flood series. Review of many long-term flood series in the region can help in determining whether the underlying distribution should be one of the conventional distributions or a mixed distribution. The mixed distribution can occur because of variations in storm types, channel-floodplain interaction, and antecedent soil moisture conditions.

9. A design flood estimation methodology is presented and its use is exemplified. It uses 3 distributions: normal with power transformation, log-Pearson type III, and mixed distribution with two components. Any outliers/inliers are detected and modified sequentially from the 0.01 to the 0.4 significance level.

REFERENCES

Anderson, D.V. 1967. 'Review of Basic Statistical Concepts in Hydrology'. In: Statistical Methods in Hydrology, McGill University, Canada, p. 7.

Armentrout, C.L. and Bissell, R.B. 1970. 'Channel Slope Effect on Peak Discharge of Natural Streams'. Journal of the Hydraulics Division, American Society of Civil Engineers, Vol. **96**, No. HY2, pp. 310-311.

Barnett, V. and Lewis, T. 1978. Outliers in Statistical Data. John Wiley and Sons, New York, 365 pp.

Blom, G. 1958. Statistical Estimates and Transformed Beta Variables. John Wiley and Sons, New York, 176 pp.

Box, G.E.P. and Cox, D.R. 1964. 'An Analysis of Transformation'. Journal of Royal Statistical Society, Vol. **B26**, pp. 211-252.

Cunnane, J. 1978. 'Unbiased Plotting Position--A Review'. Journal of Hydrology, Vol. **37**, pp. 205-222.

Harter, H.L. 1971. 'Some Optimization Problems in Parameter Estimation'. In: J.S. Rustagi (Editor), Optimizing Methods in Statistics, pp. 33-62, Academic Press, New York.

Hershfield, D.M. 1961. Rainfall Frequency Atlas of the United States. Weather Bureau Technical Paper No. 40, 61 pp., U.S. Department of the Interior, Washington, D.C.

Potter, K.W. and Walker J.F. 1985. 'An Empirical Study of Flood Measurement Error'. Water Resources Research, Vol. **21**, No. 3, pp. 403-406.

Potter, W.D. 1958. 'Upper and Lower Frequency Curves for Peak Rates of Runoff'. Transactions, American Geophysical Union, Vol. **39**, pp. 100-105.

Rajaratnam, N. and Ahmadi, R.M. 1979. 'Interaction between Main Channel and Floodplain Flows'. Journal of the Hydraulics Division, American Society of Civil Engineers, Vol. **105**, No. HY5, pp. 573-578.

Singh, K.P. 1968. 'Hydrologic Distributions Resulting from Mixed Populations and Their Computer Simulation'. Publication No. 81, International Association of Scientific Hydrology, pp. 671-681.

Singh, K.P. 1980. 'Flood Frequency Analysis by Power Transformation--Discussion'. Journal of the Hydraulics Division, American Society of Civil Engineers, Vol. **106**, No. HY3, pp. 462-465.

Singh, K.P. 1982. Runoff Conditions for Converting Storm Rainfall to Runoff with SCS Curve Numbers. Illinois State Water survey CR 288, Champaign, Illinois, 62 pp.

Singh, K.P. 1983. 'Software Design for a Versatile Flood Frequency Analysis'. Advances in Engineering Software, Vol. **5**, No. 2, pp. 107-112.

Singh, K.P. and Nakashima, M. 1981. A New Methodology for Flood Frequency Analysis With Objective Detection and Modification of Outliers/Inliers. Illinois State Water Survey CR 272, Champaign, Illinois, 145 pp.

U.S. Water Resources Council. 1981. Guidelines for Determining Flood Frequency. Bulletin 17B of the Hydrology Committee, U.S. Government Printing Office, Washington, D.C.

VERY LOW PROBABILITY PRECIPITATION-FREQUENCY ESTIMATES – A PERSPECTIVE[1]

Frank Richards
Office of Hydrology
National Weather Service, NOAA
8060 13th Street
Silver Spring, MD 20910

and

Rex G. Wescott, P.E.
U.S. Nuclear Regulatory Commission
Washington, DC 20555

ABSTRACT. Attempts to quantitatively evaluate the level of risk associated with nuclear power plants have led to an increasing desire to assign probabilities to rare precipitation events. Because it was felt that purely statistical estimates of events with return periods greatly in excess of the period of record were susceptible to potentially unacceptable levels of uncertainty, an approach that combined statistical analysis with meteorological interpretation was taken to determine a 24-hr point precipitation amount judged to be associated with a 0.001 annual exceedance probability of occurrence (1000-yr event). The paper also considers the possibility of using joint probabilities associated with storm centering, depth-area reduction and storm intensity to determine the probability of average depths of precipitation over larger areas.

1. INTRODUCTION

Attempts to evaluate the level of risk associated with various structures, including nuclear power plants, have led to an increasing desire to assign probabilities to rare events. Frequently flood risk is evaluated in terms of rainfall. Relatively low risk structures have usually been designed using precipitation-frequency estimates. While a number of distributions have been suggested, the choice of the distribution is not as crucial as an adequate data sample. A study by Hershfield (1962) showed that, for readily available data, some widely used distributions give generally similar estimates over the range of exceedance probabilities up to 1% (100-yr return period). When the return periods associated with frequency estimates greatly exceed the length of record available,

[1]This work was prepared by employees of the National Weather Service and the Nuclear Regulatory Commission. It presents information that does not currently represent an agreed-upon position by either agency. The National Weather Service and the Nuclear Regulatory Commission neither approved nor disapproved of its technical content.

V. P. Singh (ed.), Hydrologic Frequency Modeling, 243–252.

discrepancies between commonly used distributions tend to increase. We felt that, given the limited length of available data records, there was no way to conclusively select a distribution that could consistently provide adequate statistical precipitation-frequency estimates for return periods much greater than the period of record.

Probability estimates, generally limited to events with a 1% annual exceedance probability, are inadequate for design of high-risk structures. Current design criteria used for such structures is generally based on probable maximum precipitation (PMP). PMP estimates are based on consideration of factors that could have contributed to precipitation greater than that actually observed. This approach precludes quantitatively assigning a probability to PMP amounts.

The preferred method for evaluation of safety of operating power plants by the Nuclear Regulatory Commission (NRC) involves use of probabilistic risk analysis. The approach requires quantifying the level of risk associated with various factors, including flooding. After consultation with NRC, the National Weather Service (NWS) agreed to undertake a study to provide low probability precipitation estimates at specific power plant sites. It was agreed that the approach would combine statistical analysis with meteorological interpretation to produce an estimate of 24-hr point rainfall at the power plant site that could be expected to have a 0.001 probability of exceedance in any given year. The general approach used in determination of this estimate was to examine available precipitation data for an area within a 50-mi (80-km) radius surrounding the power plant.

Given a fixed observation network, the chances of detecting the maximum point value in an extremely rare rainfall event decreases as the rarity of the event increases. Presumably, with the density of the available observation network, the length of records are adequate to provide reasonable point estimates of events as rare as a 1% annual chance of exceedance. For the most extreme events, such as those associated with the PMP, the record lengths of the available data sampling network are inadequate to provide statistical estimates.

One of the main features associated with our 1000-yr estimate is the use of a "hybrid" series of annual precipitation values. Surrounding each power plant site, available data from all observations sites were analyzed individually and were also pooled. One of the key factors examined was based on the largest value that occurred in a given year, selected from among all available stations within the 50-mi (80-km) radius surrounding the power plant. This series is what we have termed a "hybrid" series. It was generally felt that the regions surrounding the power plant sites were sufficiently similar meteorologically that it was reasonable to assume that an observation at any location could also have occurred at any other location if the length of record had been long enough. In other words, the area was meteorologically homogeneous. At some sites, the results assuming meteorological homogeneity were adjusted somewhat to take into account identifiable systematic spatial variation.

What follows is a discussion of the general approach we are now considering. We provide an example for the Deerfield site, located in Rowe, MA. Finally, we conclude with a consideration of some ideas being developed to extend the use of these 24-hr point estimates.

2. GENERAL DESCRIPTION

For each site, we fit the available precipitation data to the following commonly-used frequency distributions: (1) Pearson, (2) Log-Pearson, (3) Log-Normal, (4) Gamma, and (5) Fisher-Tippett Type I. The purpose of examining several distributions was twofold. First, we wanted to demonstrate the range of estimates due only to the selection of dif-

Table 1.—Summary of statistics for individual stations surrounding the Deerfield site

No. of stations	Avg. period of record (yr)	Mean (in) (mm)	Stand. dev. (in) (mm)	Skew (in) (mm)	0.001 probability of annual exceedance: Fisher-Tippett (in) (mm)	Log-Pearson (in) (mm)	Pearson (in) (mm)	Log-Normal (in) (mm)	Gamma (in) (mm)
122	39.7	2.4 (61)	0.8 (20)	1.1 (33)	6.8 (173)	5.8 (147)	6.6 (168)	7.4 (188)	5.2 (132)

ferent distribution functions. Second, the distribution selected for each site was chosen primarily on the basis of which distribution appeared to be the most consistently conservative at that site. Distributions that produced larger estimates were considered conservative. A distribution that produced a large estimate at each site was considered a consistently conservative estimator. Once a distribution was selected, it was used to analyze pooled data, including the hybrid series.

The main focus of the meteorological analysis was to assure that all factors that might have an influence on the final estimate were adequately considered. Individual stations were examined in the context of all other stations to evaluate their representativeness. We also considered the possibility of a secular trend. Usually discrimination between a "real" trend and sampling fluctuation due to natural variability ultimately required meteorological judgment. The data were also examined in conjunction with local factors such as topographic features to evaluate the assumption of meteorological homogeneity. If necessary, further meteorological analysis was performed to attempt to define systematic variation within the study area. When possible, this analysis produced a refinement to the final estimate.

3. SITE SPECIFIC EXAMPLE

The nuclear power plant site in the Deerfield River Basin is located in the town of Rowe, MA, near the Deerfield River, and is surrounded by mountainous terrain with peaks to over 3,000 ft (900 m) in the Green Mountains to the west of the study area. The Atlantic Ocean lies about 100 mi (160 km) to the south-southeast. All available stations within 50 mi (80 km) of the power plant site that had at least 20 years of precipitation data were included in the analysis (4844 station-years of data).

Table 1 contains a summary of the average statistics based on individual station estimates. The range of 1000-yr estimates at the individual stations (not shown) varied by more than a factor of 6, from 3.7 to 22.9 in. (94-582 mm), the former from the Gamma distribution and the latter from the Log-Pearson distribution. Much of the variation between stations can probably be attributed to sampling fluctuation. In particular, the data for Belchertown, MA showed an extremely large skew 5.0 in. (127 mm). This led to 1000-yr estimates of 18.7 and 22.9 in. (475 and 582 mm) for the Pearson and Log-Pearson distributions, respectively. These extreme estimates were, in large part, due to a value of 12.25 in. (311 mm) that occurred during the passage of hurricane Dianne in August 1955.

Table 2.—Normalized measures used to evaluate some of the relative characteristics of the distributions at individual stations around the power plant

	Fisher-Tippett	Log-Normal	Pearson	Log-Pearson	Gamma
Range of normalized estimates	0.89-1.19	0.56-1.08	0.89-1.35	0.89-1.65	0.54-0.95
No. of normalized estimates less than 1	9	94	66	16	122
Average of normalized estimates	1.09	0.93	1.01	1.13	0.84
No. times normalized estimate was highest	58	0	3	61	0

Extrapolation based on the information in Hershfield (1961) would suggest an annual exceedance probability of about 0.00003. Because this value is clearly an outlier and inconsistent with the sample size, it was omitted from consideration in the hybrid series.

To focus on the selection of a frequency distribution, we wanted to minimize the effect of station-to-station variability. To do this, the frequency estimates (1000-yr) were normalized by the average of all 5 distribution estimates at each station A summary of the normalized estimates is shown in table 2. The Fisher-Tippett Type I distribution produced normalized estimates below the average of the 5 distributions less often than any of the other distributions. The Log-Pearson distribution produced the highest normalized estimate (1.65), relative to the other 4 distributions. The Fisher-Tippett Type I, Pearson, and Log-Pearson distributions produced averages of the normalized estimates greater than one. The last line in table 2 shows the frequency with which a particular distribution produced the largest normalized estimate at the individual sites (total of 122 sites). The Fisher-Tippett Type I (58) and Log-Pearson (61) distributions were most conservative. The Fisher-Tippett Type I distribution displayed considerably less station-to-station fluctuation than the Log-Pearson. Our decision was to select the Fisher-Tippett Type I distribution rather than the Log-Pearson distribution. It best met our criteria of being a consistently conservative estimator.

The 24-hr Fisher-Tippett Type I 1000-yr estimates for each observation site were plotted on a map that included topography. An analysis was prepared considering slopes, barriers, elevation and distance from moisture sources. Values along the south and southeast limit of the study area were considerably larger than those at the site. These values were strongly influenced by hurricane Dianne of August 1955. This spatial analysis of the individual station data suggested a 1000-yr value of about 7.4 in. (188 mm) at Deerfield.

The hybrid series resulted in a Fisher-Tippett Type I 1000-yr value of 12.5 in. (318 mm). In an attempt to determine whether the hybrid series was biased by fewer observing locations in the early years of the data sample, data subsets were created by dropping annual maximum precipitation values, starting with the earliest years. The 1000-yr values tended to become larger as the earliest years were dropped (see fig. 1). The estimates appeared to vary around 13 in. (330 mm) after about 20 yr were dropped. Also shown in figure 1 are the number of stations available for a given year. When there was a

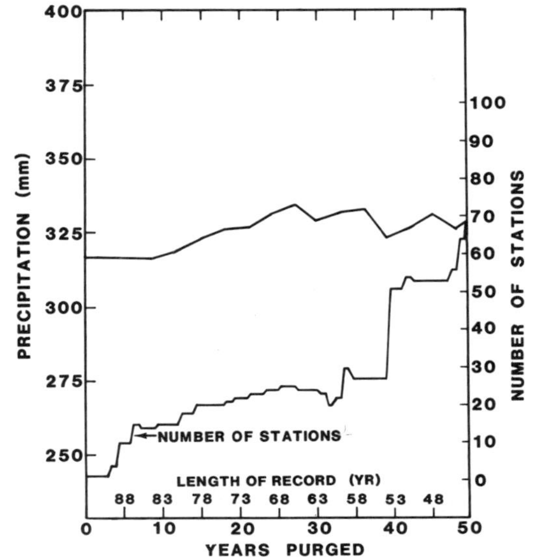

Figure 1. Variation in 1000-yr hybrid estimate as years were purged. Also shown is the number of stations that were available.

significant increase in the number of available stations, the hybrid estimate did not show a corresponding increase. We examined a number of long-record stations to determine whether the increasing hybrid estimates in the earlier years could have been caused by a secular trend. There was no indication of either short-term or long-term trends at the individual stations. While it is not possible to completely separate the effect of station density from year-to-year meteorological variability, we concluded that the lower estimates when the earlier years were included were likely to be due to inadequate station density and that a value of 13 in. (330 mm) was the most likely "true" hybrid estimate.

It is our judgment that the fully adjusted estimate at the Deerfield power plant site is 14 in. (356 mm). This estimate is based on the adopted hybrid estimate of 13 in. (330 mm). There were 79 nonrecorder station-years and 14 recorder station-years in the hybrid series. An adjustment factor of 1.11 was applied to the hybrid value of 13 in. (330 mm). to produce a corresponding 1440-min estimate of 14.4 in. (366 mm). The adjustment factor was determined by weighting the observation-day (nonrecorder) and clock-hour (recorder) adjustment factors (1.13 and 1.01, respectively) by the number of years each occurred in the hybrid series. (See Weiss (1964) for a discussion of these adjustment factors.) The resulting estimate was rounded down to 14 in. (356 mm) because the hybrid series produces estimates that are more conservative than estimates of data from fixed locations.

Although the 14 in. (356 mm) estimate is considerably larger than analysis of geographically-fixed data might suggest, it is not as large as a 24-hr observation of 18.15 in. (461 mm) at Westfield, MA. Westfield is about 55 mi (88 km) SSE of the power plant site. This precipitation was associated with the passage of decaying hurricane Dianne in August 1955. In addition to the heavy rainfall at Westfield, Dianne produced amounts of 14.0 in. (356 mm) and 13.24 in. (336 mm) of rain at Ware and Holyoke, MA, respectively. Both sites were within 65 mi (105 km) of Deerfield. This is the same storm that produced the 12.25-in. (311 mm) amount at Belchertown, MA, that was censored from the hybrid series.

The fact that excessive rainfall did not occur at Deerfield during Dianne's passage can be attributed to two factors. First, the heaviest rainfalls associated with any tropical storm are spotty. The second factor was the path of Dianne. After heading inland into the western part of southern New England, she veered to the east, limiting the northern progression of heavy rainfall. This change of course was due to large-scale meteorological patterns and, from a meteorological view point, could easily have been displaced 50-100 mi (80-160 km) further inland. There are no significant topographic barriers to impede moisture inflow as far inland as Deerfield. In fact, the elevation profile suggests that orographic enhancement of the precipitation in the vicinity of the power plant might produce values higher than those observed at Westfield. It was our judgment that an amount greater than that observed at Westfield could occur at Deerfield if a number of critical factors were present in conjunction with a hurricane, such as Dianne.

4. APPLICATION

For many applications the 0.001 annual point exceedance estimate may be used without modification. Where a less conservative estimate or an estimate for a larger area is desired, further adjustment to the values may be made. An approach presently being explored to modify point estimates is based on the assumption that the hybrid-series 1000-yr frequency estimate represents the precipitation depth that has a 0.001 per year probability of being exceeded at any point within 50 mi (80 km) of the point of interest. It is further assumed that the Fisher-Tippett Type I distribution is a reasonable representation of point precipitation potential within the 50-mi (80 km) radius.

For any storm which occurs randomly within the specified area, the probability of a given amount of precipitation being exceeded over a fixed area may be determined if the storm's areal precipitation distribution is known. This determination may be simplified by assuming the area of interest (basin catchment area) and the storm isohyets to be circular. Then the probability of a given depth of precipitation from the storm falling over the basin area may be represented as:

$$P_c(X,A_b) = (\sqrt{A_s(X)} - \sqrt{A_b})^2 / A_r \, , \quad \text{where} \qquad (1)$$

$P_c(X,A_b)$ is the probability that a basin of size A_b will be covered by an average precipitation depth of X, given the occurrence of the storm,

$A_s(X)$ is the area of coverage of precipitation depth X or greater within the storm, and

A_r is the area of the region that the storm is assumed to occur within (in this case 7,850 mi^2 [20,330 km^2]).

The probability of exceedance of a given precipitation depth over a fixed area on an annual basis [$P_t(X,A_b)$] will be a summation of joint probability combinations of storm intensity and location, or:

$$P_t(X,A_b) = \sum P_c(X,A_b) \times P_s(X,A_s) \text{ , where}$$

$P_s(X,A_s)$ is the annual probability of occurrence (discrete interval) of a storm having a precipitation depth of X over an area, A_s, within the region of interest.

The summation is over all discrete probability intervals capable of producing a precipitation depth of X or greater over an area greater than or equal to A_b.

One of the problems with trying to use this joint probability relation with point precipitation statistics is that the rate of reduction of average precipitation depth with area (depth-area reduction) for observed storms is not constant. The variation in depth-area reduction was investigated using published depth-area-duration data for 42 severe storms from Schreiner and Riedel (1978). These storms had depths of average precipitation listed for durations of from 6 to 96 hr and areas of from 10 to 50,000 mi^2 (26 to 129,500 km^2). Almost all of these storms occurred in the United States east of the 105th meridian and included hurricanes, tropical storms and extra-tropical storms.

For each of the storms which had depth information available for 24 hr and areas up to 5,000 mi^2 (12,950 km^2), a log-log regression analysis was performed such that the depth-area data could be fit with the expression:

$$X(A) = X_o A^{-K} \text{ , where}$$

X(A) is the average depth of precipitation over an area A,

X_o is the intercept of the ln(A) axis (assumed to be the maximum point precipitation of the storm), and

K is the storm decay coefficient (slope in the regression analysis).

Although the K values from the analyses showed considerable variation, it appeared that some of this variation could be accounted for by variation with X_o. Figure 2 is a plot of K versus X_o for the storms. The more intense the center of the storm, the more rapid the decrease of precipitation intensity with area. It was also determined that random variation from this relation appeared to be normally distributed in this sample of storms.

The normal distribution of the deviations from the line of best fit allowed the probability of a storm having a specified depth-area reduction relation to be defined. This distribution of possible K values was described in terms of standard deviation from the line of best fit. Each K (or K versus X_o relation) was then represented by a discrete probability interval which may be used in the joint probability computation. The lines identified as K' in figure 2 represent a deviation of one standard deviation on each side of the line of best fit.

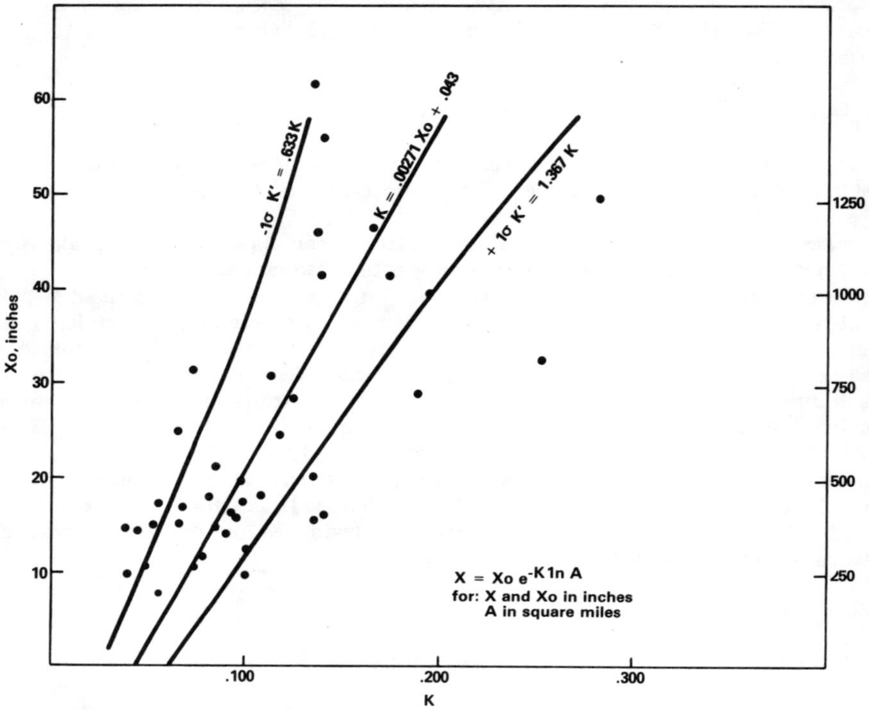

Figure 2. Realtion between decay coefficient (K) and X_o.

The relation for average depth of precipitation over area can not be used directly in the probability calculations, but can be used to derive a relation for precipitation exceedance over a specific area. This relation may be derived as follows:

$$X(A) = \frac{1}{A}\int_0^A X_e(A)\, dA \ , \quad \text{where}$$

$X_e(A)$ is the precipitation equaled or exceeded over area A.

$$\text{Then,} \quad A_s(X_e) = \left\{ \frac{X_o}{X_e}(1 - K) \right\}^{1/K} \tag{2}$$

Equation (2) is used to determine the area enclosed by the isohyet for a specified precipitation, X_e, which corresponds to the point precipitation, X_o, from the hybrid series. It is then used in equation (1) for determination of storm centering probability.

Using the relations presented, rainfall isohyets with an assumed circular shape may be developed for a desired exceedance probability. A relatively simple computer program may be used to go through all the joint probability combinations involving point rainfall frequency, depth-area reduction and centering probability. The computation, however, requires selection of the number of discrete intervals for the point rainfall probability contribution, as well as intervals representing K. A sufficient number of intervals was considered to have been reached for point rainfall probability when an increase in the

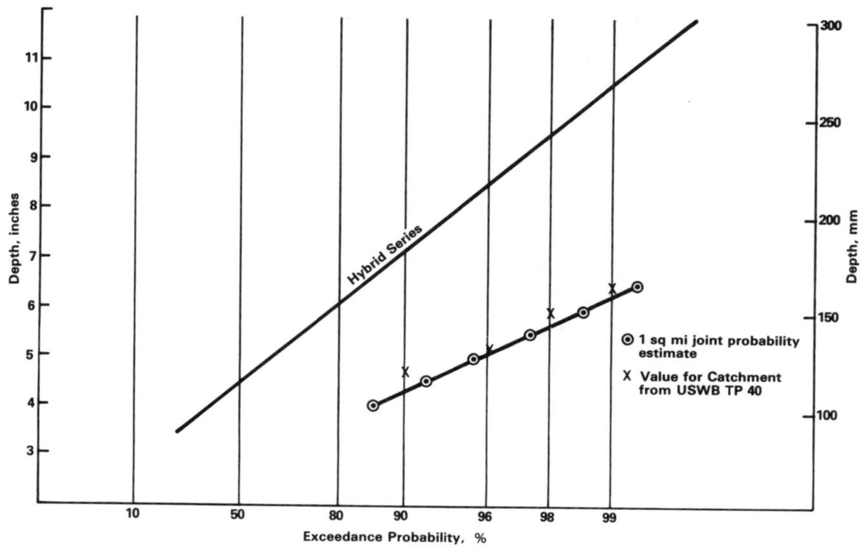

Figure 3. Precipitation amounts as a function of non-exceedance probability based on the joint probability approach [circles with dots] and, for comparison, estimates from Hershfield (1961) [X's].

number of intervals lead to minimal changes in isohyet probability. As for K, it was considered sufficient to choose five intervals, starting and ending between the limits of the observed data and one standard deviation on each side of the mean.

An example of this calculation for the Deerfield River basin in Massachusetts for a 200-mi^2 (518 km^2) catchment is given in table 3. From these calculations an average depth of 8.2 in. (208 mm) over 200 mi^2 (518 km^2) was calculated for an annual exceedance probability of 0.001. The procedure was also used to calculate point precipitation values for return periods of from 10 to 100 yr for comparison with published estimates (Hershfield, 1961). These results are shown in figure 3.

Table 3.—Probability (0.001) of enclosing a specified area with an isohyet of specified depth

Depth (in./mm)	Area (mi^2/km^2) 1 (2.6)	10 (25.9)	100 (259)	200 (518)	500 (1259)
7 (178)	4.3	4.2	4.0	3.9	3.7
8 (203)	1.3	1.2	1.2	1.2	1.1
9 (229)	0.38	0.37	0.35	0.34	0.33

5. CONCLUSIONS

Although results appear reasonable for the examples presented, the methodology is in an initial stage of development and should not be considered at this time for design applications. In particular, the distribution of storm decay coefficients may be regionally dependent. This was not taken into account, due to the scarcity of data.

6. ACKNOWLEDGMENTS

National Weather Service participation in this study was funded by the U.S. Nuclear Regulatory Commission under Interagency Agreement No. NRC-03-84-073. Roger R. Watkins provided much of the meteorological support for the Deerfield site. Typing and preparation of the report was done by Helen V. Rodgers. Roxanne Johnson and Lionel J. Watkins prepared the figures.

7. REFERENCES

Hershfield, D.M. 1961. Rainfall Frequency Atlas of the United States for Durations from 30 Minutes to 24 Hours and Return Periods from 1 to 100 Years. Technical Paper No. **40**, 115 pp., U.S. Weather Bureau, Washington, DC.

Hershfield, D.M. 1962. An Empirical Comparison of the Predictive Value of Three Extreme-Value Procedures. Journal of Geophysical Research, Vol. **67**, No. 4, pp. 1535-1542.

Schreiner, L.C. and Riedel, J.T. 1978. Probable Maximum Precipitation Estimates, United States East of the 105th Meridian. Hydrometeorological Report No. **51**, 87 pp., National Oceanic and Atmospheric Administration, National Weather Service, Washington, DC.

Weiss, L.L. 1964. Ratio of True to Fixed-Interval Maximum Rainfall. Journal of the Hydraulics Division, American Society of Civil Engineers, Vol. **90**, pp. 77-82.

SQRT-EXPONENTIAL TYPE DISTRIBUTION OF MAXIMUM

Takeharu Etoh, Dr. Eng., Prof.
Department of Civil Engineering, Kinki University
Kowakae, Higashi-Osaka, Osaka 577, JAPAN

Akira Murota, Dr. Eng., Prof.
Department of Civil Engineering, Osaka, University

Masanori Nakanishi, Research Associate
Department of Civil Engineering, Kinki University

ABSTRACT. The authors present the square-root exponential type distribution of the maximum, or SQRT-ET-max distribution, as the distribution to be applied to the annual maximum series of the total amount (or depth) of a single rainstorm. The presented distribution is theoretically derived, and is as simple in expression as conventional ones. It has two parameters and, thus, estimates them more stably than distributions with three or more parameters. Furthermore, it is as good as or superior to conventional 3-parameter distributions in overall fitness to frequency distributions of annual maximum depths of single rainstorms and rainfalls of about 24 hours.

1. INTRODUCTION

Outliers in frequency analyses of floods may be due to the following causes:

(1) Application of inappropriate distribution functions to historical data.
(2) Unstationarity of the time sequences of major floods such as long-term trends or periodicities.
(3) Diversity of the causes of rainfalls, such as typhoons, fronts, etc.,
(4) Abnormal causes of rainfalls, and
(5) Inaccuracy of measurements.

Among these, the first three produce no outliers if proper statistical techniques or models are applied to historical data. The fourth implies abnormally large rainfalls of completely different origin from those triggering several-year-probability rainfalls. Outliers caused by (3) to (5) are called contaminants. In the present paper, the first cause is examined, i.e., whether or not the applied distribution functions are proper.

V. P. Singh (ed.), Hydrologic Frequency Modeling, 253–264.

The authors present the square-root exponential type distribution of the maximum, or SQRT-ET-max distribution, as the distribution to be applied to the annual maximum series of the total amount (or depth) of a single rainstorm. Its applicability is examined to the annual maxima of daily rainfall depths at major meterological observatories throughout Japan. Five criteria are proposed to judge the goodness of distribution functions in practical applications, i.e., overall fitness, stability of the estimated parameters, reasonability of outliers, simplicity in expression, and theoretical background.

The presented distribution is theoretically derived, and is as simple in expression as conventional ones. It has two parameters and, thus, estimates them more stably than distributions with three or more parameters. Furthermore, it is as good as or superior to conventional 3-parameter distributions in overall fitness to frequency distributions of annual maximum depths of single rainstorms and rainfalls of about 24 hours. Finally, it is shown that the SQRT-ET-max distribution is a highly plausible solution to the outlier problem of rainfalls of several to 48 hours duration.

2. SQRT-ET-MAX DISTRIBUTION

The authors derived the SQRT-K distribution for the total amount of rainfall of a single storm, assuming:

(1) The duration and the peak intensity of a single rainstorm are independent of each other,

(2) One distributes exponentially; the other follows the Gamma distribution, and

(3) The total depth is proportional to (half) their product.

These may be acceptable assumptions, as shown in literature [e.g., Etoh and Murota, 1984]. Thus, the SQRT-K distribution probably fits well to the frequency distribution of the depth of a single storm. In practice, it is troublesome to handle this distribution because it is expressed in terms of the second kind modified Bessel function. Thus, we shall derive more practically convenient distributions based on analytical approximations of the SQRT-K distribution (appendix 1).

Since our focus is distribution of large rainfalls, it is necessary to simplify the function so that it provides more accurate approximations for large variables. We shall therefore employ the asymptotic expansion of the second kind Bessel function. It is, then, deduced that the depths of major rainstorms follow the square-root exponential type distribution, or SQRT-ET distribution for short. By introducing an additional assumption, that the occurrence of major rainstorms follows the Poisson distribution, the following SQRT-ET-max distribution is derived for the annual maxima of their depths:

$$F(x) = \begin{cases} 0 & (x < 0) \\ \exp\left[-\lambda \{(1 + \sqrt{\beta x}) \exp(-\sqrt{\beta x})\} \right] & (x \geqq 0) \end{cases} \tag{1}$$

where β is the scale parameter, and λ is the frequency parameter, representing the average annual frequency of major rainfalls. The shape of the distribution is also represented by λ, i.e., if λ takes the same value, the shape is similar. The scale is determined with the value of β. As observed in Eq. (1), the SQRT-ET-max distribution has a distribution function with two characteristic parameters and is as simple as the Gumbel distribution; however, it tails much longer than Gumbel's, since the negative argument of the second exponential of the function is expressed in terms of the square root of the original variable, x, while the Gumbel's argument is linear. The long tail implies the occurrence of much larger rainfall depths than predicted with the Gumbel distribution, or seeming outliers.

Consequently, it is concluded that:

(1) The frequency distribution of the annual maxima of the depths of single storms may be well approximated with the SQRT-ET-max distribution, and,

(2) It may theoretically explain the occurrence of seeming outliers.

As is well known, the combination of the exponential distribution for the intensity of rainfall of each storm and the Poisson distribution for the annual frequency provides the Gumbel distribution for the annual maximum series. As it is reported that peak intensities distribute exponentially, the annual maxima are theoretically expected to follow the Gumbel distribution instead of the SQRT-ET-max distribution.

3. APPLICABILITY

3.1. Guideline of the Examination

To objectively examine the applicability, the following rules are introduced as guideline.

Only parametric statistical estimation techniques which are usually based on assumed probability distribution functions (p.d.f.) are employed as alternatives to be compared with the SQRT-ET-max distribution. The fitness of each p.d.f. is measured by likelihood. Generally, the more parameters, the better it fits to the sample frequency distribution, though less stably these parameters are estimated. Stability also depends on sample size. Thus, an optimum number of parameters should exist for the type of p.d.f. and sample size. Hydrologic analysis usually aims at evaluating very large or very small samples. Therefore, in addition to overall fitness, one of the measures of the goodness of a distribution is how well the p.d.f. explains the tail properties of the distribution of historical records; more specifically, how well it encompasses the outliers. There are some additional standards. For example, p.d.f.'s with sound theoretical bases, such as the Gumbel distribution, are naturally better than those proposed just by intuition. However, those which are cumbersome due to the complexity of their expression should not be employed, regardless of their theoretical bases.

Consequently, we shall employ the following criteria:
(1) overall fitness, (2) stability of the estimated parameters, (3) reasonability of outliers, (4) simplicity in expression, and (5)

theoretical background.

These conceptual criteria are translated one by one into specific numerical indices for more objective evaluation.

The overall fitness is measured by likelihood. Thus, the fitting is for the most part made by the maximum likelihood method.

The stability of the estimated parameters, or the data-dependency, is evaluated as follows: Estimate the parameters using samples with and without the largest one; then evaluate the frequency of the largest sample based on both sets of estimated parameters; the stability may be measured by the ratio (or the difference) of the frequencies.

The reasonability of outliers is evaluated by the estimated frequency of the largest sample. For example, if there are many observatories where the average recurrence interval, i.e. the inverse of the annual frequency, exceeds one thousand years, the assumed p.d.f. is not acceptable. A more objective index can be constructed based on the theoretical distribution of the average recurrence interval of the largest (Appendix 4). For example, suppose we have an N-year annual maximum sample sequence of an observatory, and have estimated the N-year probability rainfall by applying a distribution function to the sequence, i.e., the estimate whose recurrence interval T is equal to N; then, if we have reasonably estimated the N-year rainfall depth, the historical largest of the N-year sample sequence should almost evenly distribute around the estimated N-year depth; theoretically, the probability that the historical N-year largest sample is less than the estimate is e^{-1} (36.8%) when N is large; therefore, the ratio of the number of observations whose historical largests are less than the estimated N-year depth should be around 37%. Thus, deviation from that value can be a good measure of the reasonability of outliers.

We can safely evaluate criteria (4) and (5) by intuition.

The above numerical indices are highly dependent on the number of parameters. Therefore, it is recommended that the distribution functions be classified into the groups by the number of parameters, and the goodness of the distribution functions of each group be compared at the first stage.

In this paper, distributions with two and three parameters are examined, since this covers most conventional distributions. Each of them is applied to the following observed annual maxima, and the indices of goodness are compared with those of the SQRT-ET-max distribution.

First, for detailed examination, they are applied to annual maxima of 24-hour depth observed from June to October of 1900 to 1983 at the Osaka Regional Meteorological Bureau. Then, to confirm the results obtained for the rainfalls of Osaka, the two selected distributions, i.e., the SQRT-ET-max and Gumbel distributions, are applied to the annual maxima of daily precipitations recorded at the major 56 meteorological observatories of Japan from the beginning of observation (before 1900) to 1980.

3.2. Application to Rainfalls of Osaka

3.2.1. Data. In 1957, record-breaking large rainfalls poured in Osaka; rainfalls with a peak intensity over 50mm/hour occurred three times in only three weeks. From June 26 to 27, the 24-hour precipitation reached 283.7mm, and at Yao, near Osaka, it amounted to 311.2mm. The next largest records in Osaka are 209.6, 188.5, and 183.2mm per 24 hours. The 24-hour record is evaluated as a several-hundred or several-thousand-year rainfall when conventional distributions with two parameters are applied. The authors have been engaged in the planning of the urban river network in Osaka, and have found a great difficulty in explaining and dealing, in the design process, with the exceptionally large record, which motivated them to initiate the present study. Thus, one key to judge the goodness of distributions is how reasonably they explain this possible outlier in Osaka.

3.2.2. <u>Stability and reasonability.</u> The evaluation of the largest 24-hour rainfall record is given in TABLE I. The largest at Yao is also evaluated, with the parameters estimated using historical records in Osaka. When the Gumbel distribution is applied, the average recurrence interval T becomes 955 to 2500 years, which is unrealistic for the length of the original record, i.e., 84 years. The same is observed when the 2-parameter log-normal distribution is applied. On the other hand, the SQRT-ET-max distribution evaluates the average recurrence interval as 214 years in Osaka and 380 years at Yao, which is reasonable since the probability of the occurrence of a rainfall depth with a recurrence interval more than 214 years (or 380 years) is 33% (or 20%) for 84-year observation (Appendix 4).

TABLE I Average recurrence interval of the historical largest 24-hour rainfall in Osaka estimated with some distribution functions (June 26-27, 1957)

Distribution		Sample size	Osaka 283.7mm	Yao 311.2mm
2-parameter	Gumbel	84	955	2500
		83*	1808	5140
	SQRT-ET-max	84	214	380
		83*	315	653
3-parameter log-normal		84	249	442
		83*	2381	7143

* Evaluated with parameters estimated from samples without the 1957-largest

When the 3-parameter log-normal distribution is employed, T becomes 249 to 442 years, which is not unrealistic.

Subsequently, to examine the stability or the data-dependency, the parameters of the distributions are estimated based on the maximum sequence without the largest one of 1957, and, then, the recurrence interval T of the largest one is evaluated based on the estimated parameters. The result is also tabulated in Table I. The estimates of T are naturally longer than the original ones. It is noteworthy that, while estimates of T by means of 2-parameter distributions almost doubled, those obtained with the 3-parameter log-normal distribution are up to ten to twenty times greater than the original ones. It is clearly observed in Figure 1, in which the solid and the broken lines representing the c.d.f.'s estimated from the respective records, with and without the 1957-largest, are compared with the Weibul-plot of the historical records. The curves of the two 2-parameter distributions (left: SQRT-ET-max, right: Gumbel) seem to be quite stably estimated, ignoring the largest record, while those of the 3-parameter log-normal distribution with or without the largest record obviously deviate from each other.

Consequently, it is concluded that:

(1) It is difficult to reason seeming outliers such as the 1957-largest in Osaka, by means of conventional 2-parameter distribution functions,

(2) Though it is possible to give realistic estimations of the frequency of exceptionally large records by means of 3-parameter distributions, such estimates fluctuate greatly depending on whether the large records are considered or ignored.

(3) Since the SQRT-ET-max distribution is a 2-parameter distribution, its estimates are relatively stable; having long tails thus explaining the occurrence of exceptionally large rainfalls, or seeming outliers.

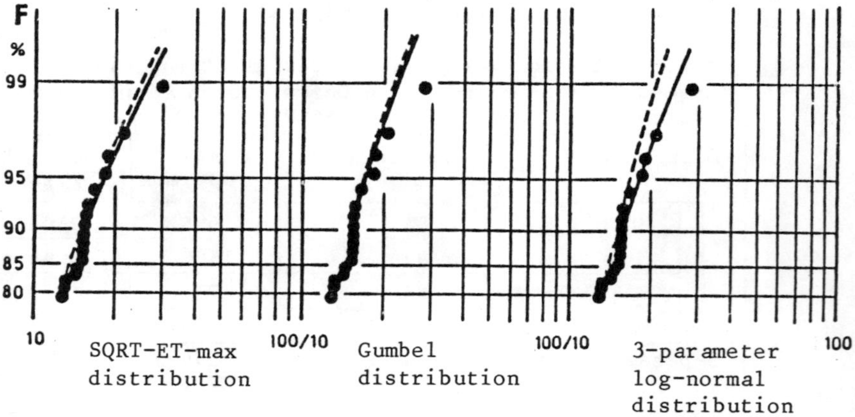

Solid and broken lines respectively represent the c.d.f.'s estimated with and without the historical largest.

Figure 1. Stability of estimation with or without historical largest record (Osaka, annual 24-hour maxima)

3.3. Applicability to Daily Precipitation in Japan

So far, we have used only the 24-hour annual maxima in Osaka. From now on, we will employ various annual maximum series; as for distributions, we shall confine ourselves to comparing the SQRT-ET-max distribution with the Gumbel distribution.

Two distributions are applied to the annual maxima of daily precipitation series at 56 meteorological observatories throughout Japan, mainly to examine their reasonability of outliers. The results are given in TABLE II, in which the likelihood is compared in the extreme right column and the historical maximum of each observatory, i.e., the candidate for outliers, is compared with the estimated N-year probability rainfall in the central two columns, where N is the observation period.

Circles mark cases where the historical maximum is less than the estimated N-year rainfall. As stated previously, the ratio of the number of circles to that of all observatories, i.e., 56, is theoretically equal to e^{-1} (=37%).

As shown in TABLE II, if we employ the Gumbel distribution the ratio is about 7%, which leads us to the unacceptable conclusion that daily rainfalls over the N-year probability rainfalls have already poured at 93% of the observatories; similarly, rainfalls with more than 100-year values have been observed at fifty (90%) of the observatories, notwithstanding that the average observation period is only 90.7 years.

The SQRT-ET-max distribution may give a full explanation of the problem. Namely, the ratio of circles is 34% for the distribution,

TABLE II Application of SQRT-ET-max and Gumbel distributions to annual maxima of daily precipitation series (at 56 meterological observatories in Japan, from beginning of observation to 1980)

Observatory No.	Sample size	Comparison of historical maximum (a) and estimated N-year probability rainfall (b)		Comparison of likelihood
		Gumbel	SQRT-ET-max	
407	83	o	x	E
409	91	o	o	G
412	94	x	o	E
~~	~~~	~~~	~~~	~~~
890	88	x	o	E
893	94	x	x	E
895	89	x	x	E
		o: 4	o: 19	E : G = 44 : 12
Total: 56 observatories Summary and remark	81-105 yrs. Average: 90.7 yrs. Excluding years without data	o: a is smaller than b. x: a is larger than b. Rate of o to total 4/56 =7.14%	19/56 =33.9%	E: Likelihood of SQRT-ET-max is larger than that of Gumbel. G: Likelihood of SQRT-ET-max is smaller than that of Gumbel.

which deviates only 3% from the theoretical prediction. Even if we employ the SQRT-ET-max distribution, the historical largest at four observatories are evaluated as more-than-1000-year rainfalls. For T=1000, theoretically, the probability that the largest record exceeds the N-year rainfall is 8.7%, where N is the length of observation; therefore, the expected number of observatories in which this case may happen is etimated to be 4.9 by multiplying the number of all observatories, i.e., 56, by the percentage (Appendix 4). A splendid consistency is obtained; the difference between 4 observatories from observation, and 4.9 from theoretical prediction, is practically negligible. We can thus conclude that even the rainfalls evaluated to be larger than 1000-year ones are not abnormally large, but quite natural from the statistical point of view with the SQRT-ET-max distribution.

Consequently, the SQRT-ET-max distribution has solved the problem of outliers in daily precipitation in Japan. As shown in TABLE II, the SQRT-ET-max distribution fits better to historical records at 80% of observatories, while, at one of five observatories, the likelihood of the Gumbel distribution is higher. This may be due to random fluctuation in the large rainfall series.

4. CONCLUDING REMARKS

The SQRT-ET-max distribution is presented for the derived distribution of annual maximum depths of single rainstorms. It is applied to various annual maxima made from the observed hourly rainfall sequence in Osaka, and those of daily rainfalls at major meteorological observatories throughout Japan, compared with conventional distribution functions.

Five criteria are proposed to judge the goodness of distribution functions in practical applications, i.e., overall fitness, stability of the estimated parameters, reasonability of outliers, simplicity in expression, and theoretical background.

The presented distribution is theoretically derived, and is as simple in expression as conventional ones. It has two parameters and, thus, estimates them more stably than distribution with three or more parameters. Furthermore, it is as good as or superior to conventional 3-parameter distributions in the overall fitness to frequency distributions of annual maximum depths of single rainstorms and rainfalls of about 24 hours. Finally, it is shown that the SQRT-ET-max distribution is a highly plausible solution to the outlier problem of rainfalls of several to 48 hours durations.

5. REFERENCES

Periodicals

Etoh, T., and Murota, A. 1984. 'A Propbabilistic Model of Rainfall of a Single Storm.' Proc. of JSCE (Japanese Society of Civil Engineers), No. 345, Vol. II-1, pp.101-109 (in Japanese).

Etoh, T., and Murota, A. 1986. 'Probabilistic Model of Rainfall of a Single Storm' Journal of Hydroscience and Hydraulic Engineering, Vol. 4, No. 1, pp. 65-77, April 1986 (English version of the above paper).

APPENDIX 1

SQRT-ET Distribution

The p.d.f. of the SQRT-K distribution is expressed as [Etoh and Murota, 1984]:

$$h(x) = c\cdot(\sqrt{\beta x})^{-\nu} k_\nu (\sqrt{\beta x}) \tag{2}$$

$$c = 2^{\nu-1}\beta/\Gamma(1-\nu) \tag{3}$$

where ν and β are the shape and the scale parameters, and k (s) is a second kind modified Bessel function, i.e.,

$$k_\nu(s) = \frac{1}{2}\left(\frac{s}{2}\right)^\nu \int_0^\infty t^{-\nu-1}\exp(-t-\frac{s^2}{4t})\,dt \tag{4}$$

As we aim at the tail property, we shall apply the asymptotic expansion of K (s) and employ the first term.
Then,

$$k_\nu(s) \sim \sqrt{\pi/(2s)}\, e^{-s} \tag{5}$$

Substituting (5) for (2), and considering that the variation of the exponential function is dominant over that of the power of s, Eq. (2) is reduced to

$$h(x) \sim \exp(-\sqrt{\beta x}) \tag{6}$$

The coefficient is determined by the condition that the integration of a p.d.f. over the whole region is unity.
Thus,

$$h(x) = \beta/2\cdot\exp(-\sqrt{\beta x}) \quad (x \geqq 0) \tag{7}$$

The c.d.f. is expressed as:

$$H(x) = 1-(1+\sqrt{\beta x})\exp(-\sqrt{\beta x}) \tag{8}$$

The distribution is represented by Eq. (7) or (8), and is named the square root exponential type, or SQRT-ET distribution.

This distribution has the following characteristics:

(1) It has only one parameter β,
(2) The c.d.f. is expressed by a simple function without integral form,
(3) It tails longer than the exponential distribution, and implies the occurrence of larger rainfall than predicted by the exponential distribution, and
(4) It is theoretically derived for the total depth of a single rainstorm.

The second feature is not only practically important, but vital in the following derivation of the distribution of the maxima. The c.d.f. of the maxima is expressed by the exponential function of the integral, i.e. the c.d.f. of the original variable as shown later {Eq. (12)}. Thus, if the second condition is not satisfied, the c.d.f. of the maxima takes a complicated form with an integral as the exponent, which is obviously impractical.

SQRT-ET-Max Distribution

As is well known, the c.d.f. of the maximum value of a compound Poisson process is expressed by Eq. (12) [Appendix 2]. The c.d.f. of the depth of a single rainstorm is deduced by substituting (8) for (12). Then,

$$F(x) = \begin{cases} 0 & (x < 0) \\ \exp\left[-\lambda \{(1+\sqrt{\beta x}) \exp(-\sqrt{\beta x})\} \right] & (x \geqq 0) \end{cases} \qquad (9) \text{ or } (1)$$

We shall call the distribution the SQRT-ET-max distribution, which is short for the square root exponential type distribution of the maximum.

The probability density conditioned by $(x \geqq 0)$ is

$$f(x) = 1/(1-e^{-\lambda}) \cdot \lambda\ h(x)\ F(x) \qquad (10)$$

where h(x) and F(x) are expressed by Eqs. (7) and (9).

The SQRT-ET-max distribution has characteristics similar to those of the SQRT-ET distribution, though it has two parameters, λ and β. The condition that x is large is automatically satisfied, since the distribution function is derived as for the maximum.
When Eq. (8) is replaced by the exponential distribution function, the Gumbel distribution is derived instead. The SQRT-ET-max distribution may possibly be approximated by the Gumbel distribution for large λ. However, the former tails much longer than the latter for medium values of λ.

It is recommended to use the maximum likelihood method for fitting. A practical, i.e., relatively simple and stable, fitting technique is proposed in Appendix 3.

APPENDIX 2

Assume the frequency of an event during a certain period follows the Poisson distribution. Then,

$$P(k) = \lambda^k / k!\ e^{-\lambda} \qquad (11)$$

where P(k) is the probability that the number of the occurrence of the event is equal to k, and λ is the average frequency of the event.
The probability that the magnitude of no event exceeds x is $\{H(x)\}^k$, where H(x) is the c.d.f. of the magnitude. In view of the randomness of k, the probability that no event exceeds, x, i.e., the cumulative

distribution of the maximum, is deduced by,

$$F(x) = \sum_{k=0}^{\infty} \lambda^k / k!\ e^{-\lambda} \cdot \{H(x)\}^k = \exp\left[-\lambda\ \{1-H(x)\}\right] \tag{12}$$

It is noted that, as x is defined in the region ($x \geqq 0$), $F(x)=0$ for $x<0$.

APPENDIX 3

The standard procedure of the maximum likelihood method may be explained as follows:

(1) Differentiate the logarithmic likelihood functions by the parameters and equate them to zero, and

(2) Solve the equations to get the values of the parameters.

However, the following procedure is more convenient and stable.

(1) Express λ in terms of β, solving the equation given by differentiating the logarithmic likelihood function ℓ by β and equating it to zero.

(2) Substitute the resulting equation for λ in ℓ, to reduce the number of the parameters to one, i.e. only β.

(3) Determine a value of β to maximize ℓ.

(4) Calculate the value of λ by substituting the estimated value of β to the expression of λ derived in (1).

The expressions of ℓ and λ are:

$$\ell = \sum_{i=1}^{N} \log_e f(x_i) \tag{13}$$

$$\lambda = \left(\Sigma \sqrt{\beta x_i} - 2N\right) / \{\Sigma\ (\beta x_i)\ \exp\ (-\sqrt{\beta x_i})\} \tag{14}$$

where f(x) is expressed by Eq. (10). x_i is the i-th sample, and N is the sample size.

APPENDIX 4

Let F(x) denote the c.d.f. of the annual maximum. The probability that the largest of the maximum sequence during N years is less than x is $\{F(x)\}^N$. When T is the average recurrence interval,

$$F(x) = 1 - 1/T \tag{15}$$

Let P and W respectively represent the probabilities that a value larger than the T-year probability does not occur at all during N-year observation and that such a value takes place at least once. Then, substituting (15) to {F(x)},

$$P = (1 - 1/T)^N, \quad W = 1 - P \tag{16}$$

It is obvious that P and W represent the probabilities that the historical largest during N years does not and does exceed the T-year probability value, respectively.

The average duration of observation at the 56 observatories, is 90.7 years. Substituting 90.7 for N and 1000, 5000, and 7800 years for T, then, W=8.7, 1.8, and 1.2%. Multiplying W by 56, the predicted number of observatories is 4.9, 1.0, and 0.7.

When T=N and N is sufficiently large (more than 50),

$$P = e^{-1} = 0.368 \tag{17}$$

For the historical largest 24-hour rainfall in Osaka in 1957, substituting 84 for N and 380 and 214 for T, W becomes 20-33%.

APPENDIX 5

For larger observation intervals, i.e., for a week to a month, two or more single rainfalls may take place in a unit interval. For the sufficiently long duration, the distribution of the total depth may be approximated by the normal distribution following the central limit theorem, and the sequence of the depths cannot be approximated by the marked point process. Thus, the distribution of the annual maxima may not be expressed either by the Gumbel or by the SQRT-ET-max distribution.

ON THE PROBABILISTIC CHARACTERISTICS OF POINT AND AREAL RAINFALL

U. Matsubayashi
Dept. of Civil Engineering
Nagoya University
Chikusa, Nagoya, 464
Japan

F. Takagi
Dept. of Civil Engineering
Nagoya University
Chikusa, Nagoya, 464
Japan

ABSTRACT. The time and spatial concentration of heavy rainfall results in the differences between point and areal averaging rainfall. The rainfall which is required for the flood control design in a drainage system should be discussed in its areal properties. If the probabilistic relationships between point and areal rainfall are clearly established probabilistic characteristics of areal rainfall can be estimated from the probability density function of point rainfall at a key-gauging station.

From this point of view, the present paper aims to establish the relationship between the probabilistic characteristics of point and areal rainfalls. For this purpose, the simulation technique of the rainfall event over a certain region has been used, after discussing the basic properties of rainfall. It is then demonstrated that the shift of the probability density function of point rainfall towards a lower rainfall intensity may give a good approximation to that of areal rainfall. The factors which govern the shift have been also discussed.

1. INTRODUCTION

Heavy rainfall occurs with high concentration in a certain time and space due to the movement of meteorological disturbances. To determine the design rainfall for a drainage system the temporal and spatial properties of rainfall should be investigated (Matsubayashi, et al,1984). Numerous research works on the temporal distributions of rainfall have been published (Talbot,1891,Sherman,1905). There are, however, many problems as yet unsolved regarding spatial distributions. Since the analyses require rainfall data from a large number of gauging stations within a region. Flood control projects are usually planned based on the statistical properties of the point rainfall without taking into account the concentration, i.e. the spatial distributions of each rainfall event. This situation results in an over-estimation of design rainfall for a drainage system.

To solve the problem DAD-analyses have been essentially performed (Horton,1924,Fletcher,1940). But, since DAD-ralationships are affected

V. P. Singh (ed.), Hydrologic Frequency Modeling, 265–275.

data, DAD-analyses also present difficulty in their application to all watersheds.

On the other hand, there is usually at least one key-gauging station in any watershed, and the rainfall data at the key station contains the local and statistical(in time) properties of rainfall at that point. In other words,local and statistical properties may be represented to some extent by the data obtained at the key-station. Therefore, if we describe the relationship in general form between the probabilistic characteristics of the point rainfall and those of areal rainfall, the relationship may be combined with the characteristics of rainfall at a certain key station to evaluate the probabilistic characteristics of the areal average rainfall even in a watershed with few gauging stations.

From this point of view, the present paper aims to investigate the relationship between the probabilistic characteristics of point and areal average rainfall. At the first step of the research, we focus our attention on problems concerned with depth-area relationship, and the duration problems are not discussed.

2. CHARACTERISTICS OF OBSERVED RAINFALL

2.1. Rainfall Data

Rainfall data analysed here are hourly data observed at 34 rain gauges in Nagoya City, Central Japan and at 11 rain gauges in its suburbs during the 13 years from 1971 to 1983. In this period, the rainfall events which brought about inundation in the urban region of Nagoya City have been dealt with. Figure 1 shows the locations of those rain gauges. The 11 rain gauges in the suburbs have been used to determine the rainfall properties in the surrounding regions of the city.

In order to deal with the statistical properties of spatial rainfall, we consider at first the statistical properties of peak rainfall* at a location where the rainfall intensity has its maximum value in a space and then the spatial distributions of rainfall around the location.

2.2. Uniformity and Isotropy of Rainfall

Firstly, let us examine the uniformity and isotropy of the rainfall events here concerned as the basic problem of discussion. Figure 2 shows the location of rainfall peaks on each rainfall occasion. In the figure, the symbols • and o express rainfall events with a peak intensity large than 30 mm/hr and 20 mm/hr at the center of the rainfall region, respectively. As will be seen in Fig.2, rainfall peak appears uniformly.

---

* The term "peak rainfall" expresses rainfall peak at the location where the rainfall intensity has its maximum value in a space,and the term "peak rainfall in time" is used for the peak of rainfall time series at a certain rainfall gauge, i.e. of point rainfall.

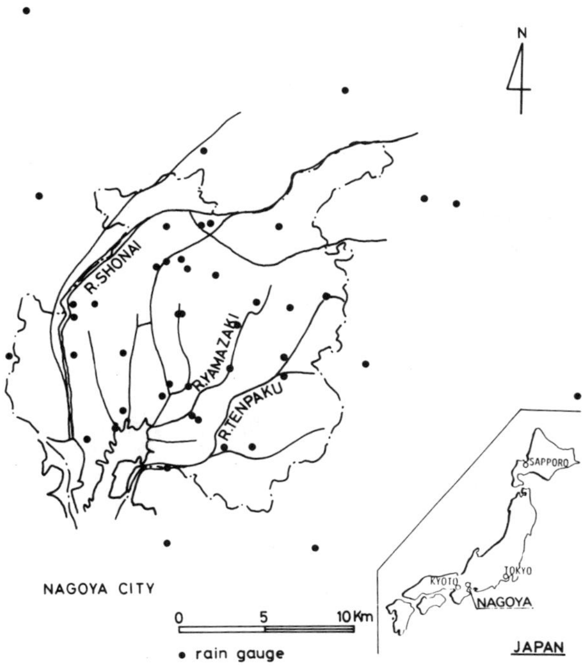

Fig. 1 Site of rain gauges.

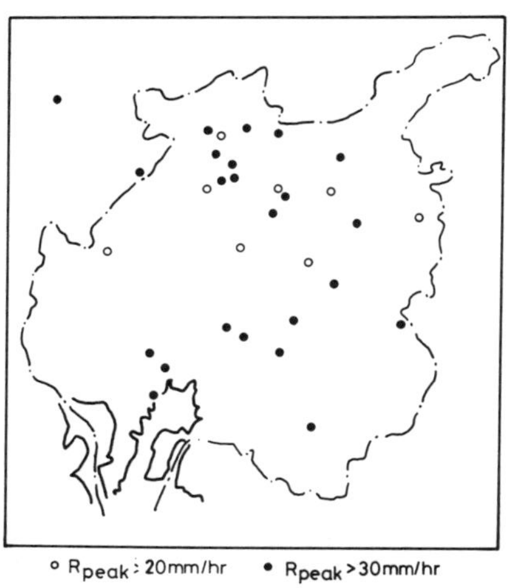

Fig. 2 Location of rainfall peak in a space.

by the local character of rainfall and we have not sufficient rainfall

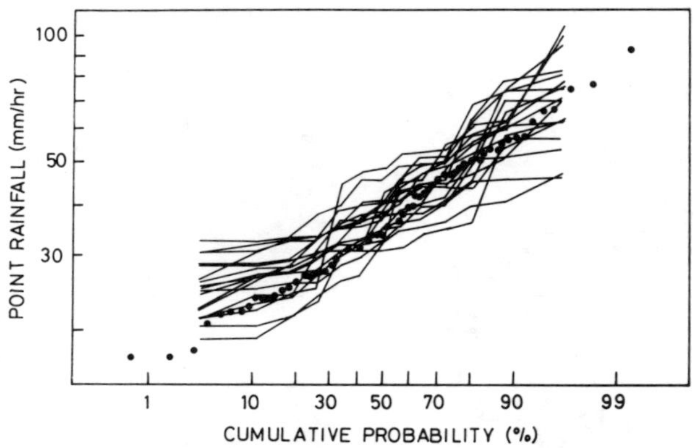

Fig. 3 Cumulative distribution function of point rainfall at each rain gauge (solid line) and Nagoya Meteorolgical Observatory (dotted line).

The statistical characteristics at each rain gauge have also been evaluated to examine uniformity. The logarithmic-normal distribution Eq.1 is applied to the upper 13 rainfalls in the period 1971-1983 at each rain gauge.

$$F(R) = \left(\frac{1}{\pi}\right)^{1/2} \int_0^{\xi} \exp(-\xi^2)\, d\xi \qquad (1)$$
$$\xi = a \{ \log (R-b) - \log (R_0-b) \}$$

The distribution of rainfall in time for each rain gauge is shown in Fig. 3. In the figure, the dotted lines express the distribution of rainfall events during a period of 76 years at the key-station (Nagoya Meteorological Observatory). Although the distribution station differ slightly from each other, the figure suggests the same type of distribution of peak rainfall in time at any gauging station,thus assuring the uniformity of rainfall events in the region.

In respect of the isotropy, Fig.4 shows the correlation coefficient of hourly rainfall of each of two gauging stations versus the distance between them. The coefficient is calculated by hourly rainfall within a few hours around the peak and is shown in Figures 4a-4d classified by the direction of the line combining the corresponding two gauges. We may call the rainfall event in a region isotropic, if the lines are same regardless of direction. In the figure, we find that decreases in the correlation with the distance vary depending with the direction but the difference may be perceived as insignificant.

The isotropy of the rainfall event in a certain instance may be examined directly by means of an iso-hyetal map. Figure 5 shows some typical examples of iso-hyetal maps at the time at which the hourly rainfall has its peak. Many of the maps confirm the isotropy of rainfall in the region, but some show more or less an elliptic shape with the major axis from north to north-east. The reason why this tendency is observed may lie in the fact that the prevailing wind direction is from south to south-west during the rainy season in Nagoya.

As stated above, although the rainfall is not isotropic in the strict sense of the word, the rainfall event in the region concerned may be considered as isotropic in the macroscopic sense. Thus for simplifying the procedure, the rainfall is assumed isotropic in the following simulation. We mention here that this assumption is not essential. In other words, if anisotropy is clearly established in the future, it is relatively easy to consider the effect of anisotropic properties in a simulation.

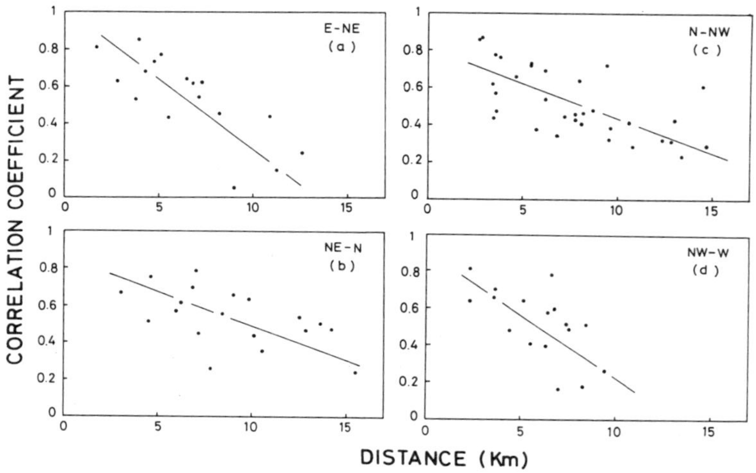

Fig. 4 Correlation coefficient of point rainfall versus the distance between gauging station.

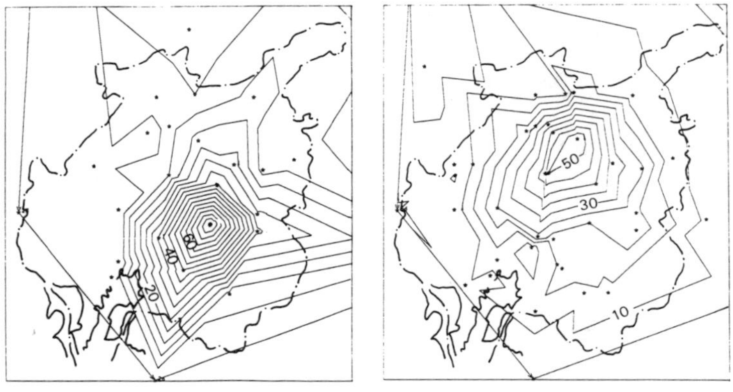

Fig. 5 Iso-hyetal maps. (unit : mm/hr)

2.3. Probability Density Function of Peak Rainfall

The cumulative probability function of peak rainfall at various locations is shown in Fig. 6 with respect to rainfall with its peak intensity larger than 20 mm/hr in Nagoya City during 13 years. The figure implies that the logarithmic normal distribution Eq. 1 provides a good approximation to the distributions of peak rainfall in a space as well as those of point rainfall.

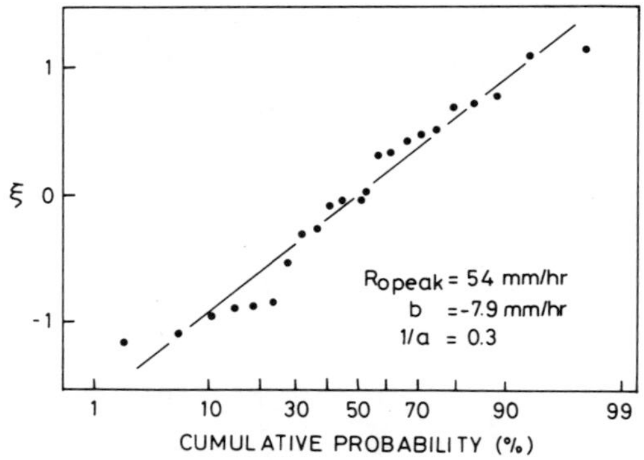

Fig. 6 Cumulative distribution function of peak rainfall.

2.4. Spatial Distribution of Rainfall

Rainfall intensity decrease with the horizontal distance from the location of the rainfall peak as observed in Fig. 5. Figure 7 shows the rainfall intensity $R(x)$ at any location normalized by the peak rainfall R_{peak} versus the distance x from the center of the rainfall event. Distance x was defined by the radius of a circle the area of which was

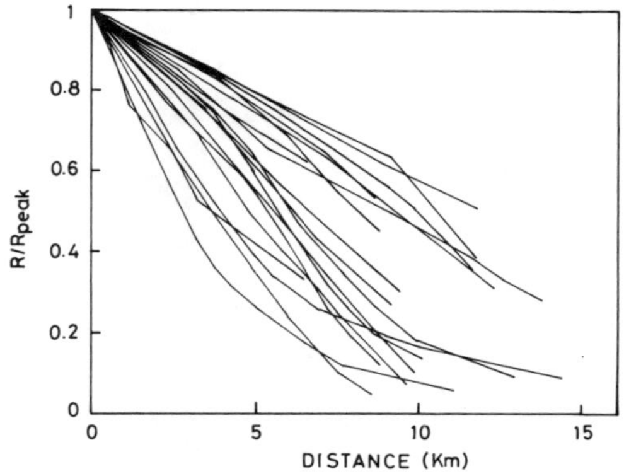

Fig. 7 Spatial recession of rainfall intensity with the distance from the location of peak rainfall.

equivalent to that encompassed by the corresponding iso-hyetal line of rainfall intensity R. From the figure we can find various decreasing patterns of R(x) with distance x. That is, the liner recession type or exponential one depending on the meteorological conditions. However, for the design rainfall, spatial emphasis should be laid on the neighbourhood with a small x of peak rainfall. Therefore we may approximately describe the rainfall distribution as

$$R(x) = R_{peak} \left(1- \frac{x}{L_o} \right) \qquad (2)$$

in which L_o denotes the distance characterizing the decrease. It is pointed out that the factor L_o is distributed uniformly over a range from 5 km to 25 km of the region concerned.

3. RAINFALL SIMULATION AND EVALUATION OF POINT AND AREAL RAINFALL

In this section, a rainfall event is simulated with respect to its peak rainfall and the spatial distribution based on the discussions in previous sections. The simulated results are then transformed into a relationship between the statistical characteristics of point rainfall and areal average rainfalls with various scales.

The simulation process is summarized as follows:

(1) First, the location of rainfall peak is chosen at random from the square simulation region (LxL) schematically shown in Fig. 8.

(2) The peak rainfall intensity is generated following the logarithmic normal distribution Eq. 1.

(3) The parameter L_o is statistically defined according to uniform distribution. Then, the spatial distribution of rainfall is estimated by Eq. 2.

(4) As the "point rainfall" the rainfall at the center of simulation region (LxL) is adopted. The "areal rainfall" is defined as the average value of rainfall over the square region (SxS) located at the center of the simulation region.

(5) Several thousand rainfall events have been generated by the above simulation procedure and the probability density functions of point rainfall and of the areal rainfall have been evaluated.

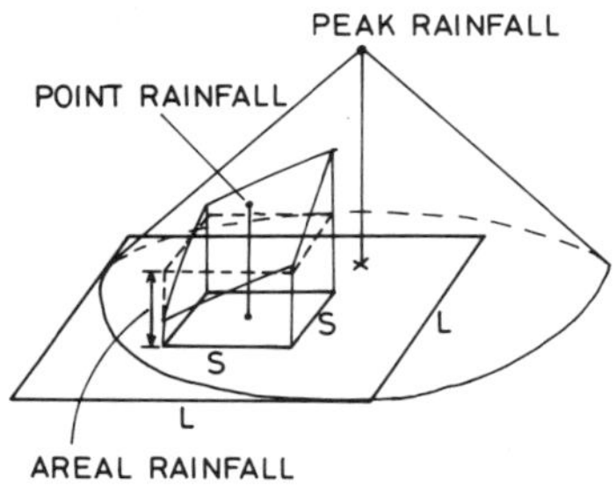

Fig. 8 Schematical figure of rainfall simulation.

To deduce the probability density function in a time ensemble from the simulated results, it should be noted as to how the probability density function of rainfall in a time ensemble is related to the number of simulated cases N, the annual frequency ρ of rainfall events per unit area and the scale of the simulation region LxL. For example, let us compare two cases with the same number of simulation N_1 over $(L_1 x L_1)$ region and over $(2L_1 x 2L_1)$ region. If the former case corresponds to the simulation of the rainfall event during M years, then the data of latter case represent data during (1/4)M years, since the simulation region is 4 times the former one. That is, generally, N-times simulation over the region (LxL) generates the data from $N/(\rho L^2)$ years. The probability density functions of point and areal rainfall have then been calculated with the upper $N/(\rho L^2)$ data among the total simulated rainfalls.

The results obtained through the simulation are expressed in terms of the following non-dimensional factors, that is,

$$R_o'area = R_o\ area / R_o\ point,$$
$$b' = b / R_o\ point,$$
$$L' = L / \overline{L}_o,$$
$$S' = S / \overline{L}_o$$

in which R_o area is the parameter R_o in the logarithmic normal distribution in Eq. 1 for areal rainfall and $\overline{L}_o$: the mean value of the parameter L_o in Eq. 1.

4. RESULT OF SIMULATION AND DISCUSSIONS

To discuss the statistical relationships between point rainfall and areal rainfall, the parameters in their logarithmic normal distribution Eq. 1 should be considered. The change of factor $R_o'_{area}$, b' and 1/a according to non-dimensional area A' (=S'xS') are shown in Figs. 9 to 11 with the parameter $1/(\rho\overline{L}_o^2)$. The parameter $1/(\rho\overline{L}_o^2)$is the expression of the value $1/\rho$ normalized only by area and has a dimension of time (year).

It may be seen from Fig. 9 that the factor $R_o'_{area}$ decreases linearly with an increase of non-dimensional area A' and reaches about 85 % of the point rainfall $R_{o\ point}$ for the non-dimensional area A'=0.6. Moreover, $R_o'_{area}$ is independent of the parameter $1/(\rho\overline{L}_o^2)$.

As seen in Fig. 10, the lower limit b' of the probability density function also decreases linearly with area A'. This factor though significantly depends to a considerable extent on the parameter $1/(\rho\overline{L}_o^2)$.

Figure 11 shows the value of 1/a corresponds to the standard deviation of the variable log (R-b) in Eq. 1. The figure clearly shows that this value changes little for area A'.

As discussed above, the changes of these three factors define the changes of probability density function with respect to the area over which the areal rainfall is considered. However, as easily understood in Eq. 1, factor b plays a role in adjusting the distribution only in the lower range of rainfall. Since factor 1/a is kept at a constant for the change of area, the decrease of $R_o'_{area}$ results in a change of the

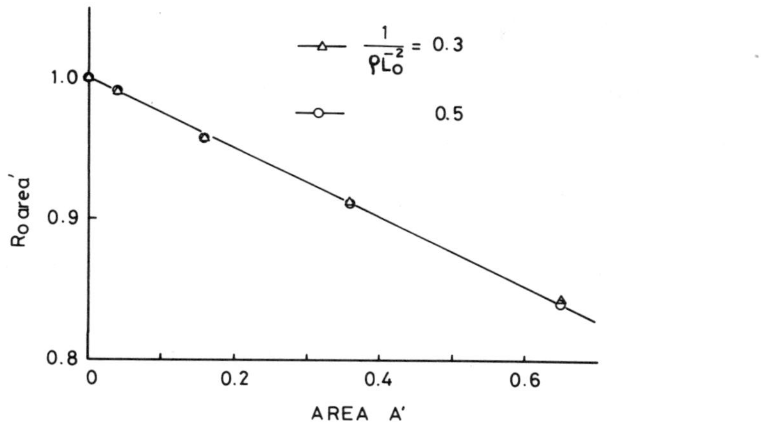

Fig. 9 Change of R_o'area (=R_o area/R_o point) with non-dimensional area A' (=S'xS')

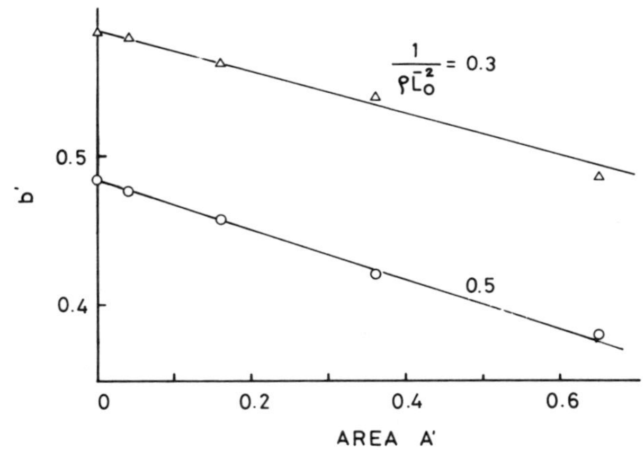

Fig. 10 Change of b' with non-dimensional area A'.

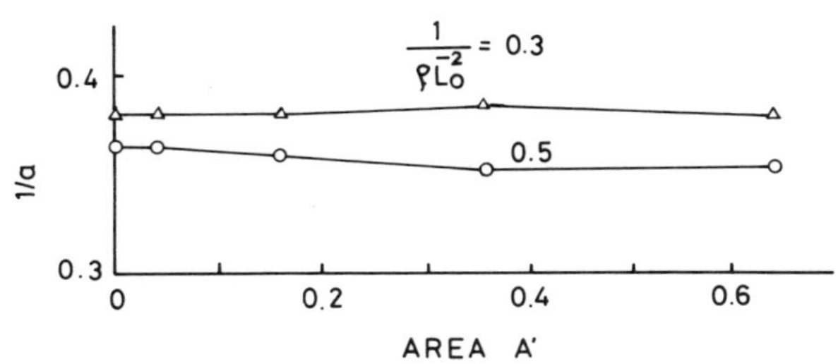

Fig. 11 Change of 1/a with non-dimensional area A'.

probability density function with an increase of area. That is, in a macroscopic sense, the probability density function of point rainfall is shifted towards the direction of a lower value of rainfall with an increasing area A' to the probability density function of areal average rainfall. Moreover, what we would also like to emphasize here is the fact that the shift defined by the $R_o'_{area}$ is independent of the parameter $1/(\rho \bar{L}_0^2)$. In this manner, we can transform the probability density function of point rainfall at the key-gauging station into the probability density function of areal average rainfall.

At the end of the present paper, we would like to mention the rainfall intensity with return period T. Figure 12 shows the change of areal average rainfall of return period T with an increment of area A'. It may be pointed out that the decreasing gradient of rainfall with area A' is almost constant regardless of of return period T. The figure also suggests the fact that point rainfall will over estimate the areal rainfall over a watershed, for example a 15 % overestimation for area A'=0.6.

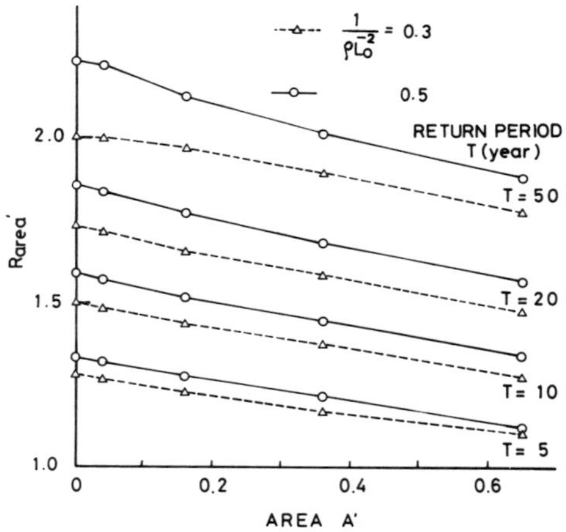

Fig. 12 Change of areal rainfall for return period T according to non-dimensional area A'.

5. CONCLUSION

The conclusions obtained through the present paper may be summarized as fallows:

(1) Logarithmic-normal distribution provides a good approximation to the distribution of peak rainfall in a space.

(2) The rainfall intensity decreases linearly with the distance from the location of peak rainfall.

(3) Rainfall events have been simulated over a certain region, and the results introduce the relationship between statistical characteristics of point rainfall and areal average rainfall. To deal with the problem in general, non-dimensional parameters have been introduced.

The relationship so obtained demonstrates the possibility of estimating the probability density function of areal average rainfall based on that of point rainfall. If we use point rainfall itself for an over all drainage system, the design rainfall can be generally over-estimated compared to that obtained based on areal rainfall.

REFERENCES

1) Fletcher, R.D. 1940. A Relation Between Maximum Observed Point and Areal Rainfall Values, Trans. AGU,31.

2) Horton, R.E. 1924. Discussion on Distribution of Intense Rainfall, Trans., ASCE, 87

3) Matsubayashi, U., F. Takagi and A.Tonomura 1984. The Probability Density Function of Areal Average Rainfall, J. of Hydroscience and Hydraulic Engineering. JSCE, 2-1 pp. 63-71.

4) Sherman, C.W. 1905. Maximum Rates of Rainfall at Boston, Trans. ASCE, 54.

5) Talbot, A.N. 1891. Rate of Maximum Rainfall, Technograph, 5, Univ. of Illinois.

COMPARISON OF THREE METHODS OF ESTIMATING RAINFALL FREQUENCY PARAMETERS ACCORDING TO THE DURATION OF ACCUMULATION

M. Slimani
Institut de Mecanique de Grenoble.
BP 68X
38402 Saint Martin d'Heres Cedex.
France

T. Lebel
ORSTOM-Hydrologie.
ZOLAD, MiniParc
34100 Montpellier.
France

ABSTRACT. The 'Gradex' method estimates high return period discharges by assuming that the two marginal increases in rainfall and flood volumes are equal for return periods greater than a hundred years. This requires an asymptotically exponential decay of the rainfall distribution, the gradex being the gradient of this exponential. Among several statistical relationships satisfying this requirement, the Gumbel distribution is widely acknowledged to properly fit the extreme value distribution of observed daily rainfalls. However, difficulties may arise for small time steps (from less than one hour up to twelve hours).

Among the many reasons for these difficulties, seasonal variation of the distribution parameters and the mixing of rainstorms issuing from different weather conditions are the most frequent. Rainfall distributions should therefore be analyzed using data from a homogeneous season. However, the mixing of weather patterns remains whatever the season considered, and it is important to find a way to estimate the gradex parameter even when the experimental distribution of extreme values does not allow a good Gumbel model fit. One solution is to fit a two component negative exponential distribution to the complete data set. Such a theoretical distribution is asymptotically parallel to a Gumbel distribution, thus providing an estimate of the gradex.

In this communication, three gradex estimators are compared from bias and efficiency standpoints : the moment and maximum likelihood methods used to fit a Gumbel distribution to an extreme value data set, and the moment method used to fit a two component exponential distribution to the complete data set of observed rainfalls. This comparison can be used to determine the most suitable estimator for subsequent regional regression and mapping of rainfall frequency parameters(see : linear relations between rainfall and morphometric parameters by Slimani and Obled, this symposium).

V. P. Singh (ed.), Hydrologic Frequency Modeling, 277–291.

1. INTRODUCTION

This communication is one outcome of a continuous effort aimed at providing engineers in relevant disciplines with means of estimating high return period floods in the **CEVENNES** (fig.1), France's most problematic region in this respect. The upland watersheds in this region are relatively small (a few hundred square kilometers), but rugged topography and complex meteorological phenomena combine to produce flash-floods threatening human life and property, as already observed in October 1958 (33 casualties) and September 1980 (a dozen casualties and damage amounting to more than one billion francs). Despite the avaibility of only one or two long series of stream flows in this region, the **GRADEX** method is believed to give reliable estimates of the risk. The method relies on the analysis of rainfall data assumed to display an asymptotically exponential decay behavior. The GRADEX is the gradient of this exponential decay. The ultimate goal of the study is to provide those in charge of hydraulic development projects with a map of the gradex (g) and rainfalls for given return periods at various durations of accumulation (see communication by Slimani and Obled, this symposium). For this, reliable estimates of g are needed, and the present paper is devoted to comparing some such possible estimates.

Accepting the exponential decay assumption, an estimation of g may be reached by studying either the parent distribution series (PDS) or the series of maxima computed over a given period (month, season, year). Under the hypothesis that the compound Poisson process (CPP) (eg see Todorovic and Zelenhasic, 1970) is a valid model for successive rainfall depths accumulated over a given duration, the asymptotic exponential decay of the PDS involves the asymptotic exponential decay of the extreme value series (EVS). While it is more convenient to deal with EVS, since the samples are smaller and complete rainfall data processing is not necessary, the heavy right hand tail of some EVS sometimes casts doubt on the validity of the classical Gumbel distribution (EV1) as a suitable model of rainfall extreme value distributions. This is particularly the case for short duration rainfall all around the Mediterranean basin since the weather patterns accounting for monthly or seasonal maxima vary greatly

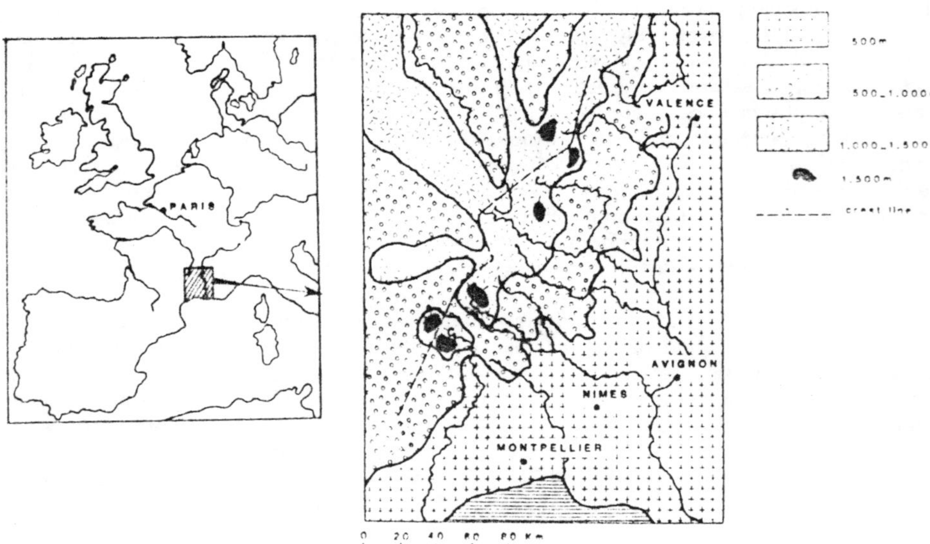

Figure 1. The Cevennes Region of France.

throughout the year. Accordingly, the year must be carefully divided into homogeneous seasons before any attempt to analyze such series. In this part of France, the large majority of severe weather patterns occurs in fall, which essentially means September, October, November. Of course, the distribution of interest is that of annual rainfall, but when almost every yearly maximum occurs during the same season, clearly the distribution of yearly maxima rapidly becomes parallel to that of the concerned season maxima(Guillot and Duband, 1967)

However, large skewness in EVS may be found, even when working on homogeneous season data, and in such cases, the EV1 distribution does not properly fit observed data. The estimation of the skewness coefficient and the detection of outliers has been given great attention by hydrologists for the past twenty years, mainly with respect to flood series analysis (e.g. Bernier,1967; Todorovic and Rousselle 1971; Matalas et al.,1975; Gupta et al., 1976; Kottegoda, 1984). By comparison, relatively few papers have been concerned with the question of how to deal with outliers in rainfall EVS, using PDS information. To our knowledge, D.Duband (1967) was the first to consider a two component exponential distribution as a model of the underlying distribution of daily rainfall depths, in order to make better use of the information contained within the rainfall data. Since the coefficient of one component is greater than that of the other, the asymptotic behavior of the model is similar to that of a single exponential; as a consequence, the derived distribution of extremes is an EV1 distribution, following Gumbel's asymptotic theory (e.g. Gumbel, 1958). It must therefore be emphasized that, if the underlying distribution is truly a two component exponential distribution, the study of PDS or EVS should lead to the same estimates of g, not taking into account model and sampling errors. Such errors are of dramatic importance, since g conditions the extrapolation of flood series to high return period values. Hence, a comparative study of three estimates was carried out, two of them applying to EVS and the other to PDS.

2. THE GRADEX METHOD OF ESTIMATING HIGH RETURN PERIOD FLOODS

Pointing out that very often available flood series are inadequate for reliable extrapolation of the experimental distribution towards extreme flood flows, Guillot and Duband (1967) proposed a method using rainfall data analysis to extrapolate flood series distributions. In flood conditions, precipitation and flood volumes computed over a common integration time tg :

$$R_g = \int_{t_1}^{t_1 + t_g} R(t)\,dt \quad and \quad Q_g = \int_{t_2}^{t_2 + t_g} Q(t)\,dt$$

can be plotted on a graph as in figure 2. If the integration time is close to the base time of the overland flow hydrograph (tb), the portion of the rainfall Dg = Rg-Qg not reaching the main stream is related to the overall basin retention during the total time (t1, t2 + tg). This retention Dg is a random quantity tending towards an upper limit as long as the rainfall intensity is large compared to the percolation rate to the ground water tables.

Most hydrologists now agree that for moderately permeable watersheds of a few hundred to a few thousand square kilometers, it is physically sound and reasonably safe to assume that, as the total basin saturation is approached, any additional precipitation Pg tends to give rise to a nearly equal increase in the discharge Qg. Thus the asymptotes of the quantile curves of the joint distribution (Rg, Qg) (figure 2) are straight lines with slopes of one (Qg = Rg) , whatever the quantile considered. The total amount of precipitation needed to

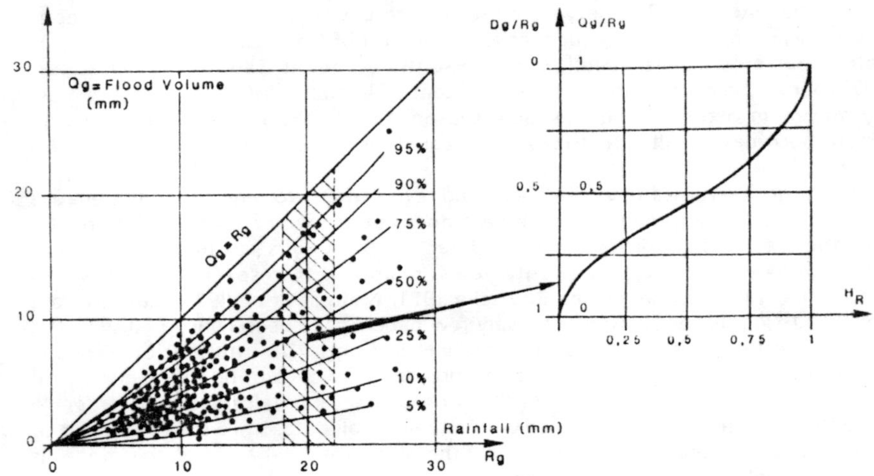

Figure 2. Conditional distribution of flood volumes with respect to rainfall depth. Dg = Qg- Pg is the retention.

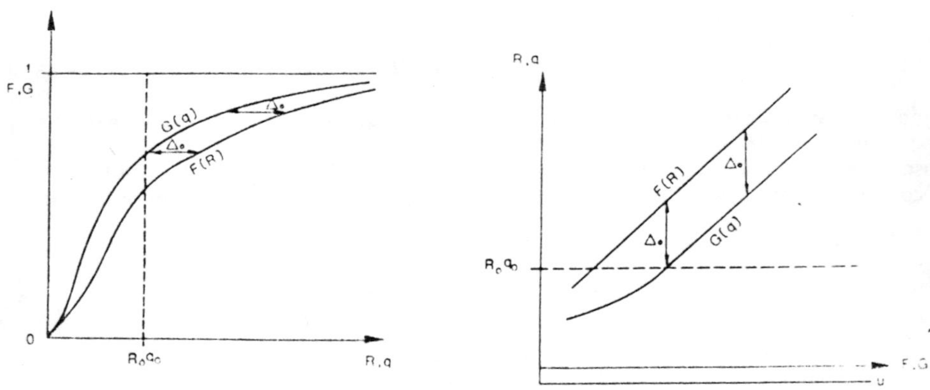

Figure 3. Parallelism of rainfall and discharge distributions, when plotted on a Gumbel probability paper. The equidistance of the distributions G(q) and F(R), beyond the threshold (Ro, qo) in arithmetic coordinates(left) involves the parallelism of the asymptotes in Gumbel coordinates(right).

reach saturation depends on the initial dryness of the soil, as does the distance between each asymptote and the curve Q = R.

When large samples of rainfall-flood data are available, the conditional probability is obtained from the experimental cumulative distributions Hp(Qg), and the theoretical flood volume cumulative distribution is given by:

$$G(Q_g) = \int_{q=0}^{q=Qg} \int_{R=0}^{R=\infty} h_p(q)\ dF(R)\ dq$$

Furthermore, a simple and operationally important consequence results from the assumptions of the method. **If the rainfall distribution is asymptotically an exponential**, the extrapolated asymptote of G(Qg) is translated from the asymptote of F(Rg) in a direction parallel to the variate (Rg,Qg) axis. In particular, on conventional Gumbel probability paper, these two asymptotes are parallel straight lines (fig. 3).To sum up, while it is very unlikely that a few decades of flood measurements allow a reliable estimation of the direction of extrapolation, the gradex method assumes that the numerous long series of rainfall data available almost anywhere in the world can provide a safe extrapolation means. Indeed, extrapolation of flood series using a model fitted to flood data only can lead to tremendous overestimation as it occurs when fitting a Frechet or a Galton model to the flood data sample. The use of log space (Z = Log(Q)) distributions has also been shown to be dangerous and counterproductive by Landwher et al.(1978); the fact that a log-type distribution is often a better description of the observed reality does not imply that it is suitable for extrapolation purposes. Accordingly, Waylen et al.(1982) proposed a two component exponential distribution to fit annual flood series (AFS) accounting for the mixture of snowmelt and rain generated floods throughout the year. Rossi et al.(1984) proposed a similar distribution for annual flood series of central and southern Italy, presenting a large skewness. Although this distribution closely reproduces the observed skewness, Rossi et al. emphasized the great uncertainty related to the parameter estimates of the outlying component (i.e. the one determining the extrapolation) when a single 40 year AFS is used, which means that in many cases the available discharge measurements can only to be used to give a good estimate of the starting point (Q0) of the extrapolation(figure 3).

For all its advantages when applied to relevant watersheds, the gradex method presents some practical difficulties which are not central to the basic concept, but which need some clarification and further study, as pointed out by Beran(1981). Some are related to the time of integration and the peak to volume adjustment factor. Others are rainfall related problems, among which the transformation of point rainfall statistics into areal rainfall is crucial. An initial procedure was presented by Lebel (1984), yielding good results for small upland catchments, but not yet extended to larger watersheds. It is also important to ascertain the best way to estimate the point gradex from rainfall series. Indeed an important characteristic of g is its spatial structure (derived from the spatial stucture of rainfall) allowing the mapping of this parameter. Such maps are needed by hydraulics engineers as a basic design tool and their preparation requires the interpolation of homogeneous point values of g, which in turn means a regionally dependable estimation procedure. Also, gradex maps are needed for several rainfall durations, since the time of integration tg is related to watershed areas. This requirement gives rise to another set of difficulties since recording raingage series are fewer and often shorter than their raindepth counterparts, and outliers are frequently more numerous in hourly rainfall than in daily rainfall series.

3. ESTIMATING THE GRADEX FROM EXTREME VALUE SERIES

A physically based explanation was given by Giraud (1959) to support the idea that, at least for durations equal to or smaller than 24 hours,the asymptotic behavior of rainfall distributions is exponential. While this has been widely accepted for decades, discussion continues as to which procedure provides the best estimate of the slope g of this exponential. The simplest method is to assume that the consecutive rainfalls PDS has a single exponential distribution, from which it is a straightforward task to derive the associated EVS distribution. Obviously, the single exponential cannot hold for every PDS, and it may be more general to write the equation for values above a threshold S0; the number of events $\{R>r\}$ for N independent observations of R is then writen as :

$$N_R(r) = \exp(-\alpha *(r-r_o)) \qquad \text{for all } r>S0 \qquad (1)$$

At the threshold:

$$N_R(S0) = \exp(-\alpha *(S0-r_o)) \qquad (2)$$

and below the threshold, $N_R(r)$ is unknown. Further assuming that :

$$P_1 = P(r<S0) \qquad (3a)$$

$$P_2 = 1-P_1 = P(r>S0), \qquad (3b)$$

NP_2 values above S0 are observed on the average for N independent observations of R. Thus, $NP_2 = N_R(S0)$, and from (2), we obtain:

$$r_o = S0 + \mathrm{Log}(NP_2)/\alpha \qquad (4)$$

Now let X be the maximum of the N independent observations. Then:

$$F_x(r) = \{F_R(r)\}^N = \{1- N_R(r)/N\}^N \qquad r>S0 \qquad (5a)$$

$$F_x(S0) = P_1^N \qquad r = S0 \qquad (5b)$$

Assuming N to be large with respect to N_R values, we obtain :

$$F_x(r) \# \exp(-N_R(r)) \qquad \text{for } S0<r, \qquad (6)$$

which substituting (1) and (4) for $N_R(r)$, gives:

$$F_x(r) = \exp\{-\exp(-\alpha * (r-r_o)\} \qquad (7)$$

with r related to S0 and N by expression (4). Expression (7) is the EV1 distribution, and $1/\alpha = g$ is the gradex. It holds for $r>S0$, while for $r<S0$ the distribution is unknown, but very likely different from that given by (6 and 7).

If the basic model (1) is plausible, it has certain practical consequences when fitting an EV1 model to EVS:

1- There is a ($100*P_1^N$) per cent chance that a given value of the EVS does not belong to the EV1 distribution ($r<S0$).

2- The upper part of the distribution is EV1, and a separation appears at S0. The gradex g is equal to $1/\alpha$.

3- As N increases, the probability of having an EVS value below the threshold S0 decreases, and the better is approximation (6).

To illustrate these statements, consider a PDS of daily rainfalls and an EVS of monthly maxima (N1 = 30), with a recurrence rate of 93% for the event {r<5 mm}(S0 = 5mm). Thus P = $(.93)^{30}$ = .142, and hence, on the average, the fourteen lowest values out of a hundred do not belong to the EV1 distribution. To get rid of this left hand tail, an EVS of season maxima (N2 = 90) may be chosen. The probability of a low value outlier then falls to .001, and the EV1 model fits the sample data very well (fig.4). Of course, the recording period must be long enough to allow the reliable fitting of any model. Thirty values is widely acknowledged to be a minimum. Therefore, short rainfall series displaying a break in the point swarm of monthly maxima call for another kind of modeling if a dependable estimate of g is to be computed. This is also the case for distributions with several high value outliers. This generally denotes the mixing of two underlying populations (R1 and R2), the recurrence rate of R2 being small compared to that of R1, but R2 yielding much heavier rainfalls. If the two distributions of the maxima X1 and X2 of each population are denoted respectively F_{X1} and F_{X2} and if they are independent, the resulting cumulation function of the rainfll X is given by:

$$F(X) = pF_{X1}(X) + (1-p)F_{X2}(X),$$

where p is the probability of occurrence of X1. In the CEVENNES region, frontal and convective rains are often mixed in the same rainfall event in fall and it proves difficult to separate data originating in the first distribution from that originating in the second distribution. On the one hand, **table 1** clearly shows that long and intense rainfalls are not always associated (compare event # 31 with event # 28). On the other hand, events #2 and 23, among others, were long and had strong hourly maxima around 60 mm. Hence, while some events belong to only one of the above distributions, others result from the addition of two physical processes.To get around this difficulty, it was decided to test the suitability of a direct model of the complete PDS, when the series was available.

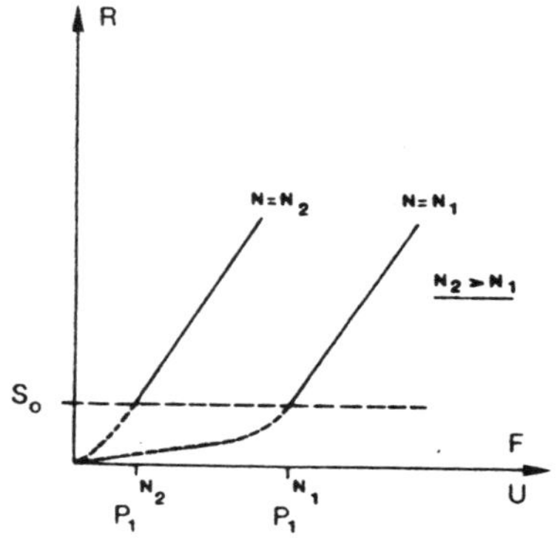

Figure 4 Elimination of low value outliers. When the number of events over which the maxima are calculated is increased, the break in the lower part of the EVS distribution disappears.

4. MODELING OF PARENT DISTRIBUTION SERIES

Let the PDS be a sequence of N consecutive rainfalls, Ri, i = 1,N , for a given duration of accumulation, including zero values. Duband(1967) assessed that the sum of two exponentials represented a good model for fitting daily rainfall distributions of the central part of France :

$$F_R(r) = 1-\gamma \exp(-r/g)- \beta\exp(-r/c), \quad r>=0, \tag{8}$$

$$\text{with } g>c. \tag{9}$$

Table.1 Main rainfall events of the ten past ten years. Rainfall depth in mm.

Event number	Beginning Date	Duration (approximate)	Mean Depth° for Event	Event Maximum	Hourly Maximum
1	19 sept 1971	30 h	23.1	68.4	28.5
2	4 sept. 1972	66 h	85.0	227.5	57.0
3	10 oct. 1972	80 h	21.4	241.7	90.0
4	25 oct. 1972	36 h	100.8	246.3	37.0
5	20 sept. 1973	34 h	34.6	272.0	80.0
6	1 oct. 1973	40 h	135.2	183.0	35.0
7	2 nov. 1973	90 h	94.2	352.0	36.5
8	16 sept. 1974	90 h	148.6	371.8	65.0
9	14 nov. 1974	103 h	64.8	219.0	19.0
10	17 nov. 1974	40 h	35.3	82.5	40.0
11	14 sept. 1975	56 h	63.4	176.0	70.0
12	29 sept. 1975	30 h	61.9	161.5	57.0
13	28 aug. 1976	90 h	148.0	392.0	60.0
14	11 sept. 1976	40 h	110.7	400.5	85.0
15	27 sept. 1976	52 h	57.7	145.5	40.5
16	30 sept. 1976	53 h	46.8	120.7	33.0
17	10 oct. 1976	92 h	78.8	178.8	34.5
18	23 oct. 1976	160 h	175.1	404.9	23.0
19	8 nov. 1976	65 h	180.2	371.4	42.5
20	28 jul. 1977	70 h	81.8	131.0	19.0
21	25 aug. 1977	48 h	87.9	266.5	49.0
22	5 oct. 1977	75 h	121.5	355.7	35.2
23	19 oct. 1977	120 h	273.3	596.5	61.0
24	25 oct. 1977	72 h	43.7	117.5	30.0
25	4 oct. 1979	36 h	50.7	118.0	71.0
26	6 oct. 1979	27 h	51.3	121.0	26.0
27	10 oct. 1979	160 h	198.6	566.0	38.5
28	22 oct. 1979	160 h	228.0	375.0	16.5
29	15 aug. 1980	22 h	35.7	86.7	51.0
30	26 aug. 1980	30 h	49.5	90.7	45.0
31	20 sept.1980	32 h	176.8	591.0	100.0
32	15 oct. 1980	46 h	141.1	204.2	60.0

° Mean value of 59 raingages.

Expressions (8) and (9) imply that, for large values of r,

$$F(r) \# 1- \gamma \exp(-r/g),$$

thus, the EVS are asymptotically EV1 distributed. This function (8) was fitted using a method similar to the method of moments, involving the estimation of four parameters γ, β, g,c, assuming at first the invariance of the coefficient of variation σ/μ, and fixing the relation:

$$\gamma = \mu^2 / \sigma^2 ,$$

where μ and σ are the mean and the standard deviation of the sample. This gives:

$$g = (\mu\gamma + (\gamma\beta (K\theta - \mu^2))^{1/2})/\gamma \qquad (10a)$$

$$c = (\mu - \gamma a)/ \beta , \qquad (10b)$$

with $\beta = 1- F(0)- \gamma;$ $\theta = 1- F(0)$; $K = (\mu^2 + \sigma^2)/2$ (10c)

5. COMPARISON OF THREE GRADEX ESTIMATES

Three estimates of the fall gradex were first compared:

* g_1: slope of the conventional moment fitting of the EV1 model to EVS.
* g_2: slope of the maximum likelihood fitting of the EV1 model.
* g_3: parameter of the model (8) computed using formula (10a) above.

Among the 54 recording raingages available in 1982, 25 were selected to carry out this comparison.

5.1. Gumbel Distribution Fitted to Period Maxima by the Method of Moments.

The method was tested on rainfall maxima over periods of 10 days, 15 days, one month, one year (i.e. one season) and two years (two seasons), with special attention paid to possible changes at any breaks in the point swarm. The aim was to determine the minimum sampling period required for correct estimation of the Gumbel parameters by the moments. As a matter of fact, when the empirical distribution does not correspond perfectly to a Gumbel distribution and when a curve or break occurs in the point swarm, this method gives a gradex which is biased by the values at the bottom of the distribution in the second order moment. Several stations offering long duration records were used, such as Montpellier Bel-Air (1922-1980), allowing the determination of maxima over periods of up to three years.The resulting lines of fit for maxima calculated over 5-day, 15-day, 1-month, 1-year, 2-year periods are shown for hourly and daily rainfalls in figure 5, revealing a break starting from the first 5-day period. This break becomes less and less distinct as the period length increases, becoming virtually negligible for annual maxima. For some other stations, monthly maxima series are almost as good as annual maxima series. In figure 5, each EVS displays a straight line asymptpotic behavior, the lines being roughly parallel to one another. Expression (4) allows computation of the theoretical distance between the asymptotes; if one asymptote is obtained with a sample of N1 independent observations, and the other with a sample of N2 independent observations, then:

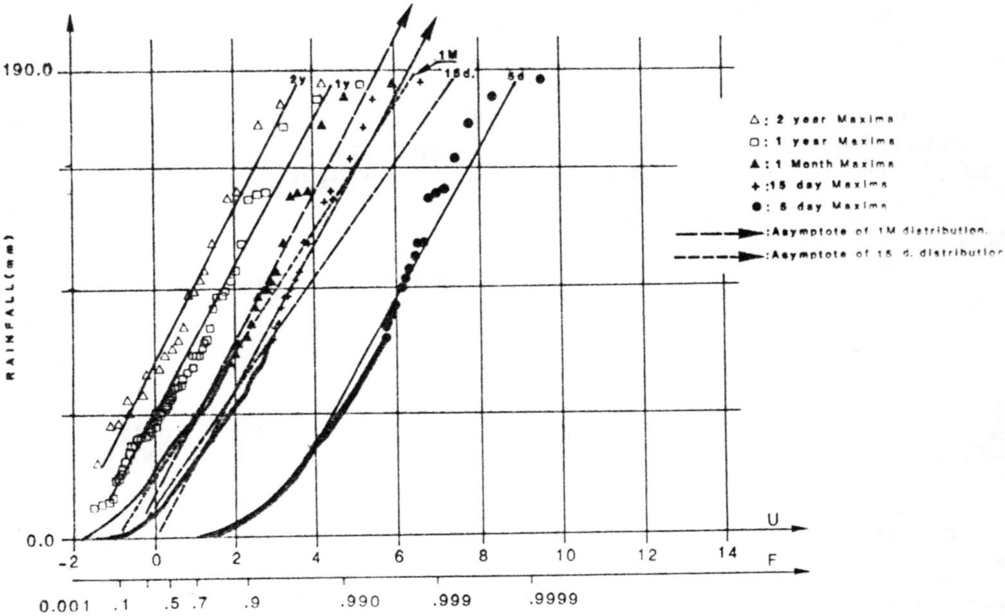

Figure 5. EVS distributions of daily rainfalls at Montpellier Bel-air. The maxima are calculated over 5-day, 15-day, 1-month, 1-year, and 2-year periods.

$$r_o(N1) - r_o(N2) = (\text{Log } N1 - \text{Log } N2)/\alpha = g * \text{Log}(N1/N2),$$

which is the distance along the rainfall axis. The ratio N1/N2 is equal to 3 when the period P1 is the month and the period P2 the season. The experimental distances measured in figure 6 are very close to these theoretical values. Even though the general asymptotic behavior is the same for each EVS, the EV1 model appears to be suitable only for periods greater than a season in the case of Montpellier Bel-Air and computing g_1 from the monthly maximum series would lead to an underestimation of the gradex.

5.2. Gumbel Distribution Fitted by the Method of Maximum Likelihood.

The estimate is given by the numerical resolution of the following system:

$$\bar{x} = (g_2 + \sum_{i=1}^{n} x_i \, exp(-y_i)) / \sum_{i=1}^{n} exp(-y_i)$$

$$r_o = -g_2 \, Log\left((\sum_{i=1}^{n} exp(-y_i))/n \right) \qquad (11)$$

where x is the mean of the sample x_i and $y_i = x_i/2$.

The fit is only good if the sample corresponds perfectly to a Gumbel distribution. Indeed, as soon as there is a slight break, the line fitted by this method shows relatively large deviations with respect to high values (fig.6) and the distribution parameters are underestimated. The same finding prompted Fiorentino and Gabriele (1984) to propose a correction. If g_2 is the gradex estimated by the maximum likelihood method, then the corrected parameters may be expressed as:

$$g'_2 = g_2/(1-.8/n), \text{ and}$$

$$r'_o = g'_2 * \text{Log}(n/\sum_{i=1} \exp[-y_i]) - .7 * g'_2/n$$

This correction is based on the results of tests on samples simulated according to the Gumbel distribution with $g = 1$ and $r_o = 0$. To evaluate its performance, we tested it on experimental samples. The correction unfortunately never exceeds 2%, which is insufficient to correct the bias of the M.L. method on our rainfall samples. This was to be expected, since the correction is based on the hypothesis that the population is exactly EV1 distributed, while we are searching for reliable estimates of g for approximately EV1 distributed EVS .

Another approach was taken by Revfeim (1983). In the procedure described in section 3 it was shown that the EV1 model is the limiting form of the cumulative distribution function $[F(R)]^N$, when N tends to infinity, if the PDS model is an exponential. Alternatively, the rain may be defined as a succession of rainfall events, the number of which in a fixed

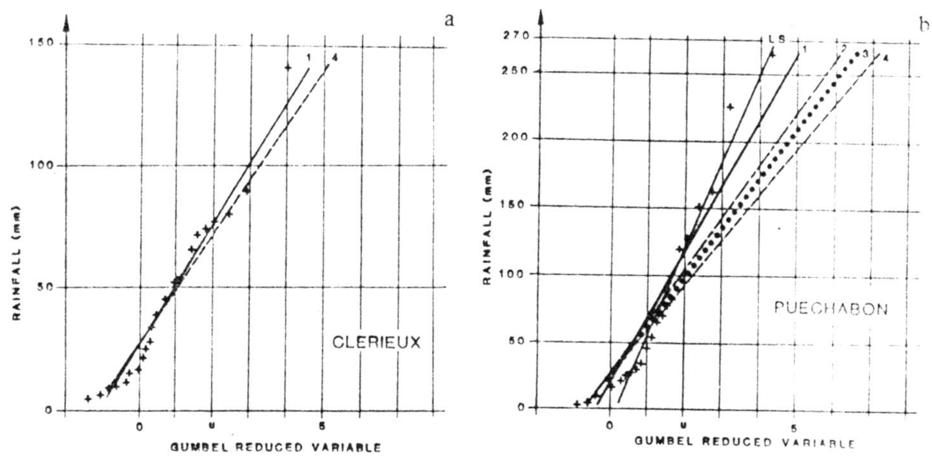

Figure 6: Fitting the EV1 distribution to monthly maximum distributions.
a: twelve-hour rainfall; Clerieux raingage.
b: two-hour rainfall; Puechabon raingage.
LS: least square fitting to the upper part of the observed distribution.
1: moment fitting; 4,2,3: maximum likelihood fitting using expreesion 11 and corrections proposed by Revfeim (1983) and Fiorentino(1984) respectively.period of time has a Poisson

distribution. If the rainfall depth of each event has an exponential distribution, the extreme value model is also an EV1 distribution, the coefficients of which can be related to the parameters of the underlying process, i.e the mean of the process (μ) and the recurrence rate(ρ). Refveim (1983) proposed maximum likelihood estimates ($\hat{\mu}$, $\hat{\rho}$) of these parameters, from which the estimated parameters of the EV1 distribution are readily obtained as:

$$g''_2 = .78\, S_x = \hat{\mu},$$

where S_x is the standard deviation of the EVS, and:

$$r''_o(\rho) = \mu \,\mathrm{Log}\, \hat{\rho} = g''_2 \,\mathrm{Log}\, \hat{\rho}$$

Note that the gradex g''_2 is the mean of the underlying process, but this process is not the sequence of every consecutive rainfall considered in the first approach described above. Hence, the method proposes new estimates of the EV1 coefficients, but no additional mean to test the suitability of the model itself for a given EVS. Figure 6 shows how a few low value outliers would lead to a biased estimate of g, especially for methods sensitive to these low values, as maximum likelihood methods generally are.

After devoting a great deal of time and effort to M.L. methods, it appears that they can be ruled out for fitting an EV1 distribution to rainfall data in the Mediterranean region, at least for time steps ranging from 1 to 24 hours.

5.3. Two Component Exponential Function Fitted to all Successive Rainfalls.

Figure 5 shows that increasing the period over which the maxima are determined eliminates the break; however the sample size is reduced (divided by 3 in the case of one season instead of one month). The risk is therefore to have insufficient sample size when using seasonal maxima, even though the sample follows a virtual Gumbel distribution. On the other hand, if the period is decreased, the break becomes increasingly distinct, leading to a limiting case consisting of two lines. This is in particular the case when the period taken is equal to the rainfall duration considered (i.e all consecutive rainfalls are taken). On Gumbel paper, these two lines require a special distribution function corresponding to the sum of two exponentials. This function accounts for both tails of the distribution with the low values represented by the second exponential and the values above a certain threshold dominated by the first exponential. The quality of the fitting of this function to maxima at low durations using expressions 10 a,b,c was poor. In order to improve short time step estimates, an experimental study of the functions $\gamma = f(t)$ and $\gamma = f(F(0))$ was carried out, allowing a function CK(t,F(0)) to be established, by which $\gamma = \mu^2 / \sigma^2$ can be multiplied :

$$CK(t,F(0)) = 1/(1-\mathrm{Log}(t/24)) \qquad (12)$$

For t= 24 hours, CK(t,F(0)) = 1 and its application is not possible for durations greater than 24 hours since it has been established empirically.

With this correction the fit of the two component exponential model was improved for all 25 stations, especially at short time steps (figure 7). For Montpellier Bel-air (52 years of recording available over the period 1922-1980), it is noteworthy that the gradex obtained by the moment fitting of the EV1 model to seasonal maxima and by the corrected fitting of the two component exponential model are very close. It was also observed that on most EVS series displaying a break in the low values,a least square fitting to the upper part of the point swarm (figure 6) provided better gradex estimations than the moment fitting.

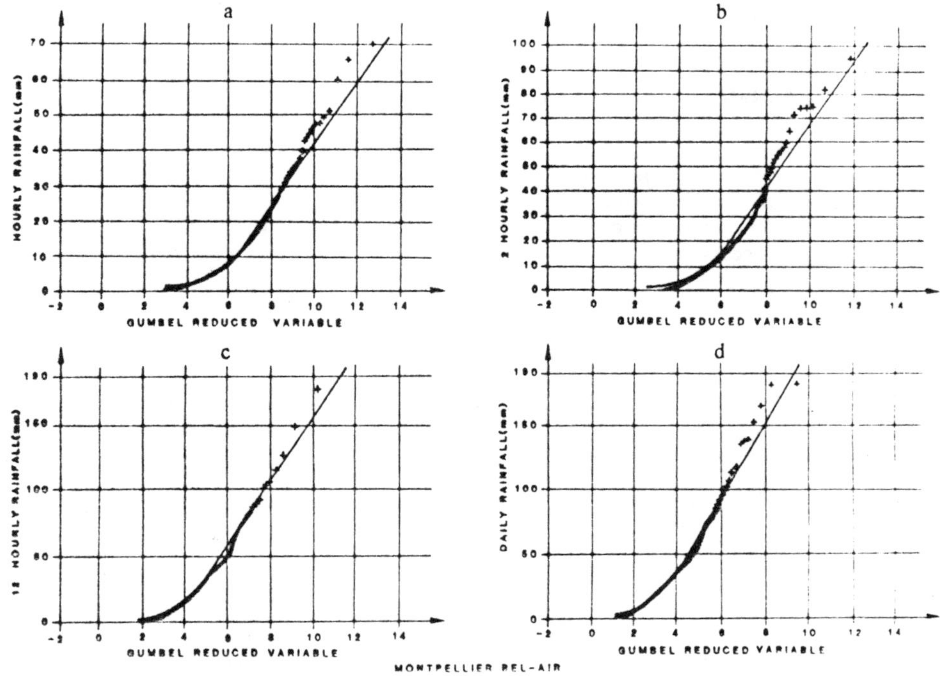

Figure 7 Fitting the two component exponential model to the rainfalls of Montpellier Bel-air. The parameters of the model are computed using the correction function CK(t,F(0)).

Table 2 Average ratios of the three estimates, computed on 25 stations. g_1 is the conventional moment estimate, while g_2 g_3 and g'_3 are calculated using expressions 11 10a and 12 respectively.

Duration (hours)		1	2	4	6	12	24
	g_1/g_2	1.15	1.15	1.20	1.20	1.30	1.40
Ratios	g_1/g_3	1.30	1.25	1.20	1.05	0.85	0.80
	g_1/g'_3	1.12	1.08	1.04	1.03	0.84	0.80

6. COMPARISON OF THE THREE METHODS

Table 2 shows that:

a) Maximum likelihood estimates (computed with expression (11)) give for all time steps consistently lower estimations of g than the other two estimates.

b) For time steps greater than 6 hours, the EV1 model gives lower estimates than the two component exponential model. Experiments performed on long duration series indicate that the estimates provided by the later model are more reliable.

c) For short time steps, none of the method is really satisfactory. The estimate g_1 is good when no break occurs in the point swarm, but underestimates the gradex in cases such as the one in figure 6.

7. CONCLUSION

The fitting of a two component exponential model to all consecutive rainfalls is the method that uses the most hydrometeorological information and appears to be the best as long as the necessary data is available. The method proved to be especially suitable when corresponding extreme value series cannot be well fitted by a Gumbel distribution. In cases where only extreme value series can be used, it was observed that maximum likelihood estimates could not compete with less sophisticated but more efficient moment fitting or least square fitting estimates.

When a break or a curve occurs in the point swarm, the least squares fit accounts for this break, and the upper part of the empirical distribution corresponds to a Gumbel distribution. If this part contains enough values, this method yields gradex estimates close to those of the two component exponential model. However, the robustness of the least squares estimate is poor when there are several high value outliers. The main advantage of working on parent distribution series rather than on extreme value series is therefore the minimization of the influence of these high value outliers, observed in rainfall in the Mediterranean area.

ACKNOWLEDGEMENTS

The study was funded by the Gard "Direction Departementale de l'Equipement" (DDE) . The Division Technique Generale of Electricite De France and the Service Regionale d'Amenagement des Eaux of the Languedoc Roussillon Region provided indispensable support.

REFERENCES

Beran,M. 1981. The "gradex" method, letter to the editor. IAHS bulletin Vol. 26, 2 pp. 223-224.

Bernier,J., 1967. Sur la theorie du renouvellement et son application en hydrologie. Bulletin de la Direction des Etudes et Recherches, Electri-cite de France, Serie B Vol.10.

Duband,D., 1967. Distribution of the sum of daily precipitations. in: International Hydrology Symposium, vol. 1 pp. 516-526. Colorado State University, Fort Collins, Colorado.

Fiorentino, M., Gabriele, S. 1984. A Correction for the Bias of Maximum-Likelihood Estimators of Gumbel Parameters. Journal of Hydrology. Vol. 73, pp.39-49.

Giraud, M. 1959. Initiation aux Processus Aleatoires. Dunod Editor, Paris.

Guillot,P.,Duband,D., 1967. The Gradex to Compute Extreme Flood Probability from the Observation of Rainfall. in: International Hydrology Symposium, vol. 1 pp. 506-515. Colorado State University, Fort Collins, Colorado.

Guillot,P., Duband,D., 1968. La methode du Gradex pour le calcul de la probabilite des crues a partir des pluies. Xmes journees de l'hydraulique, Societe Hydrotech. de France, Paris 1968.

Gumbel,E.J., 1958. Statistics of Extremes, Columbia University Press, New York.

Gupta,V.K., Duckstein,L., andPeebles,R.W. 1976. On the Joint Distribution of the Largest Flood and its Time Occurrence. Water Resources Research, Vol. 12(2), pp.295-304.

Kottegoda, N.T., 1984. Investigation of Outliers in Annual Maximum Flow Series. Journal of Hydrology, Vol. 72, pp. 105-137.

Landwher, J.M., Matalas, N.C., and Wallis, J.R. 1978. Some Comparison of Flood Statistics in Real and Log Space. Water Resources Research, Vol. 14(5), pp.902-920.

Lebel,T., 1984. Moyenne Spatiale de la Pluie sur un Bassin Versant: Estimation Optimale, Generation stochastique et Gradex des Valeurs Extremes. Thesis, Institut National Polytechnique de Grenoble, France.

Matalas, N.C., Slack, J.R., and Wallis, J.R. 1975. Regional Skew in Search of a Parent. Water Resources Research, Vol. 11(6), pp.815-826.

Revfeim, K.J.A., 1983. On the Analysis of Extreme Rainfalls. Journal of Hydrology, Vol. 62 pp. 107-117.

Rossi,F., Fiorentino, M., and Versace, P. 1984. Two Component Extreme Value Distribution for Flood Frequency Analysis. Water Resources Research, Vol. 20(7), pp. 847-856.

Todorovic,P. and Zelenhasic,E. 1970. A Stochastic Model for Flood Analysis. Water Resources Research, Vol. 6(6), pp. 1641-1648.

Todorovic,P. and Rousselle, J. 1971. Some Problems of Flood AnalysisWater Resources Research, Vol. 7(5), pp. 1144-1150.

Waylen,P.,and Woo,M.K. 1982. Prediction of annual floods generatedby mixed processes. Water Resources Research Vol.18(4), pp.1283-1286

FREQUENCY ANALYSIS OF AUSTRALIAN RAINFALL DATA AS USED FOR FLOOD ANALYSIS AND DESIGN

R. P. CANTERFORD, N. R. PESCOD, H. J. PEARCE,
L. H. TURNER AND R. J. ATKINSON

HYDROLOGY BRANCH, BUREAU OF METEOROLOGY, P.O. BOX 1289K,
MELBOURNE, AUSTRALIA, 3001.

ABSTRACT. The procedures used to obtain accurate, temporally and spatially consistent, intensity-frequency-duration (IFD) design rainfall data for Australia are discussed. These IFD design curves are used by design engineers and scientists as input to a wide range of design flood models and other environmental studies. The basic annual maximum rainfall data for durations of 6 minutes to 72 hours are fitted using a log-Pearson Type III distribution with a small positive regional skewness up to 0.7. Most of Australia is close to zero skewness ie. a lognormal distribution. Due to the sparsity of recording raingauges, various regression techniques were used to estimate short duration data at daily read raingauge sites. These procedures are discussed, along with the production of six master charts of rainfall intensity for various durations and average recurrence intervals (ARI) covering all Australia. From these six charts, plus a map of regionalized skewness, a full set of IFD curves can be obtained for any location using appropriate extrapolation and interpolation procedures. The IFD design curves extend from six minutes to 72 hours and ARI from one year to 100 years. Comparisons are made with USA work in this field. The paper also outlines the steps taken to automate this development work and produce CDIRS (Computerized Design IFD Rainfall System). CDIRS allows automatic determination of a full set of IFD curves (also in tabular form) for any location simply by supplying its latitude and longitude.

1. INTRODUCTION

The frequency analysis of rainfall data is an important part of hydrological design procedures. It is used as input to flood models for a wide range of hydraulic structures to determine design sizes and risks for overtopping or failure. In order that the rainfall design data be practical for design purposes, it is necessary that accurate, consistent intensity-frequency-duration (IFD) rainfall design curves be available for any location. An example of the IFD design curves used in Australia is shown in Figure 1.

V. P. Singh (ed.), Hydrologic Frequency Modeling, 293–302.

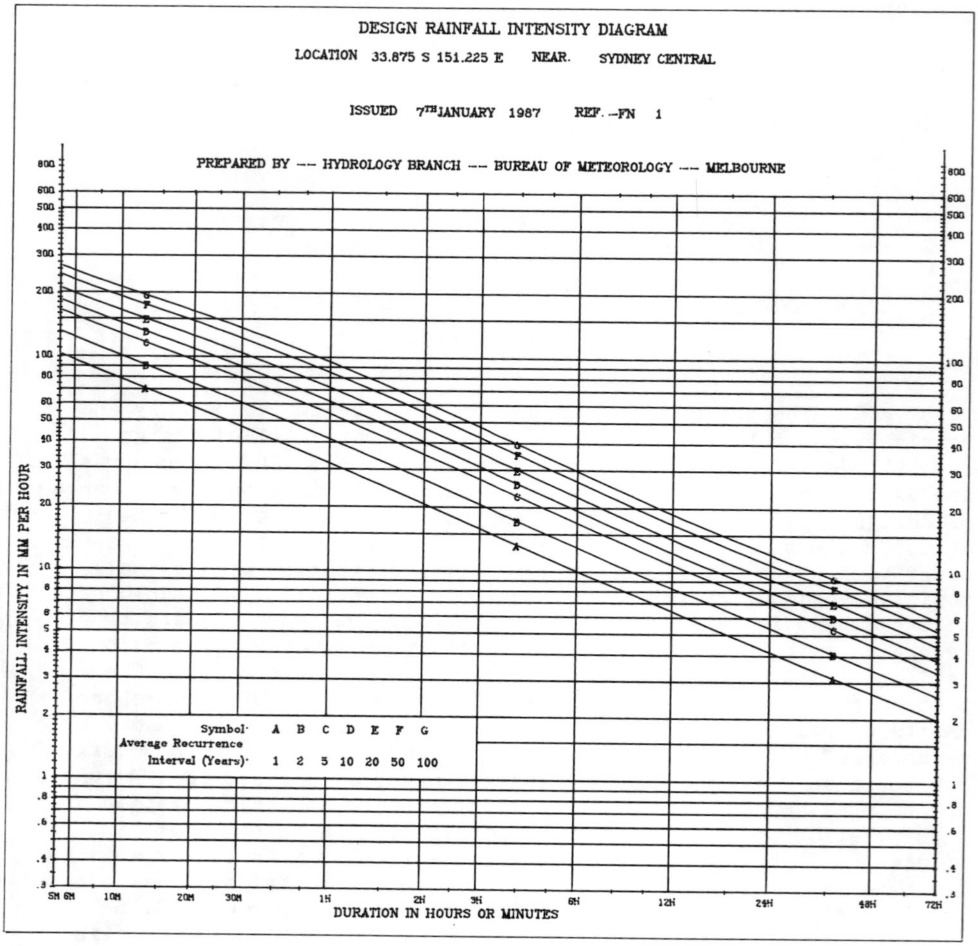

Figure 1. Rainfall intensity-frequency-duration rainfall design curves: an example.

This paper describes the derivation of the IFD design curves which are used by hydrological designers in Australia. The Australian Bureau of Meteorology commenced the project in 1983 in order to update existing design information which had developed in a restricted manner (lack of data and resources), and which was inadequate in many areas of the country. The aim of the IFD rainfall design project was to provide a full coverage of Australia with IFD data for probabilities of exceedence up to the one percent level, or average recurrence intervals (ARI) of 100 years, and for durations from six minutes to 72 hours. This IFD information will be included in a revised edition of "Australian Rainfall and Runoff", a manual of recommended hydrological design procedures, to be published in 1987 by the Institution of Engineers, Australia.

An end product of the project is CDIRS (Computerized Design IFD Rainfall System) which enables the user to obtain a full set of IFD curves automatically for any given latitude and longitude from a large data base which covers Australia at a resolution of 0.05 degrees latitude/longitude, with high rainfall gradient areas down to a resolution of 0.025 degrees. With CDIRS the user no longer has to read data from charts and use graphical techniques to produce curves; although this procedure is still available as a complementary service for users without access to the CDIRS programs and who may only require limited design information.

In the following sections the various stages of research and development that were necessary to arrive at a practical operational service for designers are described. They give an outline of the extensive work that was required to produce geographically consistent and spatially and temporally smooth sets of IFD curves; termed the generalized technique. This cannot be achieved by publishing single station analyses of rainfall data (as has also been found in the U.S.A.-J. Miller,personal communication,1983). The Australian Bureau of Meteorology has adopted similar generalized techniques to those used in the U.S.A. (Miller et al, 1973) as a basis for the project; although there are a number of major differences, including the use of more sophisticated statistical methods on short duration data, and different frequency distributions for rainfall data. This was partly due to the sparsity of short duration data. In addition the procedures were developed along lines such that the final result was completely automated.

The basis of the generalized technique are charts of rainfall intensity for six specific combinations of duration and ARI: durations of 1 hour, 12 hours and 72 hours and ARI's of 2 years and 50 years. These are termed the master charts. For any given location, the six specific intensities are obtained from these master charts, along with a regionalized skewness (discussed later), and interpolation procedures used to obtain a full set of IFD curves.

2. DATA ANALYSIS

Daily rainfall data from approximately 7,500 stations with greater than 30 years of record, and short duration (down to 6 minutes) data from approximately 600 recording rain gauges with more than 6 years of analysed record formed the basic data set for the analysis procedures. Most of these recording rain gauges, or pluviometers, are chart recorders. Their charts have been digitized and stored on computer. Again, most were daily read, allowing data to be obtained down to durations of six minutes. Figure 2 illustrates the geographical spread of these pluviometers, which are mainly located in the highly populated coastal areas. Daily stations are not shown due to their high density. Also shown on Figure 2 are the State and territory boundaries and cities referred to in this paper.

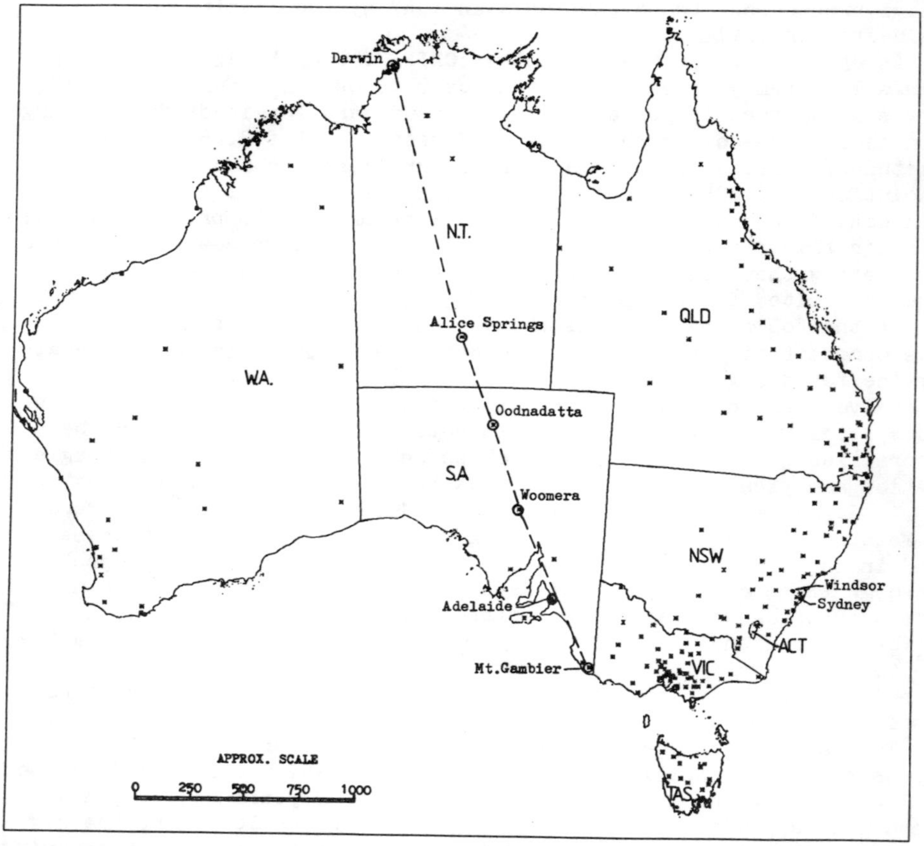

Figure 2. Bureau of Meteorology pluviometer locations with greater than 12 years of record. States and territories, and cities referred to in the text are also shown.

The primary rainfall series data of the daily stations and the pluviometers were analysed using a lognormal distribution (discussed in the next section) and various statistical and fitted parameters stored on computer. These parameters included the mean and standard deviation of the logarithms of the primary rainfall series, and the ARI 2 years to 100 years estimates. The ARI 2 years to 10 year estimates were adjusted to partial-duration series estimates using similar factors to those of Miller et al (1973). These parameters were derived for specific durations: for daily data one, two, three and four days and for the pluviometers 6,12,18,30 minutes and 1,2,3,6,12,24,48 and 72 hours. Obviously, since the pluviometers are co-located with daily stations, the 24,48 and 72 hour durations coincide if allowance is made for the restricted 9am to 9am daily totals. The restricted to unrestricted corrections used are the same as those of Miller et al (1973).

The co-located pluviometers and daily stations allowed statistical relations between the short duration data and the longer duration data and other meteorological and geographical features to be developed. This is necessary in order to increase the spatial density of rainfall intensity data for analysis of the 1 hour and 12 hour master charts. The procedures developed are described in section 3. Clearly the higher density and longer records of the daily stations allows more accurate 72 hour (ARI 2 years and 50 years) charts to be derived.

In order to handle the large amount of point data (several thousand locations for 6 specific durations and ARI's), computer plotting programs were used to plot the rainfall intensities, both raw and statistically derived (for short durations). Additionally, as will be discussed in the next section, many regions for which statistical relationships were derived were selected to allow overlap so as to ensure spatial consistency between adjacent regions. Hence for some areas two or more regression estimates were plotted at the same location; plus raw pluviometer data if available. These plotted charts were produced at various scales from 1:250,000 to 1:2.5 million, depending on the density of stations and degree of complexity in the isopluvials that needed to be analysed.

3. STATISTICAL ANALYSIS

3.1. Frequency Distributions for Primary Series Data.

For the 12 specific durations from 6 minutes to 72 hours, the lognormal distribution is fitted to the primary rainfall series. Several investigations of Australian primary rainfall series (e.g. Pierrehumbert,1972 and Pierrehumbert and Canterford, 1975) have shown that for site specific analyses, the lognormal distribution (fitted by moments) was the most suitable and convenient basic distribution after comparison with many other distributions and fitting techniques. Subsequent studies undertaken as part of the current project, however, showed areas of Australia where the logarithms of the primary rainfalls had small, but significant positive skewness (up to 0.7). Because of this, and because recent design work has placed greater emphasis on longer ARI's , the LPIII, with a smooth regionalized skewness was found to be the most appropriate distribution in the regions of positive skewness.

Since the lognormal distribution is a special case of LPIII with zero skewness, large areas were still covered by the lognormal distribution, with the main LPIII regions of Australia being the southern coastal high topographic areas and the northwest coast. The northwest coast primary rainfall series data had been found in previous work (Canterford and Pierrehumbert,1977 and 1978) to be best described by a complex double distribution, with tropical cyclones (hurricanes) forming a distinct population against the background population. However the LPIII gave similar estimates up to 200 years

ARI.

In essence, the lognormal distribution is used for the primary rainfall series, with "fine tuning" to the LPIII in some regions via a smoothed regionalized skewness (an average about 0.2 but ranging to 0.7 in limited areas). This results in percentage increases (over the lognormal) in these areas of about 5 to 15% at ARI's of 50 to 100 years. The skewness was temporally as well as spatially smoothed since no significant variation over durations was found.

The six master charts (durations 1,12 and 72 hours and ARI's of 2 years and 50 years) were plotted and analysed using lognormal intensity estimates. An additional chart represents the regionalized skewness. This allows the user to obtain quickly, lognormal estimates as a first approximation, and then adjust them, if necessary, using the appropriate skewness. This is particularly important when extrapolating to ARI's of 500 years, a design requirement for some hydraulic structures in Australia.

3.2. Determination of Short Duration Design Figures

In order to obtain more accurate 1 hour and 12 hour duration (2 and 50 year ARI) master charts, the sparse network of pluviometers must be supplemented with estimates of rainfall intensity for these durations at the daily rainfall stations. This allows interpolation between the pluviometer station rainfall data and extrapolation to areas where there are no pluviometer stations.

Principal components were derived using rainfall intensities for eight durations (1,2,3,6,12,24,48 and 72 hours) for 23 long term pluviometers around Australia. It was found that 99.8% of the variance of the ARI two year intensities, 90.6% of the variance of the standard deviations and 99.0% of the variance of the ARI 50 year intensities could be explained by the first two principal components (PC). Hence the original eight "variables" which are obviously highly correlated can be represented by two independent PC's. Further to this, a simple oblique rotation of these principal components (Harris-Kaiser method, SAS Users Guide, Statistics, 1982) was carried out in order to "link" these PC's to the one to four calendar day data from the co-located daily stations. This rotation results in one rotated PC (PC1) representing the long duration data and the other (PC2) the short duration data. In this way, PC1 could be obtained directly from daily data and PC2 found by multiple regression techniques.

A major benefit of representing, and temporally smoothing, the pluviometer data by use of principal components is that all short duration data contributes to PC2. Hence important data for other durations e.g. 6 hours, 12 hours etc. are taken into account. The factor patterns derived from the Australia-wide 23 pluviometers were then applied to the PC1 and PC2 for each daily station (calculated from daily data and the regressions), in order to create the intensities for the eight standard durations. For mathematical details, see for example Afifi and Azen (1979). The above procedures were adopted for the ARI 2 year intensities and the standard deviation of the logarithms, from which the ARI 50 year intensities were calculated. The regression estimates are used as spatial interpolation guidance, since other information is available to the analyst (section 4).

3.3. Interpolation and Extrapolation to all Durations

Sections 3.1 and 3.2 describe how the key rainfall design intensities are determined for the one hour, 12 hour, and 72 hour master charts for 2 year and 50 year ARI's. In order to obtain estimates for durations apart from these standard ones, several methods were again compared.

For durations between one hour and 72 hours the best interpolation procedures were found from an investigation of principal components applied to the geometric mean of the ARI 2 year and 50 year intensities of the 23 Australia-wide pluviometers. The "plotting position" for various durations was then obtained by regression against duration.

To extrapolate to durations less than 1 hour, ratios of short duration intensities to one hour intensities were obtained from the 23 long term pluviometers for ARI's of 2 and 50 years. The average values of the ratios were then used as multiplicative factors to apply to the one hour intensities at all locations and for all ARI's. These ratios were compared with USA data (Miller et al, 1973) and found to be similar, as shown in Table 2.

TABLE 2. Average ratios of short duration rainfall to one hour rainfall.

Duration (minutes)	5	10	15	30	60
USA	0.29	0.45	0.57	0.79	1.00
AUSTRALIA	0.29	0.43	0.53	0.75	1.00

It has been found, however that significant variation of the ratios within Australia can be explained on a geographical, meteorological and ARI basis. Due to the importance of short duration data for design purposes this variation must be taken into account. As an example, the ratios are too large in tropical and subtropical coastal areas and too small for the south-west coast. Similar variation has been found in the USA (Richards,private communication,1985). By studying more than 200 pluviometers around Australia, a relationship was developed that is dependent on (i) the one hour intensity raised to the power 0.9 for ARI 2 years, and 0.6 for ARI 50 years and (ii) a geographical factor. This will be reported elsewhere.

4. METEOROLOGICAL ANALYSIS

A major part of the production of the 6 master charts is their analysis, that is the drawing of the contours or isopleths of design rainfall intensity. In analysing the charts many items have to be taken into account. These include the various raw and regression estimates of rainfall intensities for the local area and surrounding areas, the influence of local topography, broad scale topography, distance and direction of moisture sources, primary and secondary barriers etc. Furthermore, the weight given to these aspects will vary depending on the particular chart being analysed, ie. the ARI and duration. Consistency must also be maintained among charts in terms of both return period and duration. Similar methods of analysis are used by the U.S.A. National Weather Service (Frederick et al, 1977). The general trend is for gradients of rainfall intensity to reduce from those of the 72 hour ARI 2 years chart as duration decreases and return period increases. One reason for this is that rainfall intensites for short durations and long ARI's are not affected by topography to the same extent as the longer duration and more frequent rainfalls.

As part of the meteorological analysis, the major meteorological phenomena of the regions were considered in determining moisture sources and variation with ARI eg. the depth of penetration south of a tropical cyclone or a tropical moist airmass will depend on the ARI being analysed. As an example, a study of the variation of 1 hour intensities for a cross-section through Australia from Darwin in the north to Mount Gambier on the south coast (see Figure 2.) illustrates the above feature. For an ARI of 2 years the intensities drop quickly to Alice Springs and remain fairly constant further south, while for ARI 50 years the reduction is gradual and does not reach a minumum even at Mount Gambier. This indicates that for rarer events there is further penetration south of tropical moist air. Similar studies were undertaken in other areas to determine local features, including variation with elevation.

Since more reliance can be placed on the intensity estimates for 72 hours and ARI 2 years, this chart was always analysed first. Shorter durations and longer durations were analysed next, each time referring back to earlier charts to ensure consistency. In this way basic features of the analysis could be followed.

5. FINAL IFD CURVES AND CDIRS

5.1. Determination of Final Design Curves - Graphical and Analytical

Final IFD design curves are derived for a given location by:

(i) Reading from the six master charts the lognormal design rainfall intensities for 1, 12 and 72 hour for ARI's of 2 and 50 years.

(ii) Reading from the regionalized skewness map the appropriate average skewness.

(iii) Determining LPIII distribution estimates from the lognormal ARI 2 and 50 year estimates for 1, 12 and 72 hours using the average skewness. This can be done graphically or algebraically (Australian Rainfall and Runoff, 1987)

(iv) Interpolating between 1 and 72 hours and extrapolating

to 6 minutes using the procedures of the preceding section.

5.2. C.D.I.R.S.

The Computerized Design IFD Rainfall System (CDIRS) is essentially an end product of the IFD development work described above. It allows a full set of IFD curves to be obtained by simply specifying the latitude and longitude of the required location.

The six rainfall intensity base maps (1,12 and 72 hours and ARI 2 years and 50 years) have been digitized and gridded to a resolution of 0.05 degrees latitude/longitude over most of Australia, with high gradient areas along the east coast gridded to a resolution of 0.025 degrees. In addition the skewness map was digitized and gridded to the same resolution. At each grid point these seven data values (6 rainfall and 1 skewness) are stored on magnetic tape. An accessing program reads this analysed data base and applies the interpolation procedures discussed above to obtain intensities for all durations (five minutes to 72 hours) and all ARI's (one to 100 years). This system is particularly useful for hydrologists, especially where the variation of design rainfall estimates is required across a large catchment area. It also allows the Australian Bureau of Meteorology to process client requests extremely efficiently, since charts, graphs etc are not required. It is entirely compatible with the procedure in section 5.1, which enables complete standardization of IFD design data.

6. CONCLUSIONS AND SUMMARY

A method of providing accurate, and spatially and temporally consistent, intensity-frequency-duration rainfall design data has been developed. Approximate 500,000 years of daily station annual maximum rainfalls (7,500 stations, average record length 65 years) and 12,000 years of pluviometer station data (600 stations, 20 years average record) for 12 durations were used in the development of the generalized IFD curves for Australia.

Additional studies on the determination of rainfall temporal patterns for various climatological zones, which give the distribution in time of the IFD design rainfall intensites, have also been carried out and will be published elsewhere. Further work on depth-area relationships would be useful but lack of dense networks, both in Australia and the U.S.A., limits these studies. The studies on Australian short duration data (less than 1 hour) show similarities to U.S.A. experience and the techniques developed in this project could be tested/modified for use in the USA. The methods for the determination of IFD design curves described in this paper were developed in order to provide a "package" of techniques that could be applied to many meteorological and geographical regions with varying densities of data. Again, they could be tested/modified for use in other countries. The complete automation of the system is also unique and opens up a range of possibilities for modern hydrologic design procedures.

7. ACKNOWLEDGEMENTS

Dr M. F. Hutchinson is gratefully acknowledged for the provision of specialized computer programs for gridding and advice on their application. Also Dr A. Miller and Dr G. Kuczera are thanked for their helpful comments on some statistical aspects of the work. Mr A.J. Hall is thanked for his support of the project and advice. Mr G. J. Smith provided valuable technical support and is especially thanked for his excellent attention to detail. Many other officers provided input to the study over the last 2 to 3 years and they are also thanked. The staff involved in the important task of processing pluviometer chart data and rainfall records, as well as the observers, are also acknowledged.

8. REFERENCES

Afifi, A. A. and Azen, S. P. 1979. Statistical Analysis: A Computer Oriented Approach.Second Edition. Academic Press, New York.

Australian Rainfall and Runoff:Flood Analysis and Design. 1977, 1987 Institution of Engineers, Australia. Third ed., 1987, in preparation.

Canterford, R. P. and Pierrehumbert, C. L. 1977. Frequency Distributions for Heavy Rainfalls in Tropical Australia. Hydrology Symposium, Brisbane, p119. Institution of Engineers, Australia.

Canterford, R.P. and Pierrehumbert, C.L. 1978. Unpublished Australian Bureau of Meteorology Study of 360 Australian rainfall stations.

Frederick, R. H., Myers, V.A. and Auciello, E. P. 1977. Five to 60 Minute Precipitation Frequency Atlas of the Western United States. NOAA Tech. Memo. NWS HYDRO-35.

Miller,J. F., Frederick, R. H. and Tracy, R.J. 1973. Precipitation Frequency Atlas of the Western United States Vol. III: Colorado. NOAA Atlas 2 National Weather Service, US dept. of Commerce.

Pierrehumbert, C. L. 1972. Short Period Rainfall Intensity Analysis. Working Paper 156 Bureau of Meteorology, Melbourne, Australia.

Pierrehumbert, C.L. and Canterford, R. P. 1975. Frequency Distributions for Queensland Heavy Rainfalls. Technical paper no. 2 "The Frequency of Floods in Queensland", Institution of Engineers, Australia, Queensland Division.

SAS Users Guide. 1982 Edition. SAS Institute, North Carolina, USA.

STOCHASTIC FORMULATION OF STORM PATTERN AND RAINFALL INTENSITY-DURATION CURVE FOR DESIGN FLOOD

Michio Hashino
Department of Construction Engineering
Faculty of Engineering, The University of Tokushima
2-1, Minami-josanjima, Tokushima 770
Japan

ABSTRACT. A stochastic single storm pattern, which preserves stochastic properties of actual storm rainfall, is theoretically derived from Freund's bivariate probability density function. Two typical design storm patterns, namely, the last and the central peaked types are defined by the three parameters: the reduced variate y_p of the peak rainfall intensity, the autocorrelation index k related to the autocorrelation coefficient of the rainfall intensities, and the conditional probability F. Integration of the given stochastic design hyetograph gives a new 'conditional probability' intensity-duration formula. Furthermore, a practical estimation method of the three parameters: k, F, and y_p is clearly shown. The conditional probability intensity-duration curve is demonstrated by using actual hourly rainfall data. Design intensities for shorter duration than 1 hr can be easily estimated from the available hourly data.

1. INTRODUCTION

In order to estimate the magnitudes of design floods in urban and rural areas, many designers of drainage systems have used the concept of a single design storm which is used in conjunction with an appropriate runoff model for the objective drainage system. The time distribution of rainfall intensities is determined from a rainfall intensity-duration-frequency relationship in the form:

$$r = a / (t^b + c) \tag{1}$$

where r = average rainfall intensity for duration t and a, b, c = empirical constants depending upon return period. Although several forms including Eq. 1 were empirically proposed, they have been hardly supported by the theory of probability that represents stochastic properties of actual storm patterns.

The important objections to the use of the design storms determined from the empirical intensity-duration formulas is that they attempt to summarize 'various' storm patterns in a single hyetograph

V. P. Singh (ed.), Hydrologic Frequency Modeling, 303–314.

shape and assemble components of equal individual probability of occurrence which have a quite different joint probability (Delleur and Dendrou, 1980). Frederick (1978) demonstrated that the joint occurrence of large rainfall totals within different durations of the same storm had a lower probability than the rainfall intensity-duration curve from which they were derived.

In this paper, a stochastic single storm pattern, which preserves stochastic properties of actual storm hyetograph, is theoretically derived from the bivariate probability density function defined by Freund (1961). Integration of the stochastic single storm hyetograph gives a new 'conditional probability' intensity-duration curve. The stochastic storm pattern is prescribed by three parameters: the reduced variate y_p of the peak rainfall intensity, the autocorrelation index k, and the conditional probability F that represents the conditional probability of a rainfall intensity y_i given the advanced intensity y_{i-1}. A practical estimation method of the three pattern parameters is presented, and the conditional probability intensity-duration curve is demonstrated by using actual hourly data.

2. STOCHASTIC FORMULATION OF STORM PATTERN

2.1. A Single Storm Pattern Repretented by Freund's Distribution

Freund (1961) defined a bivariate exponential density function $f(\xi,\eta)$ of a bivariate (ξ,η) with four parameters: α_1, β_1, α_2, and $\beta_2(>0)$. For the case of $\alpha_1=\beta_1=\alpha$ and $\alpha_2=\beta_2=\beta$, the variables ξ and η have identical marginal distributions. Consider a single storm pattern with a peak of rainfall intensity x_p as shown in Fig. 1.

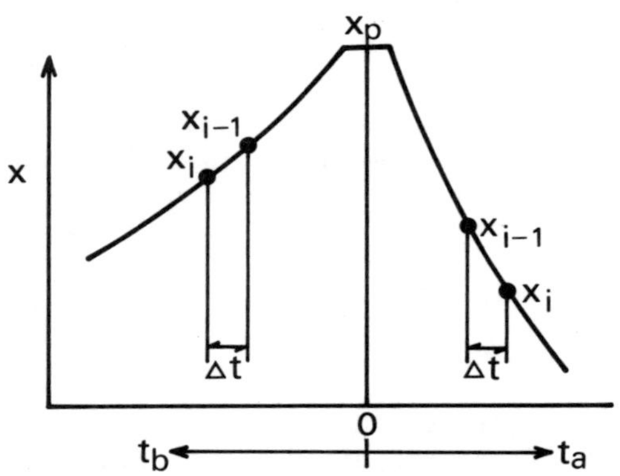

Fig. 1 A single storm pattern with a peak

Let the intensities before and after time unit Δt be x_{i-1} and x_i, respectively. By putting $\xi=x_i$ and $\eta=x_{i-1}$ and assuming that x_i and x_{i-1} have identical marginal distributions, we obtain the following density function.

$$f(x_i, x_{i-1}) = \begin{cases} \alpha\beta\exp\{-\beta x_{i-1}-(2\alpha-\beta)x_i\} & (0<x_i<x_{i-1}) \\ \alpha\beta\exp\{-\beta x_i-(2\alpha-\beta)x_{i-1}\} & (0<x_{i-1}<x_i) \end{cases} \tag{2}$$

The variance σ_x^2, and the autocorrelation coefficient ρ of the rainfall intensity x can be given as

$$\sigma_x^2 = (3\alpha^2+\beta^2)/(4\alpha^2\beta^2) \tag{3}$$

$$\rho = (1-k^2)/(1+3k^2) \tag{4}$$

where

$$k \equiv \alpha/\beta \tag{5}$$

It is clear from Eq. 4 that $\rho=0$ for $k=1$, $\rho\to1$ for $k\to0$, and $\rho\to-1/3$ for $k\to\infty$. In practice, an appropriate value of ρ in the range of $0\leqq\rho<1$ will be adopted, so that the parameter k may be supposed to be in the range of $0<k\leqq1$. We call k the autocorrelation index.

By the definition of a single storm with a peak, the rainfall intensity decreases monotonously away from the peak; that is, $x_i<x_{i-1}$. For this case, the conditional probability distribution function $F(x_i|x_{i-1})$ can be derived from Eq. 2 as

$$F(x_i|x_{i-1}) = \begin{cases} \dfrac{1-\exp\{-(2\alpha-\beta)x_i\}}{1-2\{1-(\alpha/\beta)\}\exp\{-(2\alpha-\beta)x_{i-1}\}} & (2\alpha\neq\beta) \\ \beta x_i/(1+\beta x_{i-1}) & (2\alpha=\beta) \end{cases} \tag{6}$$

The rainfall intensity x is transformed into a reduced variate y as

$$y = (x-u)/\sigma_u \qquad (0\leqq u<x) \tag{7}$$

where u = base intensity level and σ_u = standard deviation of the exceedance (x-u). Therefore, we can define the same density function as Eq. 2 for the reduced bivariate (y_i, y_{i-1}). In this case, the variance of y equals to unity, so that the combination of Eqs. 3 and 5 leads to

$$\alpha = \sqrt{3k^2+1}/2 \ ; \qquad \beta = \sqrt{3k^2+1}/(2k) \tag{8}$$

Replacing x_i and x_{i-1} into y_i and y_{i-1} in Eq. 6, substituting Eq. 8 into Eq. 6, and rewriting Eq. 6 with respect to y_i, we have the follow-equation.

$$\begin{aligned} \exp(-\lambda y_i) &= 2F\cdot(1-k)\exp(-\lambda y_{i-1})+(1-F) && (0<k\leqq1,\ k\neq1/2) \\ y_i &= F\cdot y_{i-1} + 2\sqrt{7}\cdot F/7 && (k=1/2) \end{aligned} \tag{9}$$

in which

$$\lambda \equiv (2k-1)\sqrt{3k^2+1}/(2k) \tag{10}$$

with F denoting the conditional probability $F(y_i|y_{i-1})$.

From the viewpoint of stochastic formulation of design storm, the conditional probability F may be assumed to be time-invariant before and after the peak. Hence, the values of F before and after the peak are denoted as F_b and F_a, respectively. If the autocorrelation index k, the conditional probability F, and the reduced variate y_p of the peak intensity x_p are given, a sequence of y_i or a discrete hyetograph in time unit Δt can be calculated from Eq. 9.

2.2. Conditional Probability Storm Pattern

It is inconvenient for the formulation of a conditional probability intensity-duration curve mentioned later that the time distribution of y_i calculated by Eq. 9 is discrete. Thus, we transform Eq. 9 into an equivalent equation with respect to continuous time t. Consider a first order ordinary linear differential equation: $dY/dt = A + B \cdot Y$. The general solution can be written in the form:

$$Y = C \cdot \exp(Bt) - (A/B) \tag{11}$$

where Y = variable; t = time; and A, B, C = constants. Setting $Y=Y_{i-1}$ at $t=(i-1)\Delta t$ and $Y=Y_i$ at $t=i\Delta t$, respectively, in Eq. 11, and eliminating the constant C, we obtain the following equation.

$$Y_i = \exp(B\Delta t) \cdot Y_{i-1} - (A/B) \cdot \{1-\exp(B\Delta t)\} \tag{12}$$

For 0<k<1, by equating the corresponding terms in Eqs. 9 and 12, the following conditions of equivalence are obtained.

$$\begin{aligned} &Y \equiv \exp(-\lambda y) \ ; \quad B \equiv (1/\Delta t) \cdot \ln\{2F(1-k)\} \neq 0 \\ &A/B \equiv (1-F)/\{2F(1-k)-1\} \qquad \text{for } k \neq 1/2 \text{ and } 0<k<1 \end{aligned} \tag{13}$$

$$\begin{aligned} &Y \equiv y \ ; \quad B \equiv (1/\Delta t) \cdot \ln F \neq 0 \\ &A/B \equiv -2\sqrt{7} \cdot F/\{7(1-F)\} \qquad \text{for } k=1/2 \end{aligned} \tag{14}$$

In order to determine the constant C in Eq. 11, we may set the following initial condition at $t=\Delta t/24$.

$$Y = Y_p = \begin{cases} \exp(-\lambda y_p) & (k \neq 1/2) \\ y_p & (k=1/2) \end{cases} \tag{15}$$

For k≠1/2, substitution of Eq. 15 into Eq. 11 leads to

$$\begin{aligned} &Y = \{Y_p+(A/B)\}\exp\{Bt-B(\Delta t/24)\}-(A/B) \\ &Y \equiv \exp(-\lambda y) \ ; \quad Y_p \equiv \exp(-\lambda y_p) \end{aligned} \tag{16}$$

From the definition of a single storm pattern, the intensity y at

$t>\Delta t/24$ must be always smaller than y_p at $t=\Delta t/24$. Consequently, the following conditions are necessary for Eq. 16 to be valid.

$$\begin{aligned} &2F\cdot(1-k)<1 \ ; \ -(1/\lambda)\cdot\ln(-A/B)<y_p \qquad (0<k<1/2) \\ &-(1/\lambda)\cdot\ln(-A/B)<y_p \qquad (1/2<k<1) \end{aligned} \tag{17}$$

where A/B is given by Eq. 13.

The equations for the continuous one-sided pattern before or after the peak, and the corresponding constraint conditions are summarized as follows.

(a) For $k\neq 1/2$ ($0<\rho<1$ and $\rho\neq 3/7$),

$$\begin{aligned} &Y-\delta = (Y_p-\delta)\exp\{Bt-B(\Delta t/24)\} \\ &Y \equiv \exp(-\lambda y) \ ; \quad Y_p \equiv \exp(-\lambda y_p) \\ &\delta \equiv (1-F)/\{1-2F(1-k)\} \ ; \quad B \equiv (1/\Delta t)\ln\{2F(1-k)\} \\ &\lambda \equiv (2k-1)\sqrt{3k^2+1}/(2k) \ ; \quad -(1/\lambda)\ln\delta<y<y_p \quad (\Delta t/24<t) \\ &0<F\leqq 1/\{2(1-k)\} \quad (0<k<1/2) \ ; \qquad 0<F<1 \quad (1/2<k<1) \end{aligned} \tag{18a}$$

(b) For $k=1/2$ ($\rho=3/7$),

$$\begin{aligned} &y-\delta = (y_p-\delta)\exp\{Bt-B(\Delta t/24)\} \\ &\delta \equiv 2\sqrt{7}\cdot F/\{7(1-F)\} \ ; \quad B \equiv (1/\Delta t)\cdot\ln F \quad (\Delta t/24<t) \\ &\delta<y<y_p \ ; \qquad 0<F<1 \end{aligned} \tag{18b}$$

(c) For $k=1$ ($\rho=0$),

$$y = \begin{cases} y_p & (t=\Delta t/24) \\ -\ln(1-F) & (t>\Delta t/24) \end{cases} \tag{18c}$$

$$-\ln(1-F)<y_p \ ; \qquad 0<F<1$$

The continuous storm patterns, which are given by Eq. 18, are called conditional probability storm patterns. The three parameters: k, F, and y_p are generically called as pattern parameters.

Figure 2 shows one-sided conditional probability storm patterns before or after the peak for the case of $F=0.5$ and $y_p=10$. It is clearly found from Fig. 2 that the storm pattern becomes sharp with increasing k; that is, with decreasing . For the case of a certain fixed values of k and y_p, we can easily infer that the storm pattern becomes sharp with decreasing F.

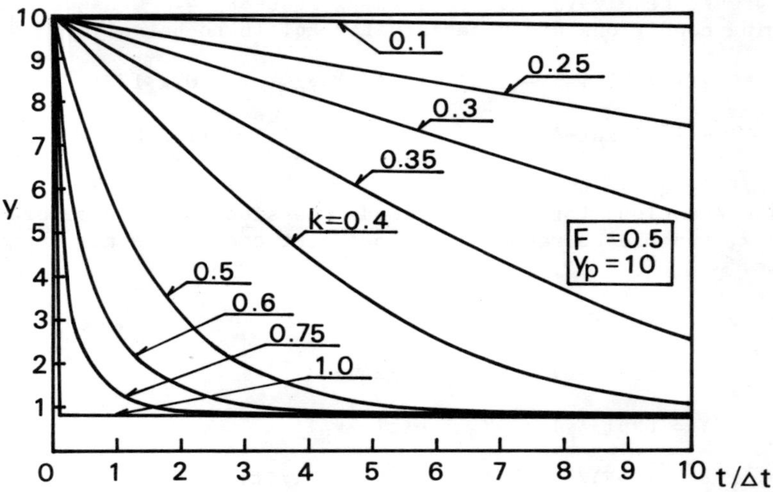

Fig. 2 One-sided conditional probability storm patterns before or after the peak for the case of F=0.5 and y_p=10

3. CONDITIONAL PROBABILITY INTENSITY-DURATION CURVE

Since the technical terms of the last peaked type (LPT) and the central peaked type (CPT) patterns are often used in engineering practice in Japan, both types of the conditional probability storm pattern are defined as shown in Fig. 3. The latter is obtained by combining two one-sided patterns back to back with the same peak intensity y_p and the same

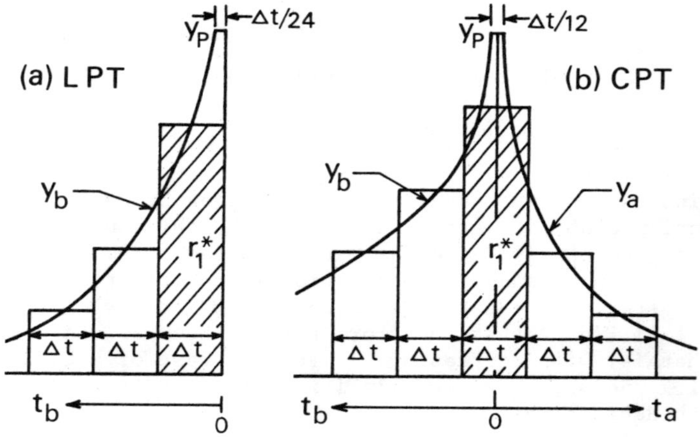

Fig. 3 Definition of two typical storm patterns

autocorrelation index k. Although CPT patterns generally have different values of the conditional probabilities F_a and F_b, the special case of a symmetrical pattern; that is, $F_a=F_b$ may be easy to use.

As shown in Fig. 3, t_b is the time before the peak, while t_a is the time after the peak. The durations of the peak intensity y_p are set to be $\Delta t/24$ and $\Delta t/12$ for LPT and CPT, respectively. The intensities y_b and y_a are the values of y at the times t_b and t_a, respectively, calculated by Eq. 18.

The reduced average intensity r* for the duration τ is given as

$$r^* = (y_p\Delta t/24+\int_{\Delta t/24}^{\tau} y_b dt)/\tau \qquad \text{(LPT)}$$

$$r^* = (y_p\Delta t/12+\int_{\Delta t/24}^{\tau_a} y_a dt+\int_{\Delta t/24}^{\tau_b} y_b dt)/\tau \qquad \text{(CPT)} \qquad (19)$$

where $\tau = \tau_a+\tau_b$ and τ_b, τ_a = sub-durations before and after the peak satisfying $y_b=y_a$. The integral of Eq. 19 for $0<k<1/2$ and $1/2<k<1$ cannot be expressed by elementary functions, so that it is numerically integrated by Simpson's formula. In Fig. 3, let r_1^* represent the average of the intensities y for time unit Δt containing the peak.

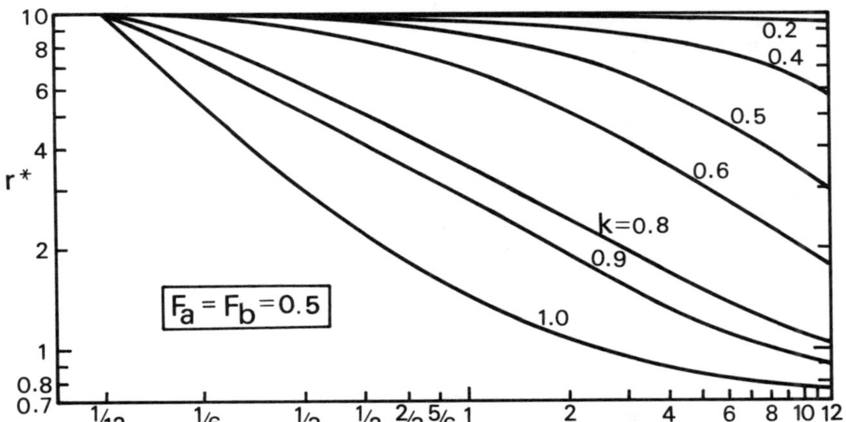

Fig. 4 Conditional probability intensity-duration curves for CPT storms and for the case of $y_p=10$ and F=0.5

Figure 4 shows conditional probability intensity-duration curves corresponding to central peaked type storms for the case of $y_p=10$ and $F_a=F_b=0.5$. It is seen from Fig. 4 that these curve shapes are quite similar to the shapes obtained from the usual analysis of intensity-duration-frequency relationship (Gilman, 1964, p.9-50).

4. APPLICATION

4.1. Fitting of Freund's Distribution to Hourly Data

Hourly data of heavy storms, which have about 100 mm or more amounts of daily rainfall, are collected. If a consective heavy storm occurrs continuously not only on a day, but also on the days before and/or after, the consective hourly data are available. Let us select a certain base level u, and consider a sequence of exceedances (x-u). A couple of exceedances (x-u) at $t=i\Delta t$ and $t=(i-1)\Delta t$, respectively, is adopted as a bivariate (X_i,Z_i); that is, $X_i=x_i-u$ and $Z_i=x_{i-1}-u$. Assuming that the bivariate follows Freund's distribution defined by Eq. 2, we can estimate the parameters: α and β by the maximum likelihood method (Freund, 1961). The theoretical marginal distribution curve with parameters: α and β is plotted on semilogarithmic paper, and the goodness of fit to the empirical distribution is examined. If a good fit is obtained, and if the sample correlation coefficient $\hat{\rho}$ agree with the autocorrelation coefficient ρ by Eq. 4, then we can employ the very value of u. Otherwise, different values of u should be reset.

The examinations of goodness of fit for Freund's distribution at five observation stations: Tokushima, Anabuki, Kito in Tokushima prefecture, Motoyama in Kochi prefecture, and Osaka, have led to satisfactory results.

4.2. Data Analysis for Pattern Parameters

We carry out fitting of the conditional probability storm patterns,that can be calculated by Eq. 18, to the actual heavy storm hyetographs with the maximum 1-hr average intensities x_c; that is, the reduced intensities r_1^* over about 3. Hourly rainfall intensities x are transformed into the reduced variates: $y=(x-u)/\sigma_u$ by using the estimated base level

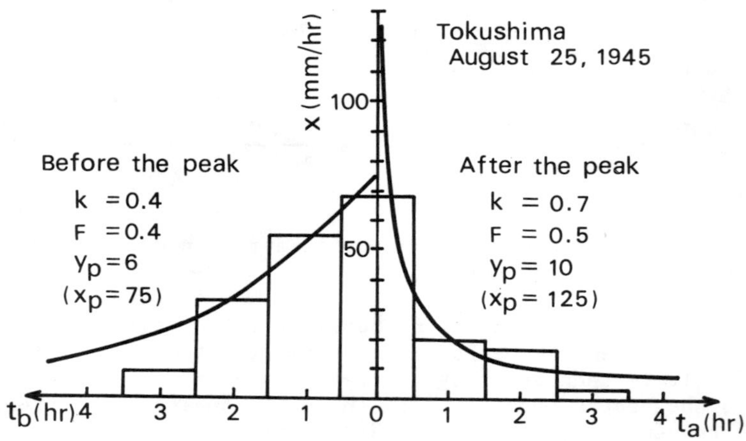

Fig. 5 Fitting of CPT patterns to an hyetograph

u and the standard deviation σ_u of exceedances (x-u). The optimum combinations of pattern parameters: k, F, and y_p, which gives the minimum sum of squares of residuals for the 3-hr duration the peak of a hyetograph, is searched among 2592 combinations of k=0.1, 0.2, ···, 0.9; F= 0.1, 0.2, ···, 0.9; and y_p=4.0, 4.5, 5.0, 5.5, ···, 19.0, 19.5.

For the case of CPT patterns, an actual hyetograph with the reduced intensity r_1^* over about 3 is divided into two portions before and after the peak, where rainfall data are independently used for analysis of the pattern parameters as shown in Fig. 5.

As for the autocorrelation index k, Table 1 summarizes the mean $\bar{k}$, the standard deviation σ_k of k, and the sample size n, obtained from the parameter optimization for each observation station. Although, in reality, the sample autocorrelation coefficients vary with time, we regard the autocorrelation index k as a constant for a design storm pattern. Then, we adopt the mean $\bar{k}$ as the value of k to determine a design storm.

Table 1 $\bar{k}$ and σ_k of the autocorrelation index k

Station \ Design type	Last peaked			Central peaked		
	$\bar{k}$	σ_k	n	$\bar{k}$	σ_k	n
Tokushima	.559	.123	22	.679	.132	19
Anabuki	.540	.156	10	.640	.174	10
Kito	.544	.137	16	.679	.161	14
Motoyama	.577	.144	22	.732	.145	19
Osaka	.518	.083	11	.655	.099	11

The sample crosscorrelation coefficients between F and k are found to be 0.79 and 0.63 for LPT and CPT patterns, respectively, so that F may be approximately determined from the following regression equation.

$$\hat{F} = \begin{cases} 1.42\ k - 0.42 & \text{(LPT)} \\ 1.11\ k - 0.36 & \text{(CPT)} \end{cases} \tag{20}$$

Assuming the reduced peak intensity y_p to be expressed by a linear regression model with three independent variables: k, F, and r_1^*, we have the following equation.

$$\hat{y}_p = \begin{cases} 1.99\ r_1^* + 24.89\ k - 11.83\ F - 9.87 & \text{(LPT)} \\ 2.38\ r_1^* + 25.03\ k - 12.18\ F - 13.67 & \text{(CPT)} \end{cases} \tag{21}$$

where the multiple correlation coefficients are 0.85 and 0.87 for LPT and CPT, respectively. However, the estimated value $\hat{r}_1^*$, which can be calculated by substituting the estimate $\hat{y}_p$ from Eq. 21, k, and F into Eqs. 18 and 19, fairly differs from the given value of r_1^* in Eq. 21.

Therefore, we use the estimate $\hat{y}_p$ obtained from Eq. 21 as a preliminary estimate for solving Eqs. 18 and 19 with respect to y_p, assuming that k, F, and the reduced 1-hr average intensity r_1^* are given, as mentioned later.

4.3. Estimation Procedure of Pattern Parameters

Based on available data of hourly rainfall intensities, a practical procedure for estimating three pattern parameters: k, F, and y_p against a given return period at a site, is summarized as follows.

1) Collect hourly rainfall data of heavy storms with about 100 mm or more amounts of daily rainfall. Fit Freund's distribution to the empirical distribution of the exceedances (x-u). If a good fit is obtained, and if the sample correlation coefficient $\hat{\rho}$ agrees with the autocorrelation coefficient ρ calculated by Eq. 4, then the value of u is adopted. The standard deviation σ_u of the exceedances (x-u) is calculated.
2) Determine the 1-hr rainfall intensity x_c against a given return period by means of a frequency analysis. Then, x_c is transformed into the reduced 1-hr average intensity $r_1^*=(x_c-u)/\sigma_u$.
3) Fit the conditional probability storm patterns given by Eq. 18 (for LPT and CPT) to actual heavy storm hyetographs with r_1^* over about 3. The mean $\bar{k}$ of the autocorrelation index k is adopted for the design storm pattern.
4) The mean $\bar{F}$ of the conditional probability F obtained in Step 3, or the estimate $\hat{F}$ by Eq. 20 may be adopted for the design storm pattern. For the case of the unsymmetrical CPT pattern, $\bar{F}$ or $\hat{F}$ is regarded as F_b, and a value of F_a, which is smaller than F_b, is selected together with y_p by an interation method mentioned in Step 5.
5) Using Eqs. 18 and 19, search y_p minimizing the square of residual $(\hat{r}_1^* - r_1^*)^2$ by an interation method. The estimate $\hat{y}_p$ by Eq. 20 can be used as a preliminary estimate for the iteration. For the case of the unsymmetrical CPT pattern, it is required to search F_a and y_p minimizing the square of $(\hat{r}_1^* - r_1^*)$.

The conditional probability intensity-duration curves against return periods: T=30, 50, 80, and 100 yr are obtained by using hourly rainfall data at Tokushima, and are plotted in Fig. 6. For these curves, the estimate F by Eq. 20 was used. The usual intensity-duration-frequency curves, which are based on annual maximum series for different durations, are also shown in Fig. 6 for reference.

The conditional probability and the usual intensity-duration curves are based on quite different data, procedures, and concepts from each other, so that it may be scarcely meaning to discuss the agreement of both methods. However, the latter curves are important references to examine usefulness and propriety of the former curves. As seen in Fig. 6, the difference between two type curves corresponding to LPT and CPT patterns becomes distinguished with increasing return perid T, and with decreasing duration τ in the range of $\tau<60$ min.

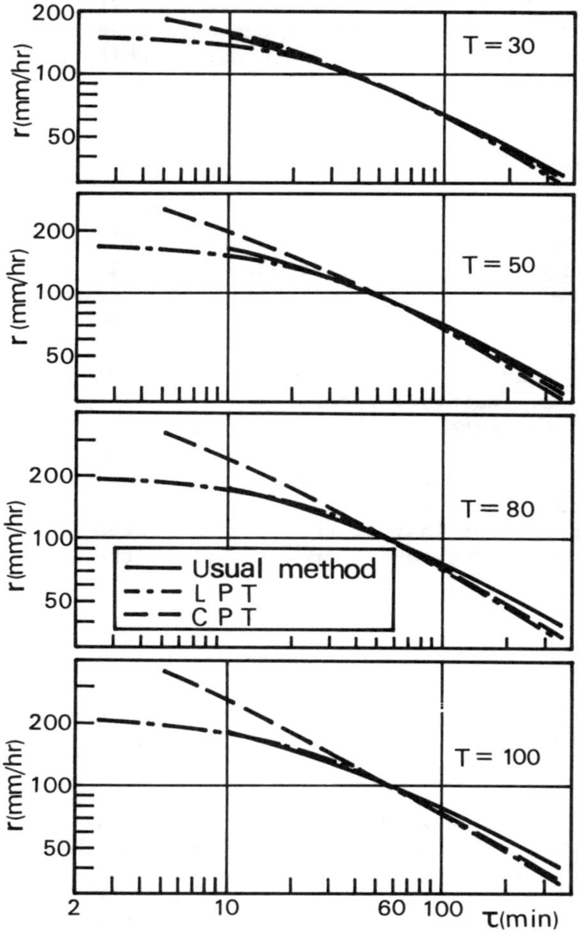

Fig. 6 Conditional probability intensity-duration curves at Tokushima for return periods T= 30, 50, 80, and 100 yr

Table 2 Pattern parameters at the case of LPT for Tokushima

u=7.5mm/hr, σ_u=15mm/hr, k=.55, F=.37				
T(yr)	30	50	80	100
y_p	9.42	10.67	12.11	12.79

As for the short duration range of $10<\tau<60$ min, if closer curves to the usual intensity-duration-frequency curves are selected based on Fig. 6, they almost all belong to LPT patterns, of which pattern parameters are listed in Table 2.

5. CONCLUDING REMARKS

A conditional probability storm pattern and a conditional probability intensity-duration curve for a design flood are formulated based on Freund's bivariate exponential distribution. Two typical storm patterns, namely, the last and the central peaked type storms are defined, and a practical method of estimating the three pattern parameters: k, F, and y_p is presented.

Based on hourly rainfall data, propriety of assuming Freund's distribution is confirmed. Rainfall intensities for shorter durations than 60 min can be easily estimated by the conditional probability intensity-duration curve which is obtained from only hourly data. Furthermore, if there is a reliable intensity-duration curve, which is obtained from annual maximum series of data for different durations, we can set a design storm pattern for the last or the central peaked type of the conditional probability storm pattern corresponding to the usual intensity-duration curve.

REFERENCES

Delleur, J.W. and Dendrou, S.A. 1980. 'Modeling the runoff process in urban areas', 10, Issue 1, CRC Critical Reviews in Environmental Control.

Frederick, R.H. 1978. 'Interstorm Relations in Pacific Northwest', Journal of the Hydraulic Division, ASCE, 104, No. HY12, pp. 1577-1586.

Freund, J.E. 1961. 'A Bivariate Extension of the Exponential Distribution', Journal of the American Statistical Association, 56, pp. 971-977.

Gilman, C.S. 1964. 'Rainfall', in: V.T. Chow(Editor), Handbook of Applied Hydrology, p. 9-50, MecGraw-Hill Book Co., New York.

RAINFALL FREQUENCY STUDIES FOR CENTRAL SAUDI ARABIA

Uygur Sendil
Assoc. Professor
Dept. of Civil Engineering
King Saud University
P.O. Box 800
Riyadh, Saudi Arabia

Abdin M. A. Salih
Professor
Dept. of Civil Engineering
University of Khartoum
Khartoum, Sudan

ABSTRACT. A frequency analysis of rainfall data for six stations in Central Saudi Arabia is presented. The Gumbel type-1 distribution is used to predict design storms for particular durations (10-min to 12-hr) and for particular return periods (2 to 50-yr). Analysis of data for each individual station as well as regional analysis are carried out using annual series. The area under study is divided into two zones by considering the rainfall and topographical characteristics, and depth-duration-frequency relationships are derived for each zone. These relationships are recommended for the design of hydraulic structures (i.e. storm sewers and drainage channels) in the area.

1. INTRODUCTION

Rainfall depth-duration-frequency relationships of high intensity rainfalls are needed for estimating floods in inadequately gaged streams, and for providing design criteria for hydraulic structures such as storm sewers for urban developments, drainage works for airports and water regulating structures on small basins.

In order to have reliable rainfall frequency relationships for a certain area, autographic point rainfall data over long years are needed. However, in many developing countries such data are either unavailable in some areas or insufficient in length. In Central Saudi Arabia, for example, rainfall depth-duration-frequency relationships are scarce mainly due to limited length of autographic data. In the present study relatively longer periods of rainfall records were available. The objective of this study is hence to obtain depth-duration-frequency relationships for Central Saudi Arabia using the longer periods of

V. P. Singh (ed.), Hydrologic Frequency Modeling, 315–326.

records available at present.

According to the Meigs (1953) classification of climate, most of Saudi Arabia including the central part is arid, with meager rainfall in a period less than half a year. A recent study by Sendil and Salih (1984) also shows that the area under study has an arid climate with very limited number of storms, i.e. less than ten storms per year, on average, with 90% of the annual rainfall depth occuring in the period between December to April. Such limited number of storms of short duration and high intensity are typical characteristics of arid and semi-arid regions (Slatyer and Mabbut, 1964).

2. RAINFALL DATA

Historic records of maximum rainfall dapths in millimeters with durations of 10-, 20-, 30-, 60-, 120-, 180-, 360-, and 720- minutes were available for the study region from six recording stations selected mainly on the basis of their lengths of record. These lengths varied from 10 to 22 years and the rain gage network covered an area of about 90,000 sq. km. Table 1 gives a summary of the particulars of these stations. For some years records were missing due to the breakdown of the recording instrument. Table 1 also shows that the relative variability of annual rainfall is often beyond 100 percent, which is typical of true desert conditions. The index of variability is based on the percent deviation of mean rainfall from the most extreme value divided by the mean value.

3. ANALYSIS OF INDIVIDUAL STATIONS

The rainfall data in each station were investigated for homogeneity by using the maximum F-ratio test of Hartley (1950). The results of the test indicate that all the stations used in the present study have homogeneity of variance.

A robust check of independence of data was made by using the empirical equation suggested by Restrepo and Eagleson (1982). The average length of rainy season for the study area ranged between 4 to 6 months, and the mean annual rainfalls are given in Table 1. The minimum time between independent storms were found to range from 69 to 96 hours for the six stations, which satisfies the independence of the present data.

A selection of frequency distribution model could be made as described by Gupta (1970). However, a recent study by Al-Shaikh (1985) showed that, among the four distributions (EV1, normal, log-normal, Pearson type III) considered, EV1 distribution gave relatively better fit for the rainfall data from 19 stations in Saudi Arabia. Hence, frequency analysis of the present data were carried out for annual series using EV1 Gumbel's distribution. The goodness of fit test for the rainfall data was carried out by dividing each rainfall series into four class intervals which yielded k-3 = 1 degree of freedom. For all the series tested the calculated chi-square values did not fall in the critical region, and hence the data fitted the EV1 distribution satis-

Table 1: Particulars of the stations used in the study

Station	R001 Riyadh	R002 Kharj	R003 Aflaj	R004 Zilfi	R005 Sudair	U214 Unayzah
Latitude	24°34'	24°10'	22°17'	26°18'	25°32'	26°04'
Longitude	46°43'	47°24'	46°44'	44°48'	45°37'	43°59'
Altitude (m)	564	430	539	605	665	650
Range of annual rainfall (mm)	13-177	8-146	0-138	36-256	67-228	25-222
Mean annual rainfall (mm)	83	67	56	122	116	93
Variability (%)	113	118	146	105	97	134
Period of graphic data	1963-84	1970-83	1973-83	1970-84	1971-84	1965-84
Total period (years)	22	14	10	15	10	20

factorily at the 95 percent confidence level.

The parameters of the EV1 distribution and the standard error of estimates are obtained by using the maximum likelihood method, which is one of the most reliable and perhaps the most widely used technique. The methods of deriving the maximum likelihood parameter estimates are available in Jenkinson (1969), Kite (1977), and other references.

Using the procedures described above rainfall depth-duration-frequency relationships for each station were determined and rainfall estimates were computed for 2-, 5-, 10-, 25-, 50-year return periods, and then 95% confidence limits were obtained for these estimates. Due to space limitation only few examples comprising of 0.17-, 1-, 12-hour duration estimates and confidence limits are shown in Table 2.

It is clear from Table 2 that, the rainfall estimates and the errors increase with higher return periods as well as with longer durations. The errors are greater for stations with shorter length of data and higher annual rainfall variability. For the individual stations studied confidence limits ranged between 16 to 54 percent of the rainfall estimates.

4. REGIONAL ANALYSIS

Since the individual station records are relatively short, rainfall frequency relationships can not yield reliable results for high return periods. The lengths of the records were therefore extended by adopting the station-year concept which is based on the assumption; that the maximum events selected for analysis are independent and that the data are homogeneous and sufficiently varied to represent a ong period.

The zones for the study area were designated according to the following topographic and climatic characteristics. Similar criteria

Table 2: Rainfall depth estimates (mm) and confidence limits for individual stations

Station	Duration hour	Return period (years) 2	5	10	25	50
R001	0.17	5.23 ± 1.38*	8.42 ± 2.12	10.53 ± 2.71	13.19 ± 3.52	15.16 ± 4.13
	1	9.08 ± 2.12	13.96 ± 3.25	17.20 ± 4.16	21.28 ± 5.40	24.32 ± 6.34
	12	13.15 ± 2.78	19.57 ± 4.27	23.82 ± 5.47	29.19 ± 7.09	33.17 ± 8.33
R002	0.17	4.63 ± 1.71	7.78 ± 2.62	9.86 ± 3.37	12.50 ± 4.36	14.45 ± 5.12
	1	7.45 ± 2.40	11.87 ± 3.68	14.80 ± 4.72	18.50 ± 6.12	21.24 ± 7.19
	12	11.79 ± 3.94	19.05 ± 6.04	23.85 ± 7.75	29.92 ± 10.05	34.43 ± 11.80
R003	0.17	3.44 ± 1.85	6.32 ± 2.83	8.22 ± 3.64	10.63 ± 4.71	12.41 ± 5.53
	1	8.42 ± 4.37	15.22 ± 6.70	19.72 ± 8.60	25.41 ± 11.14	29.63 ± 13.08
	12	16.89 ± 8.76	30.53 ±13.45	39.57 ± 17.25	50.99 ± 22.35	59.46 ± 26.25
R004	0.17	5.65 ± 1.94	9.34 ± 2.97	11.79 ± 3.81	14.88 ± 4.94	17.17 ± 5.80
	1	10.63 ± 3.74	17.77 ± 5.74	22.49 ± 7.36	28.45 ± 9.54	32.88 ± 11.20
	12	15.06 ± 4.74	24.10 ± 7.27	30.09 ± 9.33	37.65 ± 12.09	43.26 ± 14.20
R005	0.17	6.45 ± 2.65	10.56 ± 4.05	13.28 ± 5.19	16.71 ± 6.73	19.26 ± 7.90
	1	12.59 ± 3.65	18.27 ± 5.60	22.03 ± 7.18	26.78 ± 9.31	30.31 ± 10.93
	12	22.82 ± 6.01	32.18 ± 9.23	38.38 ± 11.84	46.21 ± 15.34	52.02 ± 18.01
U214	0.17	4.59 ± 1.24	7.33 ± 1.91	9.14 ± 2.44	11.42 ± 3.17	13.12 ± 3.72
	1	10.50 ± 2.13	15.19 ± 3.27	18.29 ± 4.19	22.22 ± 5.43	25.13 ± 6.38
	12	17.43 ± 2.82	23.64 ± 4.33	27.75 ± 5.55	32.95 ± 7.19	36.80 ± 8.45

* Values with ± sign are 1.96 x standard error. Thus 95% confidence limit = rainfall estimate ± ± 1.96 x standard error.

were used by Baghirathan and Shaw (1978) and Oyebande (1982).

a) Geographical location and topographical characteristics as represented by the altitude of the station.

b) Mean annual rainfall and mean annual number of storms.

c) Rainfall depth-duration-frequency characteristics as defined by the percent ratios: A = 100 (5-yr 10-min fall / 5-yr 1-h fall), and B = 100 (25-yr 1-h fall / 25-yr 12-h fall).

A study of the information in Table 1 shows that the first three stations (R001, R002, R003) are comparable in their altitudes, range of rainfall, mean annual rainfall and percent variability, whereas the second set of three stations (R004, R005, U214) are similarly comparable among themselves. Hence, the first set of stations may form Zone I, while the second set may form Zone II.

The two zones are presented in Table 3 for further discussion. It is obvious from the table that the ranges of mean annual rainfalls and intensities ratios are rather large. However, this is not surprising for desert regions with variabilities over 100%. The computed rainfall estimates as shown in Table 2 are also comparable among the stations in each zone.

Table 3: Derived rainfall zones

Rain-fall zone	Range of mean annual rainfall (mm)	Mean annual rain-fall (mm)	Mean annual no. of storms	Intensity ratio A (%)	Intensity ratio B (%)	Years of graphic data
I	56-83	69	3-6	42-66	50-73	46
II	93-122	110	7-8	48-58	58-76	45

The rainfall data in each zone were investigated for homogeneity by using Kruskal-Wallis test (Walpole and Myers, 1978, p. 495). Annual rainfall totals were used in the test. Calculations for the Kruskal-Wallis statistics resulted in 3.13 and 2.66 for zones I and II respectively. Since these values are less than the critical chi-square value of 5.99 at a significance level of 0.05 and 2 degrees of freedom, the two zones were found to be homogeneous.

Independence of data was already discussed in the analysis of individual stations.

Rainfall depth-duration-frequency relationships for each zone were derived by using the EV1 distribution with the application of maximum likelihood method and are presented in Table 4. Using these equations the rainfall estimates were computed for durations of 0.17 to 12 hours and for return periods of 2 to 100 years. The computed rainfall estimates and the corresponding confidence limits are presented in Table 5.

It may be observed in Table 5 that the errors for the extended data are smaller and the confidence limits are narrower as compared with the results for the individual stations. Confidence limits for the zones are found to be around 20 percent of the rainfall estimates.

Table 4: Fitted rainfall depth-duration-frequency relationships for the zones

Duration (h)	Zone I	Zone II
0.17	x = 2.80y + 3.63	x = 3.00y + 4.23
0.33	x = 3.34y + 4.40	x = 3.78y + 5.58
0.50	x = 3.75y + 5.35	x = 4.51y + 7.05
1.00	x = 4.68y + 6.70	x = 5.23y + 9.09
2.00	x = 5.40y + 8.18	x = 6.09y + 10.99
3.00	x = 5.72y + 8.91	x = 6.55y + 12.27
6.00	x = 6.79y + 9.90	x = 7.15y + 14.27
12.00	x = 7.40y + 10.70	x = 7.70y + 15.02

Figs. 1 and 2 show the plots of the derived depth-duration-frequency relationships together with 95% confidence limits corresponding to only 12-h duration. For Zone I (Fig. 1), five observation points, which constitute about 11% of the observations in this zone are outside the confidence limits. Scatter is less in case of Zone II (Fig. 2) and there are no outliers. It should be remembered, however, that Zone I is relatively drier and with greater percent variability than that of Zone II.

For comparison purposes rainfall estimates of durations 10-min and 12-h are shown for both Zone I and the corresponding individual stations and similarly for Zone II as shown in Figs. 3 and 4 respectively. In case of 10-min duration, the fitted lines for the zones and their corresponding individual stations are relatively close to each other. In contrast deviations are larger in case of 12-h duration. The fitted line for station R003 (Fig. 3) is outside the 95% confidence limit of 12-h duration curve of Zone I. Such a large deviation is expected because, the length of record for station R003was only 10 years and its variability was 146%. Deviations are less in case of Zone II, because this zone receives relatively higher rainfall with less variability than Zone I.

5. CONCLUSION

The goodness of fit tests show that EV1 Gumbel's distribution fit satisfactorily to rainfall data of arid climates. The available length of data for the present study was relatively short and therefore large standard errors were involved for rainfall estimates by individual stations analysis. Confidence limits ranged between 16 to 54 percent of the rainfall estimates for individual stations whereas they reduced to about 20 percent in case of regional analysis. Therefore, more confidence should be attached to rainfall depth-duration-frequency relationships with extended length of data samples by regional analysis. However, one should use the relationships with caution for a region with high rainfall variability although the tests may show the region

Table 5: Rainfall depth estimates (mm) and confidence limits for the zones

Zone	Duration h	2	5	10	25	50	100
I	0.17	4.65±0.95*	7.82±1.46	9.92±1.87	12.58±2.42	14.55±2.84	16.50±3.27
	0.33	5.63±1.13	9.41±1.74	11.92±2.23	15.09±2.89	17.44±3.40	19.78±3.90
	0.50	6.73±1.27	10.98±1.95	13.79±2.50	17.35±3.25	19.98±3.81	22.60±4.38
	1.00	8.42±1.59	13.72±2.44	17.23±3.13	21.67±4.05	24.96±4.76	28.23±5.46
	2.00	10.15±1.83	16.27±2.81	20.32±3.61	25.44±4.67	29.23±5.45	33.00±6.30
	3.00	11.00±1.94	17.49±2.98	21.78±3.82	27.21±4.95	31.23±5.82	35.23±6.68
	6.00	12.38±2.30	20.08±3.53	25.17±4.53	31.60±5.87	36.38±6.90	41.11±7.93
	12.00	13.41±2.51	21.80±3.85	27.35±4.94	34.37±6.40	39.57±7.52	44.74±8.64
II	0.17	5.33±1.03	8.73±1.58	10.98±2.03	13.83±2.63	15.94±3.08	18.03±3.54
	0.33	6.96±1.30	11.25±1.99	14.08±2.55	17.67±3.31	20.33±3.88	22.97±4.46
	0.50	8.70±1.55	13.81±2.37	17.19±3.04	21.46±3.94	24.63±4.63	27.78±5.32
	1.00	11.01±1.80	16.94±2.76	20.87±3.53	25.83±4.58	29.51±5.38	33.16±6.18
	2.00	13.22±2.09	20.12±3.21	24.69±4.11	30.46±5.33	34.75±6.26	40.00±7.19
	3.00	14.67±2.25	22.09±3.45	27.01±4.42	33.22±5.73	37.82±6.73	42.40±7.73
	6.00	16.89±2.45	25.00±3.77	30.36±4.83	37.14±6.26	42.18±7.35	47.17±8.45
	12.00	17.84±2.64	26.57±4.05	32.35±5.20	39.65±6.74	45.06±7.91	50.44±9.09

* Values with ± sign are 1.96 x standard error. Thus 95% confidence limit = rainfall estimate ± ± 1.96 x standard error.

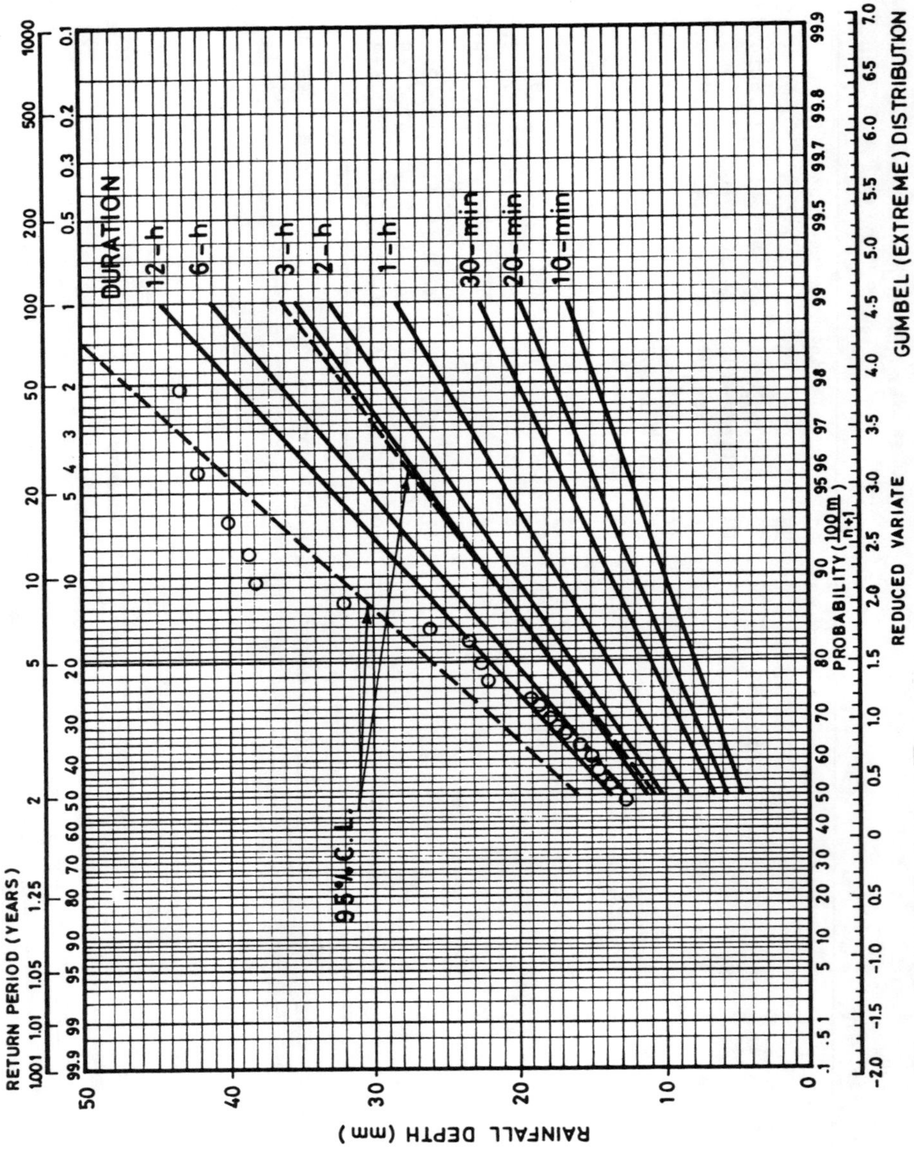

Fig. 1 _ RAINFALL DEPTH _ DURATION _ FREQUENCY RELATIONSHIPS (ZONE I)

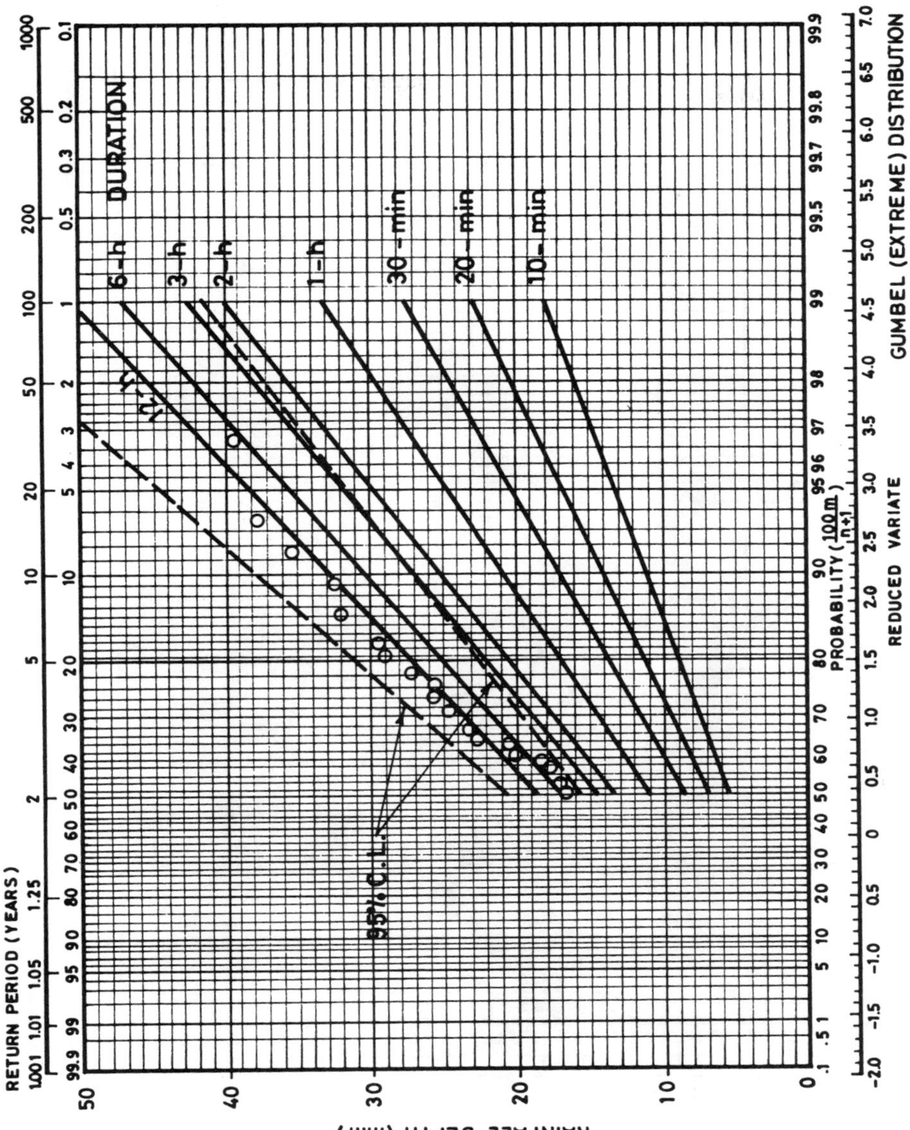

Fig. 2 - RAINFALL DEPTH - DURATION - FREQUENCY RELATIONSHIPS (ZONE II)

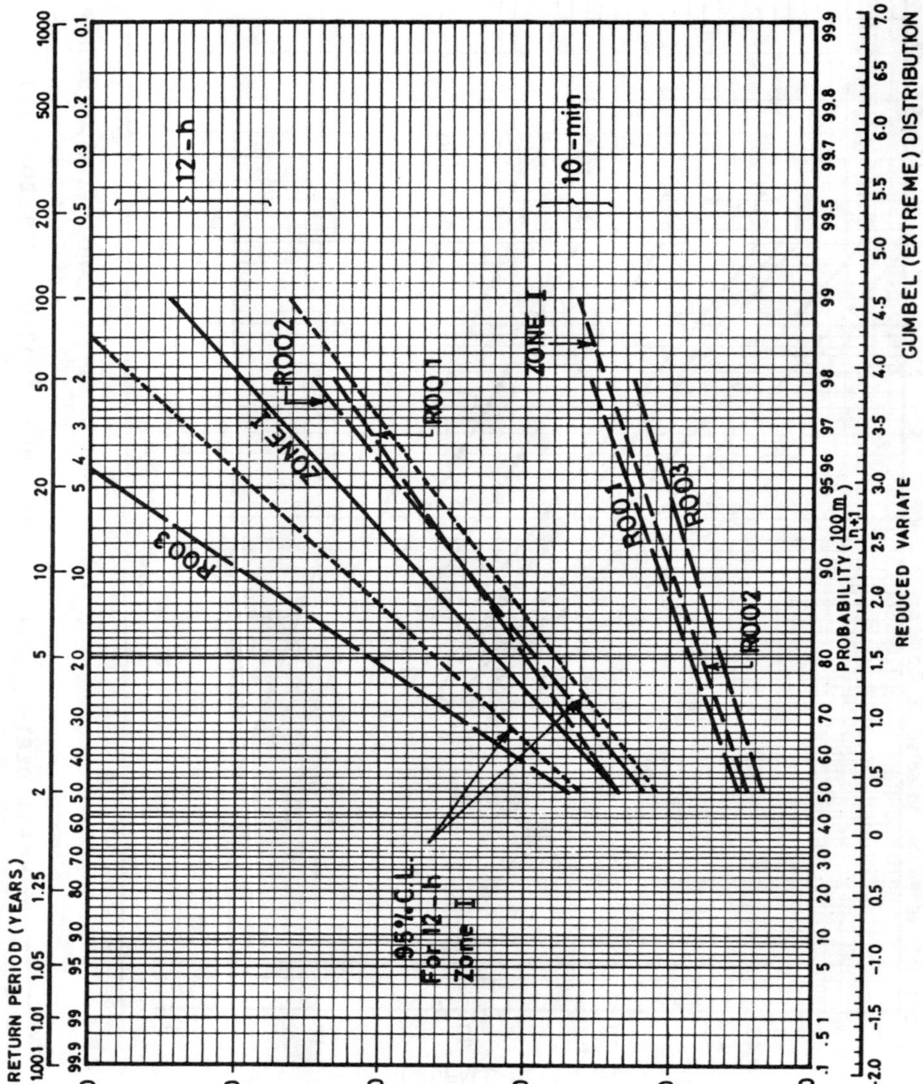

Fig. 3 – RAINFALL DDF RELATIONSHIPS FOR TWO DURATIONS (ZONE I & INDIVIDUAL STATIONS ROO1, ROO2, ROO3)

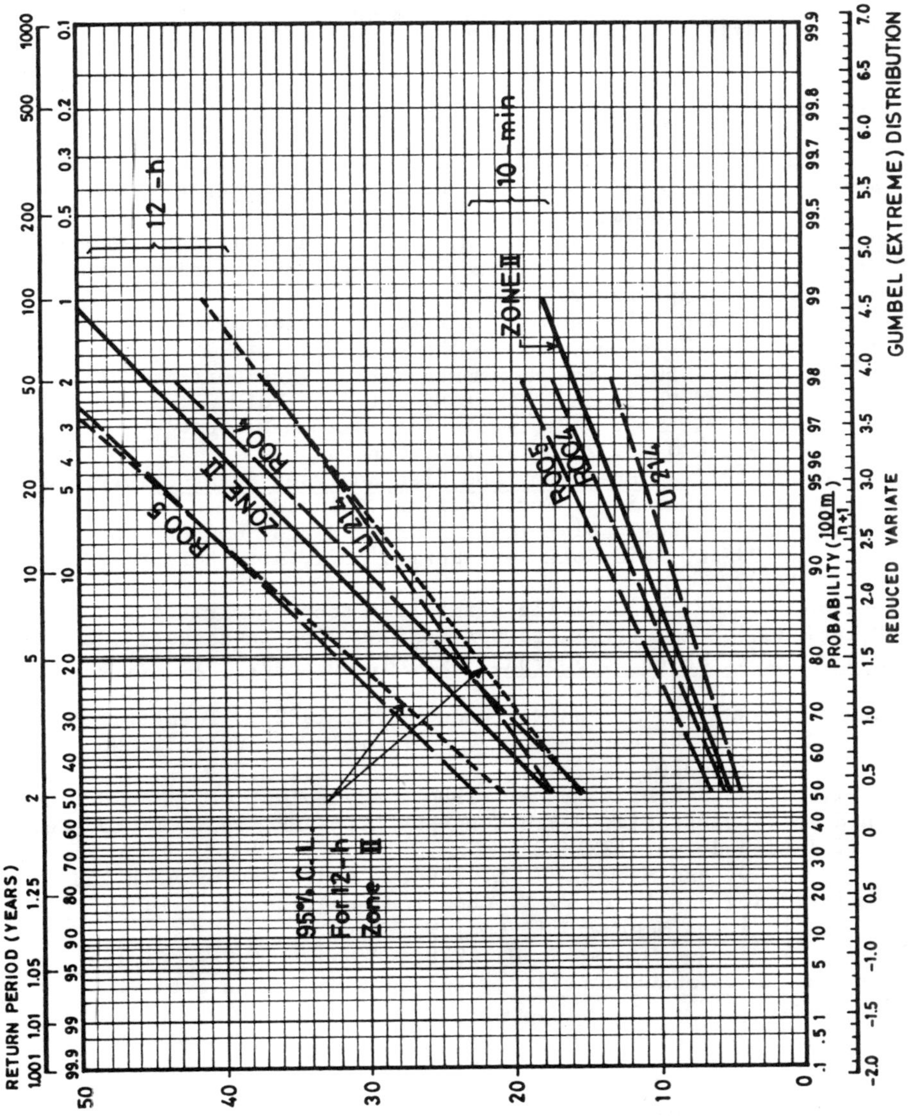

Fig. 4 – RAINFALL DDF RELATIONSHIPS FOR TWO DURATIONS (ZONE II & INDIVIDUAL STATIONS ROO 4 , ROO 5 , U 214)

to be a homogeneous one, and for return periods greater than the total length of record.

It is hoped that the derived relationships are useful for engineers designing hydraulic structures in Central Saudi Arabia.

6. REFERENCES

1. Al-Shaikh, A. A. 1985. 'Rainfall Frequency Studies for Saudi Arabia.' M.S. Thesis, Dept. of C.E., King Saud University, Riyadh, Saudi Arabia, pp. 156.
2. Baghirathan, V. R. and Shaw, E. M. 1978. 'Rainfall Depth-Duration-Frequency Studies for Sri Lanka.' *Journal of Hydrology*. Vol. 37, No. 3/4, May, pp. 223-239.
3. Gupta, V. L. 1970. 'Selection of Frequency Distribution Models.' *J. Water Res. Research*, Vol. 6, No.4, Aug., pp. 1193-1198.
4. Hartley, H. O. 1950. 'The Maximum F-Ratio as a Short-Cut Test for Homogeneity of Variance.' *J. of Biometrica*, Vol. 37, pp. 308-312.
5. Jenkinson, A. F. 1969. 'Estimation of Maximum Floods.' *WMO Tech. Note No. 98*, WMO No. 233, TP. 126, Geneva.
6. Kite, G. W. 1977. *Frequency and Risk Analyses in Hydrology*. Water Resources Publications, Fort Collins, Colorado.
7. Meigs, P. 1953. 'World Distribution of Arid and Semi-Arid Homoclimates.' In *UNESCO Arid Zone Res. Series No.1, Arid Zone Hydrology*, pp. 203-209.
8. Oyebande, L. 1982. 'Deriving Rainfall Intensity-Duration-Frequency Relationships and Estimates for Regions with Inadequate Data.' *Hydrological Sciences Journal*, Oxford, England, Vol. 27, No.3, Sept., pp. 353-367.
9. Restrepo-Posada, P. J. and Eagleson, P. S. 1982. 'Identification of Independent Rainstorms.' *J. of Hydrology*. Vol. 55, No. 1/4, Feb., pp. 303-319.
10. Sendil, U. and Salih, A. M. A. 1984. 'Hydrology of an Arid Watershed in Saudi Arabia.' 4th Congress APD-IAHR, Sept. 11-13, Chiang Mai, Thailand, pp. 1221-1234.
11. Slatyer, R. O. and Mabbut, J. A. 1964. 'Hydrology of Arid and Semiarid Regions.' In V. T. Chow (editor), *Handbook of Hydrology*. pp. 24.1-24.46, Mcgraw-Hill Co., New York.
12. Walpole, R. E. and Myers, R. H. 1978. *Probability and Statistics for Engineers and Scientists*. 2nd Ed., MacMillan Publ. Co. Inc.Ğ New York.

ANALYSIS OF FLOOD OCCURRENCE THROUGH CHARACTERIZATION OF PRECIPITATION PATTERNS

A. Kawamura, K. Jinno, T. Ueda and R. R. Medina
Department of Civil Engineering Hydraulics
Kyushu University
6-10-1 Hakozaki Higashi-ku, Fukuoka 812
Japan

ABSTRACT. This study aims to investigate the occurrences of floods by analyzing the monthly precipitation time sequence utilizing ordinary Kalman filter (OKF) and adaptive Kalman filter (AKF). OKF identifies the abnormal precipitation periods in the sequence by comparing the observed and average precipitation patterns. AKF detects the changes in the precipitation pattern by associating them with the abrupt changes in the parameters of the periodic model of the precipitation time sequence. The 92-year long precipitation record at Fukuoka City, Japan shows three types of abnormal precipitation periods, which exhibit different degrees of possibility of flood occurrence. In addition, it shows nine precipitation epochs with different precipitation patterns. The model parameters estimated by AKF in one epoch characterize its precipitation pattern and describe the occurrences of the abnormal precipitation periods, revealing whether the risk of flood occurrence is high or not in the epoch.

1. INTRODUCTION

Floods like droughts are often caused by adverse weather (Nemoto, 1974), and their occurrences usually result in losses of lives and properties. The need to limit these undesirable consequences of floods has increased society's interest on long-term prediction of adverse weather. Unfortunately, unlike daily weather forecasting, long-term weather prediction is up to now a guessing game. However we believe that the evaluation of the possibility of occurrence of adverse weather that would eventually result to flood can be improved if variations in meteorological and climatological parameters are fully understood.

One of the two most common parameters used to define weather is precipitation (the other is temperature), and its occurrence in abnormally high amounts is the major cause of flooding. Hence this study aims to investigate the occurrences of floods by characterizing the long-term variations in precipitation pattern utilizing ordinary Kalman filter (OKF) and adaptive Kalman filter (AKF). For this purpose, we use the 92-year long (1890-1981) monthly average precipitation sequence at

V. P. Singh (ed.), Hydrologic Frequency Modeling, 327–336.

Fukuoka City, Japan. In this investigation, OKF is used to detect the abnormal precipitation periods in the sequence and to determine quantitatively the periods' magnitude of abnormality. We classify the periods into three types and evaluate the degree of possibility of flood occurrence in each type. On the other hand, AKF is used to detect changes in the precipitation pattern by directly linking them with the abrupt changes in the parameters of the periodic model of the precipitation time sequence. These abrupt changes in the model parameters divide the sequence into several precipitation epochs, where each epoch is characterized uniquely by one precipitation pattern. The model parameters estimated by AKF in one epoch describe the precipitation pattern and the occurrences of the abnormal precipitation periods, where the occurrences of these periods reveal the risk of flood occurrence in the epoch.

2. METHODOLOGY

The use of the monthly average precipitation data provides a suitable approach to the investigation of long-term variations in precipitation pattern. However logarithmic transformation of the 1104 (92 years) monthly average precipitation data (in mm/day) is done to make them follow normal distribution. Also the log-transformed precipitation data is smoothed using the recursive low-pass filter shown in eq.1 (Bendat and Piersol,1976).

$$y(k) = (1-\alpha)\log\{z(k)\} + \alpha y(k-1) \tag{1}$$

where k is the time instant; y is the smoothed log-transformed monthly average precipitation (log (mm/day)); α is the smoothing coefficient and equal to 0.6; and z is the monthly average precipitation(mm/day). The smoothed log-transformed monthly average precipitation $y(k)$ at time instant k is modeled by a periodic function as follows.

$$y(k) = M_y + \sum_{i=1}^{q}(A_i\sin 2\pi f_i k + B_i\cos 2\pi f_i k) + w(k) \tag{2}$$

where M_y is the mean of the smoothed log-transformed sequence; q is the number of significant frequency components; f_i is the frequency component; A_i and B_i are the amplitudes of the frequency component; and $w(k)$ is the stochastic component which is assumed to be white Gaussian noise with zero mean and variance $W(k)$. The reasons for smoothing, choosing α=0.6 and modeling the sequence by a periodic function are mentioned in detail by Kawamura *et al.* (1985). In this study, q = 5, and the dominant frequency components used are f_1 = 1/48, f_2 = 5/72, f_3 = 1/12, f_4 = 1/3 and f_5 = 5/12 in cycles/month, having periods of 4-year, 1.2-year, one-year, 3-month and 2.4-month respectively. These periodicities are obtained by MEM spectral analysis of the smoothed-transformed data (Kawamura *et al.* 1985). Comparison of the power spectra of $z(k)$ and $y(k)$ shows similar dominant peaks, indicating that the periodic properties of $z(k)$ are retained in $y(k)$.

We now consider the problem of identifying the mean M_y and amplitudes A_i and B_i by OKF and AKF. The system equations for OKF and AKF are given respectively in eqs. 3 and 4, while the observation equation for both OKF and AKF is given in eq.5.

$$x(k+1) = \Phi(k)x(k) + \Gamma(k)u(k) \tag{3}$$

$$x(k+1) = \Phi(k)x(k) + \Gamma(k)u(k) + \delta_{k\theta}G(k) \tag{4}$$

$$y(k) = H(k)x(k) + w(k) \tag{5}$$

where x is the $(n\times 1)$ system vector; Φ is the known $(n\times n)$ state transition matrix; Γ is the known $(n\times p)$ system matrix; u is the independent, zero mean, white Gaussian $(p\times 1)$ system noise vector with known covariance matrix U; G is the unknown $(n\times 1)$ vector expressing the magnitude of abrupt change; θ is the unknown time instant when the abrupt change occurred; $\delta_{k\theta}$ is the Kronecker's delta ($\delta_{k\theta}=1$ if $k=\theta$ and $\delta_{k\theta}=0$ if $k\neq\theta$); y is the $(m\times 1)$ observation vector $(m \leqq n)$; H is the known $(m\times n)$ observation matrix; and w is the independent, zero mean, white Gaussian $(m\times 1)$ observation noise vector with known covariance matrix W. Here the dimension n of the state vector is 11 and the dimension m of the observation vector is one; $x=[M_y \ \ A_1 \ \ B_1 \ \cdot\cdot\cdot \ A_5 \ \ B_5]^T$; $\Phi(k)=I$; the observation eq.5 corresponds to eq.2; and $H(k)=[\,1 \quad \sin 2\pi f_1 k \quad \cos 2\pi f_1 k \quad \cdot\cdot\cdot \quad \sin 2\pi f_5 k \quad \cos 2\pi f_5 k\,]$. The state variables are the model parameters M_y, A_i and B_i. Details of the parameter estimations by OKF and AKF and the various properties of the use of the two filters in the analysis of time series expressed by a periodic function are given by Ueda, *et al.* (1984), Kawamura, *et al.* (1984) and Kawamura *et al.* (1986).

OKF identifies the abnormal precipitation periods in the sequence by comparing the observed and average precipitation patterns through the abnormality detection index $\phi_*(k,l)$ defined by Ueda, *et al.* (1984). This index expresses quantitatively the magnitude of abnormality of an l-month long observed precipitation period at time instant k. Since the plot of the one-step ahead predictions by OKF is almost the same as the average precipitation pattern (Ueda *et al.* 1984 and Kawamura *et al.* 1984), $\phi_*(k,l)$ is calculated recursively using the l-month long series of residuals between the predicted and observed precipitation amounts. Consequently, the occurrence of a peak ϕ_* identifies one abnormal precipitation period. Here we set l = 15.

The shifts in precipitation pattern are associated with the abrupt changes in the model parameters M_y, A_i and B_i. In this case, AKF detects whether abrupt changes in the model parameters or state variables occur or not by evaluating the l-month long series of one-step ahead prediction residuals using generalized likelihood ratio test (GLRT). The GLRT compares the value of $\phi_*(\theta,l)$ with a threshold value η. If $\phi_*(\theta,l)$ is greater than η, the hypothesis (H_1) that an abrupt change occurred at time $k = \theta$ is accepted; otherwise, the hypothesis (H_0) that no abrupt change has occurred is accepted. When an abrupt change is detected, its time of occurrence and magnitude are estimated quantitatively, and the state variables are appropriately corrected, according to the magnitude of this abrupt change, to allow the filter to adjust to the new precipitation pattern. In this study, we detect the change in precipitation pattern, which recurs at an interval of about 10 years *on the average*; for this, we set η = 5.0.

3. RESULTS

The 92 annual precipitation totals are plotted on normal paper using Weibull plotting method as shown in Fig.1. Although the annual precipitation appears normally distributed, the abnormally wet and dry years deviate significantly from the normal line. Also result of the chi-square test shows the hypothesis that the annual totals obey normal distribution being rejected at 10 % level of significance. From these results, we may conclude that the years with abnormally high and low precipitation belong to a population different from those close to or on the normal line. On the other hand, the extent of departure from the normal line provides a qualitative description of the magnitude of abnormality. For instance, the annual precipitation total in 1980 is the most abnormal because its deviation from the normal line is the largest.

Table 1 shows the daily precipitation amounts of more than or equal to 150 mm found in the 92-year record, which are arranged in descending order. These amounts, except for one (rank 6), happened in June, July, August and September. In the next section, this data is used to explain the incidence of floods in the different precipitation patterns and types of abnormal precipitation periods.

In the 92-year sequence, OKF detects 19 abnormal precipitation periods with a 5-year recurrence interval, which are given in Table 2 where they are arranged in descending order of value of peak ϕ_*. Fig.2 illustrates the first five most abnormal precipitation periods (ranks 1-5 in Table 2).

Implementing GLRT, AKF yields eight changes in precipitation pattern in the 92-year sequence, dividing it into nine precipitation epochs (see Table 3). Table 4 presents the parameter estimates at the last time instant k by AKF for each epoch and by OKF for the whole sequence. Fig.3 exhibits the nine precipitation epochs identified by AKF and the occurrences of the 19 abnormal precipitation periods detected by OKF.

4. DISCUSSION

In Fig.2, the average precipitation pattern (full line histogram) shows three peaks, April, June-July and September, which correspond to spring rainy season, summer rainy season and typhoon season respectively. Notice also that the standard deviations of these months are greater than those of the rest of the months. As shown in Table 1, the first ten heavy daily precipitation amounts of more than 200 mm (ranks 1-10) took place in these months. We can therefore say that flood risk is high in these months.

The 19 abnormal precipitation periods detected by OKF are classi-

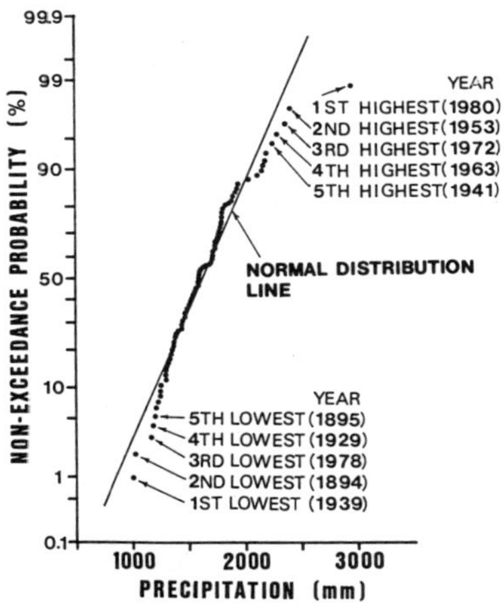

Figure 1. Plot of the annual precipitation data on the normal distribution paper.

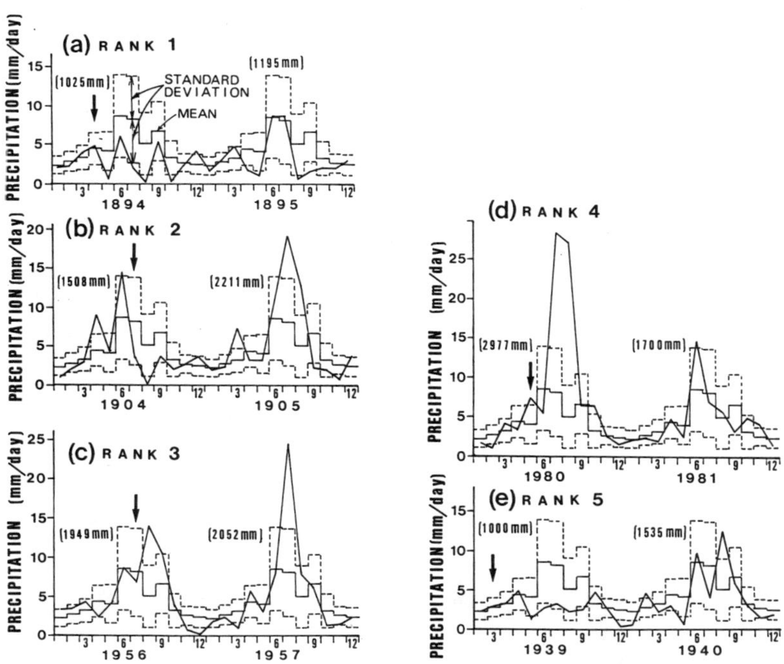

Figure 2. Abnormal precipitation periods (up to rank 5) detected by OKF. The terms in parenthesis are the annual precipitation amounts; downward-arrow indicates the time of occurrence of peak ϕ_* as detected by OKF.

TABLE 1 Daily precipitation of more than or equal to 150 mm (from 1890 to 1891) arranged in descending order.

Rank	Date	Precipitation (mm)
1	25 June 1953	307.8
2	26 July 1905	257.6
3	7 July 1948	245.0
4	29 June 1963	229.3
5	25 Sep. 1954	216.5
6	15 Apr. 1955	215.8
7	21 July 1891	208.7
8	17 Sep. 1925	207.4
9	28 June 1941	206.2
10	22 July 1891	206.1
11	30 Aug. 1980	190.5
12	26 June 1941	186.6
13	25 June 1904	185.6
14	29 Aug. 1980	185.0
15	17 Sep. 1945	182.3
16	28 June 1935	171.9
17	25 June 1941	171.6
18	11 Sep. 1948	168.2
19	10 June 1978	168.0
20	28 June 1953	168.0
21	6 June 1953	166.3
22	5 July 1948	158.6
23	30 June 1979	157.5
24	27 Aug. 1956	152.6
25	26 June 1928	150.7
26	8 Sep. 1934	150.1
27	24 July 1957	150.0

TABLE 2 The abnormal precipitation periods detected by OKF.

	Peak ϕ_*			
Rank	Magnitude	Time of occurrence	Type	Rank of daily precipitation
1	7.90	Apr. 1894	A	
2	7.85	July 1904	C	2,13
3	6.14	July 1956	C	24,27
4	6.09	May 1980	B	11,14
5	6.08	Feb. 1939	A	
6	6.02	Oct. 1944	C	15
7	5.68	Oct. 1971	C	
8	5.16	Aug. 1933	C	26
9	5.15	Oct. 1962	B	4
10	5.08	June 1902	C	
11	4.93	July 1899	C	
12	4.90	June 1948	B	3,18,22
13	4.72	May 1953	B	1,5,20,21
14	4.52	Jan. 1951	C	
15	4.45	July 1977	A	19
16	4.21	Apr. 1913	C	
17	4.19	Aug. 1936	C	
18	4.19	Apr. 1941	B	9,12,17
19	4.15	July 1976	C	

TABLE 3 Precipitation epochs identified by AKF.

Epoch			Rank of daily precipitation
Number	Inclusive dates	Duration (year, month)	
I	Jan. 1890 - Apr. 1894	4,4	7,10
II	May 1894 - May 1898	4,1	
III	June 1898 - July 1904	6,2	13
IV	Aug. 1904 - Aug. 1933	29,1	2,8,25
V	Sep. 1933 - Feb. 1939	5,6	16,26
VI	Mar. 1939 - Nor. 1947	8,9	9,12,15,17
VII	Dec. 1947 - July 1956	8,8	1,3,5,6,18,20,21,22
VIII	Aug. 1956 - Oct. 1971	15,3	4,24,27
IX	Nor. 1971 - Dec. 1981	10,2	11,14,19,23

TABLE 4 Identified system parameters by AKF and OKF.

		Mean	Frequencies (cycle/month) and their amplitudes									
			f_1 = 1/48		f_2 = 5/72		f_3 = 1/12		f_4 = 1/3		f_5 = 5/12	
	Epoch	M_y	A_1	B_1	A_2	B_2	A_3	B_3	A_4	B_4	A_5	B_5
AKF	I	0.53	0.06	0.02	0.06	-0.00	-0.10	-0.09	0.01	0.00	-0.01	0.00
	II	0.43	-0.15	-0.01	0.00	-0.12	0.01	-0.06	0.01	0.04	-0.04	0.02
	III	0.55	0.03	-0.02	-0.02	0.02	-0.12	-0.13	0.00	-0.01	-0.01	0.01
	IV	0.53	0.03	-0.01	0.03	0.02	-0.10	-0.10	0.03	0.02	-0.03	-0.00
	V	0.50	-0.03	-0.01	-0.01	-0.02	-0.18	-0.04	-0.00	0.01	-0.02	0.01
	VI	0.51	0.03	0.13	-0.05	0.02	-0.19	-0.09	0.01	0.03	-0.03	-0.00
	VII	0.61	-0.01	0.03	-0.03	0.04	-0.11	-0.12	0.02	0.02	-0.04	0.00
	VIII	0.53	0.02	0.01	0.01	-0.00	-0.14	-0.10	0.01	0.00	-0.02	-0.01
	IX	0.54	-0.12	-0.04	-0.02	-0.04	-0.14	-0.11	0.01	0.01	-0.01	-0.01
OKF	Total	0.54	-0.00	0.00	0.01	0.01	-0.12	-0.10	0.02	0.01	-0.02	-0.00

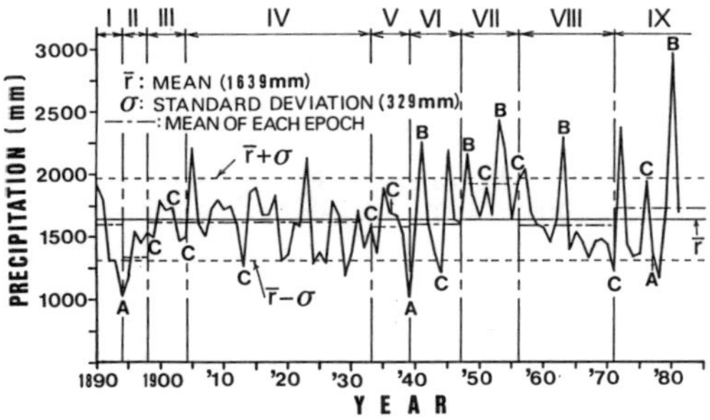

Figure 3. The annual precipitation sequence divided by AKF into nine different epochs. A, B and C are the types of the abnormal precipitation detected by OKF.

fied into three types. Type A is characterized by months with below average precipitation depths as shown in Figs.2(a) and 2(b). Type B is typified by months with above average precipitation amounts as illustrated in Fig.2 (d). As shown in Figs.2 (b) and 2 (c), Type C is characterized by both abnormally high and low precipitation amounts, which occur alternately at two or more months interval. The abnormal precipitation periods could span for one or two years. Of the 19 abnormal precipitation periods, three, five and eleven are Type A, Type B and Type C respectively, suggesting that Type C is most likely to occur whenever an increase of ϕ_* is detected. This increase of ϕ_* should warn us of the possible occurrence of both abnormally high and low precipitation.

As shown in Table 2, 20 out of 27 occasions of heavy daily precipitation occurred in the abnormal precipitation periods. Specifically the first five heavy daily precipitation amounts (ranks 1-5 in Table 1) occurred in these periods, four of them in Type B and one in Type C. This indicates that abnormally high daily precipitation is most likely to occur in Type B. Hence Type B has the highest flood risk. Heavy daily precipitation amounts tend to occur also in Type C, but they appear not as serious as those in Type B. Nevertheless Type C has high possibility of occurrence for both flood and drought. On the other hand, the severest drought ever experienced in the City took place in a Type A period (rank 15 in Table 2). Yet a heavy daily precipitation (rank 19 in Table 1) happened in this period. This suggests that flood could also occur in a Type A period. However higher ranked (1 and 5) Type A periods exhibit no trace of heavy daily precipitation. And flood in this type of abnormal precipitation period is a remote possibility.

As discussed above, we can evaluate the risk of flood occurrence in each period identified by OKF through the presence of abnormally heavy daily precipitation. In the succeeding paragraphs, we investigate the occurrence of flood in each epoch through the characteristics of its precipitation pattern.

As shown in Table 4, the average precipitation pattern defined by the parameters estimated by OKF has a very dominant one-year cycle, as indicated by amplitudes A_3 and B_3 which are much higher than the rest of the amplitudes which are almost equal to zero. Observe also in the same table that we can determine whether an epoch has much precipitation or less precipitation from the estimate of mean M_y and also which frequency component(s) is dominant in each epoch from the estimates of the amplitudes A_i and B_i of frequency component f_i. As presented in Table 4, epoch I is characterized by the one-year cycle being the most dominant. Although amplitudes A_1 and A_2 are relatively prominent (compare with the other amplitudes), this epoch's precipitation pattern is basically similar to the average precipitation pattern identified by OKF.

The second and fifth lowest annual precipitation totals shown in Fig.1 occurred in epoch II. This epoch is typified by less precipitation, as indicated by its mean being very much below the mean $(\bar{r})$ of the whole sequence (see Fig.3). In Table 4, the estimate of mean M_y of epoch II is also the lowest in the nine epochs. However the amplitudes A_1 and B_2 are the most dominant in the epoch (higher than those of the one-year cycle), suggesting that the epoch's precipitation pattern differs much from the average precipitation pattern. Table 3 illustrates no heavy daily precipitation in this epoch; hence we can safely say that flood is least expected in this kind of precipitation pattern.

The precipitation patterns in epochs III, IV, V and VIII are similar to the average precipitation pattern. And the number of occasions with heavy daily precipitation per epoch-duration is small in these epochs. In Table 3, for instance, the 29 years and one month long epoch IV has only three occasions of heavy daily precipitation. Therefore the risk of flood occurrence is not high in the precipitation patterns of these epochs.

The highest estimate of the mean M_y in Table 4 occurred in epoch VII which is characterized by much precipitation as indicated by its mean being very much above the mean $(\bar{r})$ of the sequence (see Fig.3). However the estimates of the amplitudes of the five frequency components in this epoch are practically the same as those estimated by OKF. Thus the epoch's precipitation pattern is very similar to the average precipitation pattern, except for the level of the mean. Moreover, eight of the 27 observed heavy daily precipitation amounts occurred in this short epoch (8 years and 8 months), as shown in Table 3. Four of these amounts are greater than 210 mm (see Table 1). As this epoch has the most number of heavy daily precipitation, resulting in the highest estimate of mean M_y, the risk of flood occurrence is certainly the highest among the nine epochs. Therefore, the higher the estimate of mean M_y is, the higher the flood risk is, in an epoch.

In epochs VI and IX (present epoch), the amplitudes A_1 and B_1 for the 4-year cycle are very dominant as shown in Table 4. The three types of abnormal precipitation periods (A, B and C) tend to happen at the same epoch characterized by a prominent 4-year cycle. In fact, the first lowest (1939) and the fifth highest (1941) annual precipitation totals appeared in epoch VI, whereas the third highest (1972), third lowest (the most severe drought ever experienced in the City took place in 1978), and first highest (1980) yearly precipitation totals were observed in epoch IX (see Fig.1). Hence a remarkable 4-year period is observed in these two epochs because abnormally wet and dry years took place alternately at two-year intervals as can be seen in Fig.3. However the means of these epochs appear normal because the occurrence of abnormally high precipitation is balanced by the occurrence of abnormally low precipitation. Although droughts happened in these epochs, there were also many occasions of heavy daily precipitation (see Table 3), making these two epochs next to epoch VII in terms of the number of incidence of heavy daily precipitation. Thus the risk of flood occurrence is regarded high in these epochs. However, we must pay attention to the occurrence of both flood and drought in this kind of precipitation pattern exhibited by the two epochs.

5. CONCLUSIONS

This paper has demonstrated the application of OKF and AKF in the analysis of flood occurrence through characterization of the precipitation patterns. OKF has been used successfully to identify periods with abnormal precipitation and AKF to detect changes in the precipitation pattern, where these periods and changes are usually indiscernible by visual inspection of the precipitation sequence.

Characterized by the highest estimate of mean M_y, epoch VII has therefore the highest risk of flood occurrence. In this epoch, Type B has the highest risk of flood occurrence, followed by Type C. The months with abnormally high daily precipitation amounts and with high flood risk in Type B and Type C periods (of this epoch) are April, June, July and September. Hence any combination of the epoch, period and month (e.g., epoch VII, Type B and June or epoch VII, Type C and July) has the highest risk of flood occurrence.

The epochs with the second highest flood risk have been VI and IX which are characterized by the occurrences of Types A, B and C abnormal precipitation periods. Among the three types, Type A and Type B have the lowest and highest risk of flood occurrence respectively; however, Type A has the highest risk of drought occurrence. In these epochs, the months with high risk of flood occurrence in each type of abnormal precipitation period are the same as those in epoch VII. For this epoch, we can also conclude that flood risk is high in an epoch where 4-year cycle is the most dominant.

On the other hand, the flood risk is lowest in the epoch typified by the lowest estimate of the mean M_y (epoch II). The second lowest flood risk is when the precipitation pattern of an epoch is the same as the average precipitation pattern, e.g., epochs III, IV, V and VIII. In these epochs, the low points in the plot of ϕ_* indicate periods with low flood risk. The months with low risk of flood occurrence are January-March, May, and October-December.

As presented above, the characteristics of the precipitation pattern in each epoch can be used as a basis for evaluating the possibility of flood occurrence.

REFERENCES

Bendat, J. S. and Piersol, A. G. 1971. *Random Data: Analysis and Measurement Procedures*, John Wiley & Sons, Inc., New York, pp.291-299.

Kawamura, A., Jinno, K. and Ueda, T. 1984. 'On the characteristics of the adaptive Kalman filter for detecting abnormality in a periodic time sequence' (in Japanese), *Proc. of the 28th Japan Conf. on Hydraul.*, pp.383-390.

Kawamura, A., Ueda, T. and Jinno, K. 1985. 'Analysis of long-term pattern fluctuations in a precipitation sequence' (in Japanese), *Proc. of Japan Soc. of Civ. Engrs*, 363 / II-4, pp.155-164.

Kawamura, A., Jinno, K., Ueda, T. and Medina, R. R. 1986. 'Detection of abrupt changes in water quality time series by the adaptive Kalman filter' In: *Monitoring to Detect Changes in Water Quality Series* (*Proc. Budapest Symp.*, July 1986), IAHS publ. 157, pp 285-296.

Nemoto, J. 1974. *Search for Abnormal Weather* (in Japanese), Chuokoron Book Co., Tokyo.

Ueda, T., Kawamura, A. and Jinno, K. 1984. 'Detection of abnormality by the adaptive Kalman filter' (in Japanese), *Proc. of Japan Soc. of Civ. Engrs*, 345 / II-1, pp.111-121.

THE SAMPLING THEORY OF THE BINARY RANDOM FIELD DECIDES THE RELATION ABOUT POINT AND PLANE OF THE EXTRA RAINSTORMS

Zheng-guo Pei
Engineer and Director of Yanbian Water
Conservancy and Water Electricity
Survey Designing Institute
Jilin Province China

ABSTRACT. This paper studies the theoretical relation of rainstorm point-area coefficient and raises a correlation function sampling theorem. For the white noise of low frequency band-limitedand the stochastic process of moving average mold, the conclusion to the sampling theorem of correlation function and to the cut-off frequency is entirely the same. But the cut-off frequency doesn't exist in the autoregressive mold and the like, therefore the random process of the cut-off frequency can't be used. The sampling interval obtained from the sampling theorem of correlation function can make the variance of the mean computed by the sampling value approach the real variance. In this paper, on the basis of the sampling principle, the statistic average theoretical relation of the rainstorm point area coefficient has been reasoned out with the help of correlation function of the rainfall field.

RANDOM RAINFALL FIELD AND THE RELATION OF RAINSTORM POINT AREA

The variation of a certain characteristic rainfall(for example, night or day rainfall) in a region can be regarded as a bivariate random field. Let correlation function of the random field be $B(u_1,u_2)$, then

$$B(u_1,u_2)=\overline{H(u_1)\cdot H(u_2)}$$

$$=\overline{H(u_1)\cdot H(u_1+r)} \qquad (1)$$

where, u_1 ,u_2 and r is a plane vector,$H(u_i)$ is a rainfall observation value.

For the uniform field,taking u_1 as zero vector has no real loss of generality. Then (1) becomes

V. P. Singh (ed.), Hydrologic Frequency Modeling, 337–345.

Let

$$B(r)=\overline{H(0).H(r)} \tag{2}$$

$$R(r)=\overline{H(0).H(r)} \;/\; \overline{H^2(0)} \tag{3}$$

where R(r) means the correlation function of H(r), its variance is one. In rectangular coordinates, R(r) can be represented as R(X,Y), and $H(u_i)$ as H(X,Y) respectively.

Coefficient K of storm point area denotes the ratio of average area rainfall in and the rainfall in the centre. Let the rainfall in the centre be H(0,0), then

$$K=\iint_F H(X,Y)dx.dy/F.H(0,0) \tag{4}$$

$$H(x,y)=R(x,y).H(0,0) \tag{5}$$

Putting Expression (5) in (4), we can obtain

$$K=\iint_F R(x,y)dx.dy/F \tag{6}$$

Because the rainfall variation has the obvious characteristic of regional distribution, rainfall field is generally considered as a non-uniform random field. To study the stochastic characteristics of the rainfall field in large regions. We shall transfer the observation value $H(u_i)$ of each station into N(0,1) (i.e. the random variate $Z(u_i)$ of normal distribution, which can be abbreviated to Z_i). It is evident that the random field formed by Z_i is a uniform random field. When R(r) only relates to |r|, i.e R(r) only relating to the models of vector but not relating to the direction, the isogram of R(r) forms concentric circles on the plane. We call this random field a uniform isotropic random field. Simpler maths model of the correlation function of the uniform isotropic random field can take

$$R(x,y)=e^{-\beta (X^2 + Y^2)^{1/2}} \tag{7}$$

where, β is underdetermined constant coefficient. Having put (7) in (6), we obtain

$$K= 2\pi(e^{-\beta (F/\pi)^{1/2}} (-\beta(F/\pi)^{1/2} -1)+1)/(F\cdot\beta^2) \tag{8}$$

SAMPLING PRINCIPLE OF THE STOCHASTIC PROCESS

Considering continuous stationary stochastic process X(t), what length will time interval △ take and can X(i△)(i=0,±1, ±2......) represent X(t)?

Sampling theorem shows that if the cutoff frequency f

exists in X(t), then $\Delta = 1/2f$. Sampling theorem thus has solved the sampling interval problem of the stochastic process in the case of cutoff frequency. But some of the important stochastic processes such as the autoregressive mold and so on do not have the cutoff frequency and even if some other hydrologic stochastic processes do have cutoff frequency, the uses of sampling theorem are limited because of the complicated computation and the failure in the computing precision, therefore the problem of hydrologic random sampling can hardly be solved.

1. Mean sampling theorem:

Suppose mean is zero, variance is one, which satisfies the continuous stationary stochastic process X(t) of the ergodic condition. Let

$$\hat{\mu} = \int_0^T X(t)\, dt / T \tag{9}$$

$$\hat{\mu}' = \sum_{i=1}^{n} X(i\Delta) / n \tag{10}$$

where, $n = T/\Delta$, $i = 1, 2, \ldots$.

What length will time interval Δ take to sample in order that $\hat{\mu}'$ can represent $\hat{\mu}$?

That $\hat{\mu}'$ represents $\hat{\mu}$ means that the computing precision between $\hat{\mu}'$ and $\hat{\mu}$ is equal, that is, the variance of both must be equal. According to the variance of $\hat{\mu}$ (by the study of J.S Bendat and others in 1971)

$$\mathrm{Var}[\hat{\mu}] \doteq \int_{-\infty}^{\infty} R(\tau)\, d\tau / T \tag{11}$$

where, $R(\tau)$ represents the correlation function of X(t).

$$\mathrm{Var}[\hat{\mu}'] = \sum_{j=1}^{n} \sum_{i=1}^{n} X(i\Delta) . X(j\Delta) / n^2$$

$$= (n + 2 \sum_{i=1}^{n-1} (n-i) R(i\Delta)) / n^2$$

where, $n = T/\Delta$. when $R(\Delta) = 0$, $R(i\Delta) \to 0$

$$\mathrm{Var}[\hat{\mu}'] = \Delta / T \tag{12}$$

If $\hat{\mu}'$ represents $\hat{\mu}$, then (11) must be equal to (12) . hence

$$\Delta = \int_{-\infty}^{+\infty} R(\tau)\, d\tau \tag{13}$$

When sampling $\triangle$ computed in Expression (13) satisfies $R(\triangle) = 0$ then the variance of $\hat{\mu}'$ and $\hat{\mu}$ is equal. This is the mean sampling theorem.

The correlation function of moving average mold in Example 1 can be expressed as

$$\begin{cases} R(\tau) = 1-2\,|\tau|/B & \text{when } |\tau| < B/2 \\ 0 & \text{when } |\tau| \geqslant B/2 \end{cases}$$

$$\triangle = \int_{-B/2}^{B/2} (1-2\,|\tau|/B)\,d\tau = B/2$$

$$R(B/2) = 0$$

Thus the variance of $\triangle = B/2$ sampling and the continous sampling is equal.As for the low frequency bandlimited white noise process in Example 2, when the variance is one

$$R(\tau) = (\mathrm{Sin}2\pi f\tau / 2\pi f$$

$$\triangle = \int_{-\infty}^{+\infty} (\mathrm{Sin}2\pi f\tau / 2\pi f)\,d\tau = 1/2f$$

$$R(1/2f) = 0$$

where , f is the frequency bandwidth, that is cutoff frequency. We can see from theabove that the conclusion reasoned out of the mean sampling theorem and the cutoff frequency is completely equal in the case of low frequency bandlimited white noise process.

Yet,not in all types of correlation functions, $\triangle$ being calculated with Expression (13), $R(\triangle)$ must be equal to zero. Such as the correlation function of autoregressive mold is

$$R(\tau) = e^{-\beta|\tau|}$$

$$\triangle = \int_{-\infty}^{+\infty} e^{-\beta|\tau|}\,d\tau = 2/\beta$$

$$R(2/\beta) = 0.135$$

Though $R(\triangle)$ is not zero, it is close to zero. Hence the mean sampling theorem can be used. In the general hydrologic computation, we often meet the correlation function of stochastic process, $R(\triangle) \doteq 0$ (see Table A) therefore the mean sampling theorem has got widespread significance. When $R(\triangle) \neq 0$, the variance of $\hat{\mu}'$ is larger than that of $\hat{\mu}$. And the ratio between the variance of $(\hat{\mu}'-\hat{\mu})$ and the variance of $\hat{\mu}$ is denoted by ε, ε is called the relative error of variance.

$$\varepsilon = 2\sum_{i=1}^{n-1} R(i\triangle) \tag{14}$$

Because $R(\triangle)\doteq 0$, thus the two points that are separated by $\triangle$ are independent of each other. $\triangle$ is called the step length of independent point. $n= T/\triangle$ is called the number of independent point. The mean sampling theorem takes samples on the basis of independent points, henceit is also called the sampling theorem of independent point.

$$\because \quad S(f) = \int_{-\infty}^{+\infty} R(\tau)\, e^{-j2\pi f\tau} d\tau$$

$$\triangle = S(0) \tag{15}$$

where S (f) is power spectral density function. Thus it is clear that the independent point sampling theorem is suitable for a low frequency stochastic process of $S(0)\neq 0$.

The conception of the independent point number has its great significance in application. For example, in the laying out of the rainfall station,how many observation stations should be arranged at a certain rainfall in order that the total rainfall (or the average rainfall in the area) roughly measured with the data of those observation stations can represent the real total rainfall? this number of the observation station is called the independent point number.

2. Variance sampling theorem

To Gaussian process variance must be determined except mean. So far as sampling is concerned, the precision must be maintained in both computation and variance. Let

$$\hat{\psi} = \int_0^T X^2(t)dT/T \tag{16}$$

$$\hat{\psi}' = \sum_{i=1}^{n} X^2(i\triangle')/n \tag{17}$$

For Gaussian process variance is one and the mean is zero. (16) equal to (17)

$$\triangle' = \int_{-\infty}^{+\infty} R^2(\tau)dt \tag{18}$$

and satisfies $R(\triangle')= 0$. With $\triangle'$ sampling $\hat{\psi}'$ can repressent $\hat{\psi}$, this is variance sampling theorem. Generally, $\triangle'$ is smaller than $\triangle$. But by Table A, in the case of band-limited white noise $\triangle' = \triangle$ and bandpass white noise $\triangle=0$. But $\triangle'$ changes between $1/f$ and $1/2f$.

3. Maximum value sampling theorem

For the Gaussian stochastic process ,let $X^{(i)}(t)$ denote the ith realization of stochastic process X(t), the time length T in each realization is equal. $X_{max}^{(i)}$ denotes the maximum value of ith realization. F(X) means the distribution function of X(t). The given m realization can obtain m sequences $\bar{X}_{max}^{(i)}$ and the distribution function was written as G(X). If $X^{(i)}(t)$ has n independent points,a realization is regarded as the independent experiment of n time. Based on the binomial theorem of independent experiment, probability being the extreme value of F(X), the probability occurred in the experiment of n time is

$$G = 1 - (1 - F)^n$$

$$n = \frac{\ln(1 - G)}{\ln(1 - F)}$$

where , G and F is the abbreviation of G(X) and F(X), respectively. Because F is the probability occurring from the extreme value, $F \ll 1$, hence

$$n \doteq -\ln(1 - G)/F \tag{19}$$

When F is exponential type function, the asymptotic distribution of G is

$$G(X) = 1 - e^{-e^{-Y}} \tag{20}$$

$$Y = \alpha X - \alpha u \tag{21}$$

Where, α and u can be determined by the extreme value sequence based on E.J Gumbel's least square method. When X=u,Expression (19) can be changed into

$$n = 1 / F(u) \tag{22}$$

F(u) is a normal distribution function, which can be found in the table of normal frequency curve.

Sampling interval Δ is

$$\Delta = T / n \tag{23}$$

When m realization stochastic process makes the appearance, each process takes maximum value and obtains m maximum value sequences. According to E.J Gumble's least square method, α and u are determined. With(22) the independent point number n is computed.And the sampling interval can be computed with Expression (23). This is the maximum value sampling theorem.

4. Mean sampling theorem of random field

When R(x,y) denotes its correlation,function,uniform istropic random with sampling theorem of stochastic process can lead to:

$$a= \int_{-\infty}^{+\infty} \int_{-\infty}^{+\infty} R(X.Y)\, dx.\, dy \tag{24}$$

Where ,a is an area expressed by a sampling point,an area called the controlling one of independent observation station. .For the region of area as F,n=F/a is called the independent point number. Putting Expression(7) in(24) gives

$$a= 2\pi / \beta^2 \tag{25}$$

ACTUAL ILLUSTRATIONS OF THE GREATEST RAINSTORM POINT AREA RELATION ANALYSED WITH THE SAMPLING PRINCIPLE

1. Data:

In the computation of independent station numbers in China's typhoon region"Data for Greatest Rainfall at Home and Abroad" every year's several worst rainstorms in China have been recorded.Checking the mean value of the rainstorm observation sites and Cv,Cs. Transferring the observation value of each station into the random variate Z of N(0.1)normal distribution.We list every year's maximum.Thus,by maximum value sampling theorem,in the case of having gained Z value series, the independent point number n can be determined, i.e.computing α and u based on E.J.Gumbel's least square method.By the value Zi in Table B, we find $\bar{Z}$=3.21,α=1.8922, u=2.9308, checking normal frequency curve table, obtains F(u)=0.00169, by Expression(22), n=1/0.00169=590.

The area of China's typhoon rainstorm region is 415 x 10^4 KM^2,the controlling area a of the independent station is a=7034 KM^2 . Putting Expression(25) in(8) gives

$$K=a(e^{-(2F/a)^{1/2}} (-(2F/a)^{1/2} - 1)+ 1)/F \tag{26}$$

2. The test and reasonable appraisal of theoretical curve.

Figure 1 has indicated the analysed results of several China's provinces' rainstorm point area relation. By the figure, the theoretical curve of the rainstorm point area ralation discussed in the paper has been reasonable(the imaginary line in Figure 1). But the theoretical curve in Figure 1 reflects only the statistic average numbers of the worst rainstorm in a greater region. Rainstorms often have their obvious characteristic of regions,so theoretical curve can not replace the rainstorm point area relation of all the regions.

$\triangle$ and $\triangle'$ table of the stochastic processes in common use

Table A

$R(\tau)$	$\triangle$	$\triangle'$	$\varepsilon\%$	$R(\triangle)$
$1-2\|\tau\|/B$ $\|\tau\|\leqslant B/2$; 0 $\|\tau\|\geqslant B/2$	$B/2$	$B/3$	0	0
$\sin 2\pi f\tau/2\pi f\tau$	$1/2f$	$1/2f$	0	0
$e^{-\beta\|\tau\|}$	$2/\beta$	$1/\beta$	31.3	0.135
$e^{-\alpha\tau^2}$	$(\pi/\alpha)^{1/2}$	$(\pi/2\alpha)^{1/2}$	8.64	0.0432
$(\sin\pi B\tau/\pi B\tau)\cdot\cos 2\pi f_0\tau$	0	$1/B-1/2B$	0	1.0

(Referenced in Page 5)

Maximum Z value computation table of the years' typhoon region (Referenced in Page 7)

Table B

Year	Place	$\overline{H}_{24}$ (mm)	Cv	H_{24} (mm)	Z
1953	Lushan	120	0.65	1100	4.057
1954	Wangying	95	0.65	419	2.820
1955	Qianfoshan	120	0.65	627	3.108
1956	Fengchao	160	0.65	593	2.522
1957		60	0.65	178	2.151
1958	Shitouhe	100	0.65	740	3.694
1959	Zhangjiafanzi	60	0.76	600	3.726
1960	Chaoqiao	110	0.65	653	3.323
1961	Xiakou	90	0.65	466	3.092
1962	Wanquan	50	0.78	400	3.332
1963		90	0.72	950	3.963
1964	Fugang	110	0.78	500	2.517
1965	Dafengzha	112	0.65	673	3.340
1966	Erhekou	70	0.65	297	2.753
1967	Xinliao	240	0.65	1692	3.591
1968	Shanghe	100	0.67	431	2.284
1969	Zhaoqiao	80	0.76	374	2.604
1970	Shuangjie	200	0.65	785	2.625
1971	Najin	60	0.65	330	3.218
1972	Kangbao	50	0.78	550	3.793
1973	Quzhouliting	80	o.65	557	3.590
1974	Helin	55	0.78	520	3.573
1975	LInzhuang	110	0.65	1062	4.14

Indication of rainstorm point area relation in China's typhoon region

Figure I

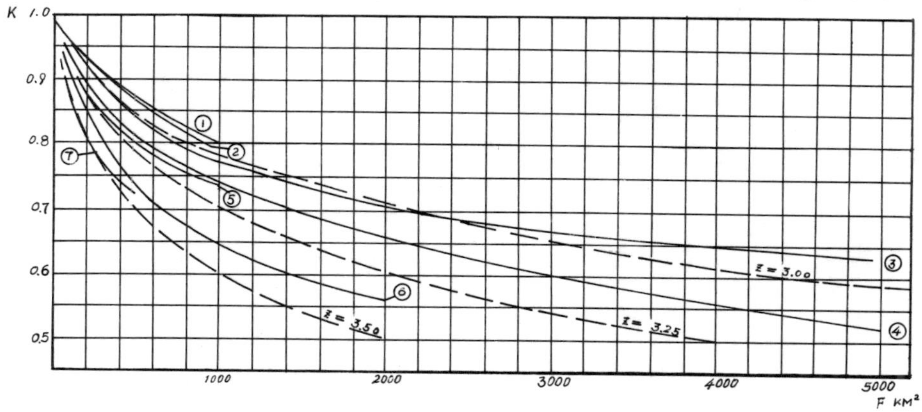

(Referenced in Page 343)

(1) Zhejiang (2) Yunnan (3) Jilin (4) Heilongjiang (5) Guangdong (6) Guizhou

REFERENCES

1. D.E New Land, 1975.
An introduction to random vibration and spectral analysis.
Long Man

2. J.S Bendat and A.G Piersol. 1971.
Random data: Analysis and Measurement Procedures.
Wiley- Interscience

SOME NEW PERSPECTIVES ON MAXIMUM ENTROPY TECHNIQUES IN WATER RESOURCES RESEARCH

A. K. Rajagopal and S. Teitler
Naval Research Laboratory
Washington, D.C. 20375

Vijay P. Singh
Department of Civil Engineering
Louisiana State University
Baton Rouge, LA 70803

ABSTRACT. After a brief expository account of the Shannon-Jaynes principle of maximum entropy (POME) for discrete and continuous variables, we give here an account of some recent research work which may be relevant to hydrology. This paper will contain discussions on: (i) a "histogram" method to contrast the discrete and continuous modes of description, which should shed some light on the practical methods of computation and the role of histogram in actual practice when dealing with continuous probability distributions. (ii) The idea of mean logarithmic decrement associated with a probability distribution is introduced and is shown to be related to the concept of differential entropy. The mean with respect to an arbitrary probability distribution of the logarithm of the ratio of the probability density function for that distribution to the probability density function for an exponential distribution is discussed in the context of hydrological investigations. Unlike the entropy of a continuous probability distribution introduced in (i), this quantity which is an example of the Kullback-Leibler (KL) Information, is always positive and invariant under coordinate transformations. (iii) The constraints entering into POME as well as a minimum K L Information are identified as a class of sufficient statistics which determine the unknown parameters in the probability density functions that occur in the most commonly used hydrological models. (iv) An example of (iii) where only the first two moments in a semi-infinite domain are given is discussed to shed light on the limitations of POME, recently recognized by Wragg and coworkers, and is made relevant to the work of Sonuga on rainfall-run off relationship. Finally, (v) a method of generating probability distributions starting from <u>one basic distribution</u> employing coordinate transformations is given. This in conjunction with (iii) leads to the notion of "physical constraints" in contrast to the "mathematical constraints" in examining parameter estimation.

V. P. Singh (ed.), Hydrologic Frequency Modeling, 347–365.

1. INTRODUCTION

The principle of maximum entropy (POME) discussed here was first introduced by Shannon (1948a,b) in communication theory and has since been extended to many fields of investigations in a general setting of POME propounded by Jaynes (1957a,b, 1961, 1982). A heuristic definition of POME has been given by Jaynes (1982): "When we make inferences based on incomplete information, we should draw them from that probability distribution that has the maximum entropy permitted by the information we do have". The latter information is succintly represented by constraints whose effect must be incorporated into the probability density function (pdf) that arises from the maximization of entropy. There have been several attempts at using POME in hydrology. See Singh et.al. (1985) for a brief review of this. However, it is fair to say that it has not as yet had much explicit impact in practice. In the present work, we show how in fact hydrological analysis can, to a large extent, be facilitated by the use of POME and that consideration of POME is useful in developing new ways to interpret hydrological data.

In Sec. 2, we give a summary of the concept of entropy and POME for discrete probabilities and continuous probability distribution functions (pdf) and exhibit in particular the difference between the two. Most importantly, the lack of invariance under coordinate transformation of the differential entropy arising from continuous distributions and its unboundedness i.e., it can take on large negative values, are often cited as disadvantages in practical use of the differential entropy concept. Since in practice, one often employs pdf for mathematical reasons while all practical data appear in discrete form, it is important to find a well-defined procedure for use in the analysis of data. For this reason, we outline a "histogram" method related to that suggested by Collins and Wragg (1977) which may clarify the situation and even provide a practical method of treating the problem in a satisfactory way. In Sec. 3, we introduce the idea of the mean logarithmic decrement of a probability distribution as a useful alternative to the entropy concept in hydrology. We define a survival probability or non-occurrence distribution as unity minus the cumulative density function (cdf) and the associated decrement rate as the ratio of the pdf to the survival probability. This decrement rate is constant for an exponential pdf. It is shown that the mean logarithmic decrement is unity minus the differential entropy associated with the continuous probability distribution so that POME becomes the principle of minimum mean logarithmic decrement. We then suggest that a theory based on Kullback-Leibler (KL) information which is invariant under coordinate transformation and is always positive unlike the differential entropy of a continuous pdf may be an alternative to POME, since a minimum KL information always exists. KL information involves specifying a reference pdf with respect to which the one under examination is measured. For hydrological applications, a specific form of KL information is suggested. It is shown that the KL information under certain conditions is numerically the mean value of the logarithm of the ratio of the decrement to a constant decrement. The specification of the known exponential pdf is via the first moment as derived from the data, for example.

It is observed that the theory of sufficient statistics is related to POME by the fact that POME results in pdf's that belong to a subset of distributions characterized by sufficient statistics. This was mentioned by Jaynes (1978), later indicated more explicitly by him (1982), shown by van Campenhout and Cover (1981) and discussed in detail by Tikochinsky et.al. (1984). Recall that a statistic is a function of one or more random variables that do not depend upon any unknown parameter. A set of statistics is said to be sufficient if a knowledge of them is sufficient to determine the unknown parameters entering into the pdf. See, for example, Hogg and Craig (1965) and Huzurbazar (1976) for accounts of this formalism. For example, the first moment is a sufficient statistic for an exponential distribution in the semi-infinite domain of the random variable and cannot by itself be a sufficient statistic on the full infinite domain. In general, once the set of sufficient statistics for an appropriate pdf is known, the latter is already determined without further recourse to the data. Further, as emphasized by Halmos and Savage (1949), if one generates an arbitrary data set based on knowledge of the set of sufficient statistics and therefore the corresponding pdf the new data set provides a representation of the random process that is just as valid as an observed data set. In short, a set of sufficient statistics extracts the whole of the relevant information in a data set about unknown parameters of an appropriate pdf. Determining the parameters of the pdf of hydrologic analyses for a given type of problem such as rainfall-runoff modelling, flood frequency analysis etc. is a major goal in the analysis of data. When that pdf is stable in the sense that the distribution with the same parameters is appropriate for different data sets for the same type of system, one is either explicitly or tacitly using pdf for which sufficient statistics exist. An important theorem in this connection is the so called Koopman-Pitman theorem first indicated by Fisher (1922), and established by Darmois (1935), Koopman (1936), Pitman (1936), and Barankin and Maitra (1963). It may be stated as follows. Under certain regularity conditions on the pdf (that can be satisfied by typical pdf's used in describing water resources problems), sufficient statistics for the pdf exist if and only if the pdf belongs to the exponential family. We will describe the nature of the exponential family in Sec. 4. We will also recall that the pdf arising from POME and the related minimum information principle (Kullback (1959)) all belong to the exponential family. With these concepts in hand, we are then able to examine closely the relationship between POME, sufficient statistics, and the analysis of water resources data. In this same section, we discuss an interesting example where some of the aspects of the Koopman-Pitman theorem become evident. This is the problem discussed by Sonuga (1972, 1976) involving a semi-infinite domain, given only the first two moments. It was pointed out by Dowson and Wragg (1973) that POME solution to this problem exists only if the relation between the two moments is constrained. We will discuss this problem in view of the above theorem of the sufficient statistics. In Sec. 5, we use coordinate transformations to derive new pdf from the old and the associated new constraints. This method in conjunction with

Sec. 4 leads to the idea that sufficient statistics in one coordinate system need not go over into sufficient statistics in another coordinate system. If we make sufficient statistics as a basic principle, then some interesting consequences follow and will be explored. In the final section, we conclude with a few remarks on future work etc.

2. BRIEF REVIEW OF THE SHANNON-JAYNES ENTROPY OF A PROBABILITY DISTRIBUTION

Following Shannon and Jaynes, we give here a brief summary of the basic definition of entropy and consequences therefrom for both the discrete probabilities and continuous probability density functions (pdf). For the most part, we shall deal with only one variable problems, since straightforward generalization to many variables is possible.

2.1 Discrete Probabilities:

If $\{p_i\}$ are probabilities of some events i occurring, (the total number of events may be finite or infinite) with $0 \leqslant p_i \leqslant 1$ and

$$\sum_i p_i = 1 \tag{1}$$

the entropy is defined as (we choose for convenience the base of logarithm to be Naperian):

$$S_D[\{p_i\}] \equiv S_D[p_1 p_2 \cdots] = -\sum_i p_i \ln p_i \tag{2}$$

(i) The <u>most certain</u> event is when $p_i = 1$ for $i=i_o$ say and zero for others and the associated entropy is zero. Also since $o \leqslant p_i \leqslant 1$ it is clear that

$$S_D \geqslant 0 \tag{3}$$

(ii) At the other extreme, the most <u>uncertain event</u> occurs when all the p_i's are equal so that if there are a total of N i's, this corresponds to $p_i = 1/N$ and $S_D = \ln N$. Or, equivalently

$$0 \leqslant S_D \leqslant \ln N \leqslant \infty \tag{4}$$

Thus S_D is always positive, has zero as its minimum value and may be taken to be a measure of how uncertain we are of the outcome of an event. The measure (2) suggested is unique under certain assumptions of (a) continuity of S_D considered as a function of the probabilities p_i; (b) if all p_i's are equal, $p_i = 1/N$, then S_D should be montonically increasing function of N which means that with equally likely events there is more choice, or uncertainty, when there are more possible events (N); (c) if a choice is broken down into two successive choices, the original S_D should be the weighted sum of the individual values of S_D (transformation of S_D under a decomposition of a choice). These assumptions are quite general for many purposes in almost all areas of investigations. There are however other measures of information more complicated in form than (2) introduced by Fisher (see Huzurbazar (1976) for a discussion) but the Shannon form (2) seems the most useful one for our purposes.

(iii) Change of labelling of p_i as occurring in a transformation of variable of the discrete type clearly does not change the value of S_D.
(iv) On the other hand, any change toward equalization of $p_1 \cdots p_n$ increases S_D. In other words, if we perform any "averaging" operation of the p_i of the form

$$q_i = \sum_j a_{ij} p_j \tag{5}$$

where $\sum_i q_i = 1$, $\sum_i a_{ij} = 1 = \sum_j a_{ij}$ and all $a_{ij} \geqslant 0$ then

$$S_D[\{q_i\}] \geqslant S_D[\{p_i\}] \tag{6}$$

The proof of this follows from the convexity property of $x \ln x$ for $x \geqslant 0$, or equivalenty $\ln(x/y) \geqslant 1-y/x$ if $x,y>0$. Thus $p_j(\ln p_j - \ln q_i) - p_j + q_i \geqslant 0$ for all q_i, $p_j > 0$. Multiplying this inequality by a_{ij}, summing on i and j, and using the above relations (5) etc., we deduce (6).

If the transformation a_{ij} is just a permutation of the p_i's then S_D does not change in value, which is a special case of (iii).

(V) POME: Given linearly independent constraints

$$\sum_i g_k(i) p_i = G_k, \quad k=1,2, \cdots K$$

maximize $S_D[p_i]$ to obtain the least biased $\{p_i\}$. This principle is supported by (6).

2.2 Continuous Variable-pdf

Again, we consider a one-variable case for simplicity. We define a dimensionless pdf p(x) in the domain of dimensionless variable x in $(0,\infty)$ which is most relevant in the hydrological contexts. (Remarks on other domains will be made when appropriate.) In most physical problems dimensional variables x_d appear (e.g. water discharge or rainfall intensity etc.) and a dimensionless version is required for defining the differential entropy, S_C. This leads to a certain innate ambiguity in choosing a dimensionless x and a corresponding dimensionless p(x) because one has always the normalization requirement

$$\int_0^\infty p_d(x_d)\, dx_d = \int_0^\infty p(x)\, dx = 1 \tag{7}$$

Thus we might consider formulations in terms of explicitly dimensional variables or those with a change of scale:

$$x \equiv x_d/u_x \equiv x_a/a \equiv x_{da}/u_x a \equiv x_d/u_{xa} a \tag{7a}$$

Here u_x is the unit choice (e.g. cm^3 or day), a is the scale factor (numerical factor such as 2 or π often introduced for convenience) and the other definitions are self-explanatory. With such a choice, (7) would imply that the corresponding $p_d(x_d)$ have compensating scale:

$$p(x) \equiv p_d(x_d)\ u_x \equiv p_d(x_a)\ a \equiv p_d(x_{da})\ u_x a \equiv p_d(x_d)\ u_{xa} a \tag{7b}$$

Define the differential entropy associated with a dimensionless pdf and variable as

$$S_C[p(x)] = - \int_0^\infty p(x) \ln p(x)\, dx \tag{8}$$

In view of (7a,b) we obtain for example,

$$S_C[p(x)] = S_C[p_d(x_d)] - \ln u_x \tag{9}$$

It is assumed that $p(x) \geqslant 0$ for all x.

(i) It is assumed that if p(x) is zero for any x, then p(x) ln p(x) is also zero there.

(ii) No comment such as (4) can be made about $S_C[p(x)]$. However as Kullback and Leibler (1951) noted, if we have two pdf's such that $p(x) \geqslant 0$ and $q(x) \geqslant 0$ for $0 \leqslant x \leqslant \infty$ and

$$\int_0^\infty p(x)\, dx = 1 = \int_0^\infty q(x) dx \tag{10}$$

then

$$I[p/q] \equiv \int_0^\infty p(x) \ln \left(\frac{p(x)}{q(x)}\right) dx \geqslant 0 \tag{11}$$

I[p/q) is often called relative entropy of p(x) with respect to a known q(x). Clearly,

$$\begin{aligned} I[p/q] &= - S_C[p] - \langle \ln q(x) \rangle_p \\ &= - S_C[p] - \int_0^\infty p(x) \ln q(x)\, dx \end{aligned} \tag{12}$$

(iii) Any coordinate transformation from a dimensionless variable x to another dimensionless variable y such that $\frac{dx}{dy}$ is positive, and maps $x:(0,\infty)$ to $y:(0,\infty)$ (or finite positive interval) implies that

$$p(x)dx = p(x) \frac{dx}{dy} dy \equiv \tilde{p}(y)\, dy \tag{13}$$

then

$$S_C[p(x)] = S_C[\tilde{p}(y)] + \int_0^\infty p(x)\, dx\, \ell n \left|\frac{dx}{dy}\right|_{x=y(x)} \tag{14}$$

Note that this transformation contains the uniform scale ambiguity mentioned in (9) as a special case and points to the non-uniqueness of S_C either in its numerical value or in its dependence on reference coordinate in terms of which all measurements are made. We see from (ii) that the KL relative entropy remains invariant under such a transformation because a compensating transformation of q(x) cancels the one that transforms p(x) in I[p/q]. We deal with the specification of the reference q(x) as required in the KL definition, later.

(iv) POME: In spite of the fact that no rigid constraints such as (6) apply for S_C, one often suggests a POME. An alternative mathematically satisfying principle is the minimum Kullback-Leibler relative entropy principle since $I[p/q] \geq 0$ for a fixed common q(x). One may rationalize POME by making a corresponding observation that with a given appropriate scale, the ambiguity in the value of $S_C[p]$ is removed and with respect to the same scale, a POME may be defined. But one should remember that this proviso in comparing two different POME estimations or in making transformations of variables. This is clearly exhibited in (14); there is a scale ambiguity in the left hand side residing in p(x) and a similar one in $\tilde{p}(y)$. This has caused much confusion and dissatisfaction among researchers. For this reason, one would like to adopt a discrete formalism throughout which has no ambiguity of the kind that appears in S_C. But the case of continuous variables is technically very powerful in establishing formal theorems etc. and hence some specific procedure which discretizes the continuous distribution in an unambiguous way ought to be sought. Before we investigate this, we state here POME and return to a discussion of it in Sec. IV. POME gives an explicit variational solution for $p_v(x)$ in terms of the given constraints:

$$p_v(x) = \exp[\sum_{i=1}^{k} \lambda_i g_i(x)]/Z[\lambda] \tag{15}$$

where λ_i are Lagrange multipliers associated with the k-constraints

$$\int_0^\infty g_i(x)\ p(x)\ dx = C_i \quad (i=1,2 \cdots k) \tag{16}$$

These are assumed to be linearly independent.

Here

$$Z[\lambda] = \int_0^\infty \exp[\sum_{i=1}^{k} \lambda_i g_i(x)]dx \tag{17}$$

to ensure that $p_v(x)$ obeys the condition of normalization, (7). Also it is important to observe that the Lagrange multipliers λ_i, must have a solution upon evaluation of (16), in terms of the constraints C_i such that the resulting $p_v(x)$ is ≥ 0 for all x as well as $Z[\lambda]$ is finite. Thus the constraints are just sufficient statistics as will be elaborated in Sec. IV. We now turn to the question of construction of a maximum entropy histogram.

2.3 Maximum Entropy Histogram

Collins and Wragg (1977) introduced a novel maximum entropy histogram. We give here a different formulation of the histogram in order to keep the formalism general and complete. Introduce the notion of cumulative probability

$$F(x) = \int_o^x p(x')dx' \tag{18}$$

It has the following important properties:

$$F(x=0) = 0 \text{ and } F(x=\infty) = 1 \tag{19}$$

i.e., F(x) is <u>positive</u>, monotonic, and goes from zero to unity continuously as x goes from 0 to ∞. Now we construct a histogram of F(x) vs. x as follows. We split the (0,1) range of the F(x) axis into N equal parts. This generates the N <u>unequal parts</u> of x-axis correspondingly and on a p(x) vs. x graph one then has a N-part histogram. In detail, we have in general,

$$F_1 = \int_o^{a_1} p(x)dx;\; F_2 = \int_{a_1}^{a_1+a_2} p(x)dx \;\cdots\; F_N = \int_{a_1+\cdots+a_{N-1}}^{a_1+\cdots+a_N} p(x)dx \tag{20}$$

F_1, F_2, $\ldots F_N$ are all positive and less than 1. Observe that

$$x_o = 0,\; x_1 = a_1,\; x_2 = a_1 + a_2 \;\cdots\cdot$$
$$x_n = a_1 + \cdots + a_n \text{ and } x_N = a_1 + \cdots + a_N = \infty \tag{21}$$

N is at our disposal but <u>not</u> the a_i's.

The following geometrical construction seems useful to note from the graph of F(x) vs. x, the slopes of the graph in the n-th interval (x_{n-1}, x_n) n=1, 2, ... N are given by:

$$\tan\theta_1 = \frac{\int_o^{x_1} p(x)dx}{x_2 a_1} = \frac{F_1}{a_1} = \frac{1}{Na_1}$$

$$\tan\theta_2 = \frac{\int_{x_1} p(x)dx}{a_2} = \frac{F_2}{a_2} = \frac{1}{Na_2}$$

$$\tan\theta_n = \frac{\int_{x_{n-1}}^{-x_n} p(x)dx}{a_n} = \frac{F_n}{a_n} = \frac{1}{Na_n} \tag{22}$$

$$\tan\theta_N = \frac{\int_{x_{N-1}}^{\infty} p(x)dx}{a_N} = \frac{F_N}{a_N} = \frac{1}{Na_N}$$

Collins and Wragg (1977) noted that their "equal area construction" meant

$$\int_{x_o}^{x_1} p(x)dx \equiv \frac{1}{N} = \int_{x_{i-1}}^{x_i} p(x)dx = \cdots = \int_{x_{N-1}}^{\infty} p(x)dx \tag{23}$$

and furthermore, they assumed that the mean value theorem can be applied in each of these intervals to deduce

$$\overline{p(\xi_1)}\ a_1 = \frac{1}{N} = \overline{p(\xi_i)}\ a_i = \cdots = \overline{p(\xi_N)} a_N \tag{24}$$

In other words, $\overline{p(\xi_i)} = \frac{1}{Na_i}$, the unknown probability densities at points ξ_i become related to the unknown $1/a_i$ which may then be treated as variational parameters in POME because

$$S_C = - \int_0^\infty p(x) \ln p(x) dx$$

$$= - \sum_{i=1}^{N} \int_{x_{i-1}}^{x_i} p(x) \ln p(x) dx \tag{25}$$

If the mean value theorem holds at least for sufficiently smooth $p(x)$ with smooth $\ln p(x)$, we then may write (25) in the form

$$S_C = - \sum_{i=1}^{N} \{\overline{p(\xi_i)} \ln \overline{p(\xi_i)}\} a_i = - \frac{1}{N} \sum_{i=1}^{N} \ln \overline{p(\xi_i)} \tag{26}$$

or equivalenty,

$$S_C = - \frac{1}{N} \sum_{i=1}^{N} \ln \left(\frac{1}{Na_i}\right) = + \frac{1}{N} \sum_{i=1}^{N} \ln a_i + \ln N \tag{27}$$

This resultant entropy is the same that would be obtained by using a uniform distribution $1/a_i$ in each interval (x_{i-1}, x_i). If we use (22) in conjunction with the Collins-Wragg prescription, we find

$$S_C[\theta_1 \cdot\cdot\ \theta_N] = - \frac{1}{N} \sum_{i=1}^{N} \ln (\tan \theta_i) \tag{28}$$

θ_i's then appear as parameters to be determined in POME. Any average value of a quantity $g_k(x)$ is then evaluated as follows, e.g. (16),

$$\int_0^\infty g_k(x)\ p(x)\ dx = \sum_{i=1}^{N} \int_{x_{i-1}}^{x_i} g_k(x)\ p(x)\ dx = \sum_{i=1}^{N} \frac{1}{Na_i} \int_{x_{i-1}}^{x_i} g_k(x)\ dx \tag{29}$$

where we have used (24) and the resulting integration $\int_{x_{i-1}}^{x_i} g_k(x)\ dx$ can be performed explicitly. For example, for $g_k(x) = x^k$, $k = 1,2,...$, the integrals can be expressed in terms of a_i's. Thus the constraint problem of POME outlined in Subsection (iv) of the last section can be formulated in terms of the unknowns $\{a_i\}$, which in turn can be determined by the variational principle which leads to a non-linear simultaneous set of equations for the a_i's. This program seems attractive in principle but has only been implemented in expository examples.

3. MEAN LOG-DECREMENT AND KL INFORMATION

The survival probability or non-occurrence distribution is defined by
1-F(x), with F(x) as cdf, (18) associated with the dimensionless pdf p(x) and the decrement rate is defined by

$$\lambda(x) = p(x)/(1-F(x)) = -\frac{d}{dx}\ln(1-F(x)). \tag{30}$$

If $p(x) = \mu e^{-\mu x}$, in (30), $\lambda(x)=\mu$, a constant. In general $\lambda(x)$ is positive for all x, and its properties are determined by the choice of distribution. A relationship between the expected value of logarithm of $\lambda(x)$ and the differential entropy exists, namely

$$1-\int_0^\infty p(x)\ln\lambda(x)\,dx = -\int_0^\infty p(x)\ln p(x)\,dx = S_C[p] \tag{31}$$

This relation is proved by noting that $dF(x) = p(x)\,dx$ and $\int_0^1 \ln(1-x)dx = -1$. It then follows that at maximum differential entropy, the expectation value of the logarithmic decrement rate, is a minimum. An interesting variation of this theorem is given here which may prove useful in the context of water resources research. Quite frequently, one introduces the concept of mean (rainfall or runoff, etc) of p(x):

$$\mu_1 = \int_0^\infty x\,p(x)dx \tag{32}$$

If we take $q(x) = \mu_1^{-1}\exp(-x\,\mu_1^{-1})$ to be the normalized reference probability in the KL information, (11), and from (30), the decrement rate associated with q(x), $\lambda(x) = \mu_1^{-1} = \lambda_o$, is independent of x, then we have from (31), (32)

$$\hat{I}[p/q] = \int_0^\infty p(x)\ln\left(\frac{\lambda(x)}{\lambda_o}\right)dx \geqslant 0. \tag{33}$$

The caret means that q(x) is chosen as above with μ_1 given by (32). The last inequality follows from the KL theorem that $I[p/q] \geqslant 0$ for any p(x), q(x), the value zero is obtained if and only if $p(x) \equiv q(x)$. The use of $q(x) = \mu_1^{-1}\exp(-x\,\mu_1^{-1})$ as a nominal standard pdf to compare with the unknown pdf p(x) is a meaningful one in the present context because this KL information is a direct measure of deviation from the single q(x) which minimizes I[p/q] given only the mean value of p(x). This follows from a simple variational calculation of $\hat{I}[p/q]$ in which the mean value of p(x) is a constraint. Note that I is only invariant under linear scale transformation. However in general the form and value of I is not an invariant in itself. Indeed it is only maintained for a linear scale transformation. Nevertheless if one determines I[p/q] by the above prescription in one coordinate frame, it coincides with the value of I[p/q] in all frames. Thus carrying out an evaluation of I[p/q] with the above procedure under two manifestly different conditions of measurements provides a test of whether the two conditions can be related by a linear scale transformation of variable.

4. POME, SUFFICIENT STATISTICS, AND EXPONENTAL FAMILY OF PDF

First we restate POME for KL information. Here POME is replaced by minimization of KL information I[p/q] subject to mutually independent constraints of the form (given before in (16))

$$E[g_i(x)] = \int_o^\infty g_i(x)\ p(x)dx = C_i \quad (i=1,2,\ldots k) \tag{16}$$

This leads to a variational estimate of pdf

$$p_v(x) = q(x)\ \exp(\sum_{i=1}^{k} \alpha_i\ g_i(x))\ /Z_I(\alpha) \tag{34}$$

where

$$Z_I(\alpha) = \int_o^\infty q(x)\ \exp(\sum_{i=1}^{k} \alpha_i g_i(x))dx \tag{35}$$

provides a normalization of $p_v(x)$. The unknown parameters α_i, i=1,2,...k, are the Lagrange multipliers as usual and are determined by the self consistency of $p_v(x)$ with the given constraints, (16). Thus, the constraints are just the sufficient statistics, by virtue of the Koopman-Pitman theorem. From (34), $p_v(x)$ is a member of an exponential family of regular pdf as will be explained presently.

With suitable alterations for our purposes, we follow Hogg and Craig (1965) in giving the following exposition. Suppose there are m-parameters $\theta_1 \cdots \theta_m$ in some known intervals, $\alpha_j < \theta_j < \beta_j$ (j=1...m). The exponential family of regular class of pdf of a continuous variable x ($0<x<\infty$) is then of the form

$$p(x;\theta_1 \cdots \theta_m) = \mathrm{Exp}\{\sum_{j=1}^{m} f_j(\theta_1 \cdots \theta_m)\ K_j(x) + S(x) + T(\theta_1 \cdots \theta_m)\} \tag{36}$$

for $0 < x < \infty$ with S(x) continuous, $dK_j(x)/dx$, j=1...m, are continuous and no one is a linear homogeneous function of the others, and $f_j(\theta_1 \cdots \theta_m)$ j=1...m are non-trival functionally independent, continuous functions of θ_j, $\alpha_j < \theta_j < \beta_j$, j=1...m. Of course $p(x;\ \theta_1 \cdots \theta_m)$ should be positive for x in $(0,\infty)$ and $\int_o^\infty p(x;\ \theta_1 \cdots \theta_m)dx = 1$ should be obeyed. Suppose $X_1 \ldots X_n$, n≥m denote a random sample from a distribution that depends on the m-parameters $\theta_1 \cdots \theta_m$ and has a pdf of the form (36). From the factorization theorem of Neyman-Pearson (see Hogg and Craig (1965)), it follows that the statistics

$$Y_1 = \sum_{i=1}^{n} K_1(X_i),\ Y_2 = \sum_{i=1}^{n} K_2(X_i)\ \ldots,\ Y_m = \sum_{i=1}^{n} K_m(X_i) \tag{37}$$

are just sufficient statistics for the m parameters $\theta_1 \ldots \theta_m$.

A comparison of these statements with (34) shows that the variational $p_v(x)$ under the conditions given above belong to a subset of the exponential family wherever the constraints C_i can be identified as sufficient statistics, and if we can identify $\theta_1 \ldots \theta_m$ above as the Lagrange multipliers in the variational principle for KL information. This is evident because the parameters α_i appear linearly whereas in (36) the parameters θ_i appear in general non-linearly, for example.

It is important to recognize that the constraints given in (16) are not arbitrary. That POME gives a subset of the exponential family of pdf will now be explained. A set of constraints determines the form of pdf. A table of differential entropies belonging to a wide class of pdf's has been given by Verdugo and Rathie (1978). A table of examples of constraints and the respective corresponding pdf's arising from POME has been given by Kagan, Linnik, and Rao (1973). This table has been extended to a much wider class of pdf's by Singh et.al. (1985, 1986). For x in $(0,\infty)$, if the only constraint is the first moment, E(x), one obtains from POME the exponential distribution. If E(x) and E(ln x) are constraints, one obtains the Gamma distribution. If $E(x^\beta)$ with β a known parameter and E(lnx) are constraints, one obtains the two-parameter Weibull distribution with the given shape parameter, β, etc.

It should be emphasized that the nature of the constraints should be distinguished from their number. For example, Dowson and Wragg (1973) have shown that the maximum entropy distributions on the range $(0,\infty)$, having prescribed first and second moments can exist i.e., with positive and normalizable pdf, only if the prescribed second moment is less than or equal to twice the square of the prescribed first moment. In detail, POME was applied to derive p(x) in $x:(0,\infty)$ such that $\mu_1 = \int_0^\infty x\, p(x)dx$ and $\mu_2 = \int_0^\infty x^2 p(x)dx$ are given. The result (15) applied directly to this problem leads to the variational answer:

$$p_v(x) = \exp(-\lambda_1 x - \lambda_2 x^2)/Z(\lambda)$$

with

$$Z(\lambda_1,\lambda_2) = \int_0^\infty \exp(-\lambda_1 x - \lambda_2 x^2)dx$$

In order that $p_v(x)$ exist such that $p_v(x) > o$ for all x in (o,∞) and $Z(\lambda_1,\lambda_2)$ remain finite, λ_1 should be real and λ_2 should be real and positive and must be consistent with the given constraints, i.e.,

$$\mu_1 = \frac{\int_0^\infty x\, e^{-\lambda_1 x - \lambda_2 x^2} dx}{Z(\lambda_1,\lambda_2)}, \quad \mu_2 = \frac{\int_0^\infty x^2 e^{-\lambda_1 x - \lambda_2 x^2} dx}{Z(\lambda_1,\lambda_2)}$$

Dowson and Wragg found that only if $\mu_2 \leq 2\mu_1^2$, one obtains $\lambda_2 > 0$ such that an appropriate $p_v(x)$ can exist. For $\mu_2 = 2\mu_1^2$, $p_v(x) \propto \exp(-\lambda_1 x)$ with $\lambda_2 = 0$ and for $\mu_2 > 2\mu_1^2$, λ_2 becomes negative and the POME solution does not exist. At an elementary level, given only the first moment μ_1 for a pdf in x in $(-\infty,\infty)$, POME solution does not exist. Only for x in (o,∞) does the POME solution lead to the linear exponential solution if $\mu_1 > o$. In the case of the two moment problem when the condition $0 < \mu_2 \leq 2\mu_1^2$ is not satisfied, trying to prescribe the first two moments is clearly not the procedure to adopt to determine a pdf with two unknown parameters. The aforementioned Gamma and Weibull distributions provide explicit examples of ways to choose two other constraints that are not the first two moments. From these examples it follows, that even if the Dowson-Wragg condition is satisfied, the first two moments may not be the appropriate sufficient statistics for the two parameter pdf pertinent to the observed data. It seems that Sonuga (1972, 1976) did not encounter the Dowson-Wragg constraint even though he examined the same problem with specific data from hydrological observations.

The thrust of this section may be summarized as follows. Under the assumption that the same pdf with the same parameters applies to different batches of the same type of system (rainfall-runoff data collected year after year, for example) an analysis should seek to identify the sufficient statistics for the parameters. If the pdf of interest is derivable from a minimum information or maximum entropy principle, the sufficient statistics are just the constraints. In practice, if one is unable to fit the data with one pdf model, one tries another. If the alternative model is also derivable from the minimum information or maximum entropy principle, one is just using a different set of constraints as sufficient statistics. The procedure of trying a different set of constraints in the maximum entropy principle when analyzing data has often been advocated by Jaynes (1978). Jaynes has repeatedly pointed out that lack of consistency with observations is just an indicator that you must change the set of constraints used in POME.

5. COORDINATE TRANSFORMATIONS AND DIFFERENTIAL ENTROPY

As mentioned earlier, the use of differential entropy as a concept is often questioned on the basis that its value changes under a transformation of the random variable. We here advocate that this transformation property may provide a new avenue to consider data analysis. The relationship between differential entropy and the expectation value of the logarithm of the decrement rate valid for all pdf's noted in Sec. 3, (31), is the key to such a development.

We may interpret a transformation in which the value of the relative entropy S[p] is changed (unchanged) to mean that the transformation describes a change (lack of change) in conditions of data acquisition (rainfall-runoff in different years or different collection stations) as revealed by $E[\ln \lambda(x)]$. This flexibility allows some relaxation of the assumption that different data sets of the system of the same type are described in terms of the same probability distribution with the same parameters. In general, the variable x may be taken to be the time or ordering parameter which is a positive cumulative function that increases monotonically from an origin. A transformation of ordering parameter is taken to mean a one-to-one relationship between two ordering parameters such that the derivative of the transformed parameter with respect to the original parameter is positive and finite everywhere, i.e., except possibly at isolated points. Implicit in our specification of the range of the ordering parameter x to be zero to infinity is the setting of the origin of data collection at zero. Transformed ordering parameters will here also be assumed to be aligned in the sense that they all share the same origin.

Consider then a transformation $u=u(x)$, $u(0)=0$ where u is a dimensionless ordering parameter. In general, such transformations are not linear. The one-to-one nature of the transformation means that

$$F(u) = F(x), \quad u=u(x) \tag{36}$$

or

$$p(u) = p(x)/[\frac{du}{dx}] \tag{37}$$

Then

$$S_u = S_x + \int p(x) \ln \left(\frac{du(x)}{dx}\right) dx \tag{38}$$

In general the integral on the right hand side of (35) does not vanish so that the differential entropy is not invariant under a transformation of the ordering parameter.

In view of (31), (38) may be rewritten as

$$E[\ln \lambda(u)] = E[\ln \lambda(x)] - \int p(x) \ln\left(\frac{du(x)}{dx}\right) dx \tag{39}$$

Thus $E[\ln \lambda(x)]$ is also not in general invariant under transformation given by (36). Changes in the conditions of data collection are a usual occurrence. Consider a simple linear transformation

$$u(x) = ax \tag{40}$$

where a is positive dimensionless constant (scale transformation). If $a > 1$, $u(x)$ is larger than x so that the process must be slowed to accommodate (36). Similarly if $a < 1$, the process must be accelerated to accommodate (36). This is revealed by considering the behavior of $E[\ln \lambda(u)]$ under the transformation (40). Thus

$$E[\ln \lambda(u)] = E[\ln \lambda(x)] - \ln a \tag{41}$$

It is clear that the process is slowed for $a > 1$ and accelerated for $a < 1$.

From the POME point of view, this is just a systematic way of modifying the constraints. The transformation of ordering parameters can then be interpreted as a similarly systematic way of modifying the constraints. Two of the most usual probability distributions are the exponential and the Weibull with respective distributions,

$$F(x) = 1 - \exp(-\lambda x) \tag{42a}$$

$$F(u) = 1 - \exp - (\eta u)^{\beta}, \quad \beta > 0 \tag{42b}$$

Here x, u are dimensionless ordering parameters and λ and η are the corresponding rate parameters. For the exponential distribution, λ is the dimensionless constant decrement rate. Consider now the transformation

$$x = bu^{\beta} \tag{43}$$

where b is a dimensionless number. From (36), it follows that

$$b = \eta^{\beta}/\lambda \tag{44}$$

The difference between the differential entropies calculated respectively for the exponential and Weibull distributions is just

$$S_x - S_u = \Delta S_{x,u} \equiv \ln(\beta\eta/\lambda) - \gamma(\beta-1)/\beta \tag{45}$$

Here γ is the Euler constant = 0.5772 ••• . Clearly, depending on the values of β and the ratio η/λ, $\Delta S_{x,u}$ can either be positive or negative corresponding respectively to decelerated or accelerated processes. We should remark that the Weibull distribution is <u>not</u> of the exponential class when the shape parameter is not fixed.

When $\Delta S_{x,u} = 0$, the differential entropy remains invariant under the transformation from an exponential to a Weibull distribution. We can similarly consider transformations of the exponential distributions to other types of distributions. Explicitly, from the exponential p.d.f. $p(x) = \lambda e^{-\lambda x}$, $x:(o,\infty)$, $\lambda > 0$, with $S_x = 1 - \ln\lambda$, we obtain the extreme value distribution

$$p(y) = \beta e^{\beta y} \exp[-(e^{\beta y}-1)] \tag{46}$$

by the transformation

$$\lambda x = \exp(\beta y) - 1 \tag{47}$$

This takes $x:(o,\infty)$ to $y = (o,\infty)$ and the entropy in the y-variable becomes

$$S_y = S_x - \ln(\beta/\lambda) + e^{\lambda}\mathrm{Ei}(-\lambda) \tag{48}$$

where $\mathrm{Ei}(-\lambda)$ is the exponential integral. Two more examples will be given where $x:(o,\infty)$ is transformed to other domains. The transformation

$$\lambda x = \mu \ln(\kappa z), \quad \mu, \kappa > o \tag{49}$$

takes $x:(o,\infty)$ to $z:(\kappa^{-1},\infty)$ and p(x) transforms to the Pareto distribution

$$p(z) = \kappa\mu(\kappa z)^{-1-\mu}, \quad z:(\kappa^{-1},\infty) \tag{50}$$

and the entropy is

$$S_z = S_x + \mu^{-1} - \ln(\kappa\mu/\lambda) \quad . \tag{51}$$

A uniform distribution in t:(a,b) is obtained from the exponential distribution in x:(o,∞) by means of the transformation

$$\lambda x = \ln\left(\frac{b-a}{b-t}\right), \quad a < t < b \tag{52}$$

and the entropy is

$$S_t = S_x + \ln \lambda(b-a) - 1 \tag{53}$$

Thus from the linear exponential pdf one may deduce other known pdf's by transformation of variable and find the altered change in differential entropy explicity as shown in a few examples above.

Another example is a transformation corresponding to modifying constraints in such a way that a Weibull pdf with a particular shape parameter, β, is changed to another Weibull pdf with a different shape parameter, β'. If S_β is the differential entropy of the Weibull pdf with β, with F(u) given by (42b), then it can be shown that

$$S_\beta - S_{\beta'} \equiv \Delta S_{\beta,\beta'} = \ln\left(\frac{\beta'\eta'}{\beta\eta}\right) + \left(\frac{\beta-\beta'}{\beta\beta'}\right)\gamma \tag{54}$$

Consider now the following situation. We obtain a number of data sets for the same type of system and each, say, can be characterized by a Weibull distribution but with respectively different scale and shape parameters. Then although the form of the pdf remains the same, we do not have the two sufficient statistics necessary for a maximum entropy pdf or the assumption of reproducibility of pdf. However, let us further suppose that evaluation of the differential entropy or E[ln λ] for the respective data sets yields essentially the same result. Then we can develop a related model based on the single sufficient statistic E[ln λ]. The result is the exponential distribution. We can understand such a circumstance to mean that there exists an underlying exponential distribution with process occurring on an internal clock of the system that is revealed as various Weibull distributions with processes occurring on various measurement (stretched or contracted) clocks. It is <u>not</u> the conditions of process that are different from data set to data set but rather the conditions of measurement.

The concept of underlying exponential distribution discussed in the above example can be generalized in the following way. When given sets of data for which conditions of data collection are supposed similar (seasonal rainfall for example) but for which a full set of sufficient statistics cannot be measured (or a full set of constraints identified), use the ones you can identify to obtain an underlying distribution, and then seek a transformation of ordering parameter to obtain the measured probability density function. The hallmark of unchanging process conditions is the invariance of E[ln λ], and the hallmark for a lack of a full set of sufficient statistics is the occurrence of statistics that depend on variable parameters.

6. CONCLUDING REMARKS

We have pointed out that for continuous random variable, POME (and minimum KL information) pdf belong to the exponential family. They are therefore a subset of pdf's with sufficient statistics. The latter just correspond to the constraints that enter into POME. Specifying the constraints is tantamount to specifying the probability model to be used in the analysis of data. We may also point out as shown by Koopman (1936), if there is a sufficient statistic then the maximum likelihood estimate is a function of that statistic, but the maximum likelihood estimate is not necessarily a sufficient statistics. Essentially all the models employed in practice can be viewed as resulting from POME (see Singh et.al. (1985, 1986)). In this sense, POME provides a unifying view for the statistical analysis. Further the ideas of transformation of the random variables along with POME lends some new insight. This introduces a novel flexibility in the description of changes under transformations and in dealing with data sets characterized by parameters that vary from set to set.

REFERENCES

Barankin, E. W. and, Maitra, A. P., 1963, 'Generalization of the Fisher-Darmois-Koopman-Pitman Theorem of Sufficient Statistics,' Sankhya, Vol. 25, pp. 217-244.

Collins, R. and, Wragg, A. 1977, 'Maximum Entropy Histograms'. Journal of Physics A: Mathematical and General, Vol. 10, No. 9, pp. 1441-1464.

Davy, B. W. and Davies, T. R. H., 1979, 'Entropy Concepts in Fluvial Geomorphology: a Reevaluation', Water Resources Research, Vol. 15, No. 1, pp. 103-106.

Darmois, G., 1935, 'Sur les lois de Probatilite a Estimation Exhaustive', Camptes Rendes de l'Academic des Sciences de Paris, Vol. 2003, p. 1265.

Dowson, D. C. and, Wragg, A., 1973, 'Maximum Entropy Distributions Having Prescribed First and Second Moments', IEEE Trans. Info. Theory, Vol. IT-19, pp. 689-693.

Fisher, R. A., 1922. 'On the Mathematical Foundations of Theoretical Statistics', Philosophical Transactions of the Royal Society of London, Vol. 222A, pp. 309-368.

Hogg, R. V. and, Craig, A. T, 1965, Introduction to Mathematical Statistics, 2nd Edition, McMillan, New York.

Halmos, P. R. and, Savage, L. J, 1949, 'Application of the Radon-Nikodym Theorem to the Theory of Sufficient Statistics', Annals of Mathematical Statistics, Vol. 20, pp. 225-241.

Huzurbazar, V. S., 1976, Sufficient Statistics, (Ed. A. M. Kshirsagar), Marcel Dekkar, Inc.

Jaynes, E. T., 1957a, 'Information Theory and Statistical Mechanics, I', Physical Review, Vol. 106, pp. 620-630.

Jaynes, E. T., 1957b, 'Information Theory and Statistical Mechanics, II', Physical Review, Vol. 108, pp. 171-190.

Jaynes, E. T., 1961. Probability Theory in Science and Engineering, McGraw-Hill Book Company, New York.

Jaynes, E. T., 1978, 'Where Do We Stand on Maximum Entropy?', In: Maximum Entropy Formalism edited by R. D. Levine and M. Tribus, MIT Press pp. 15-118.

Jaynes, E. T., 1982, 'On the Rationale of Entropy Methods',. Proceedings of the IEEE, Vol. 70, No. 9, pp. 939-952.

Kagan, A. M., Linnik, Yu V. and Rao, C. R. 1973, Characterization Problems in Mathematical Statistics, John Wiley & Sons, pp. 403-410, New York.

Koopman, B. O., 1936, 'On Distributions Admitting Sufficient Statistics', Transactions of the American Mathematical Society, Vol. 39, pp. 399-409.

Pitman, E. J. G., 1936, 'Sufficient Statistics and Intrisic Accuracy', Proceedings of the Cambridge Philosophical Society, Vol. 32, pp. 567-579.

Kullback, S., 1959, Information Theory and Statistics, John Wiley & Sons, New York.

Kullback, S. and, Leibler, R. A., 1951, 'On Information and Sufficiency', Annals of Mathematical Statistics, Vol. 22, pp. 79-86.

Shannon, C. E., 1948a, 'The Mathematical Theory of Communications I and II', Bell System Technical Journal, Vol. 27, pp. 379-423.

Shannon, C. E., 1948b, 'The Mathematical Theory of Communications III and IV', Bell System Technical Journal, Vol. 27, pp. 623-656.

Singh, V. P., Rajagopal, A. K. and, Singh, K. K., 1986, 'Derivations of Some Frequency Distributions Using the Principle of Maximum Entropy (POME), Advances in Water Resources, Vol. 9, No. 2, pp. 91-106.

Singh, V. P., Singh, K. K. and, Rajagopal, A. K., 1985, 'Application of the Principle of Maximum Entropy (POME) to Hydrologic Frequency Analysis,' Technical Report, the Louisiana Water Resources Research Institute, Louisiana State University, Baton Rouge, La.

Sonuga, J. O., 1972, 'Principle of Maximum Entropy in Hydrologic Frequency analysis', Journal of Hydrology, Vol. 17, pp. 177-191.

Sonuga, J. O., 1976, 'Entropy Principle Applied to Rainfall-Run Off Process', Journal of Hydrology, Vol. 30, pp. 81-94.

Tikochinsky, Y., Tishby, N. Z. and, Levine, R. D., 1984, 'Alternative Approach to Maximum-Entropy Inference', Physical Review, Vol. A30, pp. 2638-2644.

van Campenhont, J. and, Cover, T. M., 1981, 'Maximum Entropy and Conditional Probability', IEEE Transactions on Information Theory, Vol. IT-27, pp. 483-489.

Verdugo Lazo, A. C. G and, Rathie, P. N., 1978, 'On the Entropy of Continuous Probability Distributions', IEEE Trans. Info. Theory, Vol. IT-24, pp. 120-122.

ENTROPY AND PROBABILITY DISTRIBUTIONS

Y. Li
Department of Mathematics
Lanzhou University
Lanzhou
The People's Republic of China

Vijay P. Singh
Department of Civil Engineering and
Louisiana Water Resources Research Institute
Louisiana State University
Baton Rouge, Louisiana 70803-6405, USA

S. Cong
Department of Hydrology
East China Technical University of Water Resources
Nanjing
The People's Republic of China

ABSTRACT. Entropy was employed to investigate probability distribution functions and estimation of their parameters. Two curve-fitting methods, one of which is based on entropy, were compared using Monte Carlo simulation. A procedure to compare different distributions using entropy was suggested.

1. INTRODUCTION

A multitude of frequency distributions are employed for hydrologic analyses. For example, three-parameter log normal (LN3), Pearson type (PT) 3, log Pearson type (LPT) 3, and extreme value 1 (EV1) distributions are frequently used for flood frequency determinations. Exponential and Weibull distributions are normally used for frequency analysis of precipitation characteristics. Gamma distribution is popularly used for representing the instantaneous unit hydrograph of a watershed. This study shows that all these distributions can be derived in a unified manner using the principle of maximum entropy (POME) and that the choice of a distribution can be facilitated by POME. Furthermore, POME offers a better alternative method of distribution fitting. These aspects are investigated here using Monte Carlo simulation.

V. P. Singh (ed.), Hydrologic Frequency Modeling, 367–382.

2. ENTROPY AND THE PRINCIPLE OF MAXIMUM ENTROPY (POME)

Let a random variable X take on values A_i, i = 1, 2, ..., n, out of the totality of inconsistent possible events $A = (A_1, A_2, \ldots, A_n)$; A is called the probability field. Then the probability of $A = A_i$ is $P(A_i) = P_i$, i = 1, 2, ..., n. Evidently,

$$\sum_{i=1}^{n} P_i = 1$$

Shannon (1948) has shown that the entropy of A, H(A), can be defined as

$$H(A) = H(P_1, P_2, \ldots, P_n) = -c \sum_{i=1}^{n} P_i \log P_i, \quad c > 0 \tag{1}$$

H(A) is unique subject to the following conditions: (1) H is a continuous function of $P(A_i)$, i = 1, ..., n. (2) H becomes maximum if and only if $P(A_1) = P(A_2) = \ldots = P(A_n)$. (3) $H(P_1, P_2, P_3) = H(P_1, P_2 + P_3) + (P_2 + P_3)\, H(P_2/(P_2 + P_3), P_3/(P_2 + P_3))$.

If A is a continuous probability field with the probability density function (pdf) p(x), its entropy can be defined as:

$$H(A) = -\int_M p(x) \log p(x)dx \tag{2}$$

where M is the domain of the variation of x. Using the variational calculus, it is easy to prove the following theorem: If p(x) maximizes H(A) in (2) subject to the constraints:

$$\int_M p(x)dx = 1 \tag{3}$$

$$\int_M f_k(x)\, p(x)dx = g_k, \quad k = 1, 2, \ldots, m \tag{4}$$

where $f_k(x)$, k = 1, ..., m, are integrable real functions, g_k are real constants, then

$$p(x) = \frac{1}{z(q_1, \ldots, q_m)} \exp(q_1 f_1(x) + \ldots + q_m f_m(x)) \tag{5}$$

where $z(q_1, \ldots, q_m) = \int_M \exp(q_1 f_1(x) \ldots + q_m f_m(x))dx$. According to the constraint (4), we have the following relation:

$$g_k = \frac{1}{z} \frac{\partial z}{\partial q_k}, \quad k = 1, \ldots, m$$

the real constants q_k are to be obtained.

Jaynes (1968) has shown that the direct probability calculations and the POME yield the same results for discrete A. The same can be shown for continuous A.

3. DERIVATION OF PROBABILITY DISTRIBUTIONS USING POME

Any probability distribution can be derived using POME (Singh, Rajagopal and Singh, 1986). This derivation is consistent with given information expressed in terms of constraints, and the resulting distribution maximizes entropy. Thus, it gives free rein to all possible variability of X allowed by the constraints. The procedure is illustrated using the Pearson type 3 distribution popularly used in hydrologic practice in China. If the constraints are

$$\int_{a_0}^{\infty} p(x)dx = 1 \tag{7}$$

$$\int_{a_0}^{\infty} (x - a_0)\, p(x)dx = E(x - a_0) \tag{8}$$

$$\int_{a_0}^{\infty} \log(x - a_0)\, p(x)dx = E(\log(x - a_0)) \tag{9}$$

The solution having maximum entropy, subject to (7) - (9), is

$$p(x) = \exp(- q_0 - q_1(x - a_0) + q_2 \log(x - a_0))dx \tag{10}$$

From (7),

$$\begin{aligned} \exp(q_0) &= \int_{a_0}^{\infty} \exp(- q_1(x - a_0) + q_2 \log(x - a_0))dx \\ &= q_1^{-q_2-1}\, \Gamma(q_2 + 1) \end{aligned} \tag{11}$$

Making use of (11), (10) becomes

$$p(x) = \frac{q_1^{q_2+1}}{\Gamma(q_2 + 1)} (x - a_0)^{q_2} \exp(- q_1(x - a_0)) \tag{12}$$

(12) is the Pearson type 3 probability density function, q_2+1 is the shape parameter, q_1 is the scale parameter.

Taking partial derivative of each term of (11) with respect to q_1 and q_2, then using (8) and (9), we obtain

$$\begin{aligned} \frac{\partial q_0}{\partial q_1} &= - \exp(- q_0) \int_{a_0}^{\infty} (x - a_0) \exp(- q_1(x - a_0) + \\ &\quad + q_2 \log(x - a_0))dx \\ &= E(x - a_0) = - \frac{q_2 + 1}{q_1} \end{aligned} \tag{13}$$

$$\frac{\partial q_0}{\partial q_2} = E(\log(x - a_0)) = - \log(q_1) + \frac{\Gamma'(q_2 + 1)}{\Gamma(q_2 + 1)} \tag{14}$$

Considering (13), (14) becomes

$$\log(E(x - a_0)) - E(\log(x - a_0)) = \log(q_2 + 1) - \frac{\Gamma'(q_2 + 1)}{\Gamma(q_2 + 1)} \tag{15}$$

The left hand side (LHS) of (15) can be obtained from the samples or other prior information, and the right hand side (RHS) of (15) is denoted by G(q), called the characteristic function, i.e.,

$$G(q) = \log(q + 1) - \frac{\Gamma'(q + 1)}{\Gamma(q + 1)} = \log(q + 1) + c - \\ - \int_0^1 \frac{1 + t^q}{1 - t}\, dt \tag{16}$$

where c = 0.5772166..., is Euler's constant.

Similarly, other distributions such as normal, two- and three-parameter lognormal (LN2 and LN3), Pearson type (PT) 3, log-Pearson type (LPT) 3, gamma, Weibull, exponential beta, extreme value type 1 and 3 (EV1 and EV3), Krisky-Menkel (KM) distributions, etc., can be drived from POME.

Singh, et al. (1985, 1986) have made a comprehensive investigation of these distributions using POME. More specifically, Jowitt (1979) derived EV1 distribution, Singh and Singh (1985a, 1985b, 1987a, 1987b) PT3, LPT3, LN3, and gamma distributions, Singh (1986, 1987a, 1987b) log EV1, EV3 and Weibull distributions. Zong and Li (1981, 1984) undertook similar studies independently. Table 1 derives some of the results.

4. THE ENTROPY METHOD FOR PARAMETER ESTIMATION

The entropy method for parameter estimation can be obtained from (14) and (15). First, let us study the characteristics function,

$$G(q) = \log\ (q + 1) - \frac{\Gamma'(q + 1)}{\Gamma(q + 1)} = \log(q + 1) + \\ + c - \int_0^1 \frac{1 - t^q}{1 - t}\, dt$$

$$= \log(q + 1) + c - \sum_{k=1}^{\infty} \left(\frac{1}{k} - \frac{1}{q + k}\right) \tag{16}$$

Table 1. Some distribution functions derived from the principle of maximum entropy.

Distribution	pdf	Constraint Functions	Characteristic Functions and Relations for Parameters
PT3	$\frac{b^p}{\Gamma(p)} (x-a_0)^{p-1} \exp(-b(x-a_0))$	$f_1(x) = x-a_0$ $f_2(x) = \ln(x-a_0)$	$G(q) = \ln(q+1) - \Gamma'(q+1)/\Gamma(q+1)$ $= \ln[E(x-a_0)] - E(\ln(x-a_0)$ $p = q+1,\ E(x-a_0) = p/b$
PT5	$\frac{b^p}{\Gamma(p)} (x-a_0)^{-p-1} \exp(-b/(x-a_0))$	$f_1(x) = \frac{1}{x-a_0}$ $f_2(x) = \ln(x-a_0)$	$G(q) = \ln(q+1) - \Gamma'(q+1)/\Gamma(q+1)$ $= E[\ln(x-a_0)] + \ln[E(\frac{1}{x-a_0})]$ $E(\frac{1}{x-a_0}) = p/b,\ p = q+1$
Normal	$\frac{1}{\sqrt{2\pi}\,\sigma} \exp(-\frac{(x-\mu)^2}{2\sigma^2})$	$F_1(x) = (x-\mu)^2$ $(\text{or } f_1(x) = x^2$	$u = E(x)$ $\sigma^2 = E[(x-\mu)^2]$
LN	$\frac{1}{\sqrt{2\pi}\,\sigma_y(x+b)} \exp(-\frac{(\ln(x+b)-\mu)^2}{2\sigma_y^2})$ $y = \ln(x+b)$	$f_1(x) = \ln(x+b)$ $f_2(x) = [\ln(x+b)-\mu]^2$	$\mu = E[\ln(x+b)]$ $\sigma_y^2 = E[(\ln(x+b)-\mu)^2]$

Table 1 (continued)

Distribution	pdf	Constraint Functions	Characteristic Functions and Relations for Parameters
Gumbel	$\alpha^{-1} \exp(-(\frac{x-\mu}{\alpha}) - \exp(-\frac{x-\mu}{\alpha}))$	$f_1(x) = \frac{x-\mu}{\alpha}$ $f_2(x) = \exp(-\frac{x-\mu}{\alpha})$	$\phi(\alpha) = c\alpha + \ln \Gamma(\alpha+1)$ $= E[x] + \ln(E[\exp(-x)])$ $E(\frac{x-\mu}{\alpha}) = c = 0.57721566 \ldots$
KM	$\frac{\alpha^{\alpha}}{a^{(a/b)} b\Gamma(\alpha)} x^{(\alpha/b)-1} \exp(-\alpha(\frac{x}{a})^b$	$f_1(x) = \ln x$ $f_2(x) = x^k$	$G(q) = \ln(q+1) - \Gamma'(q+1)/\Gamma(q+1)$ $= \ln(E[x^k]) - kE[\ln x]$ $\alpha = q+1,\ E(x^k) = \alpha/q$ $a = (\alpha/k)^{(1/k)},\ b = 1/k$

$$G'(q) = \frac{1}{q+1} - \sum_{k=1}^{\infty} \frac{1}{(q+k)^2}$$

Evidently, when $q > -1$, $G'(q) < 0$. Hence, $G(q)$ is a monotonic function of q. In addition, $\lim_{q \to -1} G(q) = +\infty$, $\lim_{q \to +\infty} G(q) = 0$. Then for any $u \in (0,+\infty)$, $G^{-1}(u)$ exists and G^{-1} is also a monotonic function of u on $(0,+\infty)$. Thus, for any $u > 0$, the equation $G(q) = u$ has a unique solution $q = G^{-1}(u)$. It is well known that

$$G(q) = \log(E(x - a_0)) - E(\log(x - a_0)) \tag{17}$$

For any random variable X taking values in $(a_0,+\infty)$, the RHS of (17) is greater than zero. Thus, q can be determined from equation (17).

Usually, the estimator of $E(x)$ is obtained using the method of moments. If a_0 (or C_v) is given, the coefficients of variation C_v (or a_0) and skewness C_s can be calculated as follows:

$$a_0 = \bar{x}(1 - \sqrt{q+1}\, C_v) \tag{18a}$$

$$G(q) = \log(\bar{x} - a_0) - \overline{\log(x - a_0)} \tag{18b}$$

$$C_s = 2/\sqrt{q+1} \tag{18c}$$

where $\bar{x}$ and $\overline{\log(x - a_0)}$ are the estimators of $E(x)$ and $E(\log(x - a_0))$ respectively. If a_0 is given, C_s and C_v can be directly calculated from (18a) - (18c). If C_v is given, a_0 and C_s are determined by means of iteration.

The parameters can be estimated using POME and the principle of curve-fitting (i.e., to look for some unknown parameters that minimize the criterion of curve-fitting), for example,

$$\rho = \sum_{i=1}^{n} |x_i - F^{-1}(F_n(x_i))|,$$

where $F_n(\cdot)$ is empirical distribution. The computational steps can be as follows: (1) Determine the criterion of curve-fitting ρ. (2) Choose initial value a_0, compute $\log(x - a_0) - \log(x - a_0)$, and determine C_v, C_s from (18a) - (18c). (3) Calculate the value of the criterion ρ_0. (4) Change a_0 to a_0*, repeat steps 2, 3, and find ρ^*. (5) Take $a_0 = a_0^*$, if $\rho^* < \rho_0$, then repeat steps 4 and 5 until the minimal ρ is obtained.

This method is called entropy curve-fitting method (ECP). For other distribution functions, the ECF is shown in Table 2. As an example, a sample is taken from the population of the PT 3 distribution with $\bar{x} = 100$, $C_v = 0.5$, $C_s = 1.0$, and is arranged in descending order as follows:

Table 2. The entropy method of parameter estimation.

Distribution	Calculation of Characteristic Functions	System of Parameters	Comments
PT3	$G(q) = \ln(\bar{x}-a_0) - \overline{\ln(x-a_0)}$	$C_s = 2/\sqrt{q+1}$ $a_0 = \bar{x}(1 - \sqrt{q+1}\, C_v)$	1. a_0 (or C_v) is given, find out C_v (or a_0) 2. ECF method: change a_0 until optimization
PT5	$G(q) = \overline{\ln(x-a_0)} + \ln[\overline{(\frac{1}{x-a_0})}]$	$p = q+1$ $a_0 = \bar{x}(1 - C_v\sqrt{p-2})$ $C_s = 4\sqrt{p-2}/(p-3)$	The same as for PT3
LN	$\sigma_y = E[(y-\mu)^2]$ $y = \ln(x+b)$ $\mu = \overline{\ln(x+b)}$	$C_v = [\exp(\sigma_y^2)-1]^{1/2}$ $C_s = [\exp(3\sigma_y^2) - 3(\exp(\sigma_y^2))+2]/C_v^3$ $b = \frac{C_v \bar{x}}{C_s} [\{1 + \frac{C_s^2}{2} + C_s (0.25\, C_s^2 + 1)^{0.5}\}^{0.5} + \{1 + 0.5\, C_s^2 - C_s(0.25\, C_s^2 + 1)^{0.5}\}^{1/3} + 1] - \bar{x}$	1. b (or C_v) is given, find out C_v (or b) 2. ECF method: change b until optimization

Table 2 (continued)

Distribution	Calculation of Characteristic Functions	System of Parameters	Comments
KM	$G(q) = \ln[E(x^k)] - kE(\ln x)$ $\alpha = q+1 \quad E(x^k) = \alpha/q$ $a = (\alpha/q)^{(1/k)}$ $b = k^{-1}$	$C_v = [\frac{\Gamma(\alpha)\Gamma(\alpha+2b)}{\Gamma^2(\alpha+b)} - 1]^{0.5}$ $C_s = [\frac{\Gamma^2(\alpha)\Gamma(\alpha+3b)}{\Gamma^3(\alpha+b)} -1 - 3C_v^2]/C_v^3$	ECF method: change k until optimization
Gumbel	$\phi(\alpha) = \bar{x} + \ln[E(\exp(-x))]$	$\ln(E(\exp(-x))) = -\mu + \ln\Gamma(\alpha+1)$ $(C_s = 1.139)$	See Jowitt (1979)

Note: If X is a random variable which is subject to LPT3 distribution, then y = log X is subject to PT3 distribution. Estimate parameters using Y, first, and then obtain parameters for X.

263.37, 217.49, 183.61, 159.52, 152.28, 145.55, 124.32,
123.03, 120.43, 114.21, 112.70, 104,57, 102.83, 97.28,
90.76, 89.84, 87.70, 87.66, 85.90, 85.38, 80.48,
75.04, 61.88, 59.26, 56.44, 51.06, 37.58, 32.96,
29.77, 26.06

The search procedure for the ECF method yields:

a_0: 10.20, -10.20, 1.02, 2.04, 3.06, 4.08, 5.10, 3.98

C_v: 0.511, 0.625, 0.562, 0.557, 0.511, 0.545, 0.540, 0.546

C_s: 1.189, 1.027, 1.052, 1.065, 1.079, 1.093, 1.107, 1.091

ρ: 202.0, 195.6, 187.6, 184.4, 181.3, 179.4, 180.9, 179.3

The solution of ordinary curve-fitting (OCF) method (Cong, 1981) is:

$$a_0 = 3.07,\ C_v = 0.551,\ C_s = 1.136,\ \rho = 177.5$$

The results of the two methods are almost the same. The ECF method is a one-dimensional search method, the OCF is two-dimensional, so the former is more convenient for calculations. The functional relation between q and G(q) is computed as listed in Table 3, where $E(\log(x - a_0))$ is estimated as

$$E(\log(x - a_0)) = \frac{1}{n} \sum_{i=1}^{n} \log^*(x_i - a_0))$$

where $\log^*(x - a_0) = \log(x - a_0)$, for $a_0 \le x_i + 1$, and 0, for $a_0 > x_i + 1$

The OCF method seeks parameters C_v and C_s that minimize

$$\rho = \sum_{i=1}^{n} (x_i - F^{-1}(Fn(x_i)))$$

(or other criterion), where the mean value is estimated from method of moments.

5. COMPARISON OF TWO CURVE FITTING METHODS

The Monte Carlo technique was used for estimating values of the mathematical expectations and the root-mean-square errors of estimated parameters C_v and C_s, which are denoted by mC_v, mC_s, sC_v and sC_s respectively. Let $x_1, x_2, \ldots, x_n$, be a sample arranged in descending order and drawn from the population of a specific PT3 distribution, with Fn being its empirical frequency distribution. Then, this sample

Table 3. Value of the characteristic function G(q).

q	G(q)	q	G(q)	q	G(q)	q	G(q)
-0.95	17.5021	-0.05	0.6113	3.9	0.10552	8.5	0.053683
-0.9	8.1212	0	0.5772	4.0	0.10332	8.7	0.052567
-0.89	7.2931	0.1	0.5190	4.1	0.10125	9.0	0.050832
-0.88	6.6085	0.2	0.4714	4.2	0.099248	9.2	0.049973
-0.87	6.0337	0.3	0.4316	4.3	0.097321	9.5	0.048539
-0.86	5.5446	0.4	0.3978	4.4	0.095468	9.7	0.047628
-0.85	5.1239	0.5	0.3690	4.5	0.093684	10.0	0.046143
-0.84	4.7584	0.6	0.3440	4.6	0.091966	10.5	0.044312
-0.83	4.4581	0.7	0.3221	4.7	0.090308	11.0	0.042245
-0.82	4.1554	0.8	0.3028	4.8	0.088711	11.5	0.040780
-0.81	3.9042	0.9	0.2857	4.9	0.087169	12.0	0.038954
-0.8	3.6769	1.0	0.2704	5.0	0.085642	12.5	0.037790
-0.79	3.4777	1.1	0.2566	5.1	0.084240	13.0	0.036139
-0.78	3.2953	1.2	0.2442	5.2	0.082850	13.5	0.035227
-0.77	3.1298	1.3	0.2329	5.3	0.081502	14.0	0.033704
-0.76	2.9790	1.4	0.2226	5.4	0.080200	14.5	0.033609
-0.75	2.8412	1.5	0.2131	5.5	0.078938	15.0	0.031575
-0.74	2.7148	1.6	0.2045	5.6	0.077716	16	0.029700
-0.73	2.5981	1.7	0.1965	5.7	0.076531	17	0.028035
-0.72	2.4906	1.8	0.1891	5.8	0.075381	18	0.026547
-0.71	2.2310	1.9	0.18221	5.9	0.074266	19	0.025208
-0.70	2.2986	2.0	0.17583	6.0	0.073126	20	0.023998
-0.68	2.1323	2.1	0.16988	6.1	0.072133	21	0.022899
-0.66	1.9871	2.2	0.16431	6.2	0.071112	22	0.021897
-0.64	1.8593	2.3	0.15910	6.3	0.070114	23	0.020978
-0.62	1.7461	2.4	0.15421	6.4	0.069154	24	0.020133
-0.60	1.6451	2.5	0.14961	6.5	0.068216	25	0.019354
-0.58	1.5545	2.6	0.14528	6.6	0.067302	26	0.018633
-0.56	1.4729	2.7	0.14118	6.7	0.066414	27	0.017963
-0.54	1.3990	2.8	0.13732	6.8	0.065548	28	0.017340
-0.52	1.3317	2.9	0.13366	6.9	0.064705	29	0.016759
-0.50	1.2704	3.0	0.13018	7.0	0.063800	30	0.016216
-0.45	1.1381	3.1	0.12689	7.1	0.063083	35	0.013953
-0.4	1.0298	3.2	0.12376	7.2	0.062302	40	0.012245
-0.35	0.9396	3.3	0.12077	7.3	0.061540	50	0.009836
-0.3	0.8633	3.4	0.11793	7.4	0.060798	60	0.0082191
-0.25	0.7932	3.5	0.11522	7.5	0.060073		
-0.2	0.7419	3.6	0.11263	7.7	0.058674		
-0.15	0.6928	3.7	0.11016	8.0	0.056583		
-0.10	0.6496	3.8	0.10779	8.2	0.055450		

was subjected to PT3 distribution with unknown parameters $\bar{x}$, C_v and C_s estimated by OCF and ECF methods. The criteria used for assessing goodness-of-fit were:

1. The least-absolute-value criterion

$$\rho_1 = \sum_{i=1}^{n} |x_i - F^{-1}(Fn(x_i))|$$

2. The least-square criterion

$$\rho_2 = \sum_{i=1}^{n} (x_i - F^{-1}(Fn(x_i)))^2$$

The parameters of the population were assumed to be:

$$Cso = 1,\ 1.5,\ 2,\ 3;\ Cvo = 0.5\ Cso;\ \bar{x} = 100$$

The sample size was equal to 30 and 50. For each case, 500 samples were generated. The results are given in Tables 4 and 5, which list respectively average estimated values of mC_v, mC_s and the root-mean-square errors sC_v, sC_s for the two methods.

The values of mC_v, mC_s, sC_v and sC_s by the ECF method are almost as good as the values by OCF, but the values of mC_v and mC_s by the former are slightly smaller than those by the latter. The larger the sample size n was, the more accurate the estimators were. The performance of the two curve-fitting methods with fitting criterion of the least absolute values was better than that with criterion of the least squares. The ECF method was as good as the OCF method for estimating parameters from the samples.

6. SELECTION OF A DISTRIBUTION

The ECF method can be used to compare distributions and then choose an appropriate one for use in hydrology: (1) Choose the distribution functions, for example, PT3, PT5, LPT3 distributions, etc. (2) The sample is then fitted by the chosen distribution and its parameters are estimated by the entropy method. (3) Calculate the statistics for evaluating goodness of fit. (4) The distribution is accepted if it is found to be the best for most samples by goodness of fit criteria. Or the distribution is accepted if the goodness of fit is as good as the one for the OCF method.

POME accords neatly and consistently with the probability theory and has almost the same results as this theory, if the distribution to be sought can be regarded as the one produced by a random experiment. However, we use POME in the broadest sense, giving the widest range of applicability. Even if the experiment is not random, POME represents

Table 4. Comparison between the ECF and OCF methods (the least absolute value criterion).

Parameter		C_v			C_s		
Estimation Method	Sample Size	Cvo	mCv	sCv	Cso	mCs	sCs
OCF		0.5	0.528	0.079	1.0	1.17	0.374
ECF	30		0.524	0.076		1.13	0.380
OCF		0.5	0.520	0.060	1.0	1.115	0.302
ECF	50		0.517	0.064		1.102	0.310
OCF		0.75	0.802	0.118	1.5	1.57	0.429
ECF	30		0.784	0.117		1.51	0.421
OCF		0.75	0.774	0.090	1.5	1.54	0.302
ECF	50		0.768	0.082		1.50	0.307
OCF		1.0	1.06	0.165	2.0	2.05	0.473
ECF	30		1.02	0.134		1.98	0.476
OCF		1.0	1.04	0.126	2.0	2.03	0.346
ECF	50		1.01	0.105		1.99	0.350
OCF		1.5	1.54	0.246	3.0	2.97	0.595
ECF	30		1.46	0.185		2.90	0.571
OCF		1.5	1.53	0.215	3.0	2.99	0.492
ECF	50		1.47	0.149		2.92	0.401

Table 5. Comparison between the ECF and OCF methods (the least square criterion).

Parameter		C_v			C_s		
Estimation Method	Sample Size	Cvo	mCv	sCv	Cso	mCs	sCs
OCF			0.553	0.920	1.0	1.198	0.427
ECF	30	0.5	0.546	0.821		1.109	0.407
OCF			1.183	0.291	2.0	2.208	0.751
ECF	50	1.0	1.107	0.205		1.900	0.545
OCF			1.122	0.200	2.0	2.143	0.530
ECF	30	1.0	1.070	0.148		1.937	0.436

the most objective description of the known information. Also, if the prior information does not come from the random experiment, the original direct probability methods do not apply. POME can accommodate physical characteristics of the system through constraints. Thus, it derives the probability distribution through physical as well as stochastic considerations.

If the experiment fails to support POME and this disagreement persists in spite of indefinite repetition of the experiment, then it is concluded that there are some physical characteristics which were not taken into account in the POME results. The observed deviations provide a clue as to the nature of these new constraints. Thus, the choice of a probability distribution of given hydrologic phenomenon is facilitated by POME.

7. CONCLUSIONS

The following conclusions can be drawn from this study: (1) The principle of maximum entropy provides a unified framework for derivation of a number of probability distributions used in hydrology. (2) Using Monte Carlo simulation for log-Pearson type 3 distribution, it is shown that the entropy method is better or at least comparable to the ordinary curve fitting method. (3) The entropy method can be used to choose amongst probability distributions.

8. ACKNOWLEDGEMENTS

Participation of Dr. V. P. Singh in this study was supported by the U.S. Department of Interior, through Louisiana Water Resources Research Institute, under the project, "A Multivariate Stochastic Analysis of Flood Magnitude, Duration and Volume." Mr. Y. Li was supported by Nanjing Research Institute of Hydrology and was assisted by Prof. H. Zhang.

9. REFERENCES

Cong, S., 1981, Statistics in Hydrology, Hydrological Press of China.

Jaynes, E. T., 1968, 'Prior probabilities,' IEEE Transactions on System Science and Cybernetics, Vol. (SSC)4, pp. 227-241.

Jowitt, P. W., 1979, 'Extreme-value type-I distribution and the principle of maximum entropy, Journal of Hydrology, Vol. 42, pp. 23-38.

Li, Y. and Cong, S., 1984, 'Entropy and its application in calculation of flood frequency,' Journal of Hydrology (China).

Shannon, C. E., 1948, 'The mathematical theory of communication, I and II,' Bell System Technical Journal, Vol. 27, pp. 379-423.

Singh, V. P., 1986, 'On the log-Gumbel (LG) distribution,' Hydrology Journal of the IAH, Vol. VIII, No. 4, pp. 34-42.

Singh, V. P., 1987a, 'Derivation of the Weibull distribution,' Water Resources Management, in press.

Singh, V. P., 1987b, 'On the extreme value (EV) type III distribution,' Hydrological Sciences Journal, in press.

Singh, V. P., Rajagopal, A. K. and Singh, K., 1986, 'Derivation of some frequency distributions using the principle of maximum entropy (POME),' Advances in Water Resources, Vol. 9, No. 2, pp. 91-106.

Singh, V. P. and Singh, K., 1985a, 'Derivation of the Pearson type (PT) III distribution by using the principle of maximum entropy (POME),' Journal of Hydrology, Vol. 80, pp. 197-214.

Singh, V. P. and Singh, K., 1985b, 'Derivation of the gamma distribution by using the principle of maximum entropy (POME),' Water Resources Bulletin, Vol. 21, No. 6, pp. 941-952.

Singh, V. P. and Singh, K., 1987a 'Derivation of log-Pearson type III distribution,' Submitted to ASCE Journal of Hydraulic Engineering.

Singh, V. P. and Singh, K., 1987b, 'Three parameter lognormal distribution derived from the principle of maximum entropy,' Water Resources Bulletin, in press.

Singh, V. P., Singh, K. and Rajagopal, A. K., 1985, 'Application of the principle of maximum entropy (POME) to hydrologic frequency analysis,' Completion Report 06, 144p., Louisiana Water Resources Research Institute, Louisiana State University, Baton Rouge, Louisiana.

AN EVALUATION OF SEVEN METHODS FOR ESTIMATING PARAMETERS OF EV1 DISTRIBUTION

Kishore Arora and Vijay P. Singh
Department of Civil Engineering and
Louisiana Water Resources Research Institute
Louisiana State University
Baton Rouge, LA 70803, USA

ABSTRACT. Extreme value type 1 distribution parameters and quantiles were estimated by methods of moments, maximum likelihood estimation, probability weighted moments, entropy, mixed moments, least squares and incomplete means for Monte Carlo samples generated from two sampling cases: purely random process and serially correlated process. The performance of these estimators was statistically evaluated. The methods of maximum likelihood estimation and entropy provided most efficient quantile estimates. The methods of moments and probability weighted moments were comparable in efficiency of estimating the quantiles for small samples. The methods of mixed moments and incomplete means resulted in poor estimation of parameters and the quantiles.

1. INTRODUCTION

The cumulative distribution function (cdf) of an extreme value type 1 (EV1) or Gumbel random variable x is given by:

$$F(x) = \exp[- \exp\{- a(x-b)\}] \tag{1}$$

where 1/a and b are respectively the scale and location parameters of the distribution. Traditionally, the methods of moments and maximum likelihood estimation were used to estimate the parameters a and b and consequently to estimate the quantiles. However, several relatively new methods have been developed and successfully applied over the past few years. The method of probability weighted moments, introduced by Greenwood, et al. (1979), was used by Landwehr and Matalas, et al. (1979) to estimate Gumbel parameters and quantiles. Based on the principle of maximum entropy (Jaynes, 1961), Jowitt (1979) suggested entropy estimators. The least squares estimators were used by Chow (1953), and Lowery and Nash (1970) and have since been applied to the EV1 distribution by Jain and Singh (1986). Houghton (1978) introduced the method of incomplete means, which is easy to apply for EV1 distri-

V. P. Singh (ed.), Hydrologic Frequency Modeling, 383–394.

bution. Similarly, the method of mixed moments used by Rao (1980, 1980b, 1983) is straightforward and yields closed form algebraic expressions for parameter estimators.

In this paper, the statistical properties of the aforementioned methods are evaluated using Monte Carlo simulation. These methods are then compared with methods of moment and maximum likelihood estimation with a view to recommending optimum parameter and quantile estimators for EV1.

2. ANALYSIS

The inverse form of (1) is given by

$$x(F) = b - \frac{\ln(-\ln F)}{a} \tag{2}$$

where x(F) denotes the quantile of cumulative probability F. For sample sizes n = 5, 10, 15, 20, 30, 50, 100, 1000, the parameters a and b were estimated by the methods of moments (MOM), maximum likelihood estimation (MLE), probability weighted moments (PWM), entropy (ENT), mixed moments (MIX), least squares (LEAS), and incomplete means (ICM). The quantiles x(F), for F = 0.05, . . . , 0.99, were then calculated from (2).

Let θ denote an estimate of $\theta \in (a,b,x(F))$. $\hat{\theta}$ is a random variable whose distribution function depends upon sample size, the method of parameter estimation, and of course, the distribution of the sample itself. The performance statistics of $\hat{\theta}$ are given as follows:

$$\text{Bias, BIAS } (\hat{\theta}) = \theta - E(\hat{\theta}) \tag{3}$$

$$\text{Variance, STD}^2 \ (\hat{\theta}) = E[\hat{\theta} - E(\hat{\theta})]^2 \tag{4}$$

$$\begin{aligned}\text{Mean Square Error, MSE } (\hat{\theta}) &= E(\theta - \hat{\theta})^2 \\ &= \text{BIAS}^2 \ (\hat{\theta}) + \text{STD}^2 \ (\hat{\theta})\end{aligned} \tag{5}$$

These statistics present a good picture of the relative performance of various estimators of θ. They were estimated through Monte Carlo sampling experiments, i.e., through generation of a large number of pseudo-random samples for: (1) purely random process (independent and identically distributed Gumbel random variables), and (2) serially correlated process (with the first order serial correlation coefficient of 0.5).

On the basis of N Monte Carlo samples of size n, the statistics of $\hat{\theta}$, viz, (3) - (5) were estimated as:

$$E(\hat{\theta}) \cong \mu(\hat{\theta}) = \sum_{i=1}^{N} \frac{\hat{\theta}_i}{N} \tag{6}$$

$$STD^2(\hat{\theta}) \cong \sum_{i=1}^{N} \frac{[\hat{\theta}_i - \mu(\hat{\theta})]^2}{N - 1} \tag{7}$$

$$BIAS(\hat{\theta}) \cong \theta - \mu(\hat{\theta}) \tag{8}$$

$$MSE(\hat{\theta}) \cong BIAS^2(\hat{\theta}) + STD^2(\hat{\theta}) \tag{9}$$

These estimates are expected to be very close approximations to the theoretical values owing to the large value of N (= 50,000 for n = 5, . . . ,100, and, = 10,000 for n = 1000).

The mean square error of all methods relative to that of MLE was compared using the relative efficiency defined as:

$$EFF(\hat{\theta}) = \frac{MSE(\hat{\theta}|MLE)}{MSE(\hat{\theta}|\text{other method})} \tag{10}$$

A value of $EFF(\hat{\theta}) < 1$ implies that the method under consideration is less efficient (i.e., has higher mean square error) compared to MLE and vice versa.

3. ESTIMATION OF PARAMETERS AND QUANTILES

The parameter estimation equations for various methods are summarized as follows:

(1) MOM: $\hat{a} = \pi/(\sqrt{6} \cdot \sigma_x)$

$$\hat{b} = \bar{x} - 0.57721/\hat{a} \tag{11}$$

where σ_x is standard deviation.

(2) MLE: $$\hat{a} = \frac{\Sigma \exp(-\hat{a}x_i)}{\bar{x} \cdot \Sigma \exp(-\hat{a}x_i) - \Sigma x_i \exp(-\hat{a}x_i)}$$

$$\hat{b} = \frac{1}{\hat{a}} \cdot \ln \left[\frac{n}{\Sigma \exp(-\hat{a}x_i)}\right] \tag{12}$$

(3) PWM: $$\hat{a} = \frac{\ln(2)}{\bar{x} - 2 \cdot \Sigma x_i(n-i)/n(n-1)}$$

$$\hat{b} = x - 0.57721/\hat{a} \quad (13)$$

(4) ENT:

$$\hat{a} = \frac{0.55721 + \ln[n/\Sigma \exp(-\hat{a}x_i)]}{\bar{x}}$$

$$\hat{b} = \text{same formula as for MLE} \quad (14)$$

(5) LEAS:

$$\hat{a} = \frac{n \cdot \Sigma z_i x_i - \Sigma x_i \Sigma z_i}{(\Sigma x_i)^2 - n \cdot \Sigma x_i^2}$$

$$\hat{b} = \bar{x} + \Sigma z_i/(n \cdot \hat{a}) \quad (15)$$

where $z_i = \ln[-\ln(F_p(x_i)]$ is obtained from the plotting position formula which defines the cumulative probability of non-exceedance $F_p(x_i)$ for each data point x_i.

(6) MIX:

$$\hat{a} = 1.28255/\sigma_x$$

$$\hat{b} = \frac{1}{\hat{a}} \cdot \ln[1 + \hat{a} \cdot \bar{x} + \frac{\hat{a}^2}{2} \cdot \frac{\Sigma x_i^2}{n}] \quad (16)$$

(8) ICM:

$$\bar{x}_i = \hat{b} - \frac{n}{\hat{a}(n-n_i)} \cdot [J \ln J - J^2 \ln J/2$$

$$+ J^3 \ln J/6 - J^4 \ln j/24 + \ldots], \; i = 1, 2 \quad (17)$$

where $J = \ln(n/n_i)$, n is the sample size, and n_i is the number of observations on which the incomplete mean $\bar{x}_i$ is calculated.

All the summations above were performed over i = 1 to n and the sample statistics were calculated as:

$$\bar{x} = \Sigma x_i/n \quad (18)$$

$$\sigma_x^2 = \Sigma (x_i - \bar{x})^2/(n-1) \quad (19)$$

Landwehr and Matalas (1979) performed their calculations with the biased expression of σ_x^2, i.e., with n replacing n-1 in (19).

Having estimated parameters a and b, the respective quantiles can be calculated from (2) as:

$$\hat{x} = \hat{b} - \ln(-\ln f)/\hat{a} \quad (20)$$

4. COMPARISON OF PERFORMANCE STATISTICS OF THE PARAMETER AND QUANTILE ESTIMATORS

4.1 Case 1: Purely Random Process

The results of this case are shown in Table 1-4.

4.1.1 Parameter estimates. The MIX and ICM can be prima-facie rejected as unreliable estimators of Gumbel parameters. MIX, while performing reasonably efficiently for estimation of a, failed in providing even moderately biased estimate of b, and thus resulted in highly inefficient estimate of b. ICM failed to provide satisfactory estimate for either a or b as reflected by its very high bias and standard deviation of a and high standard deviation of b.

Of the remaining five methods, the bias of a showed the following trend:

MLE > ENT > MOM > PWM > LEAS, for n = 5, 10

MOM > MLE > ENT > PWN > LEAS, for n = 15 - 50

MOM > MLE > ENT > LEAS > PWM, for n = 75, 100

LEAS > others , for n = 1000

The standard deviation of a compared as:

MLE > ENT > MOM > PWM > LEAS, for n = 5

MOM > PWM > MLE > ENT > LEAS, for n = 10

MOM > PWM = LEAS > ENT > MLE, for n = 15

MOM > LEAS > PWM > ENT > MLE, for n > 20

The efficiency of a compared as:

LEAS > PWM > MOM > ENT > MLE, for n = 5

LEAS > PWM > ENT > MLE > MOM, for n = 10, 15

MLE > ENT > PWM > LEAS > MOM, for n > 20

where '>' means that the method on the left hand side of > has a bigger statistic (bias, standard deviation, or efficiency) then the method on the right hand side.

From the above trends, it appears that for rather small samples ($n \leq 15$), LEAS is the preferred method for estimating a. Table 1 also reveals that as n increased, the efficiency of estimating a by ENT

TABLE 1. PERFORMANCE STATISTICS OF PARAMETER ESTIMATES (RANDOM SAMPLES)

METHOD	SAMPLE SIZE	BIAS(A)	STD(A)	EFF.(A)	BIAS(B)	STD(B)	EFF.(B)
MOM	5	-0.360	0.778	1.139	-0.051	0.485	1.017
MLE		-0.434	0.806	1.000	-0.081	0.485	1.000
PWM		-0.233	0.701	1.533	0.000	0.480	1.049
ENT		-0.415	0.790	1.052	-0.083	0.485	0.997
LEA		-0.141	0.650	1.893	-0.019	0.484	1.032
MIX		-0.360	0.778	1.139	-0.834	0.497	0.257
ICM		-5.274	111.876	0.000	-0.468	0.874	0.246
MOM	10	-0.156	0.381	0.881	-0.028	0.341	0.991
MLE		-0.167	0.349	1.000	-0.040	0.338	1.000
PWM		-0.092	0.353	1.126	-0.001	0.336	1.024
ENT		-0.159	0.348	1.024	-0.040	0.338	0.996
LEA		-0.026	0.341	1.279	0.003	0.341	0.994
MIX		-0.156	0.381	0.881	-0.884	0.371	0.126
ICM		-0.532	1.489	0.060	-0.065	0.930	0.133
MOM	20	-0.076	0.249	0.739	-0.015	0.239	0.974
MLE		-0.074	0.211	1.000	-0.019	0.236	1.000
PWM		-0.041	0.225	0.953	-0.000	0.235	1.010
ENT		-0.069	0.213	0.995	-0.019	0.236	0.996
LEA		0.004	0.231	0.938	0.010	0.241	0.966
MIX		-0.076	0.249	0.739	-0.911	0.269	0.062
ICM		-0.191	0.482	0.186	0.007	0.650	0.132
MOM	50	-0.030	0.156	0.657	-0.007	0.151	0.960
MLE		-0.027	0.126	1.000	-0.008	0.148	1.000
PWM		-0.015	0.139	0.846	-0.001	0.148	1.002
ENT		-0.025	0.129	0.968	-0.008	0.148	0.995
LEA		0.012	0.151	0.722	0.008	0.152	0.947
MIX		-0.030	0.156	0.657	-0.929	0.177	0.024
ICM		-0.069	0.269	0.215	0.003	0.368	0.161
MOM	100	-0.014	0.113	0.653	-0.003	0.107	0.956
MLE		-0.012	0.091	1.000	-0.004	0.105	1.000
PWM		-0.007	0.101	0.832	-0.000	0.105	1.000
ENT		-0.012	0.093	0.966	-0.004	0.105	0.994
LEA		0.011	0.112	0.678	0.007	0.108	0.943
MIX		-0.014	0.113	0.653	-0.934	0.129	0.012
ICM		-0.032	0.184	0.243	0.002	0.252	0.174

TABLE 2. BIAS IN QUANTILE ESTIMATES : RANDOM SAMPLES

(F = PROB. OF NON EXCEEDENCE, X = QUANTILE)

METHOD	SAMPLE SIZE	F = 0.500	0.900	0.950	0.980	0.990	0.999
		X = 0.367	2.250	2.970	3.902	4.600	6.907
MOM	5	-0.017	0.161	0.229	0.315	0.378	0.586
MLE		-0.022	0.283	0.400	0.549	0.660	1.025
PWM		0.002	0.014	0.018	0.020	0.022	0.024
ENT		-0.029	0.252	0.360	0.497	0.599	0.934
LEA		-0.049	-0.198	-0.256	-0.334	-0.392	-0.589
MIX		-0.801	-0.618	-0.552	-0.468	-0.405	-0.198
ICM		-0.333	0.366	0.635	0.982	1.240	2.092
MOM	10	-0.010	0.090	0.129	0.175	0.210	0.320
MLE		-0.010	0.146	0.206	0.281	0.336	0.516
PWM		0.000	0.013	0.017	0.020	0.021	0.024
ENT		-0.014	0.126	0.179	0.246	0.295	0.456
LEA		-0.023	-0.150	-0.200	-0.266	-0.317	-0.487
MIX		-0.867	-0.762	-0.726	-0.680	-0.647	-0.537
ICM		-0.036	0.123	0.183	0.258	0.314	0.496
MOM	20	-0.004	0.054	0.076	0.102	0.121	0.180
MLE		-0.004	0.079	0.111	0.149	0.177	0.268
PWM		0.001	0.013	0.017	0.020	0.021	0.023
ENT		-0.006	0.068	0.096	0.130	0.154	0.233
LEA		-0.009	-0.098	-0.133	-0.181	-0.218	-0.341
MIX		-0.902	-0.839	-0.819	-0.794	-0.776	-0.717
ICM		0.015	0.064	0.082	0.103	0.118	0.166
MOM	50	-0.001	0.030	0.042	0.054	0.063	0.089
MLE		-0.001	0.040	0.055	0.072	0.084	0.122
PWM		0.001	0.013	0.017	0.020	0.022	0.024
ENT		-0.002	0.035	0.048	0.063	0.074	0.106
LEA		-0.002	-0.050	-0.069	-0.096	-0.117	-0.188
MIX		-0.925	-0.889	-0.879	-0.867	-0.859	-0.834
ICM		0.007	0.033	0.042	0.052	0.058	0.078
MOM	100	0.000	0.023	0.031	0.039	0.045	0.060
MLE		0.001	0.028	0.038	0.048	0.055	0.076
PWM		0.002	0.014	0.019	0.022	0.024	0.027
ENT		0.000	0.025	0.034	0.043	0.049	0.067
LEA		0.001	-0.025	-0.035	-0.051	-0.064	-0.108
MIX		-0.932	-0.905	-0.899	-0.892	-0.887	-0.872
ICM		0.005	0.024	0.030	0.037	0.041	0.052

TABLE 3. STD. DEVIATION OF QUANTILE ESTIMATES :RANDOM SAMPLES

(F = PROB. OF NON EXCEEDENCE, X = QUANTILE)

METHOD	SAMPLE SIZE	F =	0.500	0.900	0.950	0.980	0.990	0.999
		X =	0.367	2.250	2.970	3.902	4.600	6.907
MOM	5		0.530	1.122	1.406	1.780	2.057	2.960
MLE			0.530	1.017	1.252	1.562	1.797	2.533
PWM			0.526	1.167	1.470	1.869	2.150	3.123
ENT			0.533	1.041	1.285	1.605	1.848	2.609
LEA			0.536	1.282	1.627	2.072	2.389	3.506
MIX			0.620	1.387	1.685	2.059	2.327	3.264
ICM			0.805	1.059	1.339	1.746	2.068	3.151
MOM	10		0.374	0.826	1.044	1.326	1.534	2.178
MLE			0.373	0.746	0.926	1.158	1.332	1.839
PWM			0.373	0.825	1.038	1.316	1.508	2.144
ENT			0.376	0.764	0.949	1.189	1.368	1.893
LEA			0.376	0.912	1.161	1.486	1.699	2.474
MIX			0.460	1.050	1.274	1.546	1.734	2.418
ICM			0.788	0.854	1.176	1.648	2.010	3.230
MOM	20		0.262	0.609	0.778	0.993	1.148	1.592
MLE			0.262	0.547	0.686	0.861	0.990	1.314
PWM			0.262	0.594	0.752	0.953	1.091	1.498
ENT			0.263	0.559	0.702	0.882	1.014	1.352
LEA			0.263	0.654	0.841	1.078	1.225	1.750
MIX			0.331	0.795	0.964	1.159	1.282	1.776
ICM			0.541	0.629	0.887	1.251	1.522	2.421
MOM	50		0.166	0.422	0.550	0.705	0.813	1.056
MLE			0.165	0.380	0.489	0.619	0.711	0.860
PWM			0.166	0.407	0.525	0.667	0.762	0.964
ENT			0.167	0.388	0.498	0.630	0.724	0.882
LEA			0.166	0.441	0.576	0.740	0.835	1.123
MIX			0.210	0.569	0.691	0.812	0.870	1.181
ICM			0.307	0.431	0.605	0.836	1.003	1.503
MOM	100		0.118	0.331	0.439	0.566	0.650	0.773
MLE			0.118	0.304	0.401	0.510	0.584	0.631
PWM			0.118	0.319	0.421	0.539	0.614	0.700
ENT			0.119	0.309	0.406	0.518	0.592	0.646
LEA			0.118	0.342	0.452	0.582	0.654	0.801
MIX			0.143	0.459	0.557	0.636	0.654	0.863
ICM			0.210	0.336	0.472	0.644	0.765	1.064

TABLE 4. EFFICIENCY OF QUANTILE ESTIMATES : RANDOM SAMPLES

(F = PROB. OF NON EXCEEDENCE, X = QUANTILE)

METHOD	SAMPLE SIZE	F = 0.500	0.900	0.950	0.980	0.990	0.999
		X = 0.367	2.250	2.970	3.902	4.600	6.907
MOM	5	1.003	0.868	0.851	0.839	0.838	0.820
MLE		1.000	1.000	1.000	1.000	1.000	1.000
PWM		1.017	0.818	0.799	0.785	0.793	0.766
ENT		0.989	0.971	0.971	0.971	0.971	0.972
LEA		0.972	0.663	0.637	0.623	0.625	0.591
MIX		0.274	0.483	0.549	0.615	0.657	0.698
ICM		0.371	0.887	0.786	0.683	0.630	0.522
MOM	10	0.998	0.837	0.814	0.794	0.788	0.753
MLE		1.000	1.000	1.000	1.000	1.000	1.000
PWM		1.005	0.850	0.835	0.820	0.830	0.794
ENT		0.987	0.964	0.965	0.964	0.964	0.962
LEA		0.985	0.678	0.648	0.623	0.632	0.574
MIX		0.145	0.344	0.419	0.498	0.551	0.595
ICM		0.224	0.777	0.635	0.511	0.456	0.342
MOM	20	0.997	0.816	0.789	0.766	0.759	0.700
MLE		1.000	1.000	1.000	1.000	1.000	1.000
PWM		0.999	0.865	0.851	0.839	0.850	0.801
ENT		0.986	0.961	0.962	0.961	0.962	0.956
LEA		0.991	0.698	0.666	0.639	0.654	0.566
MIX		0.074	0.229	0.301	0.387	0.450	0.490
ICM		0.233	0.764	0.608	0.484	0.434	0.305
MOM	50	0.995	0.815	0.798	0.779	0.770	0.671
MLE		1.000	1.000	1.000	1.000	1.000	1.000
PWM		0.995	0.883	0.879	0.873	0.883	0.810
ENT		0.984	0.964	0.967	0.969	0.967	0.956
LEA		0.994	0.742	0.720	0.698	0.721	0.582
MIX		0.030	0.131	0.194	0.275	0.343	0.361
ICM		0.290	0.781	0.659	0.554	0.508	0.333
MOM	100	0.995	0.848	0.835	0.816	0.811	0.673
MLE		1.000	1.000	1.000	1.000	1.000	1.000
PWM		0.994	0.914	0.911	0.901	0.911	0.825
ENT		0.983	0.973	0.976	0.972	0.974	0.958
LEA		0.995	0.794	0.788	0.768	0.797	0.618
MIX		0.016	0.091	0.145	0.219	0.283	0.269
ICM		0.316	0.824	0.724	0.630	0.586	0.356

remained close to 1.00, while the efficiency from other methods reduced considerably.

In estimating b, PWM provided practically unbiased estimates. Analyzing the bias, standard deviation and the efficiency in much the same way as for a, it can be easily concluded that PWM provided superior estimates of b for the entire sample range considered.

4.1.2 Quantile estimates. PWM provided unbiased quantile estimates for all n and F. MOM provided estimates with lower bias than MLE and ENT. ENT resulted in slightly less bias than MLE, while LEAS produced more bias than MLE for all n except for n = 5. MIX and ICM again failed to provide satisfactory estimates of quantiles compared to other estimators.

MLE resulted in lowest standard deviation closely followed by ENT. MOM had slightly higher standard deviation than PWM for nearly all n and F. LEAS estimates showed higher standard deviation than MOM, although the difference reduced as n increased.

MLE estimates were most efficient, closely followed by ENT estimates. The PWM estimates proved to be more efficient than MOM estimates, though less efficient than ENT estimates. LEAS exhibits lower efficiency than MOM.

4.2 Case 2: Serially Correlated Process

Similar calculations were performed for this case and the results are highlighted below.

4.2.1 Parameter estimates. When the samples were generated from a serially correlated process but assumed to be random for the purposes of estimation, all the methods produced significantly higher bias and standard deviation in parameter estimation than the corresponding random process (case 1).

LEAS consistently produced least bias of a followed by PWM. From sample size 10 onwards, MLE, followed closely by ENT, gave least standard deviation of a. However, up to sample size n = 20, LEAS resulting in comparable standard deviation produced estimates of a with a higher efficiency than MLE. Hence, LEAS can be preferred for such sample sizes.

For b, PWM was without doubt the superior method resulting in less bias, least standard deviation and higher efficiency estimates. Although LEAS resulted in lower bias than PWM for n > 15, it showed less efficiency than PWM owing to its higher standard deviation. But it is significant to note that the effect of serial correlation was to markedly lower the performance deviation among the first five methods. In fact the first five methods performed to within 98 percent of the efficiency of MLE method for estimating b.

4.2.2 Quantile estimates. The bias in quantile estimates also increased for serially correlated samples as compared to purely random samples. LEAS provided least biased estimates of the first five methods for sample sizes 5 to 100. PWM provided the next lowest bias. MLE and ENT continued to provide very close bias estimates, although ENT produced slightly lower bias. MOM provided lower bias than MLE for up to about n = 30, beyond which MOM produced slightly higher bias than MLE. Quite in contrast with other methods, the absolute bias resulting from MIX increased with n for any given F.

MLE resulted in least standard deviation of quantile estimates among the first five methods, even for n = 5, closely followed by ENT. MOM gave lower standard deviation than PWM for n = 5 and 10 only, after which mostly PWM produced lower standard deviation estimates. LEAS provided estimates with rather high standard deviation.

Except for n = 5, and some quantiles less than 0.5 for other n, MLE turned out to be the most efficient method for serially correlated samples, followed closely again by ENT. PWM provided the next higher efficiency estimates mostly at all quantile values for all n except at n = 5. MOM came next and LEAS provided the least efficient estimates. At n = 5, all methods performed at marginally differing efficiencies except ICM.

5. CONCLUSIONS

Seven available estimators of EV1 distribution parameters and quantiles were statistically compared using Monte Carlo sampling experiments performed on two cases: a purely random process and a serially correlated process.

With regard to the inter-comparison of parameter and quantile estimators, some of the important conclusions are as follows:

(1) The methods of mixed moments and incomplete means resulted in poor estimation of the parameters and the quantiles.

(2) The method of least squares provided minimum bias and maximum efficiency estimate of the parameter 'a' for very small samples and also provided competitive estimates of the parameter 'b'.

(3) The maximum likelihood method generally provided most efficient quantile estimates followed closely by the entropy method. In fact, ENT performed practically in the same manner as MLE.

(4) For small samples, the method of probability weighted moments and the method of moments performed comparably in efficiency of estimating the quantiles. However, the efficiency of PWM somewhat improved relative to MLE with increasing sample size. PWM also resulted in nearly unbiased quantile estimates.

(5) The incorporation of serial correlation in samples resulted in deterioration of the performance of all estimators. However, all the methods performed much more similarly in this case.

6. ACKNOWLEDGEMENTS

This study was supported in part by funds provided by the U.S. Department of Interior, through the Louisiana Water Resources Research Institute, under the project, "A Multivariate Stochastic Analysis of Flood Magnitude, Duration and Volume."

7. REFERENCES

Greenwood, J. A., Landwehr, J. M., Matalas, N. C. and Wallis, J. R., 1979. 'Probability weighted moments.' Water Resources Res., 15(5), pp. 1049-1054.

Houghton, J. C., 1978b. 'The incomplete means estimation procedure applied to flood frequency analysis.' Water Resources Res., 14(6), pp. 1111-1115.

Jaynes, E. T., 1982. Papers on Probability, Statistics, and Statistical Physics. A reprint collection, Dordrecht, The Netherlands, D. Reidel.

Jowitt, P. W., 1979. 'The extreme value type-1 distribution and the principle of maximum entropy.' Journal of Hydrology, 42, pp. 23-28.

Jain, D. and Singh, V. P., 1986. 'Estimating parameters of EV1 distribution for flood frequency analysis.' submitted to Water Resources Bulletin.

Landwehr, J. M., Matalas, N. C. and Wallis, J. R., 1979b. 'Probability weighted moments compared with some traditional techniques in estimating Gumbel parameters and quantiles.' Water Resources Research, 15(5), pp. 1055-1064.

Lettenmaier, D. P. and Burges, S. J., 1982. 'Gumbel's extreme value 1 distribution: a new look.' Journal of the Hydraulics Division, Proceedings of ASCE, 108(HY4), pp. 502-514.

Lowery, M. D. and Nash, J. B., 1970. 'A comparison of methods of fitting the double exponential distribution.' Journal of Hydrology, 10, pp. 259-275.

Rao, D. V., 1980b. 'Log Pearson type 3 distribution: method of mixed moments.' Journal of Hydraulics Division, Proceedings of the ASCE, 106(HY6), pp. 999-1019.

Rao, D. V., 1983. 'Estimating log Pearson parameters by mixed moments.' Journal of Hydraulics Division, Proceedings of the ASCE, 109(HY8), pp. 1118-1131.

FITTING LOG PEARSON TYPE 3 DISTRIBUTION BY MAXIMUM LIKELIHOOD

Donthamsetti Veerabhadra Rao
St. Johns River Water Management District
Palatka, Florida 32078-1429, U.S.A.

ABSTRACT. Problems encountered in finding a maximum likelihood solution for log Pearson type 3 distribution are investigated. For a given sample at least two solutions are obtained, one with an upper bound and the other with a lower bound. In most cases the best solution is obvious from a comparison of sample statistics with the maximum likelihood estimates of statistical parameters, mean and variance. When these estimates differed greatly from the sample statistics, the sample contained either outliers or the solution is not satisfactory from a practical standpoint. Monte Carlo tests have shown that, in general, the maximum likelihood method may not give satisfactory Log Pearson fit for small samples.

1. INTRODUCTION

Log Pearson type 3 distribution (LP) may be fit to flood flow and other hydrologic samples by the method of logarithmic moments (Water Resources Council Bulletin No. 15, 1967), by the moments of real data (Bobee, 1975; Rao, 1980a), by mixing the logarithmic and real data moments (Rao, 1980b, 1983), and by the maximum likelihood (ML) method. The ML equations of LP, however, offer some difficulties in solving. Fitting LP to real data by ML is equivalent to fitting a Pearson type 3 distribution (P3) to logarithmic data by ML. In a procedure described by Matalas and Wallis (1973) the ML estimates of parameters for P3 are obtained by optimizing the location parameter (say, c) which is achieved when $|R| = f(c)$ reduces to a small value (10^{-8}). If $|R| > 10^{-8}$ after 30 iterations, Matalas and Wallis consider the solution effort a failure. A method proposed by Condie, Nix, and Boon (1979) searches for a solution within certain range of c. If no solution is reached within this range, it is considered the procedure failed to obtain ML estimates for the sample.

In this study a detailed investigation has been conducted with regard to variation of R with c. The objective was to find if 'failures' to reach an ML solution for LP could be avoided or minimized. For a given sample, it is found that $|R|$ reaches a value less than 10^{-8}, in general, for two or three values of c. A criterion

V. P. Singh (ed.), Hydrologic Frequency Modeling, 395–406.

is proposed to select 'best' among the solutions obtained. Monte Carlo tests were performed to compare the ML estimates of LP quantiles with other methods.

2. LOG PEARSON TYPE 3 DISTRIBUTION

The LP is a three-parameter probability distribution. Let a, b, and c be the parameters of LP and X the LP variable and Y = ln X. The logarithmic variable Y is distributed as P3 and the parameters a, b, and c are common for both P3 and LP. The probability density function of P3, f(y), is given by:

$$f(y) = |a| [a(y - c)]^{b-1} \exp [-a(y - c)]/\Gamma(b) \qquad (1)$$

The density function of LP distribution was derived by Bobee (1975) as:

$$f(x) = |a| \exp(ac) [a(\ln x - c)]^{b-1}/[\Gamma(b)\, x^{1 + a}] \qquad (2)$$

If $a > 0$, the P3 distribution is positively skewed, and $c \leq y \leq +\infty$. In this case, the LP distribution is also positively skewed, and $\exp(c) \leq x < +\infty$.

If $a < 0$, the P3 distribution is negatively skewed and, $-\infty < y \leq c$. In this case, the LP distribution is either positively or negatively skewed depending on the values of a and b and, $0 < x \leq \exp(c)$. At $x = 0$, $f(x)$ may be defined as zero for this case.

The distribution parameters (a, b, and c) are related to the statistical parameters, mean, variance and skew, by the following equations:

i) P3 distribution:

$$\text{Mean: } \mu_y = c + ba^{-1} \qquad (3)$$

$$\text{Variance: } \sigma_y^2 = ba^{-2} \qquad (4)$$

$$\text{Skew: } \gamma_y = \frac{a}{|a|} 2b^{-1/2} \qquad (5)$$

ii) LP distribution:

$$\text{Mean: } \mu_x = \exp(c)/(1-1/a)^b \quad \ldots\ldots (6)$$

$$\text{Variance: } \sigma_x^2 = \exp(2)[1/(1-2/a)^b - 1/(1-1/a)^{2b}] \quad \ldots\ldots (7)$$

$$\text{Skew: } \gamma_x = \exp(3c)\ \{1/(1-3/a)^b - 3/[(1-2/a)(1-1/a)]^b + 2/(1-1/a)^{3b}\}/\sigma_x^3 \quad \ldots\ldots (8)$$

3. MAXIMUM LIKELIHOOD ESTIMATION

Let $(x_1, x_2, ..x_n)$ be a sample of size n drawn from a probability distribution, with density function f(x). The ML method of estimating parameters of f(x) consists of maximizing the likelihood function, L, given by

$$L = \prod_{i=1}^{n} f(x_i) \quad \ldots\ldots (9)$$

In practice, however, the logarithmic likelihood function, ln L, is maximized, i.e., for LP, the parameters a, b and c are solved by setting the partial derivatives of ln L with respect to a, b and c to zero. This procedure yields the following equations:

$$\frac{\partial}{\partial a}(\ln L) = -nab + a^2 \sum_{i=1}^{n} (\ln x_i - c) = 0 \quad \ldots\ldots (10)$$

$$\frac{\partial}{\partial b}(\ln L) = -n\Psi(b) + \sum_{i=1}^{n} \ln[a(\ln x_i - c)] = 0 \quad \ldots\ldots (11)$$

$$\frac{\partial}{\partial c}(\ln L) = na - (b-1) \sum_{i=1}^{n} \frac{1}{(\ln x_i - c)} = 0 \quad \ldots\ldots (12)$$

in which $\psi(b)$ is the digamma or psi function, $\partial \ln\Gamma(b)/\partial b$. Application of ML procedure to P3 distribution, i.e., Eq. 1, yields relations similar to Eqs. 10-12 with $\ln x_i$ replaced by Y_i. Thus, fitting LP distribution to real data by ML is equivalent to fitting P3 distribution to logarithmic data by ML. A procedure described by Matalas and Wallis (1973) for solving Eqs. 10-12 is as follows:

Eqs. 10 and 12 can be rearranged to give

$$a = nb/S_1 \qquad (13)$$

$$b = S_1 S_2/(S_1 S_2 - n^2) \qquad (14)$$

in which $S_1 = \sum_{i=1}^{n} (\ln x_i - c)$,

and $S_2 = \sum_{i=1}^{n} 1/(\ln x_i - c)$

Assign a value for c and solve for b and a explicitly (Eqs. 14 and 13). Substitute a, b and c values in Eq. 11 to yield a value of $\partial(\ln L)/\partial b = R$. Repeat the procedure iteratively with newly assigned values of c until $|R|$ is less than a preassigned value, say 10^{-8}. The difficulties associated with this procedure are: (a) An assumption has to be made as to whether c should represent a lower bound or an upper bound which may not be apparent from data; (b) the range of c to be investigated can be too large, e.g., when the optimum solution is close to $\gamma_y = 0$; and (c) multiple solutions may exist.

Using several annual flood flow samples a detailed investigation has been conducted regarding variation of R with c. Calculations are performed with dimensionless data obtained by the transform $K_i = Q_i/\overline{Q}$ in which Q_i is the ith flood flow in a sample of size n, and $\overline{Q}$ = Sample Mean. This transform reduces the range of c to be investigated. Further, for the transformed data, the sample mean, $\overline{K} = 1$. When the population mean of LP, $\mu_x = 1$, parameter c is negative if population logarithmic skew $\gamma_y > 0$ (LP is bounded in the lower tail) and positive if $\gamma_y < 0$ (LP is bounded upwards.)

For each sample, the variation of R with c is investigated both for a positive and a negative sign c. Starting with a small value of c (determined from the lowest or the highest observed value), first the magnitude of c is gradually increased (by unity) and then rapidly increased. Three general cases of R-c relationship are observed as shown in Figure 1 qualitatively. The results obtained can be summarized as follows:

Result 1: For most samples, R has a point of inflexion at a relatively low value of c (typically less than 10). An interval for c where this point occurs can be found, and the value of c can be optimized by interval bisection method (Region 1, Figure 1).

Result 2: For all samples, $|R| < 10^{-8}$ for a large value of both positive and negative c. These two solutions (Regions 2 and 3, Figure 1) give similar flood estimates and have their ML estimates of log skew, $\gamma_y \simeq 0$.

Result 3: For samples in Result 1, the ML estimates of statistical parameters (mean, variance, and skew) in real space differ greatly from those obtained in Result 2. However, for several samples, the mean and variance estimates of one of the solution sets closely agree with the sample values.

3.1 Illustrative Examples

Results for eight annual flood samples are presented in Tables 1-3 to illustrate typical ML solutions encountered. Table 1 lists the gaging stations selected together with sample statistics in real and log spaces. Calculations are performed using the dimensionless data (K_i).

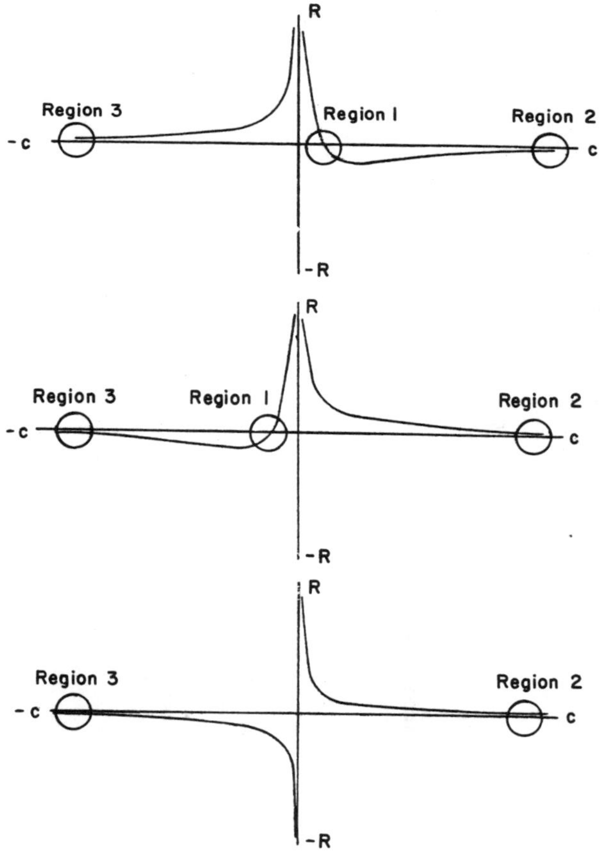

Figure 1. R-c Relationships and Regions of Possible Maximum Likelihood Solution (Note: Figure Not to Scale).

Table 1. -- Gaging Stations Selected for Illustrative Examples. (Note: 1 cfs = 0.028 m^3/sec)

				Sample statistics obtained with dimensionless data.					
				Real space			Log space		
Sample No. (1)	Gaging station (2)	Sample size Yrs. (3)	Sample Mean cfs (4)	Mean (5)	Variance (6)	Skew (7)	Mean (8)	Variance (9)	Skew (10)
1	St. Johns River near Melbourne, Florida	43	4,358	1.0000	0.7960	1.54	-0.4155	1.0161	-0.534
2	Oconee River near Greensboro, Georgia	63	15,301	1.0000	0.4851	2.52	-0.1723	0.3234	0.429
3	Wekiva River near Sanford, Florida	48	908	1.0000	0.1987	1.17	-0.0907	0.1866	-0.066
4	Oklawaha River at Eureka, Florida	17	3,487	1.0000	0.1932	0.58	-0.0934	0.2044	-0.114
5	Turkey Creek at Macclenny, Florida	25	1,030	1.0000	0.2742	1.30	-0.1191	0.2433	0.224
6	Williamson River near Chiloquin, Oregon	54	3,571	1.0000	0.5075	2.60	-0.1805	0.3381	0.502
7	Wolf Creek near Deer Park, Florida	28	1,456	1.0000	1.1442	2.72	-0.4429	1.0566	-0.692
8	Moultrie Creek near St. Augustine, Florida	25	625	1.0000	0.4259	0.35	-0.3797	1.2054	-1.224

Table 2. — Summary of Maximum Likelihood Estimates for Selected Samples.

Sample No. (1)	Solution (2)	ML estimates of parameters: a (3)	b (4)	c (5)	ML estimates of statistical parameters in real space: Mean (6)	Variance (7)	Skew (8)
1	A*	-0.28199E 01	0.80710E 01	0.24466E 01	0.9969	0.7687	1.71
2	B	0.76874E 03	0.58693E 06	-0.76391E 03	1.0849	2.0089	6.17
2	A	-0.14531E 04	0.67226E 06	0.46248E 03	0.9870	0.3646	2.06
	B*	0.86132E 01	0.23594E 02	-0.29116E 01	1.0003	0.5080	3.87
3	A*	0.88613E 03	0.14348E 06	0.16182E 03	1.0006	0.2002	1.42
	B*	0.11610E 04	0.24638E 06	-0.21230E 03	1.0008	0.2012	1.44
4	A	-0.17152E 01	0.11830E 01	0.59637E 00	1.0544	0.2092	-0.26
	B*	0.84154E 03	0.13625E 06	-0.16200E 03	1.0029	0.2139	1.49
5	A*	-0.11213E 04	0.29384E 06	0.26193E 03	0.9977	0.2615	1.67
	B	0.64160E 01	0.10082E 02	-0.16904E 01	1.0180	0.4338	4.30
6	A*	-0.13603E 04	0.61442E 06	0.45151E 03	0.9855	0.3819	2.12
	B	0.46682E 01	0.74566E 01	-0.17778E 01	1.0200	0.8100	9.38
7	A*	-0.49019E 01	0.24110E 02	0.44755E 01	0.9995	1.0174	2.67
	B	0.68954E 03	0.48491E 06	-0.70368E 03	1.0698	2.0385	6.41
8	A	-0.74618E 03	0.64338E 06	0.86185E 03	1.2184	3.2152	7.57
	B	0.74515E 03	0.64344E 06	-0.86388E 03	1.2217	3.2777	7.73

* Selected Solution

Table 3. -- The ML, MXM1, and LGMO Estimates of T-Year Events, cfs (Note: 1 cfs = 0.028 m^3/sec)

Sample No. (1)	T = 50 Yr: ML (2)	MXM1 (3)	LGMO (4)	T = 100 Yr: ML (5)	MXM1 (6)	LGMO (7)	T = 500 Yr: ML (8)	MXM1 (9)	LGMO (10)
1	15,300	15,600	16,900	17,700	18,100	20,100	22,700	23,600	27,600
2	46,300	45,600	47,000	56,600	55,400	57,700	86,600	83,100	89,100
3	1,990	1,990	1,980	2,240	2,230	2,220	2,830	2,820	2,780
4	7,830	7,450	7,810	8,830	8,300	8,750	11,300	10,100	11,000
5	2,470	2,510	2,770	2,810	2,890	3,120	3,670	3,820	4,330
6	9,730	10,800	11,400	11,400	13,100	14,200	15,600	19,800	22,700
7	5,840	6,100	5,190	7,100	7,600	6,020	10,300	11,600	7,810
			6,200*			8,020*			13,600*
8	3,880	1,400	1,920	5,200	1,430	2,060	9,400	1,460	2,290

* Results obtained with outlier adjustments recommended in Bulletin No. 17B of the U.S. Department of the Interior (1982).

For each sample two ML solutions of LP are obtained, one with a positive sign c (Solution A) and the other with a negative sign c (Solution B). These solutions yield the estimates of distribution parameters a, b, and c from which the ML estimates of statistical parameters (i.e., $\hat{\mu}_y$, $\hat{\sigma}_y$, etc.) can be calculated (Eqs. 3-8). The quantile estimates of LP, i.e., T-year events (X_T) are evaluated by the relation,

$$X_T = \text{Exp}[\,\hat{\mu}_y + K'\hat{\sigma}_y] \qquad (15)$$

in which K' = Pearson frequency factor for the T-year event. Table 2 summarizes the ML estimates of distribution parameters (a, b, and c) and the statistical parameters in real space for the selected eight samples.

As mentioned, when an ML solution is found (for a sample) with a low magnitude c, the ML estimates of statistical parameters of the two solutions, the solution with low magnitude c and the solution with large magnitude c, differed greatly from each other. The estimates of one of the solution sets closely agreed with sample statistics, especially the mean and variance. From this, it is construed that the solution which differed greatly from sample statistics (in mean and variance) is 'false.' On this basis the probable 'true' or 'best' ML solution is determined for the eight samples as follows:

Samples 1 and 2: The solution with 'smaller' magnitude of c is the best for these two samples. The ML estimates of mean and variance in real space for solution with 'large' c differed greatly from the sample values (Tables 1 and 2). Thus the solutions with 'larger' c may easily be eliminated as 'false' in similar cases.
Sample 3: No solution is found with a 'smaller' magnitude c for this sample. However, the sample mean and variance closely agreed with the corresponding ML estimates of the solution with 'large' c. Thus the solutions obtained may be accepted. For this case the solutions with both positive and negative sign c converge and the ML estimate of log skew, $\gamma_y \simeq 0$.
Samples 4 and 5: Although a solution is found with a 'smaller' magnitude c for these samples, the ML estimates of statistical parameters of these solutions more significantly differed from sample values than the solutions with 'large' magnitude c. The solutions with 'large' magnitude c appear more acceptable. This result shows that a solution with 'smaller' magnitude c may not always be the 'true' solution.
Sample 6: The estimates of statistical parameters for both solutions significantly differed from sample values. This sample has a high outlier. If it is recognized that a high outlier may have a positive bias on the sample statistics (in real space), Solution A (Table 2), with its estimates of mean, variance and skewness coefficient less than sample values, is an acceptable solution. Further, Solution B, though has a lower magnitude c, deviated more positively from the sample statistics.

Sample 7: This sample has a low outlier. Even though the ML estimate of variance differed by about 10% from the sample value, Solution A is optimal.
Sample 8: No solution is found with smaller magnitude c. The ML estimates of statistical parameters for Solutions A and B (with a large magnitude c) differed wildly from the sample values, thus the solutions may not be acceptable.

Table 3 summarizes the estimates of T-year events for T=50 yr, 100 yr, and 500 yr by the selected ML solution, Mixed Moments-I (MXM1), and logarithmic moments (LGMO) methods. The MXM1 method uses mean and variance of real data and the mean of logarithmic data for parameter estimation.

The ML estimation method proposed by Condie, Nix, and Boon (1979) searches for solution within a limited range of parameter c. The sample logarithmic skew, CSy, is used as a guidance to determine whether c should form an upper or lower bound for the solution. Let, for example, x_{min} and x_{max} be the minimum and maximum observed values in the sample. Then the range for c is determined as follows: (1) If CSy is positive, the limits for the range are ln x_{min} - 0.01 and ln x_{min} - 50; and (2) if CSy is negative, the limits are ln x_{max} + 0.01 and ln x_{max} + 50. If no solution is found within this range of c it is considered the procedure failed to obtain ML solution. Samples 3 through 6 illustrated in the foregoing demonstrate that restricting c to a limited range in finding ML solution is unwarranted.

On the basis of this study the following procedure is proposed to fit LP to hydrologic data by ML: (i) Perform calculations with dimensionless data; (ii) obtain two solutions, one with a positive, and the other with a negative sign c. Accept the solution which has its estimate of mean (in real space) more close to the sample mean; (iii) if the ML estimates of mean and variance differ significantly from the sample values, closely scrutinize the sample for the presence of outliers. If one of the solutions appears to account for the presence of outliers accept it; and (iv) if the ML estimates of statistical parameters differ wildly from the sample values, reject the solution.

It is possible to develop a Newton-Raphson iteration scheme to solve for Eq. 11. However, such procedure led to 'false' solution for several samples. Likewise, performing calculations without transforming data into dimensionless coefficients did not provide true solution for some samples.

4. MONTE CARLO TESTS

Monte Carlo tests were performed to compare the statistical properties of ML estimates of LP with LGMO and MXM1 methods. Four parameter sets were chosen with $\mu_x = 1$. The values of parameters are: (1) Set No. 1: a = -13.9502, b = 39, c = 2.7, and γ_y = -0.32; (2) Set No. 2: a = -2.1979, b = 4.0, c = 1.5 and γ_y = -1.0; (3) Set No. 3: a = 14.9058, b =

36, c = -2.5, and γ_y = 0.33; and (4) Set No. 4: a = 7.5120, b = 7, c = -1.0, and γ_y = 0.76. From each parameters set, one thousand samples of size n = 100, 50, 30, 20, and 10 were generated. The details of experimental design are given in Rao (1980b). The LP was fit to each sample by the LGMO, MXM1, and ML methods and LP quantiles were evaluated at selected frequencies. Finally, the mean and standard deviation of the quantile estimates based on 1000 samples were obtained.

For simulated samples, exploratory work has indicated that finding two ML solutions, one with a positive and the other with a negative sign c, and choosing one of them on the basis of the ML estimate of mean, in general, would give results similar to those obtained if only one solution is found with c having the sign of population c. Thus, in these tests only one ML solution is computed, with c having the sign of population c. Two cases were considered for obtaining results: (1) If an ML solution is not found with a low magnitude c (Region 1, Figure 1) the procedure proceeds to obtain the solution with large magnitude c (Region 2 or 3, Figure 1); and (2) if no ML solution is found with $|c|$ less than fifty times the population $|c|$, the sample was rejected. The following is a summary of the results:

Case 1 (No samples rejected): Table 4 presents the mean ($\bar{X}$) and standard deviation (Sx) of 100-year largest (flood) events based on 1000 simulated samples. The ML method gave satisfactory results when the population log skew (γ_y) is relatively low (Parameter Set Nos. 1 and 3), but gave very poor estimates for sample sizes $n \leq 30$ when γ_y is relatively high (Parameter Set Nos. 2 and 4). The ML estimates are inferior to MXM1 both in bias and variability when γ_y is negative (Parameter Set Nos. 1 and 2); when γ_y is positive, the two methods have somewhat balancing attributes. Compared to LGMO, the ML estimates are less variable with the exception of Parameters Set No. 2 at $n \leq 30$. Overall, the MXM1 estimates possess better statistical attributes compared to both LGMO and ML.

Case 2: To examine if inclusion of ML solutions with 'large' magnitude c unduly affected the statistical attributes of ML estimates, samples for which no ML solution was found with a 'low' magnitude c were rejected with the criterion stipulated earlier. Table 4 presents the mean and standard deviations of 100-year flood events. Rejection of the aforementioned samples did not bring an overall improvement to the ML estimates. For Parameter Set Nos. 1 and 3 the estimates became poorer. For Parameters Set No. 4, the ML gave least biased estimates, but the variability increased. For Parameters Set No. 2, the 'wild' estimates obtained earlier are not seen, but the ML estimates remain inferior to MXM1. The number of rejections varied from none (n = 100, Parameters Set No. 2) to 2430 (n = 10, Parameters Set No. 4). From these results it may be concluded: (a) the ML method may not provide

Table 4. -- Mean ($\overline{X}$) and Standard Devition (S_x) of 100 Year Flood Estimates Based on 1,000 Simulated Samples.

SAMPLE SIZE	METHOD	PARAMETERS SET NO 1 $\overline{X}$	S_x	PARAMETERS SET NO 2 $\overline{X}$	S_x	PARAMETERS SET NO 3 $\overline{X}$	S_x	PARAMETERS SET NO 4 $\overline{X}$	S_x
CASE 1: NO SAMPLES REJECTED									
100	LGMO	2.318	0.225	3.221	0.548	2.586	0.339	2.547	0.382
	MXM1	2.305	0.210	3.072	0.297	2.539	0.331	2.479	0.362
	ML	2.280	0.211	2.995	0.274	2.607	0.339	2.486	0.328
50	LGMO	2.337	0.321	3.353	0.715	2.584	0.492	2.552	0.546
	MXM1	2.305	0.287	3.086	0.417	2.509	0.446	2.445	0.476
	ML	2.247	0.289	2.989	0.555	2.612	0.489	2.434	0.446
30	LGMO	2.350	0.400	3.433	0.880	2.586	0.639	2.538	0.695
	MXM1	2.302	0.349	3.074	0.528	2.475	0.538	2.406	0.565
	ML	2.213	0.364	3.279	1.450	2.600	0.621	2.358	0.519
20	LGMO	2.372	0.496	3.542	1.052	2.597	0.815	2.533	0.845
	MXM1	2.306	0.419	3.079	0.626	2.457	0.643	2.374	0.640
	ML	2.198	0.435	3.932	2.435	2.548	0.695	2.280	0.582
10	LGMO	2.422	0.742	3.911	1.576	2.607	1.198	2.507	1.298
	MXM1	2.323	0.567	3.131	0.882	2.409	0.795	2.305	0.826
	ML	2.235	0.643	4.818	3.527	2.430	0.886	2.162	0.718
CASE 2: SAMPLES WITH $\lvert\hat{C}_{ML}\rvert$ > 50 TIMES POPULATION $\lvert C\rvert$ REJECTED									
100	LGMO	2.296	0.211	3.221	0.548	2.619	0.325	2.480	0.339
	MXM1	2.285	0.197	3.072	0.297	2.569	0.319	2.421	0.327
	ML	2.265	0.206	2.995	0.274	2.637	0.331	2.527	0.322
50	LGMO	2.281	0.289	3.340	0.706	2.645	0.476	2.437	0.469
	MXM1	2.263	0.266	3.084	0.416	2.562	0.433	2.358	0.422
	ML	2.207	0.280	2.952	0.417	2.687	0.496	2.498	0.453
30	LGMO	2.276	0.366	3.376	0.826	2.647	0.603	2.405	0.530
	MXM1	2.256	0.335	3.072	0.513	2.534	0.520	2.316	0.459
	ML	2.155	0.358	2.896	0.532	2.712	0.653	2.474	0.548
20	LGMO	2.256	0.426	3.442	1.012	2.610	0.760	2.390	0.621
	MXM1	2.235	0.376	3.098	0.628	2.480	0.625	2.294	0.515
	ML	2.085	0.389	2.867	0.644	2.679	0.808	2.453	0.628
10	LGMO	2.397	0.716	4.286	1.622	2.388	0.915	2.351	0.831
	MXM1	2.323	0.540	3.491	0.856	2.279	0.709	2.251	0.636
	ML	1.981	0.449	2.583	0.607	2.462	1.064	2.461	0.902
THEORETIC X		2.315		3.082		2.583		2.559	

satisfactory LP fit for most small size samples; and (b) the overall statistical attributes of ML estimates are inferior to MXM1.

5. SUMMARY AND CONCLUSIONS

Problems encountered in finding a maximum likelihood solution for log Pearson type 3 distribution are investigated. Performing calculations with dimensionless data obtained by the transform $K_i = Q_i/\bar{Q}$ offered several advantages. For a given sample, it is seen that at least two ML solutions of LP can be found, one with an upper bound and the other with a lower bound. For most samples the ML estimates of mean and variance (in real space) closely agreed with sample values for one of the solutions, and differed greatly for the other. From this it is concluded the solution for which the ML estimates of statistical parameters differed greatly from the corresponding sample values is 'false.' A procedure is proposed for systematically obtaining an ML solution of LP for a given hydrologic sample.

Monte Carlo tests have shown, (i) the ML method may not provide satisfactory LP fit for most small size samples, and (ii) the overall statistical attributes of ML estimates are inferior to the Mixed Moments Method-I which uses the mean and variance of real data and the mean of logarithmic data in parameter estimation.

6. REFERENCES

Bobee, B. 1975. 'The Log Pearson Type 3 Distribution and Its Applications in Hydrology,' Water Resources Research, Vol. 11, No. 5, pp. 681-689.

Condie, R., Nix, G. A., and Boone, L. G. 1979. Flood Damage Reduction Program Flood Frequency Analysis, Inland Waters Directorate, Environment Canada, Ottawa, Ontario K1A 0E7

Matalas, N.C., and Wallis, J.R. 1973. 'Eureka! It Fits a Pearson Type 3 Distribution,' Water Resources Research, Vol. 9, No. 2, pp. 281.289.

Rao, D.V. 1980a. 'Log Pearson Type 3 Distribution: A Generalized Evaluation,' Journal of the Hydraulics Division, American Society of Civil Engineers, Vol. 106, HY5, pp. 853-872.

Rao, D.V. 1980b. 'Log Pearson Type 3 Distribution: Method of Mixed Moments,' Journal of the Hydraulics Division, American Society of Civil Engineers, Vol. 106, HY6, pp. 999-1019.

Rao, D.V. 1983. 'Estimating Log Pearson Parameters by Mixed Moments,' Journal of Hydraulic Engineering, American Society of Civil Engineers, Vol. 109, No. 8, pp. 1118-1132.

U. S. Department of the Interior. 1982. Guidelines for Determining Flood Flow Frequency, Bulletin No. 17B, Geological Survey, Office of Water Data Coordination, Reston, Virginia.

U. S. Water Resources Council. 1967. A Uniform Technique for Determining Flood Flow Frequencies, Bulletin 15, Washington, D.C.

ESTIMATING THE PARAMETERS OF THE GENERALIZED GAMMA DISTRIBUTION BY MIXED MOMENTS

Huynh Ngoc Phien
Asian Institute of Technology, Bangkok, Thailand.
Van, T.V. Nguyen
Department of Civil Engineering and Applied Mechanics,
McGill University, Montreal, Quebec, Canada.
Kuo, Jung-Hua
Asian Institute of Technology, Bangkok, Thailand.

ABSTRACT. Having three parameters but many possible forms, the generalized gamma (GG) distribution can be a good candidate for flood frequency analysis. In this study, four methods of parameter estimation were introduced, and the sampling variances and covariances of the parameter estimators were analytically derived along with the variance of the T-year flood event. For 45 sets of annual flood data taken from different sources, the GG distribution was found to provide good fits when the above methods were employed. Moreover, it was also found that although the sampling variances of the estimators were high, the percent standard error of the T-year flood was relatively small. Thus use of the GG distribution and these methods would provide good design flood magnitudes at appropriately determined return periods.

INTRODUCTION

Although the log Pearson type 3 distribution has been recommended by the U.S. Water Resources Council (1977) as the base frequency curve, many hydrologists are not satisfied with the existence of an upper bound which is found to encounter in many practical applications.

Following the suggestions made by Mendel (1974) and Hoshi and Yamaoka (1980), this study proposes the generalized gamma (GG) distribution for annual flood analysis in both cases, namely *site-specific* and *regional* data (Kuczera, 1982). This proposal is based upon the following reasons:

(a) With only three parameters like the log Pearson type 3, the GG distribution is very flexible. As pointed out by Stacy and Mihram (1965), the GG distribution has many useful distributions as its special cases.

(b) There is no location parameter and consequently use of the GG distribution will rule out the existence of the upper bound.

(c) As reported by Hoshi and Yamaoka (1980), the *condition of separation* does not seem to exist for the case of the GG distribution.

While the basic properties of the GG distribution were presented elsewhere (Dilokrungterapop, 1983; Kuo, 1984), the present study concen-

V. P. Singh (ed.), Hydrologic Frequency Modeling, 407–418.

trates on the estimation of its parameters with the use of mixed moments.

METHODS OF MIXED MOMENTS

For the GG distribution, the density function is:

$$f(x;a,b,c) = |c|\ x^{bc-1} \exp[-(x/a)^c]/[a^{bc}\ \Gamma(b)] \tag{1}$$

where $x \geq 0$ is the GG variable; a,b,c are the three parameters ($c \neq 0$; $b,c > 0$; and $\Gamma(b)$ is the gamma function of b.

The (population) moment of x of order k is given by the following equation:

$$\nu_k = E(x^k) = a^k\ \Gamma(b + k/c)/\Gamma(b) \qquad b + k/c > 0 \tag{2}$$

Let

$$z = \ln\ [(x/a)^c] = c(y-\ln a) \tag{3}$$

where y = lnx, then the kth moment of z is (Stacy and Mihram, 1965):

$$E(z^k) = \Gamma^{(k)}\ (b)/\Gamma(b) \tag{3'}$$

From this equation, one can obtain the mean and variance of y easily as follows:

$$E(y) = \ln a + \psi(b)/c \tag{4}$$

$$Var(y) = \psi'(b)/c^2 \tag{5}$$

where $\psi(b)$, $\psi'(b)$ are the digamma (psi) and trigamma functions of b, respectively.

There have been three existing methods for estimating the GG parameters, namely, the method of indirect moments introduced by Stacy and Mihram (1965), methods of direct moments (see Hoshi and Yamaoka, 1980), and method of maximum likelihood (see Hager and Bain, 1970). An evaluation of these was made by Dilokrungterapop (1983). The present study concentrates on four new methods which are all based upon the first two moments of x and y = lnx. These methods, referred to as methods of mixed moments, have been found to be very useful for the log Pearson type 3 distribution (Phien and Hira, 1983; Phien and Hsu, 1985). It should be noted that only the first two moments were considered in order to avoid the large sampling errors committed in the computation of sample moments of higher orders.

Method MM1 :

This method is based upon the first two moments of x and the first moment of y = lnx. From the first moment of y, one obtains:

$$a = \exp\ [\ \bar{y} - \psi(b)/c\ ] = A(b,c') \tag{6}$$

where $c' = 1/c$ and $\bar{y}$ is the sample mean of y.

After substituting this for a in two equations obtained by setting $\nu_k = n_k$ (k=1,2) where n_k is the kth sample moment, one arrives at the following system:

$$\left.\begin{aligned} H_1(b,c') &= A(b,c')\,\Gamma(b+c') - n_1\,\Gamma(b) = 0 \\ G_1(b,c') &= A^2(b,c')\,\Gamma(b+2c') - n_2\,\Gamma(b) = 0 \end{aligned}\right| \qquad (7)$$

Newton's method may be used to solve this system for b and c'. The partial derivatives involved are readily obtained as follows:

$$\left.\begin{aligned} \partial H_1/\partial b &= A\Gamma(b+c')[\psi(b+c')-c'\psi'(b)]-n_1\Gamma(b)\psi(b) \\ \partial H_1/\partial c' &= A\Gamma(b+c')[\psi(b+c')-\psi(b)] \\ \partial G_1/\partial b &= A^2\Gamma(b+2c')[\psi(b+2c')-2c'\psi'(b)]-n_2\Gamma(b)\psi(b) \\ \partial G_1/\partial c' &= 2A^2\Gamma(b+2c')[\psi(b+2c')-\psi(b)] \end{aligned}\right| \qquad (8)$$

Having solved for b and c', one computes a using eq. 6.

Method MM2 :

This method is based upon the first two moments of x and the variance of y. Let s^2 denote the sample variance of y, then from eq. 5, one obtains:

$$s^2 = \psi'(b)/c^2$$

which, by rearranging gives:

$$H_2(b,c') = c'^2\,\psi'(b) - s^2 = 0 \qquad (9)$$

Moreover, by eliminating a in the two equations obtained by setting $\nu_k = n_k$ (k=1,2), one has:

$$G_2(b,c') = n_1^2\,\Gamma(b)\Gamma(b+2c') - n_2\,\Gamma^2(b+c') = 0 \qquad (10)$$

The partial derivatives are:

$$\left.\begin{aligned} \partial H_2/\partial b &= c'^2\,\psi''(b) \\ \partial H_2/\partial c' &= 2c'\,\psi'(b) \\ \partial G_2/\partial b &= n_1^2\Gamma(b)\Gamma(b+2c')[\psi(b)+\psi(b+2c')] \\ &\quad - 2n_2\Gamma^2(b+c')\psi(b+c') \\ \partial G_2/\partial c' &= 2[n_1^2\Gamma(b)\Gamma(b+2c')\psi(b+2c') - n_2\Gamma^2(b+c')\psi(b+c')] \end{aligned}\right| \qquad (11)$$

The system of equations 9 and 10 can be solved by Newton's method. After having obtained b and c', one computes a by:

$$a = n_1 \Gamma(b)/\Gamma(b+c') \tag{12}$$

Method MM3 :

In this method, the mean and variance of y are employed together with the mean of x. The resulting equations are:

$$a = \exp [\ \bar{y} - c'\psi(b)] \tag{13}$$

$$s^2 = \psi'(b)\ c'^2 \tag{14}$$

$$n_1\Gamma(b) = a\Gamma(b+c') \tag{15}$$

Eliminating a gives:

$$\left.\begin{aligned} H_3(b,c') &= 0 \\ G_3(b,c') &= 0 \end{aligned}\right| \tag{16}$$

where $H_3 = H_1$ and $G_3 = H_2$.

This system may be readily solved using Newton's method. All the partial derivatives needed are derived previously.

Method MM4 :

This method makes use of the mean and variance of y along with the second moment of x. Thus eqs. 13 and 14 are used again but eq. 15 is replaced by the following:

$$n_2\Gamma(b) = a^2\Gamma(b+2c')$$

Again, by eliminating a, one obtains:

$$\left.\begin{aligned} H_4(b,c') &= 0 \\ G_4(b,c') &= 0 \end{aligned}\right| \tag{17}$$

where $H_4 = G_1$ and $G_4 = H_2$.

This system may be solved by Newton's method for which all the needed partial derivatives are already obtained.

Remarks. (1) In the above systems of nonlinear equations, the psi function and its derivatives are involved. To compute these functions, algorithms similar to those of Bernado (1976) and Schmeider (1978) can readily be developed.

(2) For Newton's method to converge, good initial values of the parameters must be provided. It was found that the parameter values obtained by the method of indirect moments (using the mean, variance and skewness of y) may be used in most cases.

VARIANCE OF THE T-YEAR FLOOD

Let X_T be the magnitude of the T-year flood, then:

$$\mathrm{Prob}(x > X_T) = T^{-1} \tag{18}$$

where Prob stands for probability and T is the return period, expressed in years. Clearly X_T depends upon a, b and c; consequently, its variance may be approximately by (Kendall and Stuart, 1977):

$$\begin{aligned} \mathrm{Var}(X_T) = {} & (\partial X_T/\partial a)^2 \mathrm{Var}(a) + (\partial X_T/\partial b)^2 \mathrm{Var}(b) \\ & + (\partial X_T/\partial c)^2 \mathrm{Var}(c) + 2(\partial X_T/\partial a)(\partial X_T/\partial b)\mathrm{Cov}(a,b) \\ & + 2(\partial X_T/\partial b)(\partial X_T/\partial c)\mathrm{Cov}(b,c) \\ & + 2(\partial X_T/\partial a)(\partial X_T/\partial c)\mathrm{Cov}(a,b) \end{aligned} \tag{19}$$

In this equation, Var(a) denotes the variance of the estimator of a, and Cov(a,b) is the covariance of the estimators of a and b.

Let w be defined as:

$$w = (x/a)^c \tag{20}$$

then the density function of w is:

$$g(w) = w^{b-1} \exp(-w)/\Gamma(b) \tag{21}$$

In other words, w is the standard gamma variable with shape parameter b. So if W_T is the gamma quantile corresponding to the non-exceedance probability of $1-T^{-1}$ for $c > 0$ or T^{-1} for $c < 0$, one has:

$$X_T = aW_T^{1/c} \tag{22}$$

and consequently:

$$\begin{aligned} \partial X_T/\partial a &= W_T^{1/c} \\ \partial X_T/\partial b &= (a/c)\, W_T^{(1/c)-1} * \partial W_T/\partial b \\ \partial X_T/\partial c &= -\,(X_T/c^2) \ln W_T \end{aligned} \tag{23}$$

It becomes clear that the variance of X_T can readily be computed once the variances and covariances of the parameter estimators are obtained.

For all the methods under consideration, the estimators are obtainable from three moments, denoted by P, Q and R. As shown by Phien and Hsu (1985) once can write:

$$V_P \doteq D^{-1} V_M \tag{24}$$

where

$$V_P = \begin{bmatrix} Var(a) \\ Var(b) \\ Var(c) \\ Cov(a,b) \\ Cov(b,c) \\ Cov(a,c) \end{bmatrix} \qquad V_M = \begin{bmatrix} Var(P) \\ Var(Q) \\ Var(R) \\ Cov(P,Q) \\ Cov(Q,R) \\ Cov(P,R) \end{bmatrix}$$

D = matrix involving the partial derivatives of the moments with respect to the parameters, and

$\doteq$ indicates an approximation.

From eq. 24, it is clear that V_P is completely determined once D and V_M are obtained. The remaining part of this section deals with their derivations.

Method MM1 :

The three moments required in this method are $P = n_1$, $Q = n_2$ and $R = \bar{y}$. In this case, the components of V_M are as follows:

$$\begin{aligned} Var(n_1) &= (1/N)\ Var(x) \\ &= \frac{a^2}{N\Gamma(b)} [\Gamma(b+2/c) - \Gamma^2(b+1/c)/\Gamma(b)] \end{aligned} \tag{25}$$

$$\begin{aligned} Var(n_2) &= (1/N)\ [E(x^4) - E^2(x^2)] \\ &= \frac{a^4}{N\Gamma(b)} [\Gamma(b+4/c) - \Gamma^2(b+2/c)/\Gamma(b)] \end{aligned} \tag{26}$$

$$Var(\bar{y}) = (1/N)\ Var(y) = \psi'(b)/(Nc^2) \tag{27}$$

$$\begin{aligned} Cov(n_1,n_2) &= (1/N)\ [E(x^3) - E(x)\ E(x^2)] \\ &= \frac{a^3}{N\Gamma(b)} [\Gamma(b+3/c) - \Gamma(b+1/c)\Gamma(b+2/c)/\Gamma(b)] \end{aligned} \tag{28}$$

In the above equations, N is the sample size. For the remaining two covariances, Phien and Hsu (1985, Appendix A) gave:

$$Cov(n_1,\bar{y}) = (1/N)\ [E(xy) - E(x)\ E(y)]$$

$$Cov(n_2,\bar{y}) = (1/N)\ [E(x^2y) - E(x^2)\ E(y)]$$

After some rather tedious manipulation, the following expressions are derived:

$$E(xy) = a\ \Gamma(b+1/c)\ [\ln a + \psi(b+1/c)/c] \tag{29}$$

$$E(x^2y) = a^2\ \Gamma(b+2/c)\ [\ln a + \psi(b+2/c)/c] \tag{30}$$

All the elements of matrix D can be obtained easily:

$$\left.\begin{aligned}
\partial n_1/\partial a &= \partial E(x)/\partial a = \Gamma(b+1/c)/\Gamma(b) \\
\partial n_1/\partial b &= \partial E(x)/\partial b = a\Gamma(b+1/c)]\psi(b+1/c)-\psi(b)]/\Gamma(b) \\
\partial n_1/\partial c &= \partial E(x)/\partial c = -a\Gamma(b+1/c)\psi(b+1/c)/[c^2\Gamma(b)]
\end{aligned}\right| \tag{31}$$

$$\left.\begin{aligned}
\partial n_2/\partial a &= 2a\Gamma(b+2/c)/\Gamma(b) \\
\partial n_2/\partial b &= a^2\Gamma(b+2/c)[\psi(b+2/c)-\psi(b)]/\Gamma(b) \\
\partial n_2/\partial c &= -2a^2\Gamma(b+2/c)\psi(b+2/c)/[c^2\Gamma(b)]
\end{aligned}\right| \tag{32}$$

and

$$\partial\bar{y}/\partial a = 1/a, \qquad \partial\bar{y}/\partial b = \psi'(b)/c, \qquad \partial\bar{y}/\partial c = -\psi(b)/c^2 \tag{33}$$

Method MM2 :

This method differs from the MM1 in that it employs $R = s^2$ instead of the sample mean of y. From eqs. 3 and 3', it follows that:

$$\mathrm{Var}(s^2) \doteq [1/(Nc^4)][\psi^{(3)}(b) + 2(\psi'(b))^2] \tag{34}$$

The general expressions for $\mathrm{Cov}(n_1,s^2)$ and $\mathrm{Cov}(n_2,s^2)$ were given by Phien and Hsu (1985, Appendix B) as:

$$\mathrm{Cov}(n_1,s^2) = (1/N)[E(xy^2)-E(x)E(y^2)-2E(xy)E(y)+2E(x)E^2(y)]$$

$$\mathrm{Cov}(n_2,s^2) = [(1/N)E(x^2y^2)-E(x^2)E(y^2)-2E(x^2y)E(y)+2E(x^2)E^2(y)]$$

On the right-hand sides, all the expected values are known except $E(xy^2)$ and $E(x^2y^2)$, which can be shown, after some tedious algebra, to be:

$$\begin{aligned}
E(xy^2) = {} & a\Gamma(b+1/c)[\psi'(b+1/c)/c^2 + \psi^2(b+1/c)/c^2 \\
& + 2(\ln a)\psi(b+1/c)/c + (\ln a)^2]/\Gamma(b)
\end{aligned} \tag{35}$$

$$\begin{aligned}
E(x^2y^2) = {} & a^2\Gamma(b+2/c)[\psi'(b+2/c)/c^2 + \psi^2(b+2/c)/c^2 \\
& + 2(\ln a)\psi(b+2/c)/c + (\ln a)^2]/\Gamma(b)
\end{aligned} \tag{36}$$

This completes the determination of the components of vector V_M.

To obtain matrix D, the following derivatives are needed (in addition to those given in eqs. 31 and 32):

$$\partial s^2/\partial a = 0, \qquad \partial s^2/\partial b = \psi''(b)/c^2, \qquad \partial s^2/\partial c = -2\psi'(b)/c^3 \tag{37}$$

Method MM3 :

For this method, $P = n_1$, $Q = \bar{y}$ and $R = s^2$. All the components of the vector V_M have been derived. The elements of D have been obtained as well.

Method MM4 :

In this method, $P = \bar{y}$, $Q = s^2$ and $R = n_2$. All the components of V_M and elements of D have been obtained in the first two methods, namely MM1 and MM2.

APPLICATIONS

To evaluate the applicability of the GG distribution to flood data and the performance of the methods of mixed moments, 45 sets of annual flood data (Table 1) taken from different sources were analyzed.

Table I List of Annual Flood Data Sets Employed (unit :m³/s)

Station	River	Location	N	Period*
F1	Mekong	Vientiane, Laos	43	1923-1965
F2	Nan	Muang Nan, Thailand	51	1928-1979
F3	Mae Wang	Rasada Phisek Bridge, Lampang, Thailand	51	1928-1979
F4	Rasak	Kaeng Khol, Saraburi, Thailand	51	1928-1979
F5	Nan	Muang Phichit, Thailand	35	1945-1979
F6	Prachin	Kabinburi, Thailand	35	1945-1979
F7	Mae Yom	Muang Phrae, Thailand	35	1945-1979
F8	Nan	Tha-It, Uttaradit, Thailand	35	1945-1979
F9	River Gana	Hardwar, India	77	1901-1079
F10	Clearwater	Kamiah, U.S.A.	55	1911-1965
F11	Indus	Kotr, Pakistan	73	1901-1973
F12	Ping	Nawarat Bridge, Chiangmai, Thailand	60	1921-1980
F13	Greenbrier	Alderson, U.S.A.	66	1896-1966
F14	St. Marys	Stillwater, U.S.A.	59	1916-1965
F15	Cumberland	Cumberland Fall, U.S.A.	49	1916-1965
F16	Termpeuleau	Dodge, Wis, Canada	33	1928-1960
F17	Bow	Banff, Alberta, Canada	33	1928-1960
F18	James	Scottsville, VA., U.S.A.	33	1928-1980
F19	Benoue	Garaoua, Cameroon	32	1930-1964
F20	Oubangui	Bangui, Central Aferican Republic	38	1911-1964
F21	Fraser	Hope, Canada	53	1912-1964
F22	Red	Eamersor, Canada	53	1912-1964
F23	Assiniboine	Brandon, Canada	59	1906-1964
F24	North Sas-kaichewan	Prince Albert, Canada	53	1910-1964
F25	Saint-Maurice	Grand-Mere Power Plant, Canada	64	1901-1964
F26	St. John	Pokiok, Canada	46	1918-1964
F27	Penobscot	West Enfield, ME., U.S.A.	57	1902-1964
F28	Delaware	Trenton, NJ., U.S.A.	44	1912-1964

Table I (Continued)

Station	River	Location	N	Period*
F29	Susguen-hanna	Hairrisburg, PA., U.S.A.	67	1890-1964
F30	Missouri	Yankton, S. Dak., U.S.A.	33	1930-1964
F31	Colorado	Lees Ferry, Ariz., U.S.A.	44	1911-1964
F32	Parana	Guaira, Brazil	44	1920-1964
F33	Krishna	Vijayawada, India	60	1901-1964
F34	Ural	Kushum, Turkey	47	1915-1964
F35	Inn	Scharding, Austria	35	1930-1964
F36	Elbe	Decin, Czechoslovakia	104	1851-1964
F37	Kymijoki	Pernoo, Finland	65	1900-1964
F38	Vuoksi	Imatra, Finland	118	1847-1964
F39	Weser	Intschede, Fed., Rep. of Germany	44	1921-1964
F40	Tisza	Szeged, Hungary	44	1921-1964
F41	Po	Placenza, Italy	41	1924-1964
F42	Maas	Lith, Netherlands	54	1911-1964
F43	Tejo	Vila Velha De Rodao, Portugal	52	1913-1964
F44	Danube	Orsova., Rumania	125	1840-1964
F45	Volga	Volgograd Hydroelectric Power Plant, U.S.S.R.	85	1879-1964

(*) Some years with missing data and water years counted differently.

Fitting Results :

The results of the fitting, evaluated by the Kolmogorov-Smirnov test can be summarized as follows:

(1) The estimated values of c were negative for only two sets, namely F15 and F44. For F15 (Cumberland Fall, USA), the fitting was acceptable only at the 1% level of significance when method MM1 was used, but it is acceptable even at the 10% level when the other remaining methods were applied. For F44 (Danube at Orsova, Rumania), the fitting was accepted at the 10% level for MM2, MM3 and MM4, but it was rejected even at the 1% level when MM1 was used.

(2) When method MM1 was applied, the fitting was rejected at the 1% level for the following sets: F5, F10, F20, F21, F30, F35, F36 and F44 (mentioned before). It was rejected at the 5% level for the following additional sets: F15, F22, F24, F25 and F34. For all the remaining sets, the fitting was accepted at a level as low as 10%.

(3) When method MM2 was applied, the fitting was rejected at the 1% level of significance only for F5, rejected at the 5% level for the additional set F34. For all other sets except F22 the fitting was accepted at the 10% level.

(4) When method MM3 was applied, only two sets, manely F22 and F34, were rejected at the 10% level, and when method MM4 was used, the fitting was accepted at the 10% level for all sets.

From these observations, it is clear that fitting the GG distribution to annual flood data by means of the methods of mixed moments is applicable.

Percent Standard Errors Parameter Estimates :

The percent standard errors of the estimates of a,b,c are defined (Phien and Hsu, 1985) as:

$$\left.\begin{aligned} SE(a) &= 100\ [Var(a)]^{\frac{1}{2}}\ /\ a \\ SE(b) &= 100\ [Var(b)]^{\frac{1}{2}}\ /\ b \\ SE(c) &= 100\ [Var(c)]^{\frac{1}{2}}\ /\ |c| \end{aligned}\right| \qquad (38)$$

It should be noted that the absolute value of c is used because c can take on negative values. These percent standard errors can give insight into the order of magnitude of the error when a parameter is estimated. Here the percent standard error of a parameter estimate is used instead of its sampling variance (which was found to be very high), because it is more useful to provide the information on the error with respect to the magnitude of the parameter estimate.

For all the 45 sets of annual flood data, the means of the percent standard errors are summarized in Table 2. Except for those values corresponding to method MM3, which were very small, the tabulated values are in the same range of the percent standard errors of the parameter estimates in the log Pearson type 3 distribution (Phien and Hsu, 1985). These summarized results clearly indicate that method MM3 is the best one.

Percent Standard Error of the T-Year Flood :

The percent standard error of the T-year flood is defined in the same was as that of a parameter estimate:

$$SE(X_T) = 100\ [Var(X_T)]\ /\ X_T \qquad (39)$$

where X_T is the magnitude of the T-year flood. First, W_T is computed according to the scheme of Hoshi and Burges (1981), then X_T is obtained from eq. 22. The variance of X_T is computed according to eq. 19. The means of $SE(X_T)$ corresponding to T = 10, 50, 100, 100 and 500 years are collected in Table 3.

Although the errors committed in estimating the parameters may be high (Table 2), the percent standard error of the T-year flood is not high. Again, it can be seen that the MM3 is the best method because it yields the smallest values for $SE(X_T)$ is almost all cases.

SUMMARY AND CONCLUSIONS

The generalized gamma (GG) distribution was proposed as the base frequency curve for annual flood data analysis. This was based upon its

Table II Means of Percent Standard Errors of Parameter Estimates (in Percentage)

Method	SE(a)	SE(b)	SE(c)
MM1	415.0	144.6	79.0
MM2	653.1	216.2	112.8
MM3	22.1	1.5	12.1
MM4	457.7	184.0	98.1

Table III Means of the Percent Standard Error of the T-Year Flood

T-Years	MM1	MM2	MM3	MM4
10	7.41	6.81	6.73	6.49
50	10.98	10.45	8.59	9.82
100	12.83	12.25	9.32	11.49
200	14.69	14.08	9.99	13.16
500	17.10	16.49	10.82	15.36

flexibility in shape and was intended to get rid of the existence of the upper bound encountered in the log Pearson type 3 distribution. Since a distribution is useful only when its parameters can be reliably estimated; to this aim, four methods of parameter estimation, all based upon mixed moments, were introduced and all the needed computational schemes were provided. Moreover, the sampling variance of the T-year flood event was also investigated and needed expressions were derived.

From the application of the GG distribution and the methods of mixed moments to 45 sets of annual flood data, the following conclusions may be drawn:

(1) The GG distribution is quite appropriate for annual flood data. Even at a level as low as 10%, fitting these sets of annual flood data was accepted for at least two among the four methods of mixed moments.

(2) The percent standard errors of the parameter estimates were high except for the MM3. However, these high values are in the same range as those obtained for the log Pearson type 3 distribution.

(3) The percent standard error of the T-year flood was relatively low. This indicates that regardless of the fact that high errors may be committed in estimating the parameters, the magnitude of the T-year flood can still be determined with reasonable accuracy.

(4) Among the four methods considered, method MM3, which is based upon the mean of x, the mean and variance of $y = \ln x$, is the best one.

ACKNOWLEDGEMENTS

The authors wish to express their sincere thanks to the Asian Institute of Technology for use of its computing facilities. They are also grateful to Ms. Nantawan Nakasen for her skill in typing this manuscript.

REFERENCES

Bernado, J.M. 1976. Psi (Digamma) Function (Algorithm AS 103). *Applied Statistics*, 5 (23), pp. 315-317.

Dilokrungterapop, S. 1983. The generalized gamma distribution and its applications. *Master Thesis*, Asian Institute of Technology, Bangkok, Thailand.

Hager, W.H. and Bain, L.J. 1970. Inferential Procedure for the Generalized Gamma Distribution. *Journal of the American Statistical Association*, 65 (332), pp. 1601-1609.

Hoshi, K. and Burges, S.J. 1981. Approximate Estimation of the Derivative of a Standard Gamma Quantile for Use in Confidence Interval Estimates. *Journal of Hydrology*, 53, pp. 305-316.

Hoshi, K. and Yamaoka, I. 1980. The generalized gamma probability distribution and its application in hydrology. *Hydraulics Papers*, 133 pp., Hokkaido University, Sapporo, Japan

Kendall, M.G. and Stuart, A. 1977. *The Advanced Theory of Statistics*, 1, Charles Griffin, London, 4th ed., pp. 243-262.

Kuczera, G. 1982. Combining Site-Specific and Regional Information: an Empirical Bayes Approach. *Water Resources Research*, 18 (2), pp. 306-314.

Kuo, J.H. 1984. Evaluation of the methods of parameter estimation for the generalized gamma distribution. *Master Thesis*, Asian Institute of Technology, Bangkok, Thailand.

Mendel, H.G. 1974. A new proposal for flood frequency computation. In: *Floods and Droughts*, Proc. 2nd. Int. Symp. on Hydrology, Fort Collins, Colorado, pp. 325-333.

Phien, H.N. and Hira, M.A. 1983. Log Pearson Type 3 Distribution: Parameter Estimation. *Journal of Hydrology*, 64, pp. 25-37.

Phien, H.N. and Hsu, L.C. 1985. Variance of the T-Year Event in the Log Pearson Type 3 Distribution. *Journal of Hydrology*, 77, pp. 141-158.

Schneider, B.E. 1978. Trigamma Function (Algorithm AS 121). *Applied Statistics*, 27 (1), pp. 97-99.

Stacy, E.W. and Mihram, G.A. 1965. Parameter Estimation for a Generalized Gamma Distribution. *Technometrics*, 7 (3), pp. 349-357.

Water Resource Council, 1977. *Guidelines for determining flood flow frequency*. U.S. Water Resour. Counc., Hydrol. Comm., Washington, D.C., Bull. 17A.

ENTROPY PRINCIPLE IN THE ESTIMATION OF GUMBEL PARAMETERS

Arun Kumar
Professor of Civil Engineering
Delhi College of Engg., Delhi-6, India.

B.P. Parida
Lecturer in Civil Engg.
IITD, New Delhi-16, India.

Rema Devi
Asstt.Prof.of Civil Engg.
IITD, New Delhi-16, India.

ABSTRACT. The performance of the principle of maximum entropy (POME) in the estimation of Gumbel parameters is compared with the existing procedures, viz., the method of moments, principle of maximum likelihood, the method of least squares and the probability weighted moments (PWM). Simulation studies are performed to compare the parameters and flood quantile bias. It is shown that POME compares favourably with the PWM and performs better than the other methods. The potentialities of POME in hydrological analysis is highlighted.

INTRODUCTION

The two major problems in statistical flood frequency analysis are (i) the identification of the 'proper' distribution and (ii) selection of the most appropriate parameter estimation procedure. Presently there is no general agreement amongst hydrologists about the choice of the 'proper' distribution. The U.S. Water Resources Council (1976) have recommended the log Pearson type III distribution. Following the same guidelines, the Australian Institute of Engineers recommended the LP III distribution for Australian rivers(AIE, 1977). On the other hand, the flood studies report (FSR) of N.E.R.C., U.K. (1975) recommended the use of General Extreme Value (GEV) distribution given by Jenkinson (1969). The objective of recommending a specific distribution was to bring uniformity in the analysis undertaken by various agencies. Nevertheless, the choice of LP III distribution has generated a lot of discussion (Bobee, 1973, Landwehr et al, 1978, 1980). Recently, Wallis and Wood, (1985) using Monte Carlo simulation, have shown that GEV distribution or Wakeby (WAK) distribution fitted by PWM gives better quantile estimates than the WRC recommended procedure. Within the family of GEV distribution the EV1 or

V. P. Singh (ed.), Hydrologic Frequency Modeling, 419–431.

the Gumbel distribution has been quite popular in hydrological analysis and in this study EV1 distribution is analysed.

Once the distribution is 'identified', it remains to estimate its parameters from the observed sample and check the 'goodness of fit' to the data. Various methods have been proposed for estimating the parameters and consequently the quantiles of the distribution. The method of moments (MOM) is frequently applied primarily because of its simplicity. The method of maximum likelihood (MML) though provides minimum variance estimate, has not been frequently used because of involved computations required in many known distributions. The other popular methods are the method of least square (MLS) and the probability weighted moments (PWM). In an elaborate study undertaken by Hosking, Wallis and Wood (1984) and Wallis and Wood (1985) the PWM method suggested by Greenwood et al (1979) is recommended for estimating the GEV distribution parameters. It is shown that GEV/PWM algorithms lead to more accurate and more consistent estimation of flood quantiles than the presently recommended WRC, U.S.A. and the N.E.R.C., U.K. practices of estimating them.

A shortcoming of the PWM algorithm is that it can be applied only to those distributions that have the flexibility of being expressed in an inverse form, that is, if X is a random variable and F is the cumulative distribution function for X, the value of X may be written as a function of F; x = x(F). Many distributions like Gumbel, Weibull or Wakeby are amenable to inverse from, while many other distributions like the LP III or the Pearson type III are not. In other words,PWM method is not applicable to many distributions.

In recent years, the princple of Maximum Entropy (POME) has been used in generating least biased probability distributions from small samples (Tribus, 1969, Sonuga 1972, Jowitt, 1980, Singh, 1985). The POME can be used in solving both the problems of 'identifying' the distribution and the estimation of the unbiased parameters of the distribution. Unlike the PWM, the POME can be used in estimating the parameters of the distributions commonly used in hydrology (Gumbel, GEV, Weibull, LP III,Pearson, etc.).

The aim of this study is to compare the POME method with the method of moments (MOM), the method of maximum likelihood (MML), the method of least squares (MLS) and the probability weighted moments (PWM). Since PWM can easily be applied to obtain EV1 parameters, the EV1 distribution is chosen as the 'identified' distribution. It may however be stressed that unlike PWM, POME can also be used to obtain LP III or Pearson type III distribution parameters. Similar to the approach followed by Landwehr et al (1979), two cases were considered in comparing the five estimates, (POME, MOM, MML, MLS and PWM), namely (i) a sample from a purely random process and (ii) a sample from a serially correlated process but regarded as purely random.

Though the application of POME has tremendous potential in analysing hydrological problems, it has rarely been applied, primarily because the hydrologists have not been very familiar with this technique. Keeping this point in view, the use of POME in estimating EV1 parameters is briefly explained.

POME IN ESTIMATING EV1 PARAMETERS

Consider a probability density function, pdf, f(x) associated with a dimensionless random variable x. f(x) is a positive valued function for every x in some interval (a,b) and is normalised to unity.

$$\int_a^b f(x) \ dx = 1 \tag{1}$$

Shannon (1948) defined entropy, I(f) as a numerical measure of uncertainty associated with f(x) in describing the random variable x and defined as:-

$$I(f) = - \int_a^b f(x) \ \ln f(x) \ dx \tag{2}$$

Generally, the distribution f(x) is unknown _apriori_ although some properties of f(x) may be known. These properties are called constraints. For example, one of the constraints can be the expected value of the function $y_i(x)$ over f(x) as:-

$$C_i = \int_a^b y_i \ (x) \ f(x) \ dx \qquad i = 1,2,\ldots\ldots m \tag{3}$$

Eq. (3), C_i thus delineates the constraints or the broad information regarding the f(x).

The POME thus specifies the determination of f(x) such that I(f) in Eq. (2) is maximised subject to the constraints given in Eq. (3).

$$\text{Maximise } I(f) = - \int_a^b f(x) \ \ \ln f(x) \ dx$$

Subject to:-

$$\int_a^b f(x)dx = 1$$

$$\int_a^b y_i \ (x) \ f \ (x)dx = E \ \{ y_i \ (x) \} \tag{4}$$

$$\int_a^b y_m(x)\ f(x)dx = E\{\ y_m\ (x)\}$$

The solution of f(x) in Eq. (4) can be determined by using Lagrange's multiplier (Tribus, 1969) and is of the form:-

$$f(x) = \exp\ [\ -a_o - \sum_{i=1}^{m} a_i\ y_i(x)] \tag{5}$$

Where a_i, i = 0, 1........m are the Lagrange multipliers and can be obtained from the constraints given in Eq. (4).

A simplified approach suggested by Singh (1985) is used in deriving the parameters of EV1 distribution using POME. Exactly similar approach can be applied for deriving the parameters of other distributions.

The EV1 distribution is

$$f(x) = (1/a)\ \exp\ [\ -\ (x-b)/a\ - \{\ \exp\ \{\ -\ (x-b)/a\}\}] \tag{6}$$

and the cumulative distribution function F(x) as

$$F(x) = \exp\ [\ -\ \exp\ \{\ -\ (x-b)/a\}] \tag{7}$$

with the inverse form

$$x = b - a\ \ln\ \{\ -\ \ln\ F(x)\} \tag{8}$$

The entropy I (f) for EV1 is:-

$$I(f) = -\int_{-\infty}^{+\infty} f(x)\ \ln\ f(x)\ dx$$

$$= (\ln\ a\ -\ b/a)\ \int_{-\infty}^{+\infty} f(x)\ dx + (1/a)\ \int_{-\infty}^{\infty} x\ f(x)\ dx$$

$$+ \exp\ (b/a)\ \int_{-\infty}^{\infty} \exp\ (-b/a)\ f(x)\ dx \tag{9}$$

Comparing Eq. (8) with Eq. (3), the constraints of EV1 distribution are defined as:-

$$\int_{-\infty}^{\infty} f(x)\ dx = 1 \tag{10}$$

$$\int_{-\infty}^{\infty} x\ f(x)\ dx = E\ (x) \tag{11}$$

$$\int_{-\infty}^{\infty} \exp\ (-x/a)\ f(x)\ dx = E\ [\exp\ (-x/a)] \tag{12}$$

where E ( ) is the expected value.

Following Ea. (5), the least biased density function consistent with the constraints given in Eq. (9) to (11) is of the form

$$f(x) = \exp\ [\ -\ a_o\ -\ a_1\ x\ -\ a_2\ \exp(-x/a)] \tag{13}$$

Since $\int f(x)\ dx = 1$ and using Eq. (13)

$$\exp(a_o) = \int_{-\infty}^{\infty} \exp\ [\ -\ a_1\ x\ -\ a_2\ \exp\ (-x/a)]\ dx \tag{14}$$

which on integration results

$$\exp(a_o) = -(1/a) \exp\{-a_1 a \ln a_2\} \Gamma(a a_1) \quad (15)$$

$\Gamma(\quad)$ is the gamma function.
Thus the zeroth order Lagrange multiplier, a_o, is

$$a_o = \ln(1/a) - a_1 a \ln a_2 + \ln \Gamma(a a_1) \quad (16)$$

Substituting a_o in Eq. (12) results

$$f(x) = \exp[\ln(1/a) + a_1 a_2 \ln a_2 - \ln \Gamma(a a_1) - a_1 x - a_2 \exp(-x/a)] \quad (17)$$

Comparing Eq. (6) and Eq. (17), the values of Lagrange multipliers a_1 and a_2 in terms of EV1 parameters a and b are obtained as:

$$a_1 = 1/a, \quad a_2 = \exp(b/a) \quad (18)$$

a_1, a_2 are related to the constraints given in Eq. (10) and Eq. (11)

$$\frac{\partial a_o}{\partial a_1} = -E(x) \quad (19)$$

$$\frac{\partial a_o}{\partial a_2} = -E[\exp(-x/a)] \quad (20)$$

Hence

$$\frac{\partial a_o}{\partial a_1} = -a_1 \ln a_2 + \partial/_{\partial a}\Gamma(a_1 a) = -E(x) \quad (21)$$

Substituting the value of a_1 and a_2 given in Eq. (18) and simplifying Eq. (21).

$$E(x) = b + 0.5772\, a \quad (22)$$

Similarly from Eq. (2)

$$E[\exp(-(x-b)/a)] = 1 \quad (23)$$

Given the sample value $x_1, x_2, \ldots\ldots x_n$, a and b can be estimated from Eq. (22) and Eq. (23) using the iterative procedure given by Jowitt (1980).

In this study, the EV1 parameters estimates using POME are compared with the estimates obtained from other standard methods given as follows:

2. Method of Moments (MOM)

$$b = \bar{x} - 0.5772\ a \tag{24}$$

$$b = (6)^{\frac{1}{2}}\ \sigma/\pi = 0.7797\ \sigma \tag{25}$$

Where $\bar{x}$ and σ are the mean and the standard deviation of a given sample $(x_1, x_2, \ldots\ldots x_n)$

3. Method of Maximum likelihood (MML)

$$b = -\ a \ln \left[(1/n) \sum_{i=1}^{n} \exp(-x_1/a) \right] \tag{26}$$

$$a = \bar{x} - \frac{\left[\sum_{i=1}^{n} x_1 \exp(-\ x_1/a) \right]}{\sum_{i=1}^{n} \exp(-x_1/a)} \tag{27}$$

An iterative procedure given by Jenkinson (1969) has been used in estimating a and b.

4. Method of Least Squares (MLS)

$$b = \bar{x} + a\ (1/n) \sum_{i=1}^{n} \ln \{ -\ln F\ (x_i) \} \tag{28}$$

and

$$a = -\ \frac{(1/n) \sum_{i=1}^{n} x_i \ln\{ -\ln F(x_i) \} - \bar{x}(1/n) \sum_{i=1}^{n} \ln\{-\ln F(x_i)\}}{(1/n) \sum_{i=1}^{n} \ln^2\{-\ln F(x_i)\} - 1/n \sum_{i=1}^{n} \ln\{-\ln F(x_i)\}^2} \tag{29}$$

5. Probability Weighted Moment (PWM)

The PWM is defined by Greenwood et al, (1979)

$$M_{i,j,k} = E\ [x^i\ F^j\ (x)\ \{1-F(x)\}^k] = \int x^i F^j(x) \{1-F(x)\}^k\ dF(x) \tag{30}$$

$M_{1,0,k}$ is defined as M_k, where k is a non negative integer and is given by:-

$$M_k = (1/n) \sum_{i=1}^{n} x_i \binom{n-i}{k} \Big/ \binom{n-i}{k} \tag{31}$$

in which x_i, i = 1, 2.....n are the order statistics.

PWM estimates of a and b are given by Greenwood et al (1979) as:-

$$b = M_o - 0.5772\ a \tag{32}$$

$$a = (M_o - 2\ M_1)/\log(2) \tag{33}$$

SIMULATION STUDIES

The simulation studies are conducted on the same lines as given by Landwehr et al (1979). Two types of EV1 numbers are generated to evaluate the relative performance of various estimators.

Case 1:

In the first case, purely random numbers x following EV1 distribution were generated using Eq. (7). Without loss of generality, the values of b and a in Eq. (7) are taken as zero and one respectively. The random variable F(x) is uniformly distributed on the interval (0,1) and are generated using the algorithm given by Lewis et al (1969).

Case 2:

For the second case, serially correlated EV1 distributed numbers were generated as follows:

1. Firstly the lag 1 serially correlated normally distributed z_i are generated with zero mean and unit variance z_i is $N^i(0,1)$

$$Z_i = \rho_1 z_{i-1} + (1 - \rho_1^2)^{\frac{1}{2}}\ \varepsilon i \tag{34}$$

$_1$ is the first order serial correlation coefficient and ε is N (0,1) random generator.

2. Given a value of z_i, the corresponding $F(z_i) = P[Z < z_i]$ is computed

$$F(z_i) = \int_{-\infty}^{z_i} f_n(z)\, dz$$

$f_n(z)$ is the normally distributed density function.

3. The serially correlated EV1 distributed x_i are obtained as

$$x_i = b - a \ln \{ - \ln F(z_i) \}$$

Similar to case 1, the values of b and a are taken as zero and one respectively. The value of ρ_1 in Eq. (34) is taken as 0.3.

In both the cases 5000 EV1 distributed numbers are generated which are subdivided into N samples each of size n. The values of n are 5,10,20,25,30,40,50 and 100.

The performance of (i) MOM (ii) MML (iii) MLS (iv) PWM and (v) POME are evaluated by computing various indices to estimate the bias in parameters a, b and the quantiles x(F) where F = 0.01, 0.02, 0.05, 0.1, 0.25, 0.50, 0.75, 0.9,

0.95, 0.98, 0.99, 0.995 and 0.998.

For the sake of brevity, only limited results are discussed in the paper.

Performance Indices:

Let $\hat{\xi}$ denote an estimate of $\hat{\xi} \varepsilon \{a, b, x(F)\}$. For all the methods the bias in the estimate of ξ in N samples is calculated as follows:

The mean $\tilde{\mu}$ and the standard deviation, $\tilde{\sigma}$, of bias $(\xi - \hat{\xi})$ are:

$$\text{(i)} \quad \tilde{\mu}(\xi - \hat{\xi}) = (\xi - \frac{1}{N} \sum_{i=1}^{N} \hat{\xi}_i) \tag{35}$$

$$\text{(ii)} \quad \tilde{\sigma}(\xi - \hat{\xi}) = [\frac{1}{N} \sum_{i=1}^{N} (\xi - \hat{\xi})^2]^{\frac{1}{2}} \tag{36}$$

Where $\hat{\xi}$ is the estimate of the parameters $\hat{\xi} \varepsilon \{a, b, x(F)\}$ obtained from the ith sample. ξ is the parent value of the variable.

DISCUSSION OF RESULTS

The bias $\tilde{\mu}(\xi - \hat{\xi})$ and $\tilde{\sigma}(\hat{\xi})$ for the parameters a and b from different estimation procedures for purely random and correlated process ($\rho_1 = 0.3$) are given in Table 1. In all the methods studied, the bias in estimating parameter a is more than the bias in estimating b. For the correlated process (but taken as random in the analysis) the magnitude of bias increases. The MLS has the largest bias for both a and b for all values of n and does not seem to be a suitable method for computing the EV1 parameters. The comparative advantage of PWM decreases as n increases. For $n > 30$, all the methods except MLS give identical results. MOM, MML,and POME give nearly identical bias for all values of n. MML as expected has least bias for $n > 30$.

In terms of $\tilde{\sigma}(a)$ and $\tilde{\sigma}(b)$, it is interesting to observe that for both the uncorrelated and the correlated process and for nearly all the values of n, the POME and the MML give consistently better results than PWM, MOM and MLS procedures. MML and POME result in least $\tilde{\sigma}$ ( ), followed by PWM, MOM and MLS. In terms of the variance, POME is consistently bit more efficient than PWM. Between MOM and PWM, the MOM is bit less efficient than PWM. MLS again resulted in large $\tilde{\sigma}$.

The bias in the quantiles $\tilde{\mu}(x-\hat{x})$ for $\rho = 0$ are given in Table 2. The PWM again gives the best unbiased estimate for nearly all values of n and F. It is interesting to observe that even for very small n and for large E, PWM results in minimal bias $\tilde{\mu}(x - \hat{x})$. Compared to PWM, POME, MML and MOM have consistently a positive bias particularly for larger F. The MLS gives poor estimates and reconfirms the observation given in Table 1. Between POME and MOM, it is observed that as n and F increases, POME is bit more efficient than MOM. For example, when the sample size n is 40, F = 0.99 (100 year flood), the bias in the quantile using POME and MOM are 0.06 and 0.08 respectively, meaning thereby that POME has less bias than MOM. The MML behaves slightly more efficiently than POME.

Table 3 represents the variability $\tilde{\sigma}(\hat{x})$ of the quantile estimate obtained from different samples with respect to the 'true' value of x at different F. The comparative performance of different procedures follows the same pattern as was observed in $\tilde{\mu}(x)$ given in Table 1. For all sample sizes and for all E, POME consistently yields lesser $\tilde{\sigma}(\hat{x})$ than PWM. The effect becomes more pronounced as F increases. For example, comparing PWM and POME for n = 30 and F = 0.98 (50 year flood) the bias in $\tilde{\mu}(\hat{x})$ and $\tilde{\sigma}(\hat{x})$ are 0.007, 0.732 (PWM) and 0.08 and 0.667 (POME) respectively. However, PWM is a bit more efficient than MOM where σ is 0.790. It is evident that POME has lesser variance in the quantile estimation.

Table 4 represents $\tilde{\sigma}(\hat{x})$ for the correlated EV1 values and again POME & MML are bit more efficient than PWM, MOM, and MLS (in that order). Given the same n and F the $\tilde{\sigma}(\hat{x})$ is more in the correlated process than in the random one. For n = 30 and F = 0.98, using POME and PWM, the $\tilde{\sigma}(\hat{x})$ increases from 0.667 and 0.732 to 0.721 and 0.803, respectively.

CONCLUSION

The study highlights the principle of maximum entropy (POME) in estimating the parameters and quantiles of an EV1 distribution. Simulation studies are conducted to evaluate the comparative performance of POME with the existing methods. It is observed that for small sample size, the performance of POME is significantly better than the method of moments and the method of least squares and comparable to the method of probability weighted moments (PWM). The potentiality of POME increases as it can be used in estimating the quantiles of Pearson or log Pearson distributions which are not amenable to the PWM.

ACKNOWLEDGEMENT

The authors thank Dr. V.P. Singh, Professor of Civil Engineering, Louisiana State University, USA for useful discussions and critical comments.

Table 1 : Bias $\tilde{\mu}(\xi - \hat{\xi})$ in Parameter Estimation for $\rho = 0$ and $\rho = 0.3$

Sr. No.	Method	Sample size	$\rho = 0.0$				$\rho = 0.30$			
			$\tilde{\mu}(a-\hat{a})$	$\tilde{\sigma}(\hat{a})$	$\tilde{\mu}(b-\hat{b})$	$\tilde{\sigma}(b)$	$\tilde{\mu}(a-\hat{a})$	$\tilde{\sigma}(a)$	$\tilde{\mu}(b-\hat{b})$	$\tilde{\sigma}(b)$
1	MOM		0.03	0.22	-0.01	0.24	-0.03	0.23	0.08	0.38
	MML	20	0.03	0.18	-0.02	0.24	-0.11	0.20	0.09	0.38
	MLS		-0.20	0.26	0.09	0.23	-0.27	0.28	0.19	0.38
	PWM		0.00	0.20	0.00	0.23	-0.06	0.22	0.10	0.38
	POME		0.03	0.18	0.02	0.24	-0.09	0.20	0.11	0.38
2	MOM		0.02	0.18	-0.01	0.19	-0.05	0.20	0.09	0.31
	MML	30	0.03	0.13	-0.01	0.19	-0.15	0.16	0.11	0.31
	MLS		-0.19	0.22	0.09	0.18	-0.27	0.24	0.20	0.31
	PWM		0.00	0.16	0.00	0.18	-0.07	0.19	0.10	0.30
	POME		0.02	0.14	-0.01	0.19	-0.11	0.17	0.12	0.31
3	MOM		0.01	0.14	-0.01	0.15	-0.05	0.16	0.09	0.24
	MML	50	0.00	0.11	0.00	0.15	-0.15	0.12	0.12	0.25
	MLS		-0.18	0.17	0.00	0.15	-0.26	0.20	0.20	0.24
	PWM		0.00	0.13	0.00	0.15	-0.08	0.15	0.11	0.24
	POME		0.01	0.11	0.00	0.15	-0.13	0.13	0.14	0.24

Table 2 : Bias $\tilde{\mu}(x - \hat{x})$ in the Estimates of Quantiles F; $\rho = 0$

Sr. No.	Method	Sample size n	$\tilde{\mu}(x - \hat{x})$					
			F=0.50 x=0.37	0.90 2.25	0.95 2.97	0.98 3.90	0.99 4.60	0.995 5.29
(1)	(2)	(3)	(4)	(5)	(6)	(7)	(8)	(9)
1	MOM		0.008	0.05	0.07	0.10	0.12	
	MML	20	0.011	0.04	0.05	0.08	0.07	
	MLS		0.017	-0.36	-0.50	-0.69	-0.83	
	PWM		-0.002	0.00	0.00	0.01	0.01	
	POME		0.008	0.05	0.07	0.10	0.10	
2	MOM		-0.004	0.04	0.05	0.07	0.09	0.10
	MML	30	-0.002	0.05	0.06	0.09	0.11	0.12
	MLS		0.024	-0.33	-0.47	-0.65	-0.79	-0.90
	PWM		-0.001	0.00	0.00	0.01	0.01	0.01
	POME		0.003	0.03	0.06	0.08	0.08	0.10
3	MOM		-0.004	0.02	0.03	0.05	0.06	0.07
	MML	50	0.008	0.00	0.00	0.00	0.00	0.00
	MLS		0.020	-0.32	-0.44	-0.63	-0.76	-0.80
	PWM		0.002	0.00	0.00	0.01	0.01	0.01
	POME		-0.004	0.02	0.02	0.04	0.04	0.05

Table 3 : Standard Deviation $\tilde{\sigma}[\hat{x}]$ of the Quantiles Estimate F : $\rho = 0$

Sr. No. (1)	Method (2)	Sample size (3)	$\tilde{\sigma}[\hat{x}]$ F=0.50 (4)	0.90 (5)	0.95 (6)	0.98 (7)	0.99 (8)	0.995 (9)
1	MOM		0.27	0.60	0.75	0.95	1.10	
	MML	20	0.27	0.53	0.66	0.82	0.94	
	MLS		0.37	0.68	0.86	1.10	1.30	
	PWM		0.27	0.58	0.72	0.90	1.04	
	POME		0.27	0.54	0.66	0.82	0.95	
2	MOM		0.21	0.49	0.62	0.79	0.92	1.05
	MML	30	0.21	0.41	0.50	0.62	0.72	0.81
	MLS		0.21	0.55	0.71	0.91	1.06	1.22
	PWM		0.21	0.47	0.58	0.73	0.85	0.96
	POME		0.21	0.44	0.54	0.67	0.79	0.87
3	MOM		0.17	0.30	0.50	0.63	0.73	0.83
	MML	50	0.17	0.33	0.40	0.50	0.58	0.65
	MLS		0.17	0.44	0.56	0.71	0.84	0.95
	PWM		0.17	0.38	0.47	0.59	0.68	0.77
	POME		0.17	0.35	0.42	0.53	0.61	0.69

Table 4 : Standard Deviation $\tilde{\sigma}[\hat{x}]$ of the Quantile Estimate F : $\rho = 0.3$

Sr. No. (1)	Method (2)	Sample size (3)	$\tilde{\sigma}(\hat{x})$ F=0.50 (4)	0.90 (5)	0.95 (6)	0.98 (7)	0.99 (8)	0.995 (9)
1	MOM		0.39	0.65	0.79	0.98	1.13	
	MML	20	0.39	0.60	0.72	0.88	1.01	
	MLS		0.39	0.73	0.92	1.16	1.36	
	PWM		0.39	0.64	0.77	0.95	1.10	
	POME		0.40	0.59	0.71	0.87	1.00	
2	MOM		0.32	0.54	0.66	0.83	0.96	1.10
	MML	30	0.31	0.46	0.54	0.67	0.77	0.87
	MLS		0.36	0.61	0.77	0.98	1.14	1.31
	PWM		0.31	0.53	0.64	0.80	0.93	1.05
	POME		0.32	0.49	0.58	0.72	0.83	0.94
3	MOM		0.25	0.44	0.55	0.69	0.80	0.91
	MML		0.25	0.38	0.45	0.55	0.63	0.70
	MLS	50	0.25	0.50	0.63	0.81	0.95	1.08
	PWM		0.25	0.42	0.52	0.65	0.75	0.85
	POME		0.25	0.39	0.46	0.57	0.65	0.74

REFERENCES

1. Bobee, B. 'Sample of T-Year Events Computed of Bias in the Estimation of the Coefficient of Skewness', Water Resources Research, 1973, 9, pp. 1264-1270.

2. Greenwood, J.A., Landwehr, J.M., Metalas, N.C. and Wallis, J.R., 'Probability weighted Moments: Definition and Relation to Parameters of Several Distributions Expressable in Inverse Form', Water Resources Research, 1979, 15, pp. 1049-1054.

3. Hosking, J.R.M., Wallis, J.R. and Wood, E.F., 'The U.K. Flood Studies Report - Nine years on', Institute of Hydrology, Wallingford, U.K. (Private Communication), 1984.

4. Hydrology Committee of the Water Resources Council, 'Guidelines for Determining Flood Frequency', United States Water Resources Council, Washington, D.C., 1976, Bulletin 17.

5. Institution of Engineers, Australia, Australian Rainfall and Runoff, Flood Analysis and Design, 1977.

6. Jenkinson, A.F., 'Statistics of Extremes, Estimation of Maximum Flood', World Meteorological Organisation, Tech., 1969, 98, pp. 193-227.

7. Jowitt, P.W., 'The Extreme Value Type I Distribution and the Principle of Maximum Entropy', Journal of Hydrology, 1980, 42, pp. 23-38.

8. Landwehr, J.M., Metalas, N, N.C. and Wallis, J.R., 'Some Comparisons of Flood Statistics in Real and Log Space', Water Resources Research, 1978, 14, pp. 903-920.

9. Landwehr, J.M., Metalas, N.C. and Wallis, J.R., 'Probability Weighted Moments Compared with Some Traditional Techniques in Estimating Gumbel Parameters and Quantiles', Water Resources Research, 1979, 15, pp. 1055-1064.

10. Landwehr, J.M., Metalas, N.C. and Wallis, J.R., 'Quantile Estimation with More or Less Flood like Distribution', Water Resources Research, 1980, 16, pp. 547.

11. Lewis, P.A., Goodman, A.S. and Miller, J.M., 'A Psuedorandom Generation for the System/360', IBM System Journal 1969, 8, pp. 136-146.

12. Shannon, C.E., 'The Mathematical Theory of Communication', Bell System Technical Journal, 1948, 27, pp. 379-428.

13. Singh, V.P., 'Derivation of Some Frequency Distribution Using the Principle of Maximum Entropy (Private Communication), 1985.

14. Sonuga, J.O., 'Principle of Maximum Entropy in Hydrologic Frequency Analysis', Journal of Hydrology, 1972, 17, pp. 177-191.

15. Tribus, M., 'Rational, Descriptions, Decisions and Design', Pergamon Press, London, 1969, pp. 478.

16. Wallis, J.R. and Wood, E.F., 'Relative Accuracy of Log Pearson III Procedures', Journal of Hydraulic Engineering, American Society of Civil Engineers, 1985, 111, pp. 1043-1056.

ASSESSMENT OF USE OF AT-SITE AND REGIONAL FLOOD DATA FOR FLOOD FREQUENCY ESTIMATION

C. S. Hebson and C. Cunnane
Department of Engineering Hydrology
University College
Galway, IRELAND

ABSTRACT. The standard error of flood quantiles estimated by three different methods were obtained by simulation. The methods were based on (a) at-site data, (b) at-site and regional data combined and (c) regional data alone. Data were generated from GEV distributions. Since method (c) involves a regional regression relation an element of physical realism was preserved by basing the generating distributions on those estimated from real data of rivers in S.E. England and by using the corresponding catchment characteristic values in regional regression relations. The generated regional data set included 17 sites of 20 years and one site of variable record length n = 1, 3, 6, 10 years. Three sites (low, medium, high Cv) were investigated separately. The at-site/regional method is considerably better than either (a) or (c) especially at high return periods. For large T (c) can be better than (a) especially if n is small. In general, the mean annual flood obtained by regional regression is less precise than $\overline{Q}$ obtained from one year of record.

INTRODUCTION

In flood frequency analysis a flood magnitude Q_T of return period T is estimated from whatever suitable hydrological data are available. In the annual maximum series flood model Q_T is the quantile of exceedance probability 1/T in the probability distribution of annual maxima (AM). The data available may be

(a) at-site data (Q_1, Q_2, ...Q_N) alone
(b) at-site AM data plus AM data at a number of neighbouring sites, a total of M sites in all, giving

$$L = \sum_1^M N_j \qquad (1)$$

AM values, (Q_{ij}, i = 1, 2, ...N_j), j = 1, 2, ...M
(c) AM data at (M-1) neighbouring sites but no AM data at the site, j*, at which the flood estimate is required.

V. P. Singh (ed.), Hydrologic Frequency Modeling, 433–448.

FLOOD ESTIMATES

In case (a), a distributional form $f(q;\underline{\Theta})$ is assumed for Q, the parameters are estimated from the N values of Q and the quantile obtained from an expression such as

$$Q_T = \Theta_1 + \Theta_2 y_T(\Theta_3) \tag{2}$$

where Θ_1, Θ_2, and Θ_3 are location, scale and shape parameters respectively and y_T is a standardised variate value of return period T.

In case (b), the same procedure as in (a) might be adopted by ignoring the data at the (M-1) neighbouring sites and simply using the at-site data alone. However, recent research results suggest that the required flood quantile can be more reliably estimated by combining the site and regional information (Greis and Wood (1981), Kuczera (1982(a), (b), 1983), Lettenmaier and Potter (1985), Hosking et al. (1985a), Lettenmaier et al. (1985), Wallis and Wood (1985). If an index flood method is employed the mean annual flood $\overline{Q}$ is obtained from the at-site data and the parameters of a dimensionless distribution of $x_T = Q_T/\overline{Q}$ are obtained from the M data sets by a regional averaging procedure. The latter may take a number of forms but in this study only the index flood method based on regionally averaged PWM's (Wallis, 1980) is used. Such a procedure results in regional estimates of x_T from which the estimate

$$\hat{Q}_T = \overline{Q} \cdot \hat{x}_T \tag{3}$$

is obtained, where $\overline{Q}$ is the at-site mean.

In case (c), it is necessary to establish the dimensionless x_T-T relation from the data of the (M-1) neighbouring sites and to establish an estimate of $\overline{Q}$ from a regionally calibrated relation between $\overline{Q}$ and physically measurable catchment characteristics (Benson (1962), Nash and Shaw (1966), Thomas and Benson (1970), NERC (1975)). The form of such a relation adopted for use depends on the amount of physical and climatic data available as well as on the actual dependence between $\overline{Q}$ and the catchment characteristics. Examples are:

$$\overline{Q} = C_1 A^{0.85} \quad \text{(Benson, 1962; Cole, 1966)} \tag{4}$$

$$\overline{Q} = C_2 A^{0.85} R^{2.2} \quad \text{(Nash and Shaw, 1966)} \tag{5}$$

$$\overline{Q} = C_3 A^{0.99} S^{0.37} G^{0.98} R^{1.19}$$

$$\text{(NERC, 1975, Table I.4.13a)} \tag{6}$$

RELIABILITY OF Q_T ESTIMATES

An estimate $\hat{Q}_T$ obtained in a particular manner may differ from the true value of Q_T because of sampling effects and/or model error. The difference may be quantified in terms of bias (b), standard error (se) and root mean square error (rmse).

$$b = E[\hat{Q}_T - Q_T] \tag{7}$$

$$se = (E[\hat{Q}_T - E(\hat{Q}_T)]^2)^{0.5} \tag{8}$$

$$rmse = (E[\hat{Q}_T - Q_T]^2)^{0.5} \tag{9}$$

Site Data Available - Cases (a) and (b)

The above quantities have been obtained by many authors (see Cunnane, Table 1, 1987) for a variety of distributions commonly used for AM series analysis both for at-site estimation, method (a), and at-site/ regional estimation, method (b). Some of the studies referred to therein also examined the question of model error by considering that samples drawn from a particular distribution are used to estimate quantiles on the assumption of a different distribution. The general conclusion to be drawn from such studies is that at-site/regional methods give quantile estimates with smaller standard error than at-site methods, providing the region is reasonably homogeneous with respect to flood statistics (Lettenmaier et al. 1985; Lettenmaier and Potter 1985).

Each of the above last two mentioned studies assumes that at-site data are available to the estimation procedure. In this study, the case where no site data exist is examined and some results similar to those obtained by previous investigators are reproduced for comparison.

No Site Data Available - Case (c)

This is the least thoroughly investigated of the three cases. Central to all applications is the estimation of an index flood, such as Q, from catchment characteristics using equations of the form (4), (5) or (6). Based on the data of 57 catchments in Great Britain, Nash and Shaw (1966) concluded that a $\overline{Q}$ value obtained by an equation such as equation (4) has a standard error which is of the same order of magnitude as the standard deviation of the at-site population or equivalently an equation such as equation (4) is only as valuable as a sample of one annual maximum. A similar result was obtained by NERC (1975, I, p342) from a study based on 530 catchments. The above results are based on use of ordinary least squares (OLS) in the log-domain which, as pointed out by Stedinger and Tasker (1985), can give poor estimates of the regression relations' abilities. However, the general tenor of the above results are nevertheless true.

Thus if an index flood approach is used to estimate Q_T by equation (3), where $\overline{Q}$ is obtained from catchment characteristics, it follows that Q_T is liable to be subject to large sampling error. The simulation studies described by some authors for investigating case (b) can be designed and executed without recourse to physical catchment characteristic data and their results are capable of unambiguous interpretation. In order to study by simulation an estimator such as equation (3), with $\overline{Q}$ obtained from catchment characteristics, it is necessary to be able to generate samples from a theoretically defined population, whose relationship with catchment characteristics is similar to the relationship (such as equation (6)) found in nature between $\overline{Q}$ and the same catchment characteristics.

THE PRESENT STUDY

The major objective is to study the effectiveness of regionally-based estimators for use at a completely ungauged site. Such estimators require that $\overline{Q}$ be estimated from catchment characteristics as in equations (4), (5) or (6). In order to study such an estimator by simulation methods it is necessary to be able to specify theoretical populations, with known quantile values, whose means bear a relationship to catchment characteristics similar to the relations observed between actual $\overline{Q}$ values and actual catchment characteristics. In order to achieve this, M = 18 actual catchments from S.E. England were selected to provide the required realistic physical basis for the study. The catchment characteristics used in equation (6) as well as AM peak data of between N = 14 and N = 32 years duration were obtained for each catchment.

Simulation Strategy

Previous studies, based on the regional distribution of $x_T = Q_T/\overline{Q}$ (NERC, 1975, I(I.2.6)) had shown that EV1 and GEV distributions are acceptably good descriptors of AM flood distributions in England. Hence the GEV distribution was selected for the simulation experiements. It is defined (Jenkinson, 1969) by

$$\Pr(Q < q) = F(q) = \exp[-(1 - k(q-u)/a)^{1/k}] \ , \ k \neq 0$$

$$= \exp[-\exp(-(q-u)/a)] \ , \quad k = 0 \qquad (10)$$

Since arbitrarily chosen values of parameters (u_j, a_j, k_j, j=1,... M) might not yield random samples whose ($\overline{Q}_j$, j=1...M) values bear a relationship with actual catchment characteristics similar to what actually occurs in real data sets, it is necessary to constrain the parameter values in some way to achieve the required degree of realism. The natural way to achieve this is to estimate (u_j, a_j, k_j) from the data at site j for each j=1...M. PWM estimation (Hosking et al., 1985b) was adopted for this purpose. However, at-site estimators of the parameters a and k are extremely variable. A range of k values ($-0.316 < k < 0.483$) results which is so large as to be unrealistic.

An alternative, sensible strategy is to use a common k value (which implies a common skew coefficient Cs) for all generating populations. This was taken to be the k value of the GEV index flood distribution fitted by regional PWM's (Wallis, 1980). The (u_i, a_i) values at each site are the at-site PWM estimates, subject to this common value of k.

A further restriction can be imposed on the generating populations by forcing a common value of coefficient of variation Cv on all sites. It can be shown that the GEV Cv is a function of two quantities, the ratio (u/a) and shape parameter k. Therefore, uniform Cv requires constant (u/a) and k over all sites. GEV simulation in conjunction with regional $\overline{Q}$ estimation requires a certain internal consistency between the two processes. This can be preserved by estimating k from sample Cs, (u/a) from k and sample Cv, and then estimating the location parameter, a, from sample $\overline{Q}$, (u/a), and k.

<u>Effect of Measurement Error</u>: It is recognized that all discharge values

are subject to flow measurement error, the effect of which in flood frequency estimates has been analyzed by Potter and Walker (1985). As in most simulation studies (e.g. Wallis and Wood, 1985) the effect of measurement error is not explicitly included in the generation of random samples but its effect is nevertheless implicitly included in the relation between $\overline{Q}$ and catchment characteristics (method (c)).

Data Values and Population Parameters

The regional summary statistics for the M = 18 sites used to provide the background for the simulation study are given in Table 1. Table 2 lists the corresponding catchment characteristics and coefficients of the regression equation

$$\overline{Q} = C\, A^a\, S^s\, G^g\, R^r \tag{11}$$

where C is a constant, A is catchment area (km^2), S is mainstream 10 - 85% slope, G is a soil index, and R is annual rainfall (mm). The catchment characteristic values in Table 2 were taken from Flood Studies Report, NERC (1975) while the regression equation was calculated for this application. The independent variables in the regression were chosen somewhat arbitrarily, as it is not the goal here to produce the best possible regression relation. Rather, the regression is only used to obtain a realistic relationship between the mean annual flood and catchment characteristics to use in the simulation study. Table 3 gives the individual at-site GEV/PWM estimates. For the purposes of data generation, three separate population groups, I, II and III were used in separate experiments as outlined in Table 4. Population group III corresponds to the most homogeneous region while group I is wildly (and indeed unrealistically) heterogeneous.

Experimental Procedure

The same experimental design was used for data generated from population groups I, II, and III as follows.

1. Generate M = 18 random samples of size $N_j = 20$, $j = 1,2...M$, $j \neq j*$, and $N_{j*} = n$, where j* is a selected site of immediate interest. Initially n = 1.
2. Calculate sample means

$$\overline{Q}_j = \sum_1^{N_j} Q_{ij}/N_j \ , \ j=1 \ldots M$$

3. Estimate quantiles of T = 10-, 100-, and 1000- year return periods at site j* by

   (a) at-site GEV/PWM estimation as in Hosking et al. (1985b)
   (b) estimation of x_T by regional GEV/PWM (Hosking et al. 1985a) index flood method, and then

$$(\hat{Q}_T)_{j*} = \overline{Q_{j*}} \cdot \hat{x}_T$$

   (c) (i) calibration of the fixed form relation

$$\bar{Q} = C \cdot A^a S^s G^g R^r$$

by ordinary least squares in the (natural) log domain using $\bar{Q}$ as the dependent variable and the catchment characteristic values of Table 2 as independent variables, catchment j* excluded i.e. data for (M-1) sites only were used.

(ii) obtain the regional regression model estimate of $\bar{Q}$ at the required site j*,

$$\hat{\bar{Q}}_{j*} = C \cdot A^{\hat{a}} S^{\hat{s}} G^{\hat{g}} R^{\hat{r}} \cdot \exp(z)$$

where z is an $N(0,s_e^2)$ variate where s_e^2 is the regression model variance. The noise term z is added in every simulation because individual values of Q, rather than mean values, are being simulated. Calculate

$$(\hat{Q}_T)_{j*} = \hat{\bar{Q}}_{j*} \cdot \hat{x}_T$$

where x_T is obtained in step 3(b) but excluding the data at site j* i.e. from M-1 data sites.

Exclusion of $\bar{Q}_{j*}$ from the calibration of 3c(i) and (Q_{ij*}, $i=1,2...N_{j*}$) from the estimation of x_T makes the selected site j* completely independent of the observed regional data as would be the case for a completely ungauged catchment.

4. Store (($\hat{Q}_T)_{j*}$, $T = T_1, T_2, T_3$) results by methods (a), (b), and (c) for subsequent calculation of statistical properties of these estimates.
5. Repeat steps 1 - 4, 1000 times.
6. Calculate statistical properties of the estimators.
7. Repeat 1-6 using different values of N_{j*} (n = 3, 6, 10).
8. Repeat steps 1 - 7 for different sites j*. In all, three sites were investigated (j* = 39006, 32004, and 32003), which have respectively low, medium, and high Cv values in population group II.

RESULTS

Quantile Estimates

Standard errors of quantile estimates are presented in graphical form in Figures 1-2. There are nine such possible figures, representing three generating population groups and three chosen sites, but only a selection are shown here, the omitted ones being qualitatively similar. Other investigators (e.g. Hosking et al. 1985a) have used graphs of upper 95% confidence interval ($=b+1.96se(\hat{Q}_T)$) against T as a basis of comparison on the assumption that quantile estimates are normally distributed. However, bias in the quantile estimates is not reported here since the comparisons between the three estimators are affected by the nature of the regional regression model estimate of $\bar{Q}$. The population mean at site j* may differ systematically from $E(\bar{Q}:A,R,S,G)$ and hence in-built bias may occur in the regional estimator of $\bar{Q}$ and hence in $\hat{Q}_T$. Such bias would not be comparable with quantile bias in the other estimators. Further, se($\hat{Q}_T$) rather than 1.96se($\hat{Q}_T$) is reported because the sampling distributions of the quantile estimates display varying degrees of non-normality and the multiplier 1.96 has no particular significance. Table 5 shows the skewness and kurtosis of sampling distributions of quantile estimates as evidence of this.

Figures 1-2 show:

(i) the regional estimator (c) has considerably larger se than the at-site/regional estimator (b) at all return periods and also has considerably larger se than the at-site estimator (a) for low return periods.

(ii) the at-site estimator (a) has larger se than the at-site/regional estimator (b). The difference is very large at high T but tends to converge to zero at low T.

(iii) the se of the at-site estimator (a) increases more rapidly with T than for either of the other two methods, while se of the at-site/regional estimator increases least rapidly.

(iv) as a consequence of (iii) the at-site graph (a) intersects the regional graph (c) at some large T and this intersection value of T decreases as the at-site generating population Cv increases.

(v) as sample size decreases, se of the at-site (a) and at-site/regional (b) estimators slowly approach se of the regional (c) estimator.

(vi) as at-site sample size approaches unity the regional estimator se is still greater than se of the other two. This is consistent with results of Table 6 discussed later.

$\bar{Q}$ Estimates

When at-site data are available (cases (a) and (b)) then

$$se(\bar{Q}) = s/N^{0.5} \tag{12}$$

where s is the generating population standard deviation and N is the at-site record length. In the absence of site data $\bar{Q}$ is obtained from eq. (11) and se($\bar{Q}$) is obtainable from the simulation experiments. Knowing s, an implied record length N′ can be obtained by inverting eq. (12) to give

$$N' = s^2/se(\bar{Q})^2 \tag{13}$$

The resulting values of N′ are given in Table 6 and vary from 0.1 years to 0.6 years, indicating that the regional estimate of $\bar{Q}$ has less worth than the use of a single annual maximum as an estimate of $\bar{Q}$. This result is in keeping with those obtained by Nash and Shaw (1966) and by NERC (1975, I, p342) referred to earlier. In the latter studies the results were valid for a catchment whose characteristics values lie at the centroid of the catchment characteristics space of those catchments included in the study.

CONCLUSIONS

The goal of this research was to investigate the suitability of different mean flood estimators for quantile estimation from regional

Table 1: Regional Summary Statistics (Flows in m^3/s).

Station	N	$\bar{Q}$	s	C_v	C_s	C_k
32002	31	5.32	3.43	0.64	1.43	2.20
32003	30	7.13	4.48	0.63	0.90	0.26
32004	25	15.65	6.73	0.43	0.50	-0.45
32006	29	13.82	10.03	0.73	4.16	18.16
32007	29	16.54	7.07	0.43	0.06	-0.90
32008	23	10.63	6.94	0.65	1.26	1.16
32010	30	70.53	44.80	0.64	2.80	8.29
33005	19	21.93	4.78	0.22	-0.80	0.11
33009	18	98.06	37.31	0.38	0.80	-0.30
37001	20	23.34	8.09	0.35	0.30	-0.94
33017	14	102.51	24.89	0.24	-0.20	-0.96
38002	29	6.45	3.44	0.53	0.97	3.31
38007	19	6.27	3.13	0.50	0.91	0.04
39004	31	2.62	0.85	0.32	1.98	4.23
39006	19	13.18	3.52	0.27	0.85	1.54
39821	31	22.00	8.67	0.39	0.29	-0.66
53001	32	84.78	41.18	0.49	0.92	0.64
53003	30	159.17	73.67	0.46	1.24	0.35

Table 2: Catchment Characteristics

Station	R	A	S	G
32002	602.00	89.60	2.87	0.44
32003	617.00	74.30	3.79	0.45
32004	648.00	194.00	2.15	0.34
32006	668.00	223.00	2.35	0.45
32007	655.00	233.00	4.18	0.41
32008	673.00	107.00	4.38	0.45
32010	628.00	1530.00	0.69	0.39
33005	673.00	389.00	1.79	0.34
33009	655.00	1320.00	0.82	0.36
37001	635.00	303.00	1.22	0.41
33017	620.00	2870.00	0.53	0.35
38002	645.00	78.70	3.05	0.40
38007	640.00	21.40	7.47	0.30
39004	800.00	122.00	4.36	0.30
39006	785.00	365.00	1.90	0.21
39821	678.00	118.00	3.49	0.45
53001	815.00	666.00	0.80	0.35
53003	861.00	1600.00	0.70	0.32

Estimated Parameters of $\overline{Q}$ Relation (eqn 11)

C	r	a	s	g
0.00671	1.0728	0.4152	0.6871	0.8556

Table 3: Individual At-Site GEV/PWM Estimates and Implied Moments

Station	$\hat{k}$	$\hat{\alpha}$	$\hat{u}$	μ	C_v	C_s
32002	-0.109	2.319	3.703	5.320	0.661	2.003
32003	-0.064	3.365	4.963	7.130	0.663	1.588
32004	0.065	5.875	12.613	15.650	0.446	0.788
32006	-0.192	3.898	10.664	13.818	0.506	3.332
32007	0.223	7.063	13.764	16.540	0.442	0.176
32008	-0.144	4.589	7.229	10.631	0.698	2.435
32010	-0.316	18.919	51.127	70.529	0.691	25.279
33005	0.483	5.563	20.611	21.926	0.236	-0.586
33009	-0.158	24.899	79.111	98.064	0.423	2.661
37001	0.084	7.492	19.593	23.342	0.373	0.707
33017	0.280	27.865	92.625	102.505	0.272	-0.008
38002	0.223	3.054	5.248	6.448	0.490	0.175
38007	-0.115	2.251	4.682	6.267	0.550	2.065
39004	-0.243	0.467	2.208	2.623	0.363	5.175
39006	0.069	3.091	11.593	13.177	0.277	0.767
39821	0.137	8.114	18.297	22.002	0.408	0.489
53001	-0.028	31.918	65.444	84.776	0.501	1.343
53003	-0.251	41.860	121.325	159.170	0.550	5.698

Table 4: Method of Specifying Parameters of GEV Generating Population Groups I, II, and III.

Population Group	Method of Fixing Parameters for Generating Populations
I	(u_j, α_j, k_j) values as in Table 3.
II	$k_j = k$, all j, where k is value estimated regionally by PWM. Then, (u_j, α_j) obtained from at-site data by GEV/PWM estimator subject to fixed k.
III	$k_j = k$ and $C_{v,j} = C_v$, all j, where k and C_v are estimated regionally by PWM. Then u_j obtained from at-site data by GEV/PWM estimator subject to fixed C_v and k.

Table 5: Skewness and kurtosis (in italics) of sampling distributions of $\bar{Q}$ and quantile estimators. Each value calculated from 1000 estimate values, generating population III, site number 32003, at-site sample size = 10.

Estimator \ Statistic	$\bar{Q}$	$\hat{Q}_{10}$	$\hat{Q}_{100}$	$\hat{Q}_{1000}$
(a) At-site	0.29 *0.09*	0.53 *0.25*	1.36 *2.64*	2.27 *7.26*
(b) At-site/ Regional	0.29 *0.09*	0.29 *0.04*	0.34 *-0.01*	0.42 *-0.01*
(c) Regional	1.93 *6.04*	1.92 *5.99*	2.00 *6.67*	2.12 *7.60*

Table 6 : Worth of regional estimate of $\bar{Q}$ expressed as implied equivalent record length N' years. Here σ is standard deviation of generating population and Se($\hat{Q}$) is obtained from simulation results.

Station \ Population	I			II			III		
	σ	se($\hat{\bar{Q}}$)	N'	σ	se($\hat{\bar{Q}}$)	N'	σ	se($\hat{\bar{Q}}$)	N'
32003 Original Sample $C_v = 0.63$	4.73	5.96	0.63	4.68	9.45	0.25	3.36	5.83	0.33
32004 Original Sample $C_v = 0.43$	7.29	10.81	0.45	7.25	10.71	0.46	7.37	10.70	0.47
39006 Original Sample $C_v = 0.27$	3.65	13.47	0.08	3.81	13 50	0.08	6.21	13.75	0.20

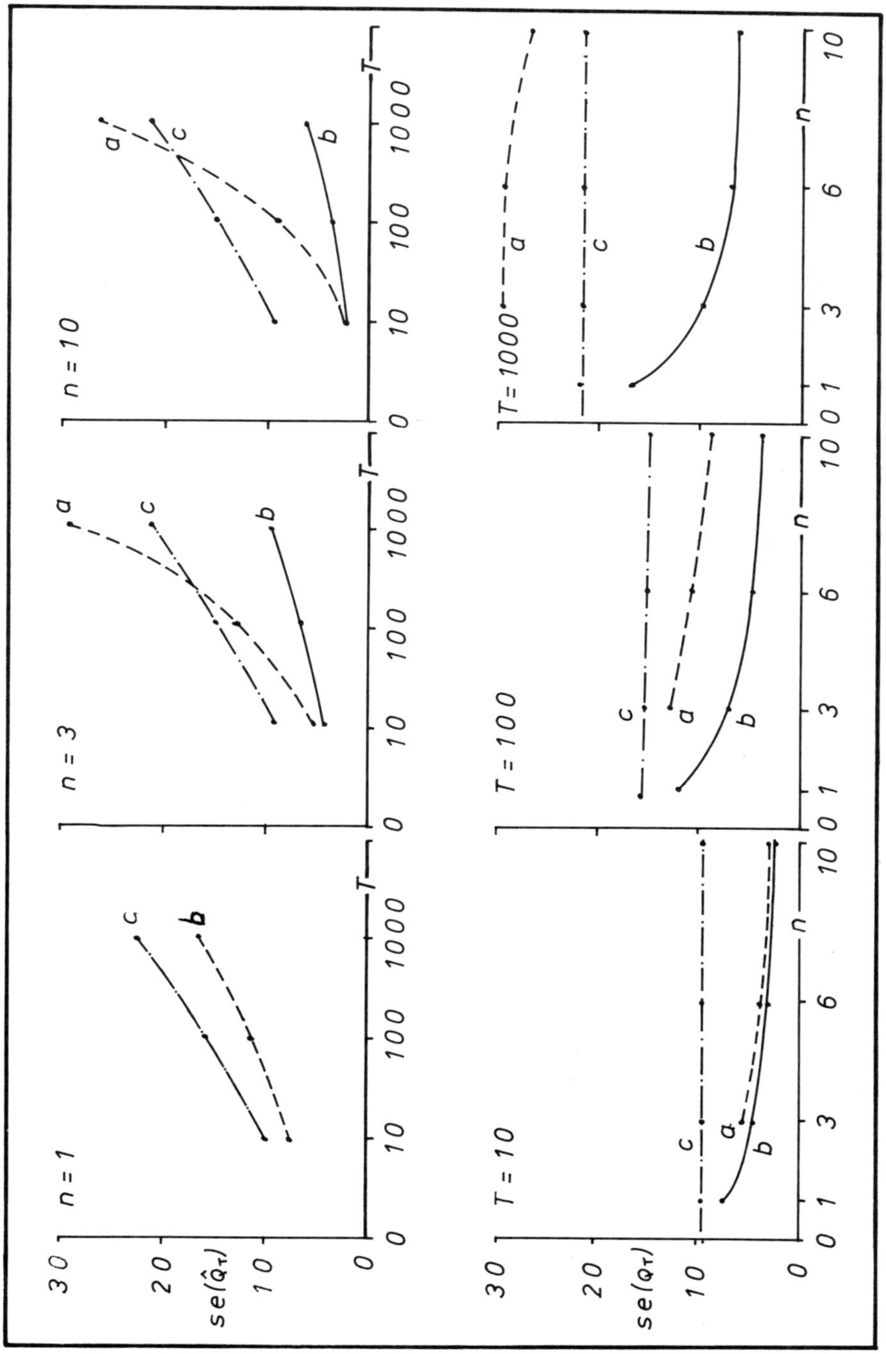

Figure 1: Dependence of standard error of quantile estimate on return period T, at-site sample site n, and estimator for population II, site 32003. True quantiles: Q_{10} = 13.11, Q_{100} = 22.54, Q_{1000} = 32.83.

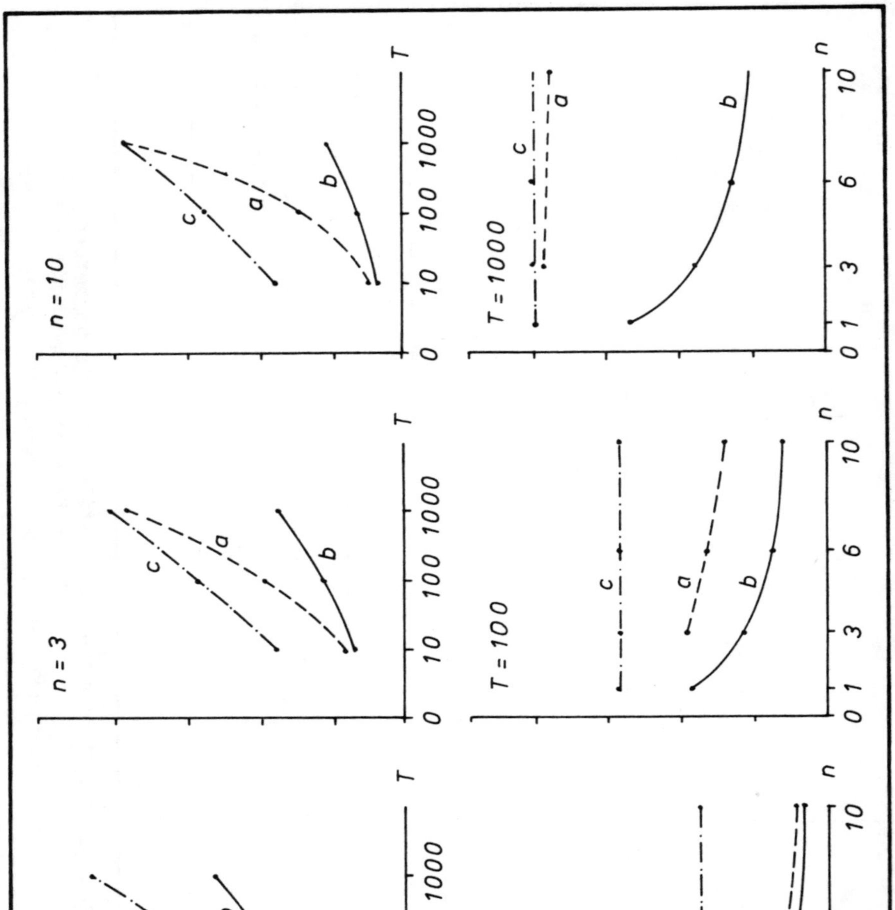

Figure 2: Dependence of standard error of quantile estimate on return period T, at-site sample size n, and estimator for population II, site 32004. True quantiles: Q_{10} = 24.93, Q_{100} = 39.54, Q_{1000} = 55.50.

frequency index flood methods. The index flood method considered here was the regionally averaged PWM technique (Wallis, 1980). Mean floods were estimated from simulated at-site data of varying record lengths (1,3,6,10 years) and from a regression model relating the mean to catchment characteristics. Three different stations (high, medium, and low Cv) and three different regions (uniform Cs and Cv, uniform Cs, and variable Cs and Cv) were considered, making for a total of nine cases. Considering the two regions of fixed uniformity, the regional/at-site quantile estimates were always superior to the purely regional estimates in terms of standard error. Thus, for the regional PWM technique even one year of data is to be preferred over a mean estimated from regional catchment characteristics. Of secondary interest here is the performance of at-site estimates. However, it should be noted that the superiority of regional/at-site estimates that was reported by previous investigators (e.g. Lettenmaier et al. 1985) was confirmed in this study. Furthermore, at-site estimates based on just a few years record were also superior to the purely regional estimates. Of course, for those very short record lengths the regional/at-site estimates would be selected for practical use. These findings are consistent with previous investigations into the worth of regional data. Nash and Shaw (1966) suggested the inadequacy of regression mean estimates for use with a regional EV1 index flood method, a finding corroborated by NERC (1975). This study extends this conclusion to the PWM/GEV treatment of regional frequency analysis.

ACKNOWLEDGEMENTS C.S. Hebson was supported by the Irish Department of Education Engineering Post-Doctoral Fellowship Program during the execution of this work.

REFERENCES

Benson, M.A., 1962. 'Factors influencing the occurrence of floods in a humid region of diverse terrain', Water Supply Paper No. 1580, 64pp., U.S. Geological Survey.

Cole, G., 1966. 'An application of the regional analysis of flood flows', in Symposium on River Flood Hydrology, 39-57, Institution of Civil Engineers, London.

Cunnane, C., 1987. 'Review of statistical models for flood frequency estimation', in International Symposium on Flood Frequency and Risk Analysis (V.P. Singh, ed.), Reidel.

Greis, N.P., and E.F. Wood, 1981. 'Regional flood frequency estimation and network design', Water Resources Research, **17**(4), 1167-1177.

Hosking, J.R.M., J.R. Wallis, and E.F. Wood, 1985a. 'An appraisal of the regional flood frequency procedure in the UK Flood Studies Report', Hydrological Sciences Journal, **30**(1), 85-109.

Hosking, J.R.M, J.R. Wallis, and E.F. Wood, 1985b.'Estimation of the generalized extreme value distribution by the method of probability weighted moments', Technometrics, **27**(3), 251-261.

Jenkinson, A.F., 1969. 'Statistics of extremes', in 'Estimation of Maximum Floods', WMO Technical Note No. 98, 183-228.

Kuczera, G., 1982a. 'Combining site-specific and regional information: an empirical Bayes approach', Water Resources Research, **18**(2), 306-314.

Kuczera, G., 1982b. 'Robust flood frequency models', Water Resources Research, **18**(2), 315-324.

Kuczera, G. 1983. 'Effect of sampling uncertainty and spatial correlation on an empirical Bayes procedure for combining site and regional information', Journal of Hydrology, **65**(4), 373-398.

Lettenmaier, D.P. and K.W. Potter, 1985. 'Testing flood frequency estimation methods using a regional flood generation model', Water Resources Research, **21**(3), 1903-1914.

Lettenmaier, D.P., J.R. Wallis, and E.F. Wood, 1985. 'Note on the comparative robustness of extreme flood quantities', AGU Spring Meeting - Baltimore, American Geophysical Union, Washington, DC.

Nash, J.E. and B.L. Shaw, 1966. 'Flood frequency as a function of catchment characteristics', in Symposium on River Flood Hydrology, 115-136, Institution of Civil Engineers, London.

NERC, 1975. Flood Studies Report, **1**:313-349, **4**, Natural Environmental Resources Council, London.

Potter, K.W., and J.F. Walker, 1985. 'An empirical study of flood measurement error, Water Resources Research, **21**(3), 403-410.

Stedinger, J.R., and G.D. Tasker, 1985. 'Regional hydrological analysis: 1. Ordinary, weighted, and generalized least squares compared.', Water Resources Research, **21**(9), 1421-1432.

Thomas, D.M., and M.A. Benson, 1970. 'Generalization of streamflow characteristics from drainage-basin characteristics', Water Supply Paper 1975, U.S. Geological Survey.

Wallis, J.R., 1980. 'Risk and uncertainties in the evaluation of flood events for the design of hydrologic structures', keynote address at Seminar on Extreme Hydrological Events - Floods and Droughts, Erice, Italy, 33pp

Wallis, J.R., and E.F. Wood, 1985. 'Relative accuracy of Log Pearson III procedures', ASCE Journal of Hydraulics Division, **111**(7), 1043-1056.

AN EMPIRICAL STUDY OF PROBABILITY DISTRIBUTIONS OF ANNUAL MAXIMUM FLOODS

A. Ramachandra Rao
School of Civil Engineering
Purdue University
West Lafayette, IN 47907 USA

P. S. Arora
URS Engineers, Dallas
Dallas, TX USA

ABSTRACT. The Wakeby distribution has been proposed for flood frequency analysis because of its attractive properties. Algorithms have been developed and tested to estimate the parameters of the Wakeby distribution from observed data.

Considerable theoretical and developmental work has been conducted on the Wakeby distribution. Besides, its parameters can be easily estimated. However, not many studies have been reported in the literature about fitting the Wakeby distribution to empirical data and comparing the results to those obtained from other well known distributions such as the extreme value or the log person type III distribution. Such studies are important in establishing the merits of Wakeby distribution in the analysis of observed data.

In view of these considerations, annual maximum data from twenty-five watersheds have been analyzed by using the log pearson type III, mixture of extreme values and Wakeby distributions and the results are discussed in this paper. The data are from watersheds in the United States and Europe and cover a variety of climatic conditions.

The parameters of the distributions are estimated by the maximum likelihood method or the method of moments and the fitted distributions are compared to the histograms of the observed data by using the goodness of fit tests. In several instances, the log pearson type III distribution estimates failed to converge. In other instances the fit of the mixture distribution to histograms of observed data were much poorer than those of log pearson type III or the Wakeby distribution. The performance of the Wakeby distribution was in general as good as that of the log pearson type III distribution and in some cases superior.

In view of these considerations, it is concluded that the Wakeby distribution should be strongly considered for flood frequency analysis, especially in view of the fact that it is so easy to use.

V. P. Singh (ed.), Hydrologic Frequency Modeling, 449–466.

1. INTRODUCTION

Methods of flood frequency analysis which started about 1914 in the U.S. have developed along several different lines. This has resulted in nonuniformity in methods of analysis and thereby in results. The Water Resources Council recommended the use of log Pearson type III distribution for flood frequency analysis for the federal agencies. In recommending this distribution, however, the Water Resources Council noted that the range of uncertainty in flood frequency analysis was still quite large regardless of the method used. Thus, they recognized a need for continued research and development in this field. New methods of flood frequency analysis have been proposed in this context. Recently proposed mixed double exponential and the Wakeby distributions for flood frequency analysis, along with the log Pearson type III distribution are investigated in this study. The first objective of the research reported herein is to compare the performance of log Pearson III, mixed double exponential and Wakeby distributions to predict floods of different frequencies. The second objective is to evaluate the computational difficulties in these methods. In order to investigate these aspects, the above three distributions are fitted to 25 sets of homogeneous and non-homogeneous annual maximum flood data from rivers around the world. The goodness of fit of the three distributions to the observed data is compared by using Chi-square and Psi tests to determine the most suitable distribution.

This paper consists of six sections. The data used in the study and their statistical properties are investigated in section II. Log Pearson type III distribution and goodness of fit tests, mixed double exponential and Wakeby distributions are discussed in sections III to V. The results are discussed and a set of conclusions are presented in section VI.

2. DATA USED IN STUDY AND STATISTICS

Data from a worldwide network of monitoring stations with long periods of record are used in this study. These data are derived from the study by Bobeé and Robitaille (1977) and include annual maximum flood data from 25 rivers. Bobeé and Robitaille classified the data as being either homogeneous or non-homogeneous. The same classification scheme is used in this study (Tables 1 and 2). Sample statistics in real and log space are estimated for the 25 data sets and are given in Tables 3 and 4. In Tables 3 and 4, the symbols μ_1', σ and C_s respectively represent the sample mean, standard deviation and coefficient of skew. The range included in Tables 3 and 4 is the difference between the largest and smallest values of the annual maximum flood data. This data set is varied enough to give a good representation of the efficiency of the three distributions to characterize them. Further details about the data and results are found in Arora (1985) or Arora and Rao (1985).

Table 1. Sources and Details of Homogeneous Data

Country	River	Location	Drainage Area, Km^2	Record	Missing Years	Years of Record
Canada	Hurricana	Amos	3,680	1915 - 1969	1932, 1933	53
Czech.	Elbe	Decin	51,104	1851 - 1968	1857, 1863, 1866, 1873 1874, 1879, 1884, 1898 1918, 1921	108
Norway	Cloma	Langnes	40,170	1902 - 1968	1984	66
Canada	Assiniboine	Headingley	162,000	1914 - 1970		57
Germany	Danube	Hofkirchen	47,495	1901 - 1968		68
Canada	Fraser	Hope	203,000	1912 - 1970		59
India	Krishna	Vijayawada	251,355	1901 - 1960		60
Canada	S. Saskatchewan	Medicine Hat	58,400	1913 - 1970		58
USSR	Neman	Smalininkai	81,200	1812 - 1969	1944, 1945, 1946 1922-1926, 1931, 1934 1939-1942, 1949, 1958 1961, 1962, 1964, 1965 1967	155
Canada	N. Saskatchewan	Prince Albert	119,500			59
Canada	S. Saskatchewan	Saskatoon	139,500	1912-1970		59
Mali	Senegal	Bakel	218,000	1903 - 1966		64
Canada	Winnipeg	Slave Falls	126,000	1908 - 1970	1909, 1911-1912, 1917	50
USA	Susquehana	Harrisburg, PA	62,400	1891 - 1967	1906, 1922, 1927 1935, 1938, 1951	70

Source — *" Use of extreme value theory in estimating flood peaks from mixed populations," Canfield et al. (1980).*

Table 2. Sources and Details of Non-Homogeneous Data

Country	River	Location	Drainage Area, Km^2	Record	Missing Years	Years of Record
Canada	Assiniboine	Brandon	92,000	1902-1970		65
Finland	Kymijoki	Pernoo	36,535	1900-1968		69
USSR	Neva	Novosaratovka	281,000	1859-1969	1942	90
Mali	Niger	Dire	340,000	1924-1968		43
Mali	Niger	Koulikoro	120,000	1907-1968		62
Poland	Oder	Gozdowice	109,365	1901-1968	1945	67
USA	Penobscot	W. Enfield	17,090	1902-1967	1913, 1928, 1944 1951, 1960, 1964	60
Canada	Red	Emerson	104,000	1913-1970		58
Canada	Saguenay	Isle-Maligne	73,000	1913-1970		58
Sweden	Vanerngota	Vanesborg	46,830	1807-1968		162
Finland	Vuoksi	Imatra	61,280	1847-1968		122

Source — *" Use of Extreme value theory in estimating flood peaks from mixed populations," Canfield et al. (1980).*

Table 3. Sample Statistics in Real Space

River	μ_1' (m^3/sec)	σ (m^3/sec)	γ_y	Range (m^3/sec)
Assiniboine	188.51	137.45	1.532	629
Assiniboine	277.12	151.36	0.634	567
Danube	1821.70	573.74	0.920	2933
Elbe	1912.59	855.57	1.081	4279
Fraser	8830.17	1676.74	0.837	10070
Gloma	2094.50	497.13	0.670	2386
Hurricana	188.04	50.21	0.737	238
Krishna	14615.00	4659.72	1.368	22578
Kymijoki	437.81	117.31	-0.427	520
N. Saskatchewan	1344.90	769.53	2.643	4813
Neman	2656.13	1141.88	1.271	5990
Neva	3325.64	476.90	0.288	2600
Niger	2347.69	197.61	-0.052	730
Niger	6284.18	1230.50	0.569	6054
Oder	1503.70	630.84	1.267	3013
Penobscot	1813.80	688.17	1.139	3509
Red	643.86	460.12	1.931	2549
S. Saskatchewan	1264.60	796.28	1.569	3850
S. Saskatchewan	1484.30	767.56	1.239	3541
Saquenay	4936.55	1467.46	0.757	6890
Senegal	4843.75	1740.59	0.492	8900
Susquehana	8396.60	2926.99	1.634	17150
Vanerngota	624.68	117.76	0.313	680
Vuoksi	709.78	146.55	0.485	837
Winnipeg	1513.88	560.50	0.610	2134

Table 4. Sample Statistics in Log Space

River	μ_y (log) m^3/sec	σ_y (log) m^3/sec	C_y (log)	Range (log) m^3/sec
Assiniboine	4.994	0.729	-0.328	3.387
Assiniboine	5.451	0.636	-0.763	2.550
Danube	7.461	0.305	0.324	1.410
Elbe	7.462	0.439	-0.093	2.184
Fraser	9.068	0.187	-0.057	1.086
Gloma	7.619	0.237	-0.019	1.119
Hurricana	5.202	0.266	-0.405	1.225
Krishna	9.546	0.287	0.054	2.421
Kymijoki	6.038	0.319	-1.148	1.562
N. Saskatchewan	7.087	0.463	0.572	2.387
Neman	7.799	0.412	0.010	2.128
Neva	8.099	0.145	-0.208	0.833
Niger	7.758	0.083	-0.265	0.318
Niger	8.727	0.197	-0.10	0.978
Oder	7.238	0.396	0.701	1.660
Penobscot	7.437	0.363	0.206	1.663
Red	6.239	0.698	-0.195	3.094
S. Saskatchewan	6.972	0.587	0.045	2.875
S. Sakatchewan	7.182	0.489	0.109	2.290
Saquenay	8.461	0.030	-0.215	1.363
Senegal	8.410	0.401	-1.030	2.257
Susquehana	8.983	0.316	0.294	1.696
Vanerngota	6.419	0.192	-0.30	1.074
Vuoksi	6.543	0.214	-0.606	1.257
Winnipeg	7.254	0.378	-0.10	1.436

3. LOG PEARSON TYPE III (LP3) DISTRIBUTION

The LP3 distribution is derived from one of the Pearson distributions. The probability density function of the Pearson type III is of the form

$$p(X) = \frac{1}{\alpha\Gamma(\beta)} \left[\frac{x-\gamma}{\alpha}\right]^{\beta-1} e^{-[(x-\gamma)/\alpha]} \tag{3.1}$$

where α, β and γ are the scale, shape and location parameters respectively. From the sample mean, standard deviation and coefficient of skewness, the parameters α, β and γ can be determined. The parameter β is always positive as it is equal to the product of the square of sample standard deviation (σ) and α. When $\alpha > 0$, the distribution is positively skewed and $\gamma \leq x \leq \infty$. When $\alpha < 0$, the distribution is negatively skewed and $-\infty \leq x \leq \gamma$. When $\gamma = 0$, a Gamma distribution is obtained.

If the natural logarithms, ln x, of a variable x are distributed as a Pearson type III variate, then the variable X is distributed as a LP3 distribution with the probability density function

$$p(X) = \frac{1}{\alpha x\Gamma(\beta)} \left[\frac{\ln x-\gamma}{\alpha}\right]^{\beta-1} e^{-[(\ln x-\gamma)/\alpha]} \tag{3.2}$$

Bobeé (1975) conducted a study of the statistical properties of the LP3 distribution. Following Bobeé, the Pearson type III distribution is expressed as

$$f(y) = \frac{|\alpha|}{\Gamma(\lambda)} [\alpha(y-m)]^{\lambda-1} e^{-\alpha(y-m)]} \tag{3.3}$$

where α, λ and m are parameters of the distribution. It should be noted that α, λ and m in eq. 3.3 corresponds to $1/\alpha$, β and γ respectively in eq. 3.1.

The variate x has thus a LP3 distribution if $y = \log_a x$ has a Pearson type III distribution. Bobee considered logarithmic transformation to base a as a general case. The density function g(x) of the LP3 distribution derived from eq. 3.3 is as follows,

$$g(x) = \frac{|\alpha|}{\Gamma(\beta)} e^{-\alpha(\log_a x - m)} [\alpha(\log_a x - m)]^{\lambda-1} \frac{k}{x} \tag{3.4}$$

where k is a constant dependent on base a. α, λ and m are parameters of the LP3 distribution. It is assumed that $a > 1$ and therefore $k > 0$, so that

$$k = \log_a e = (\ln a)^{-1}$$

The density function of the LP3 distribution may also take the form

$$g(x) = \frac{|\alpha|\ k\ e^{\alpha m}}{\Gamma(\lambda)x^{1+\alpha k}} \left[\alpha(k \ln x - m)\right]^{\lambda-1} \tag{3.5}$$

The parameters of the log Pearson type III distribution are usually estimated by the method of moments or the maximum likelihood method. It is not always possible to obtain parameter estimates by the maximum likelihood method. Matalas and Wallis (1973) have noted that for very small values of sample skew a solution may not be possible for the LP3 distribution. Similarly, if $\beta < 1$, a maximum likelihood solution is not possible. Since $\beta = (2/\gamma)^2$, if β is greater than 1 then the sample coefficient of skew, γ, cannot exceed 2. Also if the sample coefficient of skew is negative, the Pearson type III distribution becomes bounded at the upper limit which is not appropriate for analysis of extreme events.

For fitting the LP3 distribution to annual maximum flood data, a program LP3, taken from Kite (1977) is used. For the maximum likelihood method included in the program LP3, two convergence criteria are used to determine whether the method gives an acceptable solution. The first convergence criterion is the number of iterations, which is set at 25. The second convergence criterion is the difference δ between the magnitudes of values of the distribution parameter, γ obtained in the nth and $(n-1)^{th}$ iterations. δ is set at 10^{-8} in the program LP3 for the present study.

3.1 Goodness of Fit Tests

In applying various distributions to a particular data set, the question of determining the "best" fit arises. To arrive at the best distribution, goodness of fit tests are used. The chi-square test and Kolomogorov-Smirnov tests are most frequently used for this purpose. In this study the chi-square test and the Psi tests (Tribus, 1969) are used. Details of chi-square test are available in standard texts and hence are not discussed here. The χ^2-test statistic is denoted by J in the following discussion.

The Psi test is similar in structure to chi-square test. The statistic Ψ for the Psi test is given by

$$\Psi = 10n \sum_{j=1}^{k} f_j \log_{10} \frac{f_j}{p_j} \tag{3.5}$$

where n is the sample size and f_j and p_j are the observed and expected number of observations in the jth class interval. Tribus (1970) has shown that 3.5 can be written as 3.6

$$\Psi = n \sum (p_j - f_j)^2/2p_j \tag{3.6}$$

The term on the right of eq. 3.6 is analogous to the chi-square statistic J. Thus, when the ratio f_j/p_j is nearly unity, there is a close correspondence between the chi-square and psi tests. However, when there is a significant difference between the observed and expected frequencies, the results differ significantly.

Only the results from goodness of fit tests are not sufficient for acceptance of a particular probability distribution. If the goodness of fit were the only criterion of acceptance, then higher order polynomials would often provide a much better fit than any of the standard distributions. However, the polynomial or other similar methods are not used because there is no hydrologic justification for them. The important criteria in selecting a distribution are that there be a sound theory describing the phenomenon and that the distribution should abstract the maximum information from the data via good estimation techniques. The U.S. Water Resources Council Work Group in realizing this noted that "no single method of testing the computed results against the original data is acceptable to all those on the Work Group, and the statistical consultants could not offer a mathematically rigorous method." The Work Group concluded that a probability distribution cannot be chosen solely on statistical grounds, and the situation is not different today. However, if a particular distribution is rejected on the basis of goodness of fit tests, then the chances of it being an appropriate distribution are slim.

3.2 Results

Table 5 includes the results of goodness of fit tests using the chi-square and Psi tests for LP3 distribution for the data used in the study. For the chi-square test, the two significance levels of 0.05 and 0.1 are used for testing the hypothesis.

For the LP3 distribution, the chi-square test results indicate that the distribution provides a good fit for 13 of the 25 data sets. For 7 data sets the fit is marginal and for 5 data sets the fit is not acceptable at all. The Psi test indicates that the LP3 distribution provides a good fit for 10 of the 25 data sets, when the Ψ values are compared with $\chi^2_{0.95}(k-4)$ values in Table 5. For 9 data sets the difference between Ψ and $\chi^2_{0.95}(k-4)$ values ranges from 0.49 to 5, for 4 data sets the difference ranges from 5 to 10, and for 2 data sets the difference is more than 10. Typical cumulative empirical and fitted distributions are shown in Fig. 1.

4. MIXED DOUBLE EXPONENTIAL DISTRIBUTION

Canfield and Borgman (1975) used reliability theory to show that the distribution of the extremes in a sequence of mixture random variables can be written as

$$F(X) = F_i(X)^{P_1} F_{i'}(X)^{P_2} \tag{4.1}$$

where $F_i(X)$ and $F_{i'}(X)$ represent any of the EV1, EV2 or EV3 distributions. The exponents P_1 and P_2 in 4.1 can be absorbed by reparameterization, so that 4.1 can be written as

$$F(X) = F_i(X) F_{i'}(X) \tag{4.2}$$

This redues the number of parameters in the distribution and makes computations easier. As shown by Canfield and Borgman (1975) the distribution in 4.2 gives primary consideration to the shape of the right tails. For flood frequency analysis, this is important since the right tails of the distribution represent high flows. The distribution is thus a product of the tails of $F_i(X)$ and $F_{i'}(X)$. Potter (1958) has used the mixture of random variables in the statistical distribution of floods. He used the standard mixed distribution for the case of two components

$$F(X) = P_1 G_1(X) + P_2 G_2(X) \tag{4.3}$$

where $G_1(X)$ and $G_2(X)$ represent the two components of the mixture distribution. The parameters P_1 and P_2 are the probability of occurrence such that $P_1 > 0$, $P_2 > 0$, and $P_1 + P_2 = 1$. The parameters P_1 and P_2 cannot be absorbed

Table 5. Goodness of Fit Results for LP3 Distribution

River	Psi Test Ψ	J	Chi-Square Test $\chi^2_{0.90}(k-4)$	$\chi^2_{0.95}(k-4)$
Assiniboine	3.585	1.897	7.78	9.49
Assiniboine	16.99	8.485	6.25	7.81
Danube	10.34	4.88	7.78	9.49
Elbe	33.08	15.915	12.0	14.1
Fraser	11.39	6.697	6.25	7.81
Gloma	16.03	3.414	6.25	7.81
Hurricana	9.41	5.99	6.25	7.81
Krishna	14.31	6.017	9.24	11.1
Kymijoki	7.23	3.28	6.25	7.81
N. Saskatchewan	5.81	5.80	9.24	11.1
Neman	7.02	3.726	7.78	9.49
Neva	13.09	1.43	10.6	12.6
Niger	6.057	2.807	9.24	11.1
Niger	15.299	7.792	10.6	12.6
Oder	5.65	2.607	7.78	9.49
Penobscot	6.62	2.787	9.24	11.1
Red	5.76	2.564	6.25	7.81
S. Saskatchewan	6.15	4.378	9.24	11.1
S. Saskatchewan	18.186	12.164	7.78	9.49
Saquenay	8.78	5.765	6.25	7.81
Senegal	10.89	6.504	10.6	12.6
Susquehana	7.76	3.525	4.61	5.99
Vanerngota	15.95	6.62	9.24	11.1
Vuoksi	54.20	37.83	6.25	7.81
Winnipeg	17.576	9.594	10.6	12.6

Table 6. Goodness of Fit Results for mixed double exponential Distribution

River	Psi Test Ψ	J	Chi-Square Test $\chi^2_{0.9}(k-6)$	$\chi^2_{0.95}(k-6)$
Assiniboine	8.63	4.482	4.61	5.99
Assiniboine	21.28	12.56	2.71	3.84
Danube	248.87	124.4	4.61	5.99
Elbe	17.27	7.856	9.24	11.1
Fraser	4.44	2.09	2.71	3.84
Gloma	3.24	5.21	2.71	3.84
Hurricana	6.29	3.03	2.71	3.84
Krishna	14.659	6.837	6.25	7.81
Kymijoki	3.62	1.59	2.71	3.84
N. Saskatchewan	3.98	0.68	6.25	7.81
Neman	25.48	14.03	4.61	5.99
Neva	1.459	1.515	7.78	9.49
Niger	11.645	7.055	6.25	7.81
Niger	13.978	10.25	7.78	9.49
Order	7.44	3.33	4.61	5.99
Penobscot	5.0	5.227	6.25	7.81
Red	4.53	2.167	2.71	3.84
S. Saskatchewan	4.554	2.225	4.61	5.99
S. Saskatchewan	11.08	6.225	6.25	7.81
Saquenay	6.80	2.977	2.71	3.84
Senegal	10.102	1.488	7.78	9.49
Susquehana	14.245	8.007	4.61	5.99
Vanerngota	21.37	13.636	6.25	7.81
Vuoksi	24.33	13.05	2.71	3.84
Winnipeg	14.589	10.558	7.78	9.49

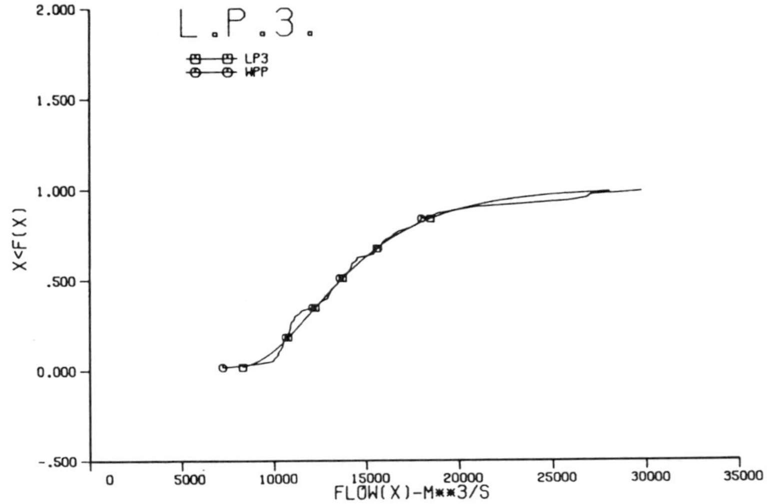

Krishna River

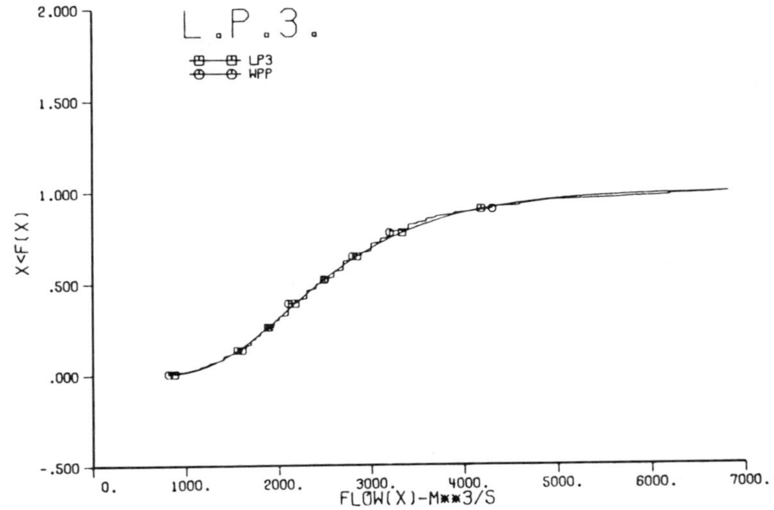

Neman River

Fig. 1. Log Pearson distributions and empirical data.

by reparameterization and thus must be estimated in addition to all the parameters of both $G_1(X)$ and $G_2(X)$. This mixture distribution represents a weighted average of the tails of the distributions $G_1(X)$ and $G_2(X)$. Thus the tail shapes may not necessarily be correct for flood frequency analysis even if two extreme value distributions are used in eq. 4.3 (Canfield et al., 1980).

Canfield et al. (1980) consider the EV3 distribution to be applicable to most of the flood frequency analysis applications. Thus the form of both $F_i(X)$ and $F_{i'}(X)$ in eq. 4.2 are selected by them in the EV3 form. Substituting the EV3 distribution for $F_i(X)$ and $F_{i'}(X)$ in eq. 4.2, we obtain the cumulative distribution function as follows:

$$F(X) = \begin{cases} 1 & , x > b \\ \exp\left[-\left(\frac{b-x}{c}\right)^{a} - \left(\frac{b-x}{c'}\right)^{a'}\right], & x \leq b, \; c > 0, \; c' > 0 \end{cases} \tag{4.4}$$

Equation 4.4 is a five-parameter distribution. The parameters are a, b, c, a′ and c′. The parameter b is the same for both extreme value components of the distribution; and, as discussed before, represents the concept of limiting flood based on the physical limitations of meteorological phenomenon and basin characteristics which control river flows. Numerically b is the most difficult parameter to estimate; however, the distribution is least sensitive to it so long as its magnitude is relatively large. Parameters a, c, and a′, c′ are similar parameters for the two components of the cumulative distribution respectively with different magnitudes.

Canfield et al. (1980) used the graphical method of estimating the parameters of eq. 4.4. They improved and used a least square estimating technique reported in Canfield and Borgman (1975) to estimate the parameters a, a′, c and c′ of the mixture distribution function. The parameter estimation method discussed in Canfield et al. (1980) is used in the present study.

Canfield et al. (1980) stated that the objective of their research was to develop and evaluate an extension of extreme value theory for flood frequency analysis for nonhomogeneous river flows. The mixed double exponential distribution has five parameters, and is therefore capable of fitting a wide variety of data. In the present study, the mixed double exponential distribution is fitted to both homogeneous and non-homogeneous annual maximum flows, and the results are compared to those given by the LP3 and Wakeby distributions.

The results of goodness of fit tests obtained in the present study are given in Table 6. For mixed double exponential distribution, the chi-square test indicates that the distribution provides good fit for 9 of the 25 data sets. The fit is marginal for 8 data sets and is not acceptable. The Psi test indicates that the distribution provides a good fit for 6 of the 25 data sets, when the Ψ values are compared with $\chi^2_{0.95}(k-6)$ values in Table 6. For 10 data sets the difference in Ψ and $\chi^2_{0.95}(k-6)$ values ranges from 0.6 to 5, for 4 data sets the difference ranges from 5 to 10, and for 5 data sets the difference is more than 10. Typical empirical cumulative distribution and fitted mixed double exponential distributions are shown in fig. 2.

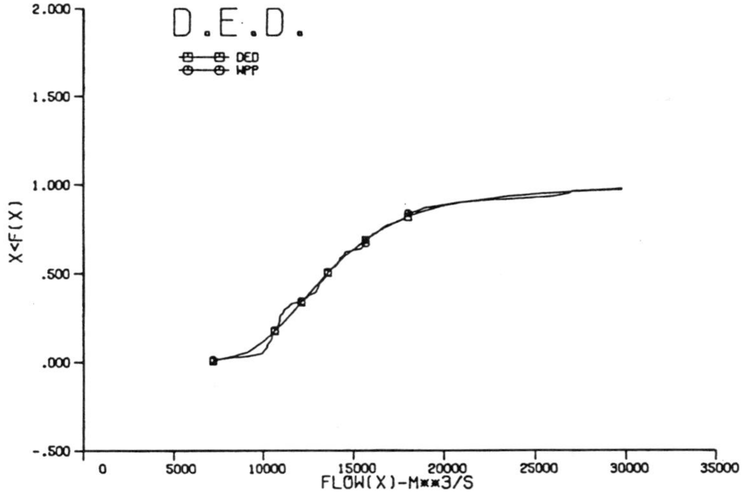

Krishna River

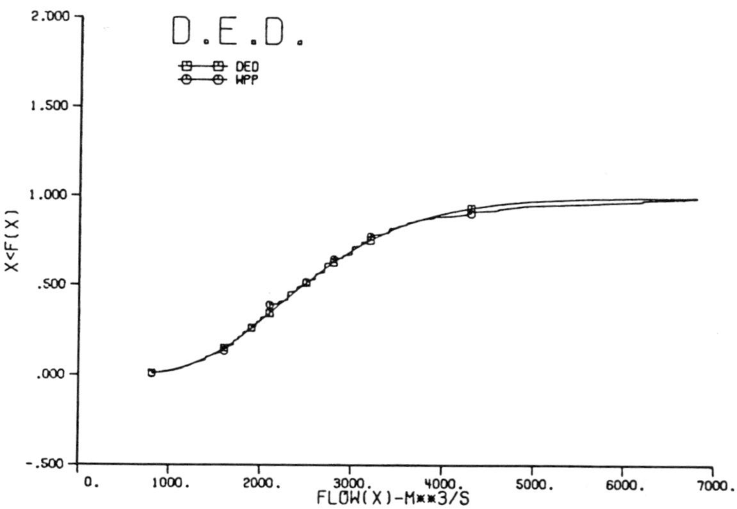

Neman River

Fig. 2. Mixed exponential distributions fitted to empirical data.

5. WAKEBY DISTRIBUTION

The Wakeby distribution was initially studied for hydrologic applications by Houghton (1978). The Wakeby distribution is a five-parameter distribution. In estimating the parameters of a distribution, the higher moments and their functions such as coefficient of skew become unstable. This makes the estimated floods unreliable. Also the smallest observations can have a substantial effect on the right-hand side (large flows) of the distribution. As discussed by Houghton, the left-hand side (small flows) does not necessarily add information to estimating the quantilies of the right-hand side. Since floods are not known to follow any particular distribution, it appears intuitively better to separate the left-hand side from the right-hand side of the distribution. The Wakeby distribution utilizes this strategy in flood frequency analysis.

The Wakeby distribution also accounts better for the "separation effect" as described by Matalas et al. (1975). Landwehr et al. (1978) studied the condition of separation in logarithmic space and arrived at the conclusion that the Pearson type III distribution is unlikely to explain the condition of separation. Landwehr et al. (1987) found that it is difficult to infer the skewness coefficient of flows from skewness coefficient of logarithm of flows. Therefore, the skew map in real space cannot be uniquely determined from the skew map in logarithmic space.

The Wakeby distribution was originally introduced to account for the "separation effect." In order to account for this effect a distribution is needed with a very thick right-hand tail and a left-hand tail thick enough to decrease the average skew. This makes the middle part of the distribution function steeper than traditional skewed curves. The Wakeby distribution has this property. The five-parameter inverse Wakeby distribution is defined as in 5.1

$$x = -a(1-F)^b + c(1-F)^{-d} + e \tag{5.1}$$

where F is the uniform (0,1) variate and represents the probability of occurrence of x. The values of the parameters a, b, c and d are always positive and e is a location parameter. The density function is defined as

$$f = \frac{dF}{dx} = [ab(1-F)^{b-1} + cd(1-F)^{-d-1}]^{-1} \tag{5.2}$$

From 5.1 it can be seen that the upper bound of x is infinity when $F = 1$. The first moment about zero (mean) is given as

$$\mu_1'(x) = e - \left[\frac{a}{b+1} - \frac{c}{1-d}\right] \tag{5.3}$$

The second moment about mean (variance) is

$$\mu_2(x) = \frac{c^2}{1-2d} - \frac{2ac}{1+b-d} + \frac{a^2}{1+2b} - \left[\frac{c}{1-d} - \frac{a}{1+b}\right]^2 \tag{5.4}$$

From 5.3 and 5.4 it is seen that for $d > 1$ the mean is infinite and for $d > 0.5$ the variance is infinite. For $F > 0.25$ the term $-a(1-F)^b$ in 5.1 has generally no effect on the value of x. This is based on values of a and b which are typical of flood records. Thus the Wakeby distribution can be identified in two parts. The right-hand tail $c(1-F)^{-d} + e$ and the left-hand tail $-a(1-F)^{b}$ which is in effect an adjustment to the graph of the right-hand tail. This property of Wakeby distribution makes it account better for the "separation effect," as it makes the middle part of the distribution function steeper.

Landwehr et al. (1979b) developed an algorithm for estimating the parameters and quantiles of the Wakeby distribution by using probability weighted moments. They considered the case where the true lower bound m was known and could be subtracted from the sample values. For example, if the random variable x is distributed as Wakeby with the lower bound m, then the random variable $z = x - m$ is also distributed as Wakeby but with lower bound zero. Thus when the lower bound m is known, it can be subtracted from the sample values and the algorithm is used with m equal to zero. Estimation of parameters with the assumption that the true lower bound m is known represents a biased estimate. Landwehr et al. (1979b) suggest that when the lower bound is zero and particularly for small sample sizes, it is preferable to use a biased estimate of the probability weighted moments rather than an unbiased one. A computer program developed by Dr. J. R. Wallis was used in the present study to estimate the parameters of the Wakeby distribution.

The goodness of fit test results for the Wakeby distribution are given in Table 7. For the Wakeby distribution, the chi-square test indicates that the distribution provides a good fit for 11 data sets. For 6 data sets the fit is marginal and for 8 data sets the fit is not acceptable at all. The Psi test indicates the distribution provides good fit for 10 data sets, when the Ψ values are compared with $\chi^2_{0.95}(k-6)$ values in Table 7. For 6 data sets the difference in Ψ and $\chi^2_{0.95}(k-6)$ values ranges from 0.22 to 5; for 6 data sets the difference ranges from 5 to 10; and for 2 data sets the difference is more than 10. Examples of empirical and fitted distributions for Wakeby distribution are shown in fig. 3.

6. DISCUSSION OF RESULTS AND CONCLUSIONS

Three distributions for of flood frequency analysis have been tested on 25 data sets from watersheds located around the world. Two goodness of fit tests are applied and magnitudes of return period floods are computed.

Results shown in Tables 5, 6 and 7 show that the results from chi-square and Psi tests are not consistent. Based on chi-square test results, the hypothesis of LP3 distribution as a good fit would be acceptable for 20 of the 25 data sets at 0.05 and 0.1 significance levels and for one data set at 0.1 significance level. For the mixed double exponential distribution, the null hypothesis would be acceptable for 12 of the 25 data sets at 0.05 and 0.1 significance levels and for 5 data sets at 0.1 significance level. For the Wakeby distribution, the null hypothesis would be acceptable for 11 of the 25 data sets at 0.05 and 0.1 significance levels and for 6 data sets at 0.1 significance level.

The results from the Psi test are markedly different. These results suggest that the Wakeby distribution provides a better fit for 13 of the 25 data sets compared to 3 for LP3 distribution and 9 for the mixed double exponential distribution (Table 8).

Table 7. Goodness of Fit Results for Wakeby Distribution

River	Psi Test Ψ	Chi-Square Test J	$\chi^2_{0.9}(k-6)$	$\chi^2_{0.95}(k-6)$
Assiniboine	3.18	1.74	4.61	5.99
Assiniboine	9.60	9.92	2.71	3.84
Danube	6.21	2.622	4.61	5.99
Elbe	10.39	4.577	9.24	11.1
Fraser	7.35	3.155	2.71	3.84
Gloma	10.48	6.075	2.71	3.84
Hurricana	2.19	1.124	2.71	3.84
Krishna	15.293	6.347	6.25	7.81
Kymijoki	13.20	6.0	2.71	3.84
N. Saskatchewan	5.63	10.068	6.25	7.81
Neman	14.48	10.58	4.61	5.99
Neva	19.851	0.809	7.78	9.49
Niger	6.828	3.318	6.25	7.81
Nigher	18.60	9.067	7.78	9.49
Oder	2.02	0.903	4.61	5.99
Penobscot	3.86	1.90	6.25	7.81
Red	6.63	3.208	2.71	3.84
S. Saskatchewan		No Solution		
S. Saskatchewan	4.12	3.41	6.25	7.81
Saquenay	6.59	4.48	2.71	3.84
Senegal	8.99	4.096	7.78	9.49
Susquehana	10.741	4.722	4.61	5.99
Vanerngota	14.74	6.376	6.25	7.81
Vuoksi	3.63	1.467	2.71	3.84
Winnipeg	19.86	10.727	7.78	9.49

Table 8. Comparison of Psi Test Goodness of Fit Test Results

River	Ψ LP3	Mixed Double Exponential Dist.	Wakeby
Assiniboine	3.585	8.63	3.18
Assiniboine	16.99	21.28	9.60
Danube	10.34	247.87	6.21
Elbe	33.08	17.27	10.39
Fraser	11.39	4.44	7.35
Gloma	16.03	3.24	10.48
Hurricana	9.41	6.29	2.19
Krishna	14.31	14.659	15.293
Kymijoki	7.23	3.62	13.20
N. Saskatchewan	5.81	3.98	5.63
Neman	7.02	25.48	14.48
Neva	13.09	1.459	19.851
Niger	6.057	11.645	6.828
Niger	15.299	13.978	18.60
Oder	5.65	7.44	2.02
Penobscot	6.62	5.0	3.86
Red	5.76	4.53	6.63
S. Saskatchewan	6.15	11.08	4.12
S. Saskatchewan	18.186	4.554	No Solution
Saquenay	8.78	6.80	6.59
Senegal	10.89	10.102	8.99
Susquehana	14.559	14.245	10.741
Vanerngota	15.95	21.37	14.74
Vuoksi	54.20	24.33	3.63
Winnipeg	17.576	14.589	19.86

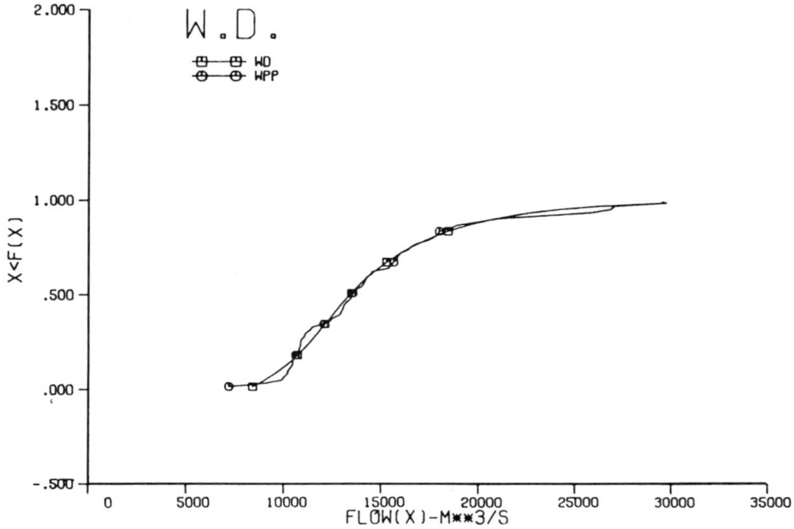

Krishna River

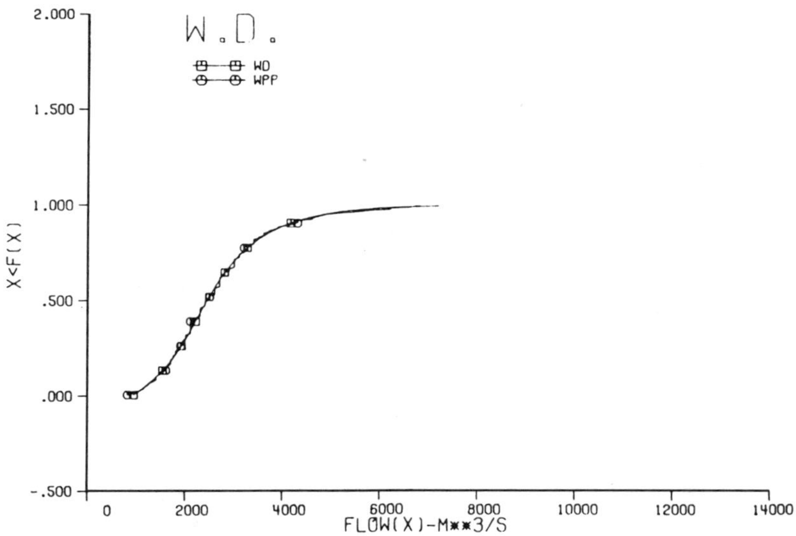

Neman River

Fig. 3. Wakeby distributions fitted to empirical data.

The chi-square goodness of fit test requires that there be 5 or more observations in each class for the results from the test to be valid. To achieve this, it is necessary to group the data into larger classes, thus guaranteeing that no class is empty. However, for several data sets it was not possible to obtain 5 or more observations in each class.

The chi-square test is a first order approximation to the Psi test; and when the Psi test and chi-square test measures diverge, the fit is usually not very good. The mixed double exponential distribution shows lower divergence between the Psi test and chi-square test statistics for 11 of the 25 data sets compared to 10 for the Wakeby distribution and 4 for the LP3 distribution.

The estimated floods for different recurrence intervals are compared to discern general trends in the prediction of flood values. Comparing the lower floods of 10 and 25 year recurrence interval, the LP3 distribution underpredicted 30 estimated values; the mixed double exponential distribution underpredicted 9 values; and the Wakeby distribution underpredicted 11 out of 50 estimated flood values. On the higher side, the mixed double exponential distribution overpredicted 29 estimated values, the Wakeby distribution, 17 values; and the LP3 distribution underpredicted 4 out of 50 estimated flood values.

Comparing the 200-year return period floods, the LP3 distribution gave lower values in 17 cases, the mixed double exponential distribution in 5 cases, and the Wakeby distribution in 3 cases out of 50. On the high side, the Wakeby distribution overpredicated 11 values, the mixed double exponential distribution 6 values, and the LP3 distribution 4 values out of 50 estimated values. Thus the LP3 distribution appears to underpredict estimated flood values for both low and high return period floods. The mixed double exponential distribution overpredicts the estimated values for low return period floods. The Wakeby distribution neither consistently underpredicts or overpredicts estimated values for either low or high return period floods.

In terms of ease of applying the different methods studied, the Wakeby and LP3 distributions are one-step applications from data input to output of usable results. The mixed double exponential distribution involves several steps to obtain usable results.

Based on the results of goodness of fit tests used in the study, it is rather difficult to choose one particular distribution over the other. The Psi test is not restrictive in its application compared to the chi-square test, which requires a minimum number of observations in each class interval. As mentioned earlier, this criterion for chi-square test could not be satisfied for some data sets. The results of Psi test for the three distributions can be compared for corresponding river flow data (see Table 8). The data in Table 8 is derived from Tables 5, 6 and 7 in order to compare the values of the parameter Ψ for the three distributions. The results in Table 8 show that the Wakeby distribution provides a better fit for 13 of the 25 data sets compared to 9 for the mixed double exponential distribution and 3 for LP3 distribution. Thus based on the Psi test results the Wakeby distribution proves to be superior to the LP3 and mixed double exponential distributions.

Based on the above discussion of the results of the study, the following conclusions are presented:

1. The LP3 distribution underpredicts estimated flood values for both the low and high return periods, compared to the Mixed double exponential and Wakeby distributions.
2. The Mixed double exponential distribution overpredicts the estimated values for low return periods, compared to the LP3 and Wakeby distributions.
3. The Wakeby distribution neither consistently underpredicts nor overpredicts estimated values for either low or high periods floods, compared to the LP3 and Mixed double exponential distributions.
4. The Wakeby distribution based on the Psi test results is superior for flood frequency analysis compared to the LP3 and Mixed double exponential distributions. The Wakeby distribution is also superior to the other two distributions in the sense that it neither underpredicts nor overpredicts the estimated flood values.

ACKNOWLEDGEMENTS

We are grateful to Dr. J.R. Wallis of IBM Research, Yorktown Heights, N.Y. for sending us a copy of his computer program to estimate the parameters the Wakeby distribution.

REFERENCES

1. Arora, P.S. 'Fitting mixture distributions to flood data,' M.S.C.E. Thesis, Purdue University, W. Lafayette, IN 47905, 1985.

2. Arora, P.S. and A. R. Rao, 'Probability distributions of flood data,' Tech. Rept. CE-HSE-84-06, School of Civil Engineering, Purdue University, W. Lafayette, IN 47907, 1985.

3. Bobeé, B.B., 1975. 'The Log Pearson Type III Distribution and Its Application in Hydrology,' Water Resources Research, **11**(5):681-689.

4. Bobeé, B.B., and R. Robitaille, 1975. 'Correction of Bias in the Estimation of the Coefficient of Skewness.' Water Resources Research, **11**(6):851-861.

5. Bobeé, B.B., and R. Robitaille, 1977. 'The Use of the Pearson Type III and Log Pearson Type III Distributions Revisited.' Water Resources Research, **13**(2):427-443.

6. Canfield, R.V., and L.E. Borgman, 1975. 'Some Distributions of Time to Failure for Reliability Applications.' Technometrics, **17**(2):263-268.

7. Canfield, R.V., D.R. Olsen, R.H. Hawkins, and T.L. Chen, 1980. 'Use of Extreme Value Theory in Estimating Flood Peaks From Mixed Populations.' Hydraulics and Hydrology Series, Utah Water Research Laboratory,

Utah State University, Logan, Utah.

8. Houghton, J.C., 1978. 'Birth of a Parent: The Wakeby Distribution for Modeling Flood Flows.' Water Resources Research, **14**(6):1105-1110.

9. Houghton, J.C., 1978. 'The Incomplete Means Estimation Procedure Applied to Flood Frequency Analysis.' Water Resources Research, **14**(6):1111-1116.

10. Kite, G.W., 1977. *Frequency and Risk Analysis in Hydrology.* Water Resources Publications, Fort Collins, Colorado.

11. Landwehr, J.M., and N. C. Matalas, 1979a. 'Estimation of Parameters and Quantile of Wakeby Distributions, 2. Unknown Lower Bounds.' Water Resources Research, **15**(6):1373-1379.

12. Landwehr, J.M., and N.C. Matalas, and J.R. Wallis 1979b. 'Estimation of Parameters and Quantile of Wakeby Distributions, 1. Known Lower Bounds.' Water Resources Research, **15**(6):1361-1372.

13. Landwehr, J.M., N.C. Matalas, and J.R. Wallis, 1978. 'Some Comparison of Floor Statistics in Real and Log Space.' Water Resources Research, **14**(5):902-920 and 15(6).

14. Matalas, N.D., and J.R. Wallis, 1973. 'Eureka! It Fits a Pearson Type III Distribution.' Water Resources Research, **9**(2):281-289.

15. Matalas, N.C., J.R. Slack, and J.R. Wallis, 1975. 'Regional Skew in Search of a parent.' Water Resources Res., **11**(6) 815-826.

16. Potter, W.D., 1958. 'Upper and Lower Frequency Curves for Peak Rates of Runoff.' Transactions of the American Geophysical Union, **39**:100-105.

17. Tribus, M., 1969. *Rational Descriptions, Decisions and Designs.* Pergamon Press, New York, N.Y.

COMPARISON OF SOME FLOOD FREQUENCY DISTRIBUTIONS USING EMPIRICAL DATA

D. Jain and Vijay P. Singh
Department of Civil Engineering and
Louisiana Water Resources Research Institute
Louisiana State University
Baton Rouge, Louisiana 70803-6405, USA

ABSTRACT. The Gumbel, 3-parameter lognormal, Pearson type 3, log-Pearson type 3, and Boughton distributions as well as power transformation were evaluated and compared using 55 flood data sets from various sizes of drainage basins representing various parts of the U.S.A. No distribution was found to fit all flood data sets accurately. Based on empirical criteria, the log-Pearson type 3 distribution emerged as the best distribution.

1. INTRODUCTION

Frequency analysis is used to determine the return periods of recorded events of known magnitudes or to estimate the magnitudes of events for design return periods beyond the recorded range. This information on magnitude and frequency relationship is used in the design of dams, culverts, highways, water supply systems, bridges, flood control systems, stream control works, hydro-electric power plants, waste disposal, etc.

Commonly used empirical distributions for frequency analysis are Gumbel (GU), normal distribution (NO), 2-parameter lognormal (LN2), 3-parameter lognormal (LN3), 2-parameter gamma distribution (G2), Pearson type 3 (PT3), log-Pearson type 3 (LPT3) and Boughton (BO). Beard (1974) compared LPT3, LN2, GU, and G2 distributions. Using the tests for consistency and reliability, he found LPT3 and LN2 distributions to be better suited for frequency analysis. The U.S. Water Resources Council (1981) published guidelines for performing flood frequency analysis. On similar grounds, the LPT3 distribution was found to be the best distribution. In Italy, Cicioni, et al. (1973) compared LN2, LN3, G2, PT3 and GU distributions on 108 data sets and found LN2 distribution to be the best. Loganathan, et al. (1985) compared GU, PT3, LPT3, BO distributions and power transformation method. Using empirical criteria on 20 long-term records, they found that BO, LPT3 and PT were suitable methods for frequency analysis. Moore and Chong (1983) compared LN2,LN3, GU, PT3 and LPT3 distribu-

V. P. Singh (ed.), Hydrologic Frequency Modeling, 467–485.

tions. For comparison they tested them on 22 long-term flood records. They found that GU was the best of the five distributions compared. Boughton (1980, 1983) disagreed with the use of 2-parameter distribution in flood frequency analysis. He argued that at least three parameters were required to describe the variations in distributions of annual floods from different watersheds, and introduced a new distribution designated as BO. This distribution has simpler algebraic equations for frequency factor than LPT3, and gave better results than LN2 and GU distributions.

Sangal and Biswas (1970) compared LN3 distribution with GU, PT3, LPT3, LN2 and NO distributions. They concluded that: (1) LN3 distribution could be successfully used for flood frequency analysis; (2) the GU distribution was a special case of LN3 distribution; and (3) it was better than LN2 distribution because if the sample skewness for the logarithms of the data was negative, the LN2 distribution gave high estimates for flood values. Burges, et al. (1975) also argued in favor of LN3 distribution for the following reasons: (1) It provides a relatively simple method for preserving the first three sample moments, and that its logarithmic transform by definition is normally distributed. (2) The normal property of the logarithm transform makes use of LN3 distribution very appealing in cases where observed flows have skew coefficients that cannot be satisfactorily modeled by a two parameter distribution.

Benson (1968) compared G2, GU, log-Gumbel (LGU), LN2, LPT3 distributions and the Hazen method for flood frequency analysis, and concluded that the LPT3 distribution should be adopted as a base method, with the proviso that departures from the base method be justified. The choice of the base method should not be considered as final. Bobee and Robitaille (1977) compared PT3 and LPT3 distributions using different fitting methods. They found that the PT3 distribution represented annual flood data generally better than LPT3 distribution. The LPT3 distribution was, however, preferable if the method of moments was applied to the observed data rather than to the logarithmically transformed data. However, correction for bias of the coefficient of skewness was needed.

Another empirical method for frequency analysis involves reconstitution of data by a suitable transformation such that the transformed data follows a normal distribution. Chander, et al. (1978) used the power transformation suggested by Box and Cox (1964) and showed that it transformed the data to a near normal series. This transformation (PT) is versatile and can incorporate the effect of kurtosis. For their data, the PT method worked better than LPT3 and GU distributions. Table 1 summarizes these comparative studies. Clearly, no method appears uniformly superior under all conditions.

The objective of this study is to compare the GU, LN3, PT3 LPT3, BO distributions and PT method using 55 long-term records of annual flood maxima, and choose the best of these methods for flood frequency analysis.

Table 1. A survey of studies on comparing empirical methods of flood frequency analysis.

Author	Distributions and Methods Used										Recommended Method
	Normal	2-Parameter Lognormal	3-Parameter Lognormal	Gumbel	2-Parameter Gamma	Pearson Type 3	Log-Pearson Type 3	Boughton	Transformation	Hazan Method	
Benson (1968		*		*	*		*			*	LPT3
Sangal and Biswas (1970)	*	*	*	*		*	*				LN3
Cicioni, et al. (1973)		*	*	*	*	*					LN2
Beard (1974)		*		*	*		*				LPT3, LN2
Bobee and Robitaille (1977)						*	*				PT3
Chander, et al. (1978)				*			*		*		PT
U.S. Water Resources Council (1981)	*										LPT3
Boughton (1980, 1983)		*		*			*	*			BO
Moore and Chong (1983)		*	*	*		*	*				GU
Loganathan, et al. (1985)				*		*	*	*	*		BO, LPT3, PT

* denotes the method included in comparison.

2. SELECTION OF DATA

Long term records of annual maximum flood flows from 55 gaging stations were selected. These stations represent drainage basins of various sizes located in different parts of the U.S.A. Each record was at least 30 years long, complete, independent and homogeneous. The drainage basins ranged from 20 to 3,000 km^2. Pertinent characteristics of the data of these stations are given in Table 2. The coefficient of variation (C_v) ranged from 0.291 to 1.22 and the coefficient of skewness (C_s) from 0.16 to 5.66 for these records.

3. METHODS OF FREQUENCY ANALYSIS

Six empirical methods of frequency analysis were compared: Gumbel, 3-parameter lognormal, Pearson type 3, log-Pearson type 3, and Boughton distributions, as well as power transformation method. Table 3 gives probability distribution functions (pdf) for the six methods and pertinent equations for estimation of their parameters. For GU, LN3, PT3, and LPT3 distributions, parameters were estimated by the method of maximum likelihood estimation (MLEM). For parameters of the BO distribution, an empirical method presented by Boughton, (1980, 1983) was used. For the PT method, the transformation parameter was obtained by an iterative method. All six methods were applied to 55 data sets. For each data set parameter estimates of each method are given in Table 4.

4. METHODS FOR COMPARISON

Two tests, which measure the relative goodness of fit, were employed for comparison of the methods of fitting the flood data. These tests have the same origin as the methods given by the Work Group on Flow Frequency Methods reported by Benson (1968) and Bobee (1977). The mean absolute deviation test (MADT) is defined as the average of relative deviations between observed and computed flood discharges for the entire sample, with algebraic sign ignored,

$$D_a = \Sigma|G(T)|/n \quad (1)$$

in which

$$G(T) = (x_0(T) - x_c(T))/x_0(T) \quad (2)$$

where x_0 and x_c are observed and computed flood values respectively for a given value of return period T, and n is sample size.

The other test is mean square deviation test (MSDT) defined as the average of squares of relative deviations,

Table 2. Pertinent characteristics of annual flood data of fifty-five USGS gaging stations. x = discharge in m^3/s.

St. No.	River	Location	Drainage Area (sq km)	Record	Years of Record	$\bar{x}$	S_x	C_s	C_k
1	Comite	Olive Branch, LA	1405	1943-1982	38	238.2	174.5	0.70	2.52
2	Comite	Comite, LA	1896	1943-1982	38	315.7	166.8	0.54	2.77
3	Amite	Amite, LA	4092	1949-1982	34	745.1	539.5	0.71	3.03
4	Amite	Magnolia, LA	1804	1949-1980	31	698.0	365.9	0.16	2.15
5	St. Mary	Still Water, Nova Scotia	1653	1915-1974	59	409.5	147.9	1.42	6.25
6	St. John	Ninemile Bridge, ME	1890	1951-1982	32	699.0	223.7	0.41	3.01
7	St. John	Dickey, ME	5085	1947-1982	36	1449.7	517.7	0.35	2.55
8	Allagash	Allagash, ME	1659	1932-1982	51	438.8	159.8	0.71	3.30
9	St. Francis	Connors, New Brunswick	1105	1952-1982	31	215.3	84.4	0.53	3.25
10	Fish	Fort Kent, ME	890	1930-1982	53	241.1	71.4	0.30	3.45
11	St. John	Fort Kent, ME	9097	1927-1982	56	2405.2	754.1	0.43	3.22
12	Machias	Ashland, ME	1105	1952-1981	29	191.3	87.9	1.19	5.54
13	Aroostook	Washburn, ME	1884	1931-1982	51	687.4	236.7	0.35	2.65
14	Meduxnekeag	Houlton, ME	589	1941-1984	42	103.4	41.3	0.78	2.90
15	Narraguagus	Cherryfield, ME	685	1948-1982	35	128.8	52.6	0.98	4.33
16	Mattawamkeag	Mattawamkeag, ME	1640	1935-1982	47	487.9	142.0	0.37	2.59
17	Piscataquis	Dover-Foxcroft, ME	1136	1903-1982	79	259.8	125.0	1.16	4.07
18	Sheepscot	North Whitefield, ME	405	1938-1982	43	63.9	32.6	1.74	6.50
19	Carrabassett	North Anson, ME	897	1925-1982	56	391.0	187.1	0.68	2.78
20	Sebasticook	Pittsfield, ME	2500	1929-1980	52	193.4	63.4	0.80	4.71
21	Diamond	Wentworth Location, NH	843	1942-1982	40	135.5	40.1	0.62	4.34
22	Swift	Roxbury, ME	753	1930-1982	51	182.8	97.9	0.93	3.47
23	Nezinscot	Turner Center, ME	733	1942-1982	41	105.2	61.2	3.03	14.95
24	Royal	Yarmouth, ME	642	1950-1982	32	110.7	54.0	2.24	10.67
25	Oyster	Durham NH	140	1935-1981	44	9.1	4.4	1.16	5.22
26	Pemigewasset	Plymouth, NH	1243	1904-1981	75	658.6	297.6	2.19	8.36
27	Smith	Bristol, NH	135	1919-1981	63	57.5	33.5	3.04	14.94
28	Soucook	Concord, NH	103	1951-1981	30	41.2	19.3	1.52	6.26

Table 2 (Continued)

St. No.	River	Location	Drainage Area (sq km)	Record	Years of Record	$\bar{x}$	S_x	C_s	C_k
29	Squannacook	West Groton, MA	585	1950-1982	32	43.8	23.0	0.95	4.40
30	Parker	Byfield, MA	50	1946-1982	37	6.5	2.8	0.91	3.88
31	Hop	Columbia, CT	65	1934-1982	49	64.9	41.1	1.73	5.81
32	Safford	Woodstock Valley, CT	58	1950-1982	31	10.6	6.0	1.39	4.81
33	Moose	Victory, VT	63	1946-1982	35	60.4	19.1	1.01	4.76
34	Moose	St. Jonsbury, VT	120	1930-1982	53	84.3	34.6	0.78	3.03
35	Ammonoosuc	Bethlehem Junction, NH	289	1927-1982	36	139.2	61.8	1.37	4.56
36	Ammonoosuc	Bath, NH	596	1941-1982	42	343.1	163.5	1.51	4.81
37	White	West Hartford, VT	1500	1918-1982	65	592.0	428.3	4.68	31.58
38	Moss Brook	Wendell Depot, MA	45	1917-1982	66	9.1	6.7	2.98	14.74
39	North	Shattuckville, MA	270	1940-1982	43	145.4	76.3	1.13	3.87
40	Hop Brook	New Salem, MA	20	1949-1982	34	4.5	2.0	0.33	2.54
41	East Branch	Hardwick, MA	116	1936-1982	47	26.1	27.5	5.06	32.38
42	Guaboag	West Brimfield, MA	125	1903-1982	70	43.5	47.6	5.55	36.09
43	Middle B Westfield	Goss Heights, MA	560	1930-1982	53	112.3	97.3	3.07	13.94
44	West Branch Westfield	Huntington, MA	782	1937-1982	46	186.4	144.4	2.14	8.20
45	Scantic	Broad Brook, CT	178	1930-1982	53	42.0	51.3	5.66	38.59
46	Burlington	Burlington, CT	92	1937-1982	46	9.7	7.3	3.35	19.00
47	Connecticut	Hartford, CT	3500	1905-1982	78	980.1	761.5	2.87	15.55
48	North Branch Park	Hartford, CT	125	1956-1982	27	45.6	50.4	4.34	23.15
49	Salmon	East Hampton, CT	575	1930-1981	52	107.4	93.6	2.59	10.77
50	Eightmile	North Plain, CT	320	1938-1982	45	30.4	26.0	3.44	18.39
51	East Branch Eightmile	North Lyme, CT	197	1937-1982	46	24.9	23.3	3.84	19.70
52	Quinnipiac	Wallingford, CT	205	1932-1982	51	62.3	37.9	2.38	10.74
53	Blackberry	Canaan, CT	215	1943-1982	30	57.8	67.8	4.79	27.34
54	Tenmile	Gaylordsville, CT	430	1921-1982	52	103.5	81.3	2.97	14.01
55	Still	Lanesville, CT	495	1935-1982	48	45.7	37.9	2.74	13.0

Table 3. Empirical methods for flood frequency analysis.

Method	pdf	Pertinent Equations for Parameter Estimation by MLEM
Gumbel Distribution	$\exp(-y)\exp(-\exp(-y))$ where $y = a(x-b)$	$b = \frac{1}{a} \ln \left(\frac{n}{\Sigma \exp(-ax_i)}\right)$ $\frac{1}{a} = \bar{x} - [\Sigma x_i \exp(-ax_i)/\Sigma \exp(-ax_i)]$
Log-normal 3 Distribution	$\frac{1}{(x-a)\ \sigma_y\ (2\pi)^{0.5}} \exp - \frac{(\ln(x-a)-\mu_y)^2}{2\sigma_y^2}$	$\mu_y = \frac{\Sigma \ln(x_i-a)}{n}$ $\sigma_y^2 = \frac{\Sigma(\ln(x_i-a)-\mu_y)^2}{n}$ $\Sigma(x_i-a)^{-1}(\mu_y-\sigma_y^2) = \Sigma(\ln(x_i-a)/(x_i-a))$
Pearson Type 3 Distribution	$\frac{1}{a\Gamma(b)} [\frac{x-c}{a}]^{b-1} \exp - (\frac{x-c}{a})$	$a = \frac{1}{n} \Sigma(x_i-c) - \frac{n}{\Sigma(x_i-c)^{-1}}$ $b = \frac{\Sigma(x_i-c)^{-1}}{[\Sigma(x_i-c)^{-1}] - n^2[\Sigma(x_i-c)]^{-1}}$ $\Sigma \ln(x_i-c) = n\psi(b) + n \ln(a)$

Table 3 (Continued)

Method	pdf	Pertinent Equations for Parameter Estimation by MLEM
Log Pearson Type 3 Distribution	$\frac{1}{ax\Gamma(b)} [\frac{\ln(x)-c}{a}]^{b-1} \exp - (\frac{\ln x-c}{a})$	$a = \frac{1}{(nb)} \Sigma(\ln(x_i)-c)$ $b = \frac{\Sigma(\ln(x_i)-c)^{-1}}{[\Sigma(\ln(x_i)-c)^{-1}] - [n^2(\Sigma(\ln(x_i)-c)^{-2}]}$ $\Sigma(\ln(x_i)-c) = nba$
Boughton Distribution		$\log Q = \bar{x} + KS$ $K = A + ((1.009\ A^{2.04})/(\ln\ \ln(T/(T-1)))-A)$
Power Transformation	$y_i = \frac{(x_i^{\lambda}-1)}{\lambda}$	

Table 4. Parameter estimates of the six methods of flood frequency analysis for fifty-five data sets.

St. No.	LPT3			BO			LN3			PT3			PT	GU	
	a	b	c	S	A	C	a	μ	σ	a	b	c	λ	a	b
1	0.062	173.5	-5.6	0.3144	2.591	7.05	-589.68	6.7	0.20	50.62	10.05	-270.78	0.15	.00779	159.0
2	0.114	32.0	1.9	0.2489	2.892	8.88	-692.18	6.9	0.16	47.69	11.34	-225.20	0.49	.00740	238.0
3	0.197	23.5	1.7	0.3470	2.452	6.30	-1794.06	7.8	0.20	191.68	6.87	-572.63	0.48	.00244	498.0
4	0.132	27.3	2.8	0.2381	2.190	4.96	-6879.27	8.9	0.05	35.87	99.97	-2888.28	0.80	.00318	522.0
5	0.029	131.9	2.1	0.1372	5.792	35.63	60.05	5.8	0.40	84.89	2.94	160.20	-0.20	.00920	345.0
6	0.067	26.8	4.7	0.1390	3.651	14.25	-1123.81	7.5	0.12	55.36	15.67	-168.64	0.45	.00526	593.0
7	0.071	31.1	5.0	0.1574	3.316	11.73	-3350.66	8.5	0.11	108.00	21.93	-918.36	0.50	.00228	1200.0
8	0.053	50.3	3.3	0.1535	4.089	17.87	-294.32	6.6	0.21	64.34	6.04	50.36	0.68	.00775	365.0
9	0.081	30.8	2.8	0.1765	3.363	12.09	-319.34	6.3	0.15	27.75	9.05	-35.84	0.60	.0141	176.0
10	0.050	46.0	3.1	0.1314	3.284	11.51	-513.59	6.6	0.09	12.36	33.03	-167.02	0.30	.0153	207.0
11	0.048	56.0	5.1	0.1383	3.458	12.78	-3138.10	8.6	0.13	188.47	16.02	-614.81	0.55	.00149	2050.0
12	0.088	32.0	2.3	0.1968	3.895	16.23	-65.38	5.5	0.32	41.11	4.38	11.10	0.05	.0149	153.0
13	0.062	39.5	4.0	0.1518	3.368	12.11	-1458.49	7.7	0.11	47.07	24.44	-462.98	-0.15	.00489	574.0
14	0.001	153230.0	-150.2	0.1631	3.620	13.99	-69.98	5.1	0.23	8.79	19.30	-66.30	0.30	.0316	84.7
15	0.032	153.9	-0.1	0.1639	4.044	17.47	-52.44	5.2	0.27	12.64	14.69	-56.83	-0.05	.0254	105.0
16	0.039	59.8	3.8	0.1215	3.574	13.64	-759.34	7.1	0.11	29.31	22.41	-169.13	-0.77	.00837	420.0
17	0.009	2594.3	-17.4	0.2526	2.612	7.35	-90.17	5.8	0.33	71.38	2.91	52.34	0.10	.0112	205.0
18	0.173	6.1	3.0	0.1754	5.573	33.01	-1.57	4.1	0.42	5.88	22.94	-70.90	-0.415	.0495	50.6
19	0.030	266.1	-2.1	0.1994	3.412	12.42	-492.98	6.8	0.20	53.62	10.75	-185.44	0.50	.00696	305.0
20	0.050	53.5	2.5	0.1418	3.788	15.35	-65.08	5.5	0.24	19.39	10.66	-13.40	-0.08	.0183	164.0
21	0.052	42.2	2.7	0.1281	3.343	11.94	-76.52	5.3	0.19	11.65	12.17	-6.18	-0.54	.0277	117.0
22	0.024	493.8	-6.6	0.2179	4.030	17.35	-160.52	5.8	0.27	28.39	9.82	-96.12	0.05	.0141	139.0
23	0.134	10.3	3.2	0.1761	4.826	24.81	26.59	4.2	0.61	36.98	1.86	36.60	0.20	.0301	83.3
24	0.005	7307.1	-31.9	0.1803	3.464	12.81	21.16	4.3	0.60	26.04	3.55	18.20	-0.60	.0274	89.1
25	0.087	34.6	-0.9	0.2045	4.280	19.58	-3.55	2.5	0.33	0.87	21.90	-9.90	-0.52	.299	7.18
26	0.090	15.8	5.0	0.1320	2.380	5.78	180.29	6.0	0.52	150.03	2.90	222.80	0.05	.00567	544.0
27	0.139	9.7	2.6	0.1744	5.145	28.17	15.46	3.5	0.63	5.28	25.53	-77.30	0.325	.0551	45.5
28	0.020	481.2	-6.0	0.1839	4.641	22.99	-3.84	3.7	0.39	9.45	3.71	6.10	0.05	.0731	33.0

Table 4 (Continued)

St. No.	LPT3			BO			LN3			PT3			PT	GU	
	a	b	c	S	A	C	a	μ	σ	a	b	c	λ	a	b
29	0.109	30.3	0.3	0.2380	3.632	14.11	-38.49	4.4	0.26	12.98	3.19	2.44	0.325	.0565	33.5
30	0.014	872.1	-10.6	0.1770	4.238	19.19	-3.71	2.3	0.26	0.61	17.98	-4.47	0.07	.469	5.25
31	0.153	12.2	2.2	0.2204	5.274	29.60	-17.08	4.3	0.42	8.49	16.64	-76.40	-0.41	.0398	4.84
32	0.111	20.8	-0.1	0.2110	5.677	34.23	-4.39	2.6	0.35	1.07	24.67	-15.70	-0.32	.254	8.09
33	0.049	36.9	2.2	0.1228	3.698	14.60	-3.31	4.1	0.28	4.19	17.74	-14.00	-0.46	.0694	51.8
34	0.006	4284.3	-0.1	0.1679	4.105	18.00	-59.37	4.9	0.23	6.80	22.87	-71.15	-0.04	.0378	68.5
35	0.132	8.6	3.7	0.1612	4.769	24.25	-15.41	5.0	0.35	16.37	10.49	-32.47	-0.62	.0245	113.0
36	0.127	10.6	4.4	0.1679	4.414	20.79	-28.42	5.8	0.38	111.99	1.91	129.40	-0.52	.00959	276.0
37	0.084	34.3	3.4	0.1909	3.507	13.10	193.63	5.8	0.38	189.67	2.65	89.60	-0.17	.00488	457.0
38	0.115	26.2	-1.0	0.2335	5.515	32.34	0.60	1.9	0.63	0.70	65.37	-36.90	-0.11	.275	6.66
39	0.077	40.2	1.8	0.2032	4.846	25.04	-81.19	5.4	0.31	21.40	9.92	-66.80	-0.27	.0189	112.0
40	0.083	44.8	-2.3	0.2001	2.441	6.26	-14.92	3.0	0.10	0.39	24.36	-4.91	0.71	.592	3.60
41	0.157	13.4	0.9	0.2144	2.720	7.75	1.43	3.0	0.62	3.20	37.46	-93.80	-0.35	.0950	18.5
42	0.179	6.5	2.4	0.0001	-0.227	-1.24	3.76	3.5	0.54	6.47	19.72	-84.20	-0.68	.0670	32.2
43	0.210	8.1	2.8	0.2431	5.549	32.72	-9.77	4.6	0.55	24.46	7.51	-71.50	-0.46	.0213	79.0
44	0.090	52.9	0.2	0.2686	5.365	30.63	-55.45	5.4	0.48	101.22	1.59	25.70	-0.10	.0121	131.0
45	0.208	8.2	1.8	0.1725	2.053	4.19	-0.69	3.5	0.60	7.39	17.70	-88.80	-0.38	.0560	29.0
46	0.015	1600.6	-2.3	0.2558	4.148	18.39	1.11	1.9	0.77	0.84	51.43	-33.70	0.04	.242	7.10
47	0.122	26.0	3.5	0.3339	2.074	4.49	-10.39	6.7	0.62	407.63	1.95	183.60	-0.30	.00229	693.0
48	0.286	3.7	2.5	0.0040	0.330	-0.85	-5.95	3.8	0.52	8.13	15.96	-84.10	-0.75	.0546	31.7
49	0.252	6.5	2.8	0.2601	6.030	38.56	-26.39	4.7	0.51	25.05	7.20	-72.90	-0.45	.0210	73.6
50	0.325	3.5	2.1	0.2399	5.912	37.09	0.09	3.2	0.60	3.53	34.12	-90.00	-0.61	.0788	21.5
51	0.259	3.4	2.2	0.0024	0.362	-0.83	-0.23	3.0	0.52	2.80	40.99	-90.00	-0.99	.112	18.0
52	0.127	16.1	1.9	0.2055	6.060	38.95	4.36	3.9	0.55	6.87	20.86	-81.00	-0.22	.0440	47.5
53	0.179	12.3	1.6	0.0406	1.159	0.87	-6.59	4.0	0.55	11.39	12.63	-86.10	-0.26	.0405	40.0
54	0.194	9.4	2.6	0.2406	5.554	32.78	-1.13	4.5	0.58	20.02	8.76	-71.80	-0.31	.0233	74.5
55	0.263	5.8	2.1	0.2549	5.948	37.53	-6.44	3.8	0.54	6.78	19.70	-87.90	-0.485	.0500	31.8

$$D_r = \Sigma(G(T))^2/n \tag{3}$$

The statistics D_a and D_r are objective indexes of the goodness of fit of each method to sample data. The observed flood discharge corresponds to a return period which was computed from the Gringorton plotting position formula,

$$T = (n + 0.12)/(m - 0.44) \tag{4}$$

in which m is the rank assigned to each data point in the sample with one for the highest flow, two for the second highest flow and so on, with n for the lowest value. Appropriate plotting position formulae were used for different distributions to maintain unbiasedness.

These criteria were computed for each method of frequency analysis for each sample. Tables 5-6 give the values of D_a and D_r. The six methods were ranked for all the gaging stations according to the values of D_a and D_r on a scale of 1 to 6, with one being the best method. Table 7 ranks the six methods according to D_a. Clearly, the LPT3 distribution was the best of all, followed by the PT, BO, GU, PT3 and LN3 methods in descending order of their performance. It was noted that the GU distribution fitted the data sets well only if the coefficient of skewness (C_s) and coefficient of kurtosis (C_k) were within one standard deviation of their theoretical values of 1.1396 and 5.4 respectively. Its efficiency of fitting diminished appreciably with large departures of C_s and C_k from their theoretical values. Similarly, LN3 distribution did not work well if the C_s for the logarithmically transformed data was more than 1.0. For PT3 distribution MLEM was not always workable for small values of skewness. Further, if b > 1 then C_s and C_k value cannot exceed 2. If C_s < 0, the PT3 distribution becomes bounded at the upper limit and is then rendered unsuitable for analysis of floods. The main difficulty in using MLEM for PT3 distribution was finding the minimum variance solution for each data set which required a careful investigation of the shape of the function. For the 55 data sets, PT3 distribution fitted only 25, a success ratio of 45% only. The BO distribution worked well as long as the parameter S was greater than 0.2, C greater than 8.0, and A greater than 3.0 (refer to stations 1, 3, 4, 17, 26, 41-43, 45, 47-49, 51, 253 in Tables 3, 4 and 5). The effect of each parameter being lower than its respective limit given above was cumulative. The maximum effect was caused by deviation in the C value. As long as the parameters were above their lower bounds, the BO distribution fitted the data well. The method in general underestimated the design discharges. The effect was very pronounced for stations which had coefficient of skewness greater than 3.0.

The PT method performed well in almost all cases. In some cases, it required a large number of iterations to find the correct value of λ. Thus, it might not be easily adaptable to a pocket calculator. The adjustment in the standard deviate K for transformed data sets having a value of the coefficient of kurtosis different from 3.0 was not significant for 42 of the 55 data sets fitted by this method.

Table 5. Mean absolute deviation test (MADT) statistic for the six methods.

St. No.	Methods					
	LPT3	BO	PT3	LN3	GU	PT
1	5.62	7.83	24.49	5.93	14.56	6.61
2	2.05	0.65	1.42	2.52	0.88	2.85
3	2.84	5.82	19.12	3.49	15.76	6.88
4	1.79	3.12	5.75	2.86	2.04	8.96
5	0.12	0.17	0.14	0.55	0.14	0.22
6	0.25	0.24	0.22	0.47	0.27	1.68
7	0.34	0.31	0.29	5.77	0.35	2.11
8	0.21	0.22	0.19	3.73	0.20	1.09
9	0.31	0.32	0.32	8.06	0.44	2.41
10	0.19	0.19	0.20	6.38	0.50	2.27
11	0.27	0.25	0.29	5.70	0.56	1.72
12	0.55	0.48	0.51	7.93	0.49	1.90
13	0.24	0.22	0.21	5.79	0.35	1.71
14	0.49	0.58	0.70	4.42	0.38	1.08
15	0.59	0.67	0.88	5.50	0.49	0.80
16	0.36	0.40	0.33	4.76	0.31	1.43
17	0.44	4.04	0.28	3.59	0.36	0.55
18	0.87	0.99	3.40	4.35	1.10	0.15
19	1.80	1.21	1.59	7.54	0.93	1.67
20	0.45	0.30	0.34	6.11	0.54	1.48
21	0.50	0.39	0.47	7.16	0.67	2.62
22	2.16	1.09	2.52	6.98	1.35	1.08
23	0.50	1.09	0.83	3.73	0.90	0.21
24	1.49	0.88	0.97	10.47	0.80	1.62
25	0.25	0.23	1.06	6.59	0.26	2.16
26	0.53	2.64	0.82	4.76	0.72	0.06
27	0.49	1.02	4.35	4.17	0.91	0.24
28	0.61	0.62	0.61	5.09	0.53	1.69

Table 5 (Continued)

St. No.	Methods					
	LPT3	BO	PT3	LN3	GU	PT
29	0.36	0.21	0.27	9.61	0.23	1.77
30	0.53	0.45	0.87	4.97	0.44	3.20
31	1.50	1.04	5.89	4.65	1.83	4.60
32	1.21	0.55	4.23	4.31	1.11	1.27
33	0.41	0.68	0.57	5.19	0.38	0.43
34	0.56	0.46	0.95	5.13	0.35	0.90
35	0.66	0.83	1.21	4.30	0.66	0.13
36	0.85	1.19	0.64	5.28	0.86	0.14
37	0.14	1.79	2.09	8.63	1.31	0.76
38	0.85	0.96	27.17	6.45	1.64	3.41
39	0.44	0.76	2.15	5.42	1.06	0.52
40	0.48	0.63	0.51	12.86	1.45	5.62
41	1.67	3.60	50.81	7.64	3.37	1.29
42	2.76	18.62	25.89	8.52	3.43	0.37
43	2.08	1.89	9.99	5.34	4.25	0.30
44	2.24	1.28	1.84	7.41	3.76	1.25
45	2.17	6.37	37.68	9.07	4.50	0.98
46	1.81	0.90	22.47	7.08	1.29	4.03
47	1.19	10.00	2.18	7.35	4.28	0.40
48	3.85	27.08	27.98	6.21	4.98	0.24
49	0.11	1.84	12.37	5.20	5.72	0.32
50	0.40	1.58	25.08	3.21	3.98	0.35
51	0.13	18.23	36.84	9.16	5.36	0.25
52	0.41	0.54	4.68	4.58	0.93	0.70
53	1.46	79.67	28.01	12.76	4.41	1.63
54	0.19	1.16	6.83	5.87	2.71	0.53
55	2.34	1.57	22.08	4.24	5.00	5.92

Table 6. Mean square deviation test (MSDT) statistic for the six methods.

St. No.	Methods					
	LPT3	BO	PT3	LN3	GU	PT
1	4.27	23.87	30.58	32.28	25.30	0.32
2	3.60	5.54	6.85	7.35	6.48	4.16
3	11.66	18.66	23.90	27.23	21.21	8.20
4	6.98	13.55	12.99	13.10	10.96	2.49
5	0.12	3.02	2.88	2.60	2.59	1.12
6	0.49	3.87	3.75	3.97	4.43	0.50
7	0.85	4.53	4.39	4.73	4.50	0.90
8	0.39	3.55	3.34	3.54	3.33	0.42
9	1.26	4.83	4.42	4.58	5.05	1.32
10	0.79	3.36	3.31	3.25	5.28	0.90
11	0.92	3.16	2.84	2.89	3.98	0.90
12	1.29	5.06	4.88	4.71	4.59	0.80
13	0.82	3.34	3.30	3.47	3.79	1.26
14	0.39	6.16	6.71	5.63	4.39	0.20
15	0.42	6.59	7.27	6.35	5.99	1.11
16	0.33	4.41	4.03	4.03	4.62	0.43
17	0.24	15.35	4.31	5.32	4.73	0.26
18	0.28	8.47	13.92	7.98	9.07	0.32
19	0.79	9.37	9.93	10.26	8.35	1.35
20	0.89	3.45	3.15	3.10	4.06	0.93
21	0.91	4.79	4.88	4.67	6.04	0.13
22	0.63	9.01	11.83	10.84	9.62	0.11
23	0.39	7.69	6.69	4.27	6.57	0.05
24	2.01	6.28	7.25	8.42	5.86	0.05
25	0.73	3.44	6.30	3.57	3.59	0.12
26	0.46	12.75	7.37	5.18	6.02	0.53
27	0.38	7.13	15.40	4.71	6.45	0.36
28	0.62	5.93	5.81	5.21	5.15	0.92

Table 6 (Continued)

St. No.	Methods					
	LPT3	BO	PT3	LN3	GU	PT
29	1.81	3.80	3.88	4.69	3.69	1.73
30	0.38	4.75	6.06	4.92	4.76	0.36
31	0.48	7.59	17.00	9.03	9.52	1.23
32	0.29	6.02	14.62	8.65	8.63	0.35
33	0.37	6.36	5.88	5.30	5.34	0.39
34	0.50	5.23	6.94	5.63	4.88	0.65
35	0.33	7.26	8.98	6.41	6.20	0.35
36	0.47	9.61	6.64	7.98	7.86	0.45
37	1.50	10.46	12.02	18.63	8.58	1.65
38	0.79	7.29	28.62	6.67	8.92	0.80
39	0.41	7.44	11.50	9.40	8.86	0.39
40	4.73	6.99	6.12	6.10	6.87	4.15
41	1.38	14.72	44.37	8.48	14.25	1.30
42	2.07	31.76	38.95	11.69	14.15	2.35
43	0.53	11.21	24.66	11.96	15.98	0.65
44	0.92	9.25	11.39	12.20	14.54	1.00
45	1.94	18.77	40.19	9.32	15.44	1.96
46	1.09	6.11	24.26	10.20	6.62	1.00
47	0.85	26.94	10.98	7.69	13.09	0.80
48	1.39	40.43	40.22	14.74	17.31	1.45
49	0.43	10.98	27.20	15.42	18.70	0.45
50	0.22	9.24	33.01	8.67	14.79	0.30
51	1.65	33.84	46.88	18.06	19.65	1.71
52	0.31	5.75	14.90	4.97	7.06	0.32
53	4.16	44.83	37.29	12.51	14.65	4.17
54	0.61	7.50	19.20	6.77	11.08	0.65
55	0.39	10.35	31.58	12.57	17.45	0.38

Table 7. Ranking of the frequency analysis methods for the 55 gaging stations by mean absolute deviation test D_a (on a scale of 1 to 6 with 1 being the best method).

Method	Number of Gaging Stations Receiving Ranking					
	1	2	3	4	5	6
BO	11	14	9	11	8	2
GU	9	9	14	17	6	0
PT	15	6	4	9	18	3
PT3	5	6	11	6	9	18
LN3	0	2	2	6	13	32
LPT3	15	18	15	6	1	0

Table 8. Ranking of the frequency analysis methods for the 55 gaging stations by mean square deviation test D_r (on a scale of 1 to 6 with 1 being the best method).

Method	Number of Gaging Stations Receiving Ranking					
	1	2	3	4	5	6
BO	0	0	13	14	14	14
GU	0	0	13	21	13	8
PT	21	34	0	0	0	0
PT3	0	0	8	7	15	25
LN3	0	0	21	13	13	8
LPT3	34	21	0	0	0	0

This adjustment was significant for the remainder of stations, but did not improve the fit materially.

The ranking of the seven methods according to D_r is given in Table 8. Again, the previous conclusions hold except that LN3 distribution was ranked higher than PT3 distribution.

5. CONCLUSIONS

On the basis of fitting the GU, PT, PT3, LN3, BO and LPT3 methods to 55 data sets following inferences can be drawn:

1. The LPT3 method was the most accurate of all, followed by PT, BO, GU, LN3 and PT3 methods.
2. If the parameters, S, A, C were greater than 0.2, 3.0 and 8.0 respectively, then the fit was good for the BO distribution.
3. The maximum effect on the fit of the BO distribution was caused by the deviation of the parameter C from its lower bound of 8.0 (minimum value required) for the fit to be good. The least effect was caused by A for BO distribution.
4. The fit was good for GU distribution if the sample C_s and C_k were near their theoretical values of $C_s = 1.1396$ and $C_k = 5.4$.
5. If the sample skewness for ln(x-a) random variable was close to zero, the LN3 distribution fitted the data well.
6. If the parameter b was less than 0.15, MLEM did not perform well for PT3 distribution.
7. All the data sets were fitted well by LPT3 distribution according to MADT and MSDT tests.
8. For approximately 67% of the stations, the MLEM fit gave a lower estimate of the 100 year flood but only by the order of a few percent for LPT3 distribution.
9. When the C_s of the logarithmically transformed data approximately approached 0.0, it was desirable to fit LN2 distribution instead of LPT3 distribution.
10. The PT method reduced the C_s and C_k of the data. The transformed data tended to be normal in general.
11. The PT method has a small drawback. It cannot control C_s and C_k simultaneously. Thus, in reducing the coefficient of skewness of the data, it may change C_k which originally satisfied the condition of normality.
12. The PT method is versatile as the transformed data can be adjusted for kurtosis.
13. The PT method is very sensitive to the value of λ. C_k was not, however, affected significantly once an approximate value of λ had been found.

6. ACKNOWLEDGMENTS

This study was supported in part by funds provided by the U.S. Geological Survey through the Louisiana Water Resources Research

Institute under the project, "A Multivariate Stochastic Analysis of Floods Magnitude, Duration and Volume."

7. REFERENCES

Beard, L. R., 1974. 'Flood flow frequency techniques.' Technical Report, University of Texas at Austin, Texas.

Benson, M. A., 1968. 'Uniform flood frequency estimating methods for federal agencies.' Water Resources Research, Vol. 4, No. 5, pp. 891-908.

Bobee, B. and Robitaille, R., 1977. 'The use of Pearson type 3 and log Pearson type 3 distributions revisited.' Water Resources Research, Vol. 13, No. 2, pp. 427-443.

Boughton, W. C., 1980. 'A frequency distribution for annual floods.' Water Resources Research, Vol. 16, No. 2, pp. 347-354.

Boughton, W. C., 1983. 'Fitting procedures for the log-Boughton distribution.' Journal of the Hydraulics Division, Proceedings of the ASCE, Vol. 109, No. 4, pp. 579-589.

Box, G. E. P. and Cox, D. R., 1964. 'An analysis of transformation.' Journal of the Royal Statistical Society, Vol. B26, pp. 211-252.

Burges, S. J., Lettenmaier, D. P. and Bates, C. L., 1975. 'Properties of the three-parameter lognormal distribution.' Water Resources Research, Vol. 11, No. 2, pp. 229-235.

Chander, S., Spolia, S. K. and Kumar, A., 1978. 'Flood frequency analysis by power transformation.' Journal of the Hydraulics Division, Proceedings of the ASCE, Vol. 104, No. HY11, pp. 1495-1504.

Cicioni, G., Giuliano, G. and Spaziani, F. M., 1979. 'Best fitting of probability functions to a set of data for flood studies.' Proceedings of the Second International Symposium on Hydrology of Floods and Droughts, Water Resources Publications, Fort Collins, Colorado, pp. 304-314.

Fiorentino, M. and Versace, P., 1984. 'Two-component extreme value distribution for flood frequency analysis.' Water Resources Research, Vol. 20, No. 7, pp. 847-856.

Loganathan, G. V., Kuo, C. Y. and McCormick, T. C., 1985. 'Methods of analyzing instream flows.' Virginia Water Resources Research Center, Bulletin 148, Blacksburg, Virginia.

Moore, M. S. and Chong, She-King, 1983. 'Flood frequency analysis for small watersheds in southern Illinois.' Water Resources Bulletin, Vol. 19, No. 2, pp. 277-282.

Sangal, B. P. and Biswas, A. K., 1970. 'The 3-parameter lognormal distribution and its applications in hydrology.' Water Resources Research, Vol. 6, No, 2, pp. 505-515.

Water Resources Council, 1981. 'A uniform technique for determining flood flow frequencies.' Bulletin 17-B, Hydrology Committee, Washington, D.C.

USE OF HISTORICAL DATA IN FLOOD-FREQUENCY ANALYSIS

Jorge Machado Damazio and Jerson Kelman
Centro de Pesquisas de Energia Eletrica, CEPEL
C.P.2754
Cidade Universitaria, Ilha do Fundao
Rio de Janeiro, Brazil

ABSTRACT. It is presented an assessment of several flood frequency procedures for the estimation of the 10,000 years flood which takes into account historical information, as for example when one knows the maximum flow in a period of 100 years. Monte Carlo experiments are employed. The conclusion favors the use of the exponential distribution.

1. INTRODUCTION

The spillway design flood for dams can be calculated through flood frequency analysis, which is a set of procedures to estimate x(T), the flow which will be exceeded in any year with probability 1/T. For major hydraulic structures T is chosen to be as large as 10,000 years. The most usual approach to estimate x(T) is to fit a probability distribution F(.) to the annual maximum flow series. Due to the short length of these series (usually less than 50 years), as compared with the required return period, these estimators have large sample variation. Research on this area has been directed to find out which is the best fitting distribution and which estimation procedure should be employed. Special consideration ought to be given to the case when historical information is available, as for example when one knows the maximum flow in a period of 100 years and the streamgage record is only 10 years long. In this paper it is presented an assessment of several flood frequency procedures which take into account the historical information. Monte Carlo experiments are employed. Special attention is given to the estimators of the ten thousand years flow. Similar investigations can be found in Tasker and Thomas (1978), Condie and Lee (1982), Cohn (1984) and Hosking and Wallis (1986).

The experiment is performed using a set of twelve Wakeby parents chosen to represent typical brazilian annual flood data. For each one of them a major sequence of 100,000 independent values is generated and then sequentially split into sequences with h elements: n of these elements simulate gauge data and (h-n) elements simulate the

V. P. Singh (ed.), Hydrologic Frequency Modeling, 487–497.

historical information. That is, only the greatest of the (h-n) elements, u, is kept.

Two-parameters distributions (Gumbel and Exponential) and three-parameter distributions (Pearson type III, log-pearson type III and Generalized extreme-value) are fitted by the method of the moments for each sequence. The two-parameter distributions are also fitted by the maximum-likelihood method. Consideration of the historical event is done in the method of the moments using a weighted-moment procedure as that of Bulletin 17B (U.S. Water Resources Council, 1982) and in the method of the maximum-likelihood using the likelihood function given by:

$$\left\{\prod_{i=1}^{n} g(x_i;\theta)\right\} \; g(u,;\theta) \; \left\{G(u;\theta)\right\}^{h-n-1} \tag{1}$$

where G(x;) and g(x;) are respectively the distribution and density functions of the fitted distribution and is the vector of unknown parameters. Each procedure is also tested ignoring the historical event. Detailed description of the procedures are given in Damazio and Kelman (1984). The procedures are assessed considering the mean absolute error of the estimate, given by:

$$MAE = k^{-1} \sum_{j=1}^{k} \left| \frac{\bar{x}_j - x}{x} \right| \tag{2}$$

where x_j is the estimate given by the j-th sequence, x is the population value and k is the number of sequences.

In the presentation and discussion of the results obtained the procedures are referred by the names listed in Table I.

TABLE I: TESTED PROCEDURES

PROCEDURE	DISTRIBUTION	FIT	HISTORICAL DATA
EPR	Exponential	moments	no
EPR*	Exponential	maximum likelihood	no
GUR	Gumbel	moments	no
GUR*	Gumbel	maximum likelihood	no
LP3R	Log-Pearson	log-moments	no
P3R	Pearson	moments	no
GEVR	Generalized extreme value	moments	no
EPM	Exponential	moments	yes
EPM*	Exponential	maximum likelihood	yes
GUM	Gumbel	moments	yes
GUM*	Gumbel	maximum likelihood	yes
LP3M	Log-Pearson	log-moments	yes
P3M	Pearson	moments	yes
GEVM	Generalized extreme value	moments	yes

2. THE MONTE-CARLO EXPERIMENT

The experiments were designed to test the procedures in a set of conditions:

a) Historical period (h): 50, 100 and 150 years.
b) Record length (n): 5, 10, 25 and 50 years.
c) Parent distribution: Twelve conditions were tested, using Wakeby distributions. Figure 1 illustrates the position of each of these parents in a kurtosis versus skewness plane.

It is also included in Figure 1 points that correspond to annual maximum flow sequences for different regions in Brazil. Table II presents the parameters and the first four moments for each of the twelve parents and presents some of the quantiles.

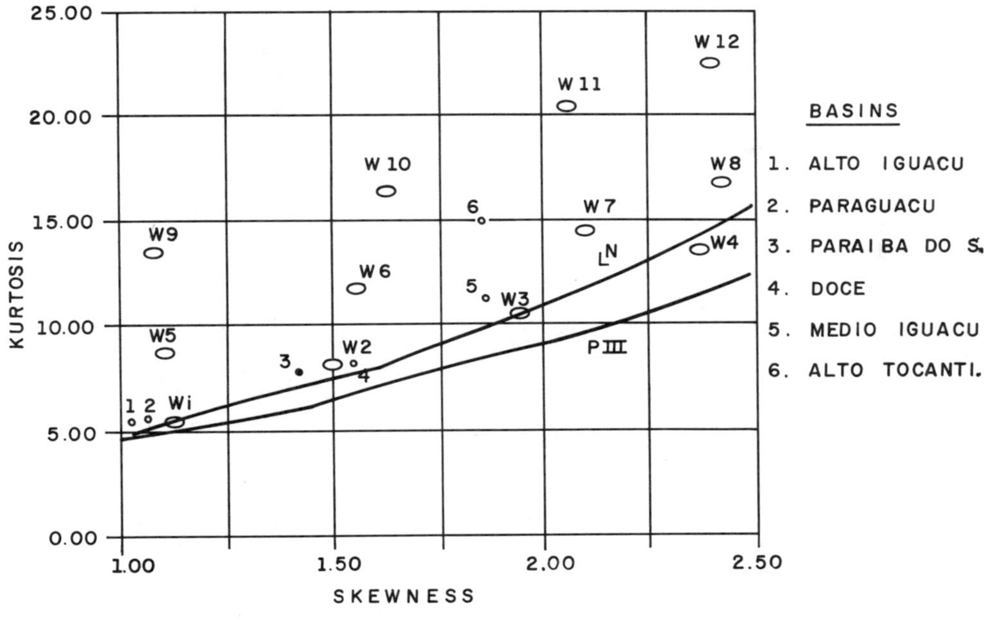

FIGURE 1. SKEWNESS AND KURTOSIS OF THE WAKEBY DISTRIBUTIONS

3. THE RESULTS

The method which gives the minimum MAE for each condition is presented in Table III, where it is also given the correspondent MAE value. It was also calculated an efficiency coefficient for each method given by the ratio between the minimum MAE along the methods and the MAE of the considered method. Table IV presents the mean of this coefficients obtained through the twelve parents for each method and pair (h,n).

Table V presents the minimum coefficient obtained through the twelve parents for each method and pair (h,n).

4. ANALYSIS OF THE RESULTS

It can be verified that the two parameter distributions yielded results better than the three parameter distributions. This is due to the fact that adjusting only two parameters one avoids the influence of the highly unstable sample skewness.

Considering the sequences generated by the parent w-1 the fit of the Gumbel distribution by the maximum likelihood method presents the lowest MAE for all pairs (h,n) that were tested. This is due to the fact that the skewness and the kurtosis of this population are very close to the correspondent values of the Gumbel distribution. Considering the sequences generated by the parent w-3, which skewness and kurtosis are very close to the exponential, the fit of this distribution by the maximum likelihood method presents the lowest MAE for all combinations (h,n) that were tested. As could be expected, the fit of the Gumbel distribution to the sequences generated by the parent w-3 and the fit of the exponential distribution to the sequences generated by the parent w-1 presented low performances. Figure 2 shows the variation of MAE when the Gumbel and the Exponential distributions are used as fitting distributions with parameters estimated by maximum likelihood. It can be noted that the MAE can increase with the length of the record when the fitting distribution is too different from the parent population.

As the identification of the population from a sample is always unrealiable one should choose the fitting distribution taking into account the performance through the twelve parents. For example, if we consider the twelve tested parents as equally probable, it can be selected the fitting distribution which has more oftenly obtained the lowest MAE. The last column of table III presents the recommended procedure using this criteria, which is equivalent to maximizing the probability of using the best procedure. For all pairs (h,n) this criteria recommends the fit of the exponential distribution by the method of the maximum likelihood. Curiously, if n = 50, the criteria recommends to abandon the historical data.

Also considering the twelve parents as equally probable, it would be rational to include a penalty to the procedures proportional to the loss of efficiency. This is done by adopting the procedure with the greatest mean efficiency. In this case it is also recommended (Table IV) the fit of the exponential distribution by the maximum likelihood for every tested pair (h,n), but the recommendation to abandon the historical information when the gauge period approaches 50 years doesn't hold anymore.

Any of the two criteria adopted above can be used in conjunction with a set of "a priori" probabilities for the twelve parents but the resultant recommendation will depend on this distribution. The definition of the "a priori" probabilities poses a problem similar to the identification of the population.

TABLE II: THE TWELVE WAKEBY DISTRIBUTION USED AS PARENT DISTRIBUTIONS

Wakeby	a	b	c	d	m	E(X)	Std.Var.(X)	Skew(X)	Kurt(X)	x(T) T=1000	x(T) T=10000
W - 1	0.55	2.00	8.24	0.04	0.29	1.00	0.49	1.12	5.46	3.46	4.51
W - 2	0.49	2.00	3.45	0.09	0.33	1.00	0.49	1.50	8.13	3.79	5.27
W - 3	0.32	1.50	3.80	0.09	0.43	1.00	0.49	1.95	10.52	4.03	5.66
W - 4	0.14	1.50	4.19	0.09	0.50	1.00	0.49	2.37	13.03	4.25	6.05
W - 5	0.89	1.50	0.89	0.19	0.25	1.00	0.49	1.11	8.76	3.56	5.37
W - 6	0.65	4.00	1.96	0.14	0.16	1.00	0.49	1.56	11.87	4.01	5.97
W - 7	0.42	2.00	2.08	0.14	0.38	1.00	0.49	2.10	14.37	4.19	6.27
W - 8	0.31	1.50	2.18	0.14	0.46	1.00	0.49	2.42	16.42	4.32	6.51
W - 9	0.93	4.00	1.06	0.19	0.00	1.00	0.49	1.07	13.50	3.81	5.97
W - 10	0.73	2.50	1.13	0.19	0.22	1.00	0.49	1.63	16.32	4.02	6.32
W - 11	0.60	2.00	1.20	0.19	0.32	1.00	0.49	2.05	20.36	4.18	6.63
W - 12	0.53	1.15	1.22	0.19	0.43	1.00	0.49	2.39	22.58	4.27	6.76

Note: A Wakeby distribution is described by $X = m + a(1 - (1 - U)^b) - c(1 - (1 - U)^{-d})$ where U is a uniform (0,1) random variable.

TABLE III : MINIMUM MAE

H	N	W - 1	W - 2	W - 3	W - 4	W - 5	W - 6	W - 7	W - 8	W - 9	W - 10	W - 11	W - 12	WINNER
150	50	GUM* (0,08)	EPM (0,10)	EPM* (0,09)	EPR* (0,17)	EPM (0,10)	EPM* (0,13)	EPR* (0,09)	EPR* (0,18)	EPM (0,18)	EPM* (0,08)	EPR* (0,08)	EPR* (0,17)	EPR*
	25	GUM* (0,09)	EPM* (0,11)	EPM* (0,11)	EPM* (0,17)	EPM* (0,10)	EPM* (0,11)	EPR* (0,13)	EPR* (0,20)	EPM* (0,12)	EPM* (0,12)	EPR* (0,10)	EPR* (0,18)	EPM*
	10	GUM* (0,12)	EPM* (0,14)	EPM* (0,16)	EPM* (0,19)	EPM* (0,15)	EPM* (0,17)	EPR* (0,20)	EPM* (0,23)	EPM* (0,17)	EPR* (0,20)	EPR* (0,18)	EPR* (0,25)	EPM*
	5	GUM* (0,13)	EPM* (0,16)	EPM* (0,17)	EPM* (0,19)	EPM* (0,18)	EPM* (0,21)	EPM* (0,24)	EPM* (0,24)	EPM* (0,22)	EPM* (0,25)	EPM* (0,25)	EPM* (0,28)	EPM*
100	50	GUM* (0,08)	EPM (0,11)	EPM* (0,09)	EPR* (0,17)	EPM* (0,10)	EPM* (0,16)	EPR* (0,09)	EPR* (0,17)	EPM (0,18)	EPM* (0,09)	EPR* (0,08)	EPR* (0,17)	EPR*
	25	GUM* (0,10)	EPM* (0,11)	EPM* (0,12)	EPM* (0,18)	EPM* (0,11)	EPM* (0,11)	EPR* (0,13)	EPR* (0,19)	EPM* (0,14)	EPM* (0,11)	EPR* (0,11)	EPR* (0,18)	EPM*
	10	GUM* (0,13)	EPM* (0,14)	EPM* (0,17)	EPM* (0,21)	EPM* (0,15)	EPM* (0,17)	EPR* (0,20)	EPM* (0,24)	EPM* (0,18)	EPR* (0,18)	EPR* (0,19)	EPR* (0,25)	EPM*
	5	GUM* (0,15)	EPM* (0,17)	EPM* (0,20)	EPM* (0,22)	EPM* (0,20)	EPM* (0,22)	EPM* (0,24)	EPM* (0,27)	EPM* (0,24)	EPM* (0,26)	EPM* (0,28)	EPM* (0,28)	EPM*
50	50	GUR* (0,09)	EPR (0,12)	EPR* (0,09)	EPR* (0,17)	EPR (0,12)	EPR (0,19)	EPR* (0,09)	EPR* (0,18)	EPR (0,19)	EPR* (0,16)	EPR* (0,08)	EPR* (0,17)	EPR*
	25	GUM* (0,10)	EPM* (0,13)	EPM* (0,12)	EPR* (0,19)	EPM (0,13)	EPM* (0,13)	EPR* (0,13)	EPR* (0,19)	EPM* (0,19)	EPM* (0,10)	EPR* (0,12)	EPR* (0,18)	EPM*
	10	GUM* (0,14)	EPM* (0,15)	EPM* (0,18)	EPM* (0,22)	EPM* (0,14)	EPM* (0,17)	EPR* (0,20)	EPR* (0,24)	EPM* (0,16)	EPM* (0,19)	EPR* (0,19)	EPR* (0,24)	EPM*
	5	GUM* (0,16)	EPM* (0,18)	EPM* (0,21)	EPM* (0,24)	EPM (0,20)	EPM* (0,23)	EPM* (0,26)	EPM* (0,28)	EPM* (0,25)	EPM* (0,26)	EPR* (0,28)	EPM* (0,31)	EPM*
H = N	25	GUR* (0,11)	EPR (0,16)	EPR* (0,12)	EPR* (0,18)	EPR (0,13)	EPR (0,21)	EPR* (0,13)	EPR* (0,20)	EPR (0,21)	EPR* (0,15)	EPR* (0,10)	EPR* (0,18)	EPR*
	10	GUR* (0,17)	EPR (0,22)	EPR* (0,20)	EPR* (0,24)	EPR (0,19)	EPR* (0,24)	EPR* (0,20)	EPR* (0,25)	EPR (0,25)	EPR* (0,20)	EPR* (0,18)	EPR* (0,25)	EPR*
	5	GUR* (0,25)	EPR (0,30)	EPR* (0,30)	EPR* (0,33)	EPR (0,26)	EPR* (0,29)	EPR* (0,29)	EPR* (0,33)	EPR (0,31)	EPR* (0,27)	EPR* (0,27)	EPR* (0,32)	EPR*

TABLE IV: MEAN EFFICIENCY COEFFICIENT

H	N	EPR	EPR*	GUR	GUR*	LP3R	P3R	GEVR	EPM	EPM*	GUM	GUM*	LP3M	P3M	GEVM
150.	50.	0.63	0.70	0.45	0.44	0.29	0.35	0.33	0.69	0.82	0.46	0.47	0.36	0.46	0.41
150.	25.	0.59	0.71	0.45	0.44	0.23	0.33	0.31	0.68	0.88	0.47	0.49	0.33	0.43	0.43
150.	10.	0.66	0.76	0.54	0.54	0.18	0.39	0.36	0.75	0.96	0.57	0.71	0.37	0.55	0.55
150.	5.	0.61	0.69	0.53	0.53	0.12	0.40	0.38	0.71	0.99	0.58	0.84	0.38	0.62	0.62
100.	50.	0.64	0.71	0.46	0.46	0.30	0.35	0.34	0.71	0.80	0.48	0.48	0.36	0.40	0.39
100.	25.	0.59	0.71	0.45	0.45	0.24	0.34	0.32	0.69	0.88	0.48	0.49	0.34	0.42	0.41
100.	10.	0.65	0.77	0.54	0.54	0.20	0.40	0.37	0.77	0.95	0.58	0.67	0.37	0.55	0.54
100.	5.	0.64	0.71	0.57	0.57	0.14	0.43	0.42	0.78	0.99	0.64	0.84	0.41	0.65	0.65
50.	50.	0.73	0.77	0.51	0.51	0.33	0.39	0.37	0.73	0.77	0.51	0.51	0.33	0.39	0.37
50.	25.	0.65	0.74	0.49	0.49	0.26	0.36	0.34	0.72	0.88	0.52	0.52	0.32	0.41	0.39
50.	10.	0.64	0.78	0.54	0.55	0.19	0.40	0.37	0.79	0.94	0.60	0.64	0.36	0.51	0.49
50.	5.	0.69	0.76	0.61	0.60	0.12	0.46	0.44	0.86	0.99	0.69	0.80	0.36	0.65	0.64
25.	25.	0.74	0.82	0.56	0.56	0.29	0.40	0.37	0.74	0.82	0.56	0.56	0.29	0.40	0.37
10.	10.	0.80	0.90	0.66	0.67	0.22	0.47	0.44	0.80	0.90	0.66	0.67	0.22	0.47	0.44
5.	5.	0.86	0.95	0.76	0.76	0.17	0.57	0.54	0.86	0.95	0.76	0.76	0.17	0.57	0.54

TABLE V: MINIMUM EFFICIENCY COEFFICIENT

H	N	EPR	EPR*	GUR	GUR*	LP3R	P3R	GEVR	EPM	EPM*	GUM	GUM*	LP3M	P3M	GEVM
150.	50.	0.30	0.17	0.22	0.21	0.17	0.18	0.17	0.31	0.25	0.22	0.22	0.19	0.20	0.20
150.	25.	0.34	0.19	0.26	0.25	0.16	0.21	0.20	0.37	0.39	0.27	0.28	0.20	0.26	0.26
150.	10.	0.56	0.28	0.45	0.44	0.08	0.32	0.30	0.60	0.75	0.46	0.56	0.27	0.40	0.38
150.	5.	0.46	0.30	0.44	0.41	0.05	0.32	0.29	0.61	0.87	0.50	0.78	0.23	0.48	0.46
100.	50.	0.30	0.16	0.22	0.21	0.21	0.19	0.18	0.31	0.22	0.22	0.21	0.22	0.21	0.21
100.	25.	0.38	0.22	0.29	0.28	0.20	0.24	0.22	0.41	0.38	0.30	0.30	0.23	0.27	0.26
100.	10.	0.56	0.31	0.46	0.45	0.13	0.33	0.30	0.63	0.76	0.49	0.54	0.28	0.40	0.38
100.	5.	0.50	0.34	0.49	0.47	0.07	0.36	0.33	0.69	0.94	0.58	0.80	0.25	0.54	0.53
50.	50.	0.30	0.19	0.22	0.21	0.21	0.19	0.18	0.30	0.19	0.22	0.21	0.21	0.19	0.18
50.	25.	0.40	0.22	0.29	0.31	0.18	0.21	0.20	0.43	0.31	0.30	0.31	0.22	0.23	0.21
50.	10.	0.56	0.33	0.45	0.45	0.12	0.29	0.28	0.63	0.70	0.49	0.51	0.28	0.34	0.32
50.	5.	0.55	0.36	0.55	0.52	0.07	0.38	0.36	0.82	0.89	0.64	0.76	0.24	0.54	0.51
25.	25.	0.34	0.23	0.26	0.26	0.17	0.21	0.20	0.34	0.23	0.26	0.26	0.17	0.21	0.20
10.	10.	0.56	0.40	0.45	0.44	0.10	0.35	0.33	0.56	0.40	0.45	0.44	0.10	0.35	0.33
5.	5.	0.71	0.58	0.61	0.60	0.09	0.48	0.46	0.71	0.58	0.61	0.60	0.09	0.48	0.46

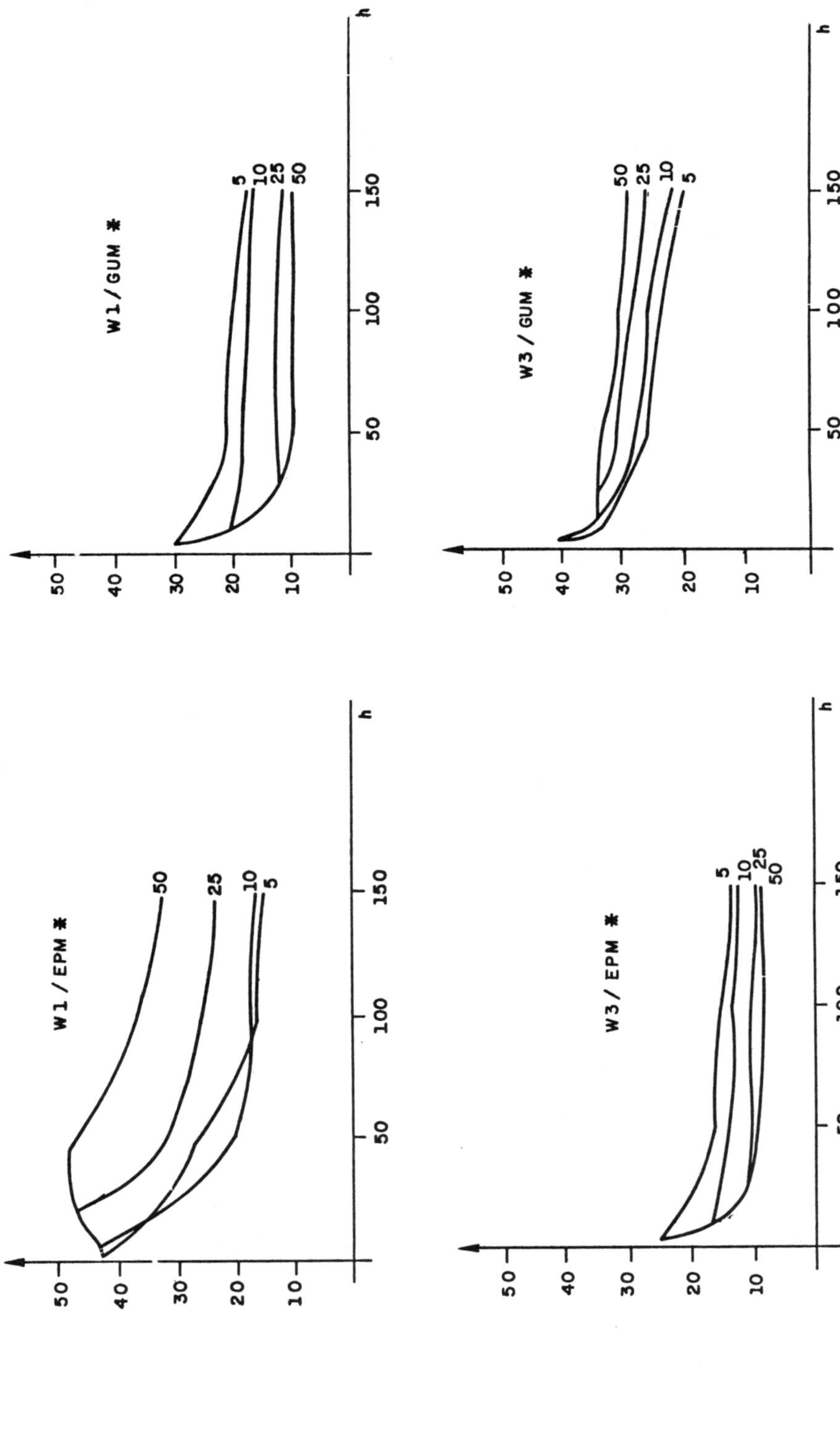

FIGURE 2. MAE FOR W-1 AND W-3

Kuczera, (1982) presents an interesting approach to the problem of the choice of the fitting distribution: he considers that a model should be capable to estimate extreme events, for every parent population, without strong loss of efficiency. The criteria of maximizing the minimum efficiency is coherent with this principle. Furthermore it independs on the "a priori" probabilities for the parents.

It is shown in Table V that the exponential distribution is recommended again by the criteria of the maximum efficiency for every tested pair (h,n). The method of fitting varies. If $h \leq 4n$ it is recommended the use of the method of moments. If $h > 4n$ it is recommended the use of the maximum likelihood. Historical data should be always used.

5. CONCLUSIONS

a) The results obtained with the three parameter distributions were always worse than the results obtained with two parameter distributions.

b) The expected errors of estimates can grow with the quantity of available information when one uses distributions that are very different from the population.

c) The identification of the population is usually precarious. The alternative is to choose "a priori" a distribution that is capable to estimate without strong loss of efficiency, disregarding the population. The maximization of the minimum efficiency indicated, for the tested parents, the use of the exponential distribution fitted by the method of the moments when $h \leq n$ and fitted by the method of the maximum likelihood when $h > 4n$.

6. REFERENCES

COHN, T., 1984. The Incorporation of Historical Information in Flood Frequency Analysis, M.Sc. Thesis, Dept. of Environmental Engineering, Cornell University.

CONDIE, R. and LEE, K., 1982. "Flood Frequency Analysis with Historic Information", J. Hydrology, 58, pp. 47-61.

DAMAZIO, J.M. and KELMAN, J., 1984. Use of Historical Data in Flood-Frequency Analysis. Technical Report No. 650/84, CEPEL, Rio de Janeiro, Brazil.

HOSKING, J.R.M. and WALLIS, J.R., 1986. "Paleoflood Hydrology and Flood Frequency Analysis", Water Resources Research, Vol. 22, No. 4, pp. 543-550.

KUCZERA, G., 1982. "Robust Flood Frequency Models", Water Resources Research, vol. 18, No. 2, pp. 315-324.

TASKER, G.D. and THOMAS, W.O., 1978. "Frequency Analysis with Pre-record Information", J. Hydraul Div. ASCE, 104, pp. 249-259.

U.S. Water Resources Council, 1982. Guidelines for Determining Flood Flow Frequency, Bulletin 17B, Hydrology Committee, Washington, D.C.

THE 1983 IGUAÇU RIVER FLOOD
EFFECT OF A RARE FLOOD ON FREQUENCY ANALYSES

Heinz D. Fill
Martha R. v. Borstel Sugai
Nelson L. de S. Pinto
CEHPAR - University of Paraná - COPEL, Caixa Postal 1309
80000 Curitiba, PR, BRAZIL

ABSTRACT. Methods of flood frequency analysis commonly used in engineering studies are evaluated and criticized on the basis of actual hydrological data which include a recent extraordinary flood. The analysis is restricted to the 67,300 km^2 Iguaçu river basin, almost entirely situated in Brazilian territory.

The criteria for selection of the best theoretical frequency distribution are analysed and verified to be clearly dependent on the extension of the series and on subjective decisions such as the selection of class intervals. Design flood estimates are found to be dramatically influenced by the extension of the hydrological series or even by the methods used to define the parameters of the best fit theoretical probability distribution function. The relativity of such a concept as the 10,000 yr design flood, either due to limitations of the existing sample or to the subjectivity of the methodology used, has become evident.

1. INTRODUCTION

The design flood for any hydraulic project is always related to a certain degree of risk, clearly or implicity accepted by the designer. Estimating that risk is one of the main purposes of the flood frequency analysis.

Two basic problems in such analyses are: (1) the criteria for selection of the best theoretical frequency distribution, and (2) the correct evaluation of the parameters of the distribution function.

Those problems are not satisfactorily solved as yet. Results depend strongly on the available sample, as will be shown in this work. It is recognized that no one of the classical extreme distribution functions can be chosen "a priori" on the basis of their own theoretical reasoning. Tests of adherence have to be carried out, but the doubt remains on how representative the distributions are in relation to the actual distribution of the population of events.

Referring to the classical Ven Te Chow formula:

V. P. Singh (ed.), Hydrologic Frequency Modeling, 499–504.

$$q_T = \bar{q} + K_T\, s \tag{1}$$

where $\bar{q}$ - temporal average of the maximum observed annual floods, s - sample standard deviation, and K_T - recurrence factor, dependent of the distribution function and the recurrence interval; it is clear the influence, on $\bar{q}$ and s, of one or two extraordinary flow values eventually included in the observed series. The use of cumulative distribution curves as suggested by Roche (1963) is of no great help either. In spite of such well known drawback, 30 to 40 yr series are frequently accepted as representative for selecting the best distribution function and to define its parameters.

The following Iguaçu river flood analyses, will refer to the 1926-83; 1937-82; and 1936-83 series, Table I. The series were so chosen as to emphasize the influence of the two major floods on the selection of the best probability function distribution and on the definition of their parameters, as well as on the design flood studies.

TABLE I
IGUAÇU RIVER MAXIMUM ANNUAL FLOODS - CATARATAS GAGING STATION ($67,300\ km^2$)

Year	Discharge (m^3/s)	Year	Discharge (m^3/s	Year	Discharge (m^3/s)	Year	Discharge (m^3/s)
1926	7,480	1941	8,550	1956	7,090	1971	10,700
1927	8,430	1942	4,870	1957	15,900	1972	15,000
1928	17,500	1943	3,890	1958	6,430	1973	7,370
1929	6,480	1944	1,880	1959	4,470	1974	4,550
1930	7,070	1945	3,600	1960	4,600	1975	9,200
1931	12,600	1946	7,710	1961	5,600	1976	5,720
1932	15,200	1947	6,880	1962	5,670	1977	3,780
1933	2,220	1948	6,080	1963	8,470	1978	4,630
1934	3,770	1949	2,920	1964	5,740	1979	11,000
1935	13,900	1950	6,800	1965	9,020	1980	4,800
1936	28,600	1951	5,410	1966	5,600	1981	9,030
1937	9,960	1952	9,790	1967	4,350	1982	16,800
1938	15,500	1953	8,090	1968	4,530	1983	35,300
1939	9,540	1954	7,870	1969	7,710		
1940	3,100	1955	12,100	1970	5,740		

2. SELECTION OF THE BEST FIT DISTRIBUTION

The selection of the best theoretical distribution followed a systematic approach in which the influence of the extension of the record and of the selected class intervals were easily detected.

Seven probability distribution functions were used: 1) two parameter log-normal; 2) three parameter log-normal; 3) exponential; 4) Goodrich; 5) Gumbel; 6) Pearson type III; 7) log-Pearson type III.

Except for the Gumbel distribution, parameters were estimated by the method of moments according to well known procedures, (Roche, 1963; Kite, 1977). The least square method was used to adjust the Gumbel distribution (Kamogawa, 1981).

Goodness of fit tests used were Chi-Square and the likelihood ratio test, the latter following the procedures indicated by Hoel (1980). To make the results of two and three parameter distributions comparable, the computed values of the logarithm of the likelihood function were divided by the corresponding degrees of freedom.

Three criteria for division in classes were used; I) class intervals of same amplitude with an average of 5 elements in each class (Roche, 1963; Snedecor and Cochran, 1980); II) number of classes defined by Sturges empirical formula:

$$k = 1 + 3{,}3 \log n$$

where n - size of the sample, k - number of classes (two extreme classes with expected frequency but no observed frequency are added (Gomide and Kaviski, 1981), and III) class intervals with at least 5 observed values in each class (Roche, 1963).

The degrees of freedom were computed for each distribution and each different class division according to

$$DF = n - 1 - p$$

where n - number of classes, and p - number of parameters estimated from the sample (two or three).

The seven theoretical distributions were classified according to the fitness tests (Chi-Square and likelihood ratio) for the three different series and class intervals, as shown in Table II.

Basic findings of this study are as follows: (1) The method of class division has a definite influence in the classification of the best fit distribution, especially when defined by the Chi-Square method. (2) Different series extensions may result in different best fit theoretical distributions. For that matter, the Chi-Square method is less stable than the likelihood ratio method. (3) It seems clearly demonstrated that a 30 to 40 yr series is not long enough to define the best fit theoretical distribution. That is more clearly so, when the Chi-Square method is used to define the goodness of fit.

3. DESIGN FLOOD ESTIMATES

After selection of the best fit probability distribution function, frequency analysis is oftenly used to estimate rare events such as the 10,000 yr flood for design purposes. The extension of the series and the method of estimating the parameters of the distribution may have a considerable effect on the results.

Table III presents the values for the 10,000 yr flood computed according to the two most promising distributions, two parameter log-normal and log-Pearson type III.

TABLE II
BEST FIT DISTRIBUTION AS INFLUENCED BY CLASS DIVISION AND EXTENSION OF THE SERIES

Series	1937 - 1982						1936 - 1983						1926 - 1986					
	Chi-Square			likelihood ratio			Chi-Square			likelihood ratio			Chi-Square			likelihood ratio		
Distribution	div I	div II	div III	div I	div II	div III	div I	div II	div III	div I	div II	div III	div I	div II	div III	div I	div II	div III
Gumbel	3	1*	5	2	2	2	7*	6*	5*	3	3	3	7	4*	6*	3	3	3
Fuller	6*	4*	7	6	3	3	6	2	4	2	1	2	4	2	3	2	1	3
Two-parameter lognormal	1	2*	1	1	1	1	2	3	2	1	2	1	1	2	1	1	2	1
Three-parameter lognormal	5	3*	4	5	7	5	4	4	3	5	6	6	3	3	2	7	6	4
Goodrich	4	2*	3	4	4	5	5	4	2	7	5	5	2	3	4	5	5	6
Pearson Type III	2	3*	2	3	5	4	3	5	6*	6	5	7	6	3	7*	6	5	7
Log-Pearson Type III	2	3*	6	3	6	6	1	1	1	4	4	4	5	1	5	4	4	5

* Rejected at a level of significance of 5%. (Mood, Graybill and Boes, 1974).

TABLE III
TEN THOUSAND YEAR FLOOD AS INFLUENCED BY THE EXTENSION OF THE RECORD

Distribution	Ten thousand year flood (m^3/s)		
	1927-82 (n=46)	1936-83 (n=48)	1926-83 (n=58)
Two-parameter lognormal	34,700	72,900	68,500
Log-Pearson type III	32,700	123,000	89,200

The two parameter distribution is somewhat more stable. The larger variations observed for the log-Pearson type III distribution results mainly from the high sensibility of the skewness coefficient to the extreme values. The results show very large differences and will have to be considered subjectively in flood design studies. Both results point out the limitations of the flood frequency analysis in estimating such events, in spite of the long historical series.

Even the method of estimating the parameters of the distribution may have a considerable effect on the results. The most commom methods currently used, the maximum likelihood method, the method of moments, and the least square method were investigated, and their influence illustrated by the results obtained for the ten thousand year flood, estimated from the 58 year record, Table IV.

TABLE IV
TEN THOUSAND YEAR FLOOD AS INFLUENCED BY THE METHOD OF ESTIMATING THE PARAMETERS OF THE DISTRIBUTION

Distribution	Ten thousand year flood (m^3/s)		
	Moments	Max. likelihood	Least square
Gumbel	48,200	38,000	51,700
Two-parameter lognormal	68,500	58,200	62,400
Galton	69,900	67,900	-
Log-Pearson type III	89,200	80,100	-

Although less spectacular than the effect due to the extension of the series, the effect shown in table IV, in which differences of up to 36% are detected, is far from being negligible.

Some practical conclusions from the analysis of the results, summarized in Tables III, and IV, are as follows: (1) The influence of the period of observation on the flood frequency results is greater than that of the type of distribution or of the method of estimating the parameters of the distribution. (2) The errors in the estimated design floods can be larger than 100% depending on the caracteristics of the historical series, even for 40 to 50 yr records. (3) Two parameter distributions are more stable than three parameter distributions for estimating rare floods.

The conclusions would not change if the 1983 flood had been treated as an outlier due to its low probability, certainly smaller than 1:58. In this case some of the differences would be less striking, but the subjectivity of the analysis would remain.

4. REFERENCES

Gomide, F.L.S., Kaviski, E. 1981. 'Poder de Testes de Adequação de Ajuste: Uma Investigação Experimental' in IV Simpósio Brasileiro de Hidrologia e Recursos Hídricos, vol. 1, pp. 551-562.

Hoel, B.J. 1980. Estatística Matemática, Guanabara, pp-334-337.

Kamogawa, L. F. 1981. 'Análise Crítica dos Fatores de Freqüência Tabelados por Leonard L. Weiss para a Lei dos Extremos tipo I' in IV Simpósio Brasileiro de Hidrologia e Recursos Hídricos, vol. 1, pp. 569-577.

Kite, G.W. 1977. 'Frequency and Risk Analyses in Hydrology', Water Power Resources Publications, Fort Collins, Colorado. 224 p.

Mood, A.M., Graybill F.A., Boes, D.C. 1974. Introduction to the Theory of Statistics, Mc Graw-Hill Kagakusha Ltd., Tokio, p. 553.

Pinto, N. L. S., Holtz, A.C.T., Martins, J.A., Gomide, F.L.S. 1976 Hidrologia Básica, Edgar Blücher, São Paulo, p. 243.

Roche, M. 1963. Hydrologie de Surface, Gauthier-Villars, Paris, pp. 17-43.

Snedecor, G.W., Cochran, W.G. 1980. Statistical Methods, The Iowa State University Press, Ames, pp. 75-78.

ANALYSIS OF FLOOD FREQUENCIES IN THE CAUVERY VALLEY

M.RAMESH
Lecturer, National Institute of Engineering, Mysore, India

M.C.SRINIVASA MURTHY
Assistant Professor, National Institute of Engineering, Mysore, India

RAMA PRASAD
Professor, Department of Civil Engineering, Indian Institute of Science, Bangalore, India

ABSTRACT. Annual peak flood records from ten subbasins in the upper reaches of the river Cauvery in south India were analysed. It was found that floods of longer return periods were correlated well with the drainage area of the subbasin, while the correlation for floods of short return periods was poor. Correlation with normal annual rainfall in the subbasin was also found to be poor. Return periods for design floods based on empirical formulae, used in the region for small reservoirs (called 'tanks'), were calculated using the fitted distributions. These are much too high in relation to the expected life of the tanks in all cases but one. This would suggest that while the majority of the tank spillways were overdesigned, some would be underdesigned. This conclusion is also supported by the fact that breach due to inadequate spillway capacity is rare.

1. INTRODUCTION

There are thousands of small reservoirs called tanks in south India, some of them a thousand years old and few hundred new ones are being built every year. In the state of Karnataka (Geographical area 192 000 km^2), there are 38 000 tanks, besides 35 bigger reservoirs. The tanks irrigate an area of about 600 000 ha, out of a total of 2.2 million ha, while the bigger reservoirs account for a million ha. The spillways for most of, if not all, these reservoirs, especially the smaller ones, were designed on the basis of empirical formulae of the type

$$Q = C\ A^{m} \tag{1}$$

Where Q is the design flood, C a coefficient, A the drainage area and m an exponent. The value of m is sometimes taken as 2/3 as recommend by Ryves or 3/4 as recommended by Dickens (Ellis, 1950), with, of

V. P. Singh (ed.), Hydrologic Frequency Modeling, 505–513.

course, different values for C. Sometimes another formula proposed by Inglis is used:

$$Q = 123A/\sqrt{A+10.4} \tag{2}$$

Where Q is in m^3/s and A in km^2. While isolated cases of breaching of tank embankments have occurred, the vast majority of spillways have easily passed the floods during the lifetime of the tanks, giving rise to the suspicion that many spillways might have been overdesigned. The present work was taken up with the main objective of studying the flood frequencies using data from ten subbasins in the upper reaches of the river Cauvery, and estimating the return periods of floods calculated from empirical formulae. A secondary objective was to compare the performance of two popular flood frequency distributions in respect of these subbasins.

2. DATA

The Cauvery is one of the most important rivers in India, irrigation in its basin being practised for the past few thousand years. It rises in the Western Ghats (elevation exceeding 1700 m at places) and flows a long course eastwards to the Bay of Bengal. Fig.1 shows the upper part of the basin along with its location in the state of Karnataka. Elevation decreases to around 850 m in the eastern part of the region

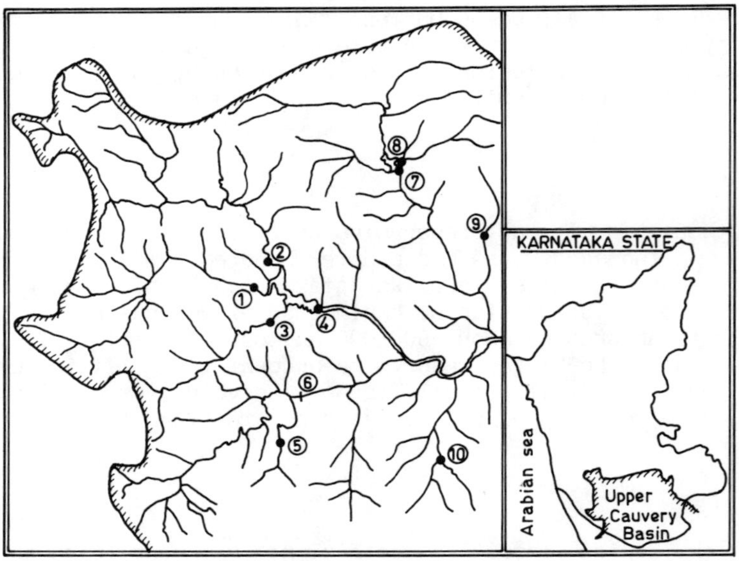

FIG. 1 - THE UPPER CAUVERY BASIN

Table 1. Gaging Stations and related details

Station no	Stream	Drainage area (km^2)	Normal annual rainfall (mm)	Data period	Peak flow (m^3/s) max	min
1	Cauvery	2968	680	1918-80	2879	624
2	Hemavathy	5198	742	1918-80	2979	369
3	Lakshmana-thirtha	1502	763	1918-80	1825	80
4	Cauvery	10619	671	1938-80	6200	1206
5	Nugu	984	920	1918-80	877	33
6	Kabini	4850	920	1918-74	4129	491
7	Shimsha	4131	764	1918-80	1952	3
8	Nagini	748	764	1918-80	707	5
9	Kanva	344	839	1918-80	998	11
10	Suvarna-vathi	1437	676	1918-80	2084	31

shown. Area of the basin within the State is 36130 km^2, and 60 steam gaging stations are in operation. Ten of these stations, chosen in this study, are shown in Fig.1. Annual peak flows for these stations were obtained from the State Water Resources Development Organization. Details on the stations, including length of record, are provided in Table 1. While the normal annual rainwall values at the gaging stations lie in the range 670-920 mm, rainfall further upstream is much higher, reaching more than 3000 mm. Peak flood occurs sometime during the monsoon months from June to October. The rainfall is mainly orographic in the upper reaches, but in the lower part of the basin cyclonic precipitation makes a significant contribution to the annual rainfall.

3. METHOD, RESULTS AND DISCUSSION

3.1. Frequency Distribution

The return periods of the floods in the data series were calculated using Weibull's definition (Linsley et al, 1975). Since the length of record is 63 years (in most cases), the probability of exceedence of the observed flood magnitudes lies in the range 1.6% to 98.4%. Two distributions used widely, namely the log Pearson Type III (LP3) and the Gumbel, were fitted to the data and the parameters of the distributions determined. Floods estimated from these distributions are shown plotted against observed values for the same probabilities in Fig.2 and 3 respectively for four representative stations. 45°-lines

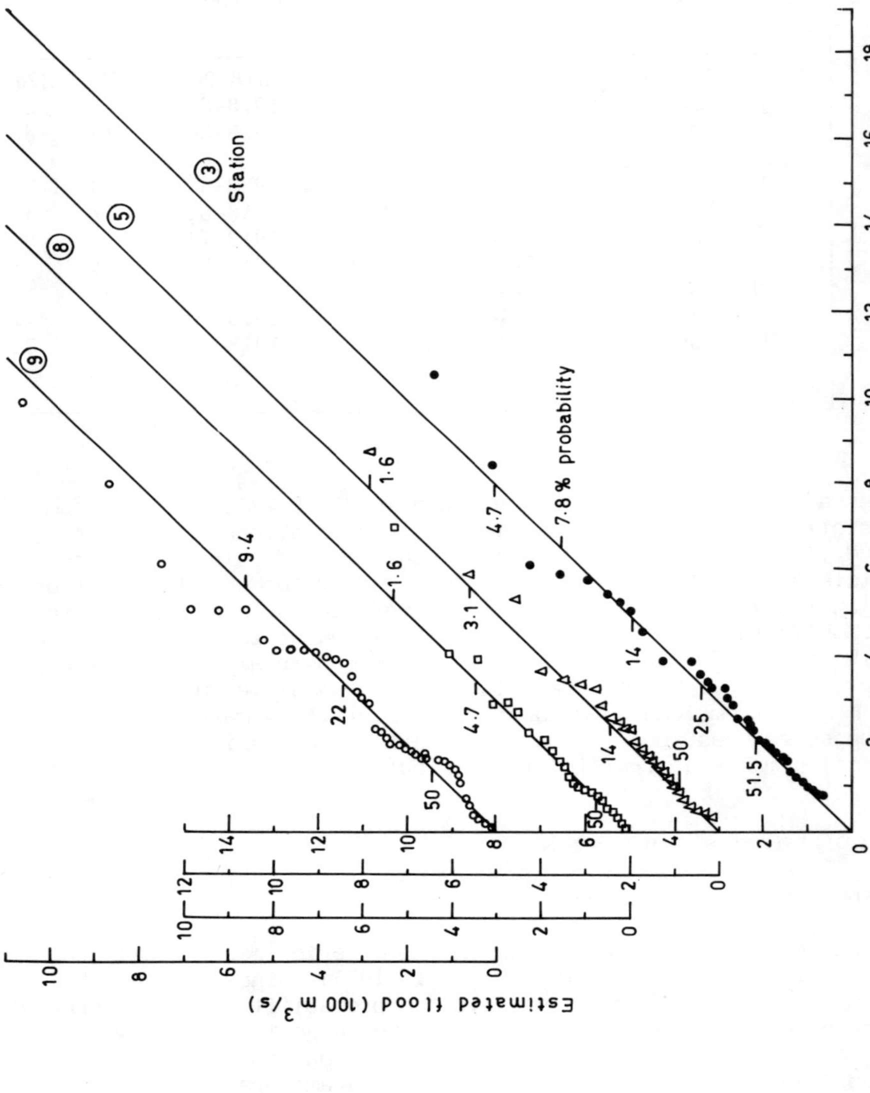

FIG. 2 – LP3 PREDICTION COMPARED WITH RECORD

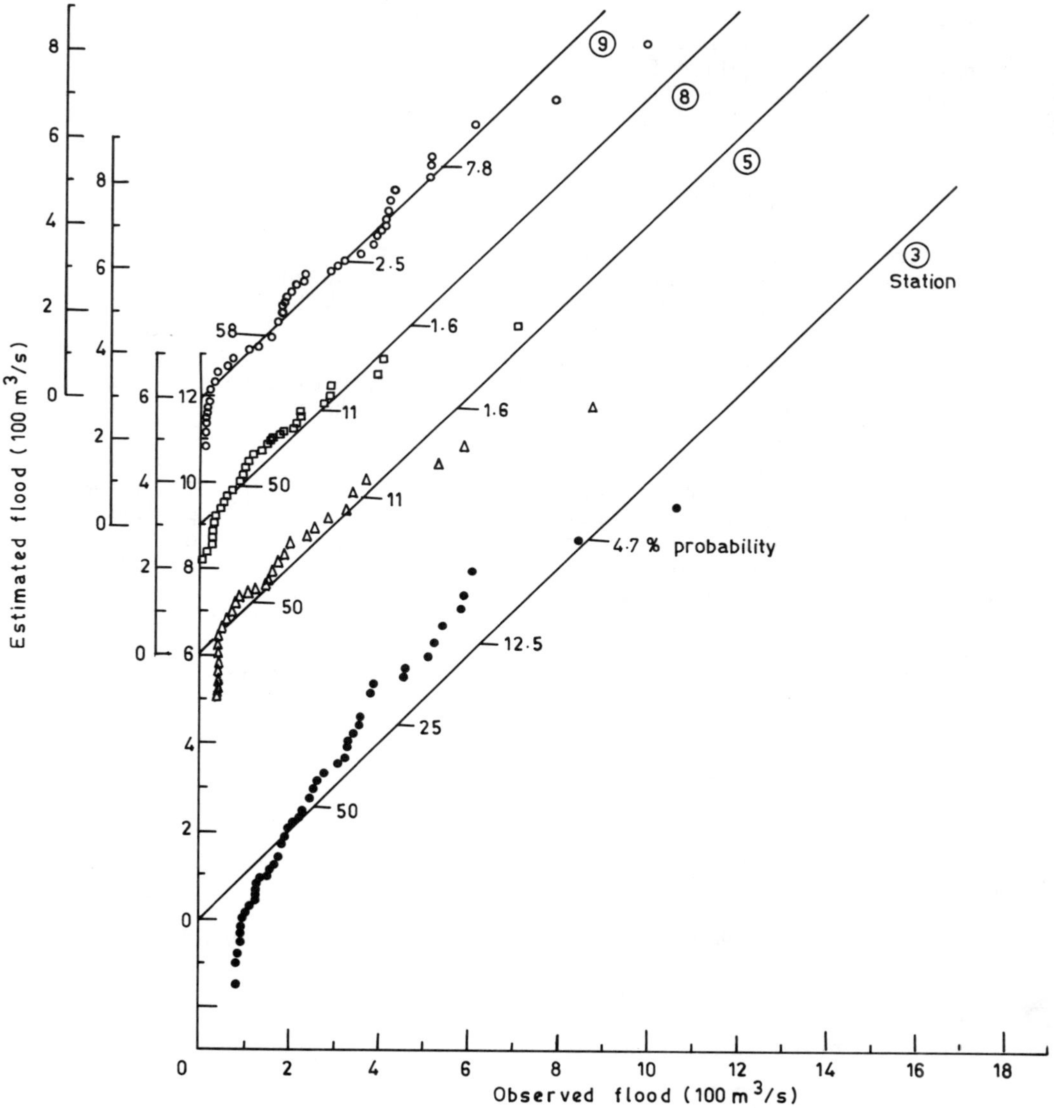

FIG. 3 - GUMBEL PREDICTION COMPARED WITH RECORD

are drawn through the origin (offset vertically for each station for clarity) to bring out deviations. The probability of exceedence is marked on the lines at selected points. Agreement is seen to be generally better for LP3 than for Gumbel. Moreover, for many subbasins, Gumbel predicts negative floods in the higher probability range, which of course is physically impossible. Among all subbasins, the best agreement is for Station 3 (LP3 distribution), and the worst for Station 7.

3.2 Regression Analysis

Since empirical formulae involving drainage area to estimate the design flood are in use for a long time in the region, an analysis of such relationships is made for each subbasin here. Peak flows have been related elsewhere too to selected characteristics of drainage basins like drainage area, mean basin elevation, vegetation cover, soil infiltration capacity, precipitation, temperature etc. (e.g. Natural Environmental Research Council, 1975). Drainage area has been found in all regions to be significantly related to peak flows. While it was the only significant variable in many cases, mean annual precipitation or mean basin elevation enters as an additional parameter (Campbell and Sidle, 1984) in several other cases. Area-based empirical formulae are therefore not to be totally ignored, but the coefficients and exponents normally used may neither be the right ones nor valid for all watersheds, and they definitely depend on the return period.

In the present case, a regression was performed between the flood Q of a given return period and (i) the drainage area A alone, and (ii) drainage area and normal annual precipitation P, respectively in the forms

$$Q = a_0 A^{a1} \qquad (3)$$

$$Q = b_0 A^{b1} P^{b2} \qquad (4)$$

Where Q is in m^3/s, A in km^2 and the rest are constants. The analysis was confined to return periods within the length of the record, from 2 to 64 years.

Table 2. Regression constants for flood frequencies

Return Period years	Eq.(3) a_0	a_1	R^2	Eq.(4) b_0	b_1	b_2	R^2
64	20.2	0.60	0.80	929	0.58	-0.55	0.81
32	6.03	0.72	0.79	6.62	0.72	-0.013	0.79
16	2.12	0.82	0.83	99.6	0.79	-0.55	0.83
8	2.11	0.78	0.81	1221	0.75	-0.92	0.82
4	1.09	0.82	0.71	1609	0.78	-1.05	0.72
2	0.326	0.92	0.66	10655	0.86	-1.50	0.68

Table 2 shows the results of the analysis. The coefficient of determination R^2 is also shown. The exponent varies from 0.6 to 0.92, increasing with decreasing return period, in both regressions. This contrasts with 0.67 and 0.75 used respectively in the formulae of Ryves and Dickens. The formulae recommended by the Natural Environment Research Council (1975) for the mean annual flood use exponents nearer unity. For sites in Ireland, for example, exponents ranging from 0.85 to 0.99 are used. A study by Hanna and Wilcock (1984) also gives an exponent of 0.99 for Northern Ireland for the mean flood.

The negative values for b_2 (eq.4) indicate that the normal annual rainfall at the low end of the watershed may be inversely related to the flood-producting rainfall in the basin. This should not, however, be regarded as conclusive, since the coefficient of determination for the two equations is about the same, and area should therefore be regarded as the principal factor involved.

The multiplier a in eq.(3) decreases approximately linearly with the return period. When a_0 and a_1 are related to the return period T, eq.(3) can be written

$$Q_T = (0.32\ T - 1.75)\ A^{1.08T^{-0.13}} \qquad (5)$$

3.3 Evaluation of the Formulae of Ryves, Dickens and Inglis

The design floods given by eqs. (1) and (2) were next examined to determine their return periods. The LP3 and Gumbel distributions with parameters obtained from observed data for each basin were used to determine T. Table 3 shows the results.

Table 3. Return periods of Floods (in centuries) by Empirical Formulae

Station no	Return period in centuries according to					
	LP3 for the formula of			Gumbel for the formula of		
	Ryves	Dickens	Inglis	Ryves	Dickens	Inglis
1	100	200	650	826	2439	14286
2	1000	2000	3000	100000	∞	∞
3	6.5	30	48	217	100000	∞
4	13	70	60	42.6	1370	794
5	6.3	15.8	28.8	25000	∞	∞
6	28	22.3	27.7	5882	2381	5556
7	2.5	3	4.5	20000	∞	∞
8	18	85	200	33333	∞	∞
9	0.45	2.7	14	1.1	55.7	2778
10	35	400	1000	31.2	5000	100000

The three empirical equations naturally provide three different estimates of flood for a given basin, as is reflected by the differing return periods calculated from a given distribution. However, these differences are very small compared to the differences in the return periods of the same flood calculated from the two distributions. In all cases except one (Station 10, Ryves formula), Gumbel gives a higher return period than LP3. In many cases, the difference amounts to several orders of magnitude. It is interesting to note that floods for some stations calculated by the Dickens and Inglis formulae are assigned infinite return period by Gumbel's distribution.

The Ryves formula is much more commonly used to design spillways for tanks than other two. It is still being used, since over the decades, spillways designed using it have by and large not failed. A rare case of breaching of the embankment does occur. If tanks are designed to pass a flood of probability p, then an average of 38000p tanks must be breached every year in Karnataka State due to overtopping. From past data, it can be taken that one tank embankment in Karnataka is overtopped each year, so that the average return period of the design floods used is 38000 years. The results shown in Table 3 are also in conformity with this observation. For example, except for station 9, the return periods of Ryves floods from LP3 are several centuries. Only for Station 9 a rather low value of 45 years is obtained. An average tank has a life of 40 to 50 years, and hence in this case the Ryves flood causes the spillway to be underdesigned. Thus while the large majority of spillways would be overdesigned by the use of the Ryves formula, a small number are underdesigned, leading to rare cases of breaching. Field experience can also therefore be taken to support indirectly the superiority of the LP3 distribution over Gumbel in addition to the smaller scatter of observed data. Eq.(5) may be adopted as a more suitable formula for ungaged watersheds in the region.

4. CONCLUSION

The analysis shows that the LP3 distribution performs better than the Gumbel distribution in predicting flood frequencies in the upper Cauvery subbasins. This is also supported indirectly by the fact that there are rare cases of breaching of tanks. The results indicate that the empirical formulae used for a long time in the region to calculate design floods in ungaged catchments are far from satisfactory. A formula, based again on watershed area, but with return period as a parameter, as been derived for use in the region.

REFERENCES

Campbell, A.J. and Sidle, R.C. 1984. 'Prediction of Peak Flows on Small Watersheds in Oregon for Use in Culvert Design'. Water Resources Bulletin, Vol.**20**, No.1, pp.9-14.

Ellis, W,M. 1950. Irrigation, College of Engineering, Madras.

Hanna, J.E. and Wilcock, D.N. 1984. 'The Prediction of Mean Annual Flood in Northern Ireland'. Proceedings of the Institution of Civil Engineers, part 2, Vol. **77**, pp.429-444.

Linsley, R.K., Kohler, M.A. and Paulhus, J.L.H. 1975. Hydrology for Engineers, McGraw Hill-Kogakusha, Tokyo, p.340.

Natural Environment Research Council. 1975. Flood Studies Report, London.

A MULTIVARIATE STOCHASTIC FLOOD ANALYSIS USING ENTROPY

P. F. Krstanovic and Vijay P. Singh
Department of Civil Engineering and
Louisiana Water Resources Research Institute
Louisiana State University
Baton Rouge, Louisiana 70803-6405, USA

ABSTRACT. The principle of maximum entropy (POME) was used to derive a multivariate stochastic model for flood analysis. By specifying appropriate constraints in terms of covariances, variances and cross-covariances, multivariate Gaussian and exponential distributions were derived. As a special case, the bivariate process of flood peaks and volumes was investigated for three cases: (1) the peaks and volumes are independent and occur the same number of times; (2) the number of peaks is more than that of volumes in the same time interval; and (3) peaks and volumes exhibit dependence. Special emphasis was given to the structure of the matrix of Lagrange multipliers in the model. Marginal distributions of flood characteristics were obtained, first with no restrictions imposed, and then with assumptions of independent occurrences and a high threshold value. The conditional distribution of flood volume given the peak was then discussed.

This multivariate stochastic model was related to maximum entropy spectral analysis (MESA). This relationship was shown by deriving power spectrum from marginal distributions of flood characteristics and cross-spectrum from the bivariate distribution of peaks and volumes. This connection has two useful practical applications: use of the derived distributions in statistical inference and use of the spectral analysis for prediction and reconstruction of historical records.

1. INTRODUCTION

Stochastic approaches to flood analysis have been developed along two major lines: (1) analysis of flood peaks using theory of probability, and (2) streamflow synthesis using time series analysis. In the first category are considered the floods exceeding some threshold value x_0 (partial duration series approach) in a given time interval, or their maxima (complete duration series). Borgman (1963), and Shane and Lynn (1964) found the distribution function of the flood peaks in a certain time interval (0,t) assuming their number to be a homogeneous Poisson

V. P. Singh (ed.), Hydrologic Frequency Modeling, 515–539.

process and their number of occurrences to follow an exponential distribution. Todorovic and Zelenhasic (1970) have considered the case where the number of exceedances follows a nonhomogeneous Poisson process, with the sequence of flood discharges kept independently identically distributed (i.i.d.) in exponential form. In 1971 Todorovic and Rouselle relaxed independency assumption while dividing the water year into different seasons and keeping the i.i.d. assumption only within the season being considered. Although this model was further generalized (Todorovic and Woolhiser 1972; Gupta, et al., 1976; Todorovic, 1978; North, 1980; Ashkar and Rousselle, 1982), its i.i.d. assumption was retained, and consequently it is valid only for a high threshold value. The i.i.d. assumption neglects seasonal variations and the choice of a high threshold value renders the model invalid for lower threshold values. A theoretically more general model was developed by Kavvas and co-workers (Kavvas and Delleur, 1975; Kavvas, 1982; Kavvas, et al., 1983) treating flooding as the clustering phenomenon and its mechanisms as centers of clusters of flood peaks. Although the disadvantages of the model developed by Todorovic and co-workers were removed, this model is not widely used because of mathematical complexity.

In the time series category, historical information is assumed as discrete time series or observations available at specific instants in time separated by the same time interval Δt. A simple model for streamflow is represented by a single probability function $f(x,\theta)$ with θ as row vector of the parameters $(\theta_1, \ldots, \theta_n)$ valid for the positions in time $t_1, t_2, \ldots, t_n$. Random variables $X_1, X_2, \ldots, X_n$, can be dependent on one another or independent. The extensions of this simple model are various ARMA models. These have been widely used for modeling long-term dependent hydrologic sequences assuming that flood mechanisms can be represented by either autoregessive (AR) or moving average (MA) components of the model, but do not simulate short-term properties satisfactorily.

The above two categories of approaches have been developed almost independently but how they are related to each other is not clear. The objective of this paper is to develop, using entropy, theoretical foundations from which these two seemingly disjointed approaches can be derived. The approach developed here is multivariate and treats various flood characteristics which have different stochastic properties.

Entropy has only recently been applied to hydrologic analyses. Sonuga (1972, 1976) used it for hydrologic frequency analysis and rainfall-runoff modeling. Amorocho and Espildora (1973) used it for assessment of uncertainty in hydrologic systems and models. Singh and associates (Singh and Krstanovic, 1985, 1986; Singh, et al., 1985) have applied it for derivation of frequency distributions, parameter estimation, and sediment and water quality modeling.

2. DEFINITION OF ENTROPY

The concept of entropy was originally developed by Shannon (1948) and successfully extended by Jaynes (1957). Entropy is defined as expectation of information or measure of uncertainty. If S is a system of events E_1, E_2, ..., E_n, and $p(E_k) = p_k$ probability of k-th event to occur then the entropy of the system S is:

$$H(S) = - \sum_{k=1}^{n} p_k \log(p_k) \tag{1}$$

This discrete representation is easily extended to the continuous case: If X and Y are two random variables in probability space Ω, then the marginal entropy H(X) and the joint entropy H(X,Y) can be defined as:

$$H(X) = - \int_\Omega p(x) \log p(x) dx \tag{2.1}$$

$$H(X,Y) = - \int_\Omega \int p(x,y) \log p(x,y) dxdy \tag{2.2}$$

An important characteristic of this concept is to assign probabilities based on prior knowledge to a given system of events. Based on this, Jaynes (1957) formulated the principle of maximum entropy (POME) when he stated: "While assigning the probabilities to various events, assign them by maximizing entropy subject to the given information." That assignment would be minimally prejudiced as it excludes all our assumptions and makes maximum use of the given information. Thus, this procedure prescribes adequate information and uses that information as constraints in order to describe the probability distribution.

Recently, POME has been applied to time series analysis by Jaynes (1982), and Skilling and Gull (1984). Assume that we have a discrete time series $\{x_0, \ldots, x_T\}$, some information about this series $g(x_i)$ at time t_i and about the mutual properties of the series at two different times t_i and t_j expressed as $g(x_i, x_j)$. The problem is to find the probability density function $p\{x_0, x_1, \ldots, x_T\}$ which has the maximum entropy:

$$H(X_0, \ldots, X_T) = - \int dx_0 \ldots \int dx_T \, p(x_0, \ldots, x_T) \cdot \log p(x_0, \ldots, x_T) \tag{3}$$

subject to the constraints $g(x_i)$ and $g(x_i, x_j)$. If these constraints take the form of mean values A_k' of m different quantities A_k then they are expressed as:

$$A_k' = \int dx_0 \ldots \int dx_T \, p(x_0, \ldots, x_T) \, A_k(x_0, \ldots, x_T), \quad 1 \le k \le m$$

The POME procedure proceeds as follows. The partition function is defined as

$$Z(\lambda_1, \ldots, \lambda_m) = \int d\vec{X} \exp(-\Lambda\vec{A}) \tag{4}$$

and the maximum entropy distribution as:

$$p(\vec{X}) = \frac{1}{Z} \exp(-\Lambda\vec{A}) \tag{5}$$

where $\vec{X}$ and $\vec{A}$ denote respectively the row vectors of random variables x_i, i = 1, 2, ..., T, and A_i, k = 1, 2, ..., m, while Λ is the column vector of Lagrange multipliers. The maximum entropy is:

$$H_{max} = H(A_1', \ldots, A_m') = \log Z + \Lambda\vec{A}' \tag{6}$$

where $\vec{A}'$ is the row vector of A_k', k = 1, 2, ..., m, and Lagrange multipliers are determined from

$$\lambda_k = \frac{\partial H}{\partial A_k'} \tag{7}$$

3. DERIVATION OF THE MULTIVARIATE STOCHASTIC MODEL

3.1 General Case

First, we consider the univariate case with multidimensional time properties. Let $\vec{X} = (x_0, \ldots, x_T)$ be a row vector of discharges adjusted for the mean measured at discrete time intervals (e.g., monthly values) for the available historical record of length T (e.g., T + 1 = n = number of monthly discharges). If the information is available in terms of variances or autocovariances then the probability density function (pdf) of $\vec{X}$ is (Singh and Krstanovic, 1986):

$$p(\vec{X}) = \frac{1}{Z(\Lambda)} \exp[-\vec{X}\Lambda\vec{X}^*] \tag{8}$$

where * denotes the row vector transpose. We note that (8) is of the multivariate Gaussian form. The constraints are defined as A_k = (T+1)/2 c_k, and the pdf takes the form:

$$p(\vec{X}) = \frac{1}{Z(\Lambda)} \exp[-\frac{1}{2}\vec{X}\Lambda\vec{X}^*] \tag{9.1}$$

where

$$Z(\Lambda) = \frac{1}{(|\Lambda|)^{0.5}} \tag{9.2}$$

and Λ is the matrix of Lagrange multipliers in Toeplitz form:

$$\Lambda = \begin{cases} \lambda_{j-i} & \ldots \ |j-i| \leq m \ (= \text{no. of constraints}) \\ 0 & \ldots \ \text{otherwise} \end{cases} \tag{9.3}$$

Rewriting (9),

$$p(\vec{X}) = |\Lambda|^{0.5} \exp[-\frac{1}{2} (\vec{X}-\vec{\mu}) \Lambda (\vec{X}-\vec{\mu})^*] \tag{10}$$

On comparing with multivariate Gaussian distribution,

$$p(\vec{X}) = \frac{1}{(2\pi)^{(T+1)/2} |\Sigma|^{0.5}} \exp[-\frac{1}{2} (\vec{X}-\vec{\mu}) \Sigma^{-1} (\vec{X}-\vec{\mu})^*] \tag{11}$$

The matrix of Lagrange multipliers Λ is inverse of the sample variance-covariance matrix of values x_i, $i = 0, \ldots, T$:

$$\Lambda = \Sigma^{-1} \tag{12}$$

Thus, the POME distribution is equivalent to the multivariate Gaussian distribution weighted by some constant value $(1/2\pi)^{(T+1)/2}$. This is the most general case assuming dependence of all or some diagonal terms of the matrix in (12). The parameters of the model are Lagrange multipliers. Every new Lagrange multiplier in the model adds one new diagonal to the Λ matrix. It is shown (Singh and Krstanovic, 1986b) that every new λ is of the form

$$\lambda_k = 1/c_k \tag{13}$$

and accounts for additional information being introduced. Thus, its role is similar to the partial autocorrelation function ϕ_{kk} of the time series model that, at lag k, accounts for the correlation not introduced by previous k-1 lags.

Note that the dimension of the Λ matrix is (T+1)x(T+1) equal to the length of the vector of the hydrologic variable (e.g., discharges). A limitation of the model is that m must be much smaller than T. Note that we not only have a multivariate distribution but also autocovariances c_k expressed as a function of Lagrange multipliers λ_k. For $0 \leq k \leq m$ the model can be used for known sample values c_k and for $\overline{k} > \overline{m}$ it can be used for prediction purposes. Further, we can use the known connection between autocovariances c_k and the power spectrum P(f) to derive the spectral representation of the flood series (Burg, 1975; Jaynes, 1982):

$$c_k = \frac{1}{2\pi} \int_0^{2\pi} \frac{\exp(ik\theta)d\theta}{\sum_{k=0}^{m} \lambda_k \exp(ik\theta)} \tag{14.1}$$

$$P(f) = \frac{1}{\sum_{k=-m}^{+m} \lambda_k \cos 2\pi f_k}, \text{ for } |f| \leq 0.5 \tag{14.2}$$

The λ_k of the model can be calculated from (14.1) as:

$$\frac{1}{2\pi} \int_0^{2\pi} \frac{\exp(ik\theta)\, d\theta}{\sum_{k=-m}^{+m} \lambda_k \exp(ik\theta)} = \frac{1}{2\pi i} \int_{|z|=1} \frac{z^{k-1}\, dz}{\sum_{k=-m}^{+m} \lambda_k z^k} \tag{15}$$

where z is variable in the complex plane. The denominator of (15) can be expressed as product of two polynomials such that the roots of the first lie outside the unit circle and the roots of the second inside the unit circle:

$$\sum_{k=-m}^{+m} \lambda_k z^k = \frac{1}{P_m} \left(\sum_{k=0}^{m} a_k z^k \right) \left(\sum_{k=0}^{m} a_k^c z^{-k} \right) \tag{16}$$

where P_m is constant > 0, and a_k^c, a_k are coefficients and z denotes polynomial roots. Note that $a_0^c = 1$. In general a_k^c is the complex conjugate of a_k. Since $a_k^c = a_k$ for real values, we can write (14) as:

$$P(f) = \frac{P_m}{\left| \sum_{k=0}^{m} a_k z^{-k} \right|^2} = \frac{P_m}{\left[\sum_{k=0}^{-m} a_k \cos(2\pi f_k) \right]^2 + \left[\sum_{k=0}^{-m} a_k \sin(2\pi f_k) \right]^2} \tag{17}$$

(17) expresses the power spectrum as a function of a_k, which can be determined using the Levinson-Burg algorithm (Burg, 1975; Singh and Krstanovic,1986b). The Lagrange multipliers can then be obtained from (16) by equating the terms of equal powers:

$$\lambda_{-j} = \lambda_j = \frac{a_j + \sum_{k=1}^{m-j} a_k a_{k+1}}{P_m} \tag{18}$$

To summarize, the following steps can be used for development of the POME distribution (8) and maximum entropy spectrum (14.2): (1) Express the known information as k autocovariances c_k. (2) Use the Levinson-Burg algorithm to determine the coefficients a_k. (3) Use (18) to determine the Lagrange multipliers λ_k. (4) Calculate multi-dimensional entropy distribution (8) and maximum entropy spectrum (17). (5) Validate the model by comparing the entropy spectrum with the unsmoothed sample spectrum. These steps validate the model in time series mode. To test the model in probability theory mode, one must obtain repeated measurements of random variable X(t). Thus, one can consider T as a cycle (e.g., hydrologic year). Then $\vec{X} = (x(t_1)$ = first discharge value that has n observations for n years record, $x(t_2)$ = second value, ...). Distribution obtained in that way is compared with observed frequency distribution and tested for goodness of fit.

3.2 Univariate Distributions of Flood Characteristics

A careful examination of (8) with Λ defined by (9.3) reveals:

(a) in case of complete independency among the flood values, Λ is a diagonal matrix since nothing new can be introduced by the information at different lags. Thus,

$$\Lambda = \begin{bmatrix} \lambda & 0 & 0 \\ 0 & \lambda & 0 \\ 0 & 0 & \lambda \end{bmatrix}_{(T+1)(T+1)} \tag{19}$$

For this case, the information needed for prescribing the constraints is simply the variance for m available lags.

(b) Assuming the distribution to be identical for all lags, the multidimensional distribution (8) becomes one-dimensional for a particular hydrological characteristic:

$$p(x_i) = \frac{1}{Z(\lambda)} \exp[- \sum_{i=1}^{m} \lambda_i x_i^2] \tag{20}$$

(20) is of univariate Gaussian form. If information is available in terms of the mean, as usually is the case, then we have:

$$p(x_i) = \frac{1}{\bar{x}} \exp[- \frac{x_i}{\bar{x}}] \tag{21}$$

which is the univariate exponential distribution.

We showed (Singh and Krstanovic, 1986b, Appendix D) that the Poisson distribution of the number of flood events in an arbitrary interval is the special case of (8). Similarly, (Singh and Krstanovic, 1986b, Appendix C) if annual maximum peaks are assumed to be i.i.d. with exponential tail, then the result is the Gumbel distribution and that it is also a special case of multidimensional distribution (8).

Marginal distributions of peak X_t and volume Y_t can also be derived as special cases. Making the basic assumption of partial duration series, we consider the peaks and volumes only above some threshold value X_0. We emphasize that this value need not be very high. For very low level we will have Λ matrix of m significant Lagrange multipliers (e.g., for m = 2),

$$\Lambda = \begin{bmatrix} \lambda_0 & \lambda_1 & \lambda_2 & 0 & 0 & 0 \\ \lambda_1 & \lambda_0 & \lambda_1 & & 0 & 0 \\ \lambda_2 & \lambda_1 & & & & 0 \\ 0 & \lambda_2 & & & & \lambda_2 \\ 0 & 0 & & & & \lambda_1 \\ 0 & 0 & 0 & \lambda_2 & \lambda_1 & \lambda_0 \end{bmatrix} \qquad (22)$$

for higher level off-diagonal elements will simply lose their significance and Λ will become diagonal matrix which will simplify the calculations. The observations of X_t or Y_t may not have the same times. While the work of Todorovic took the number of flood occurrences as a random variable, in our model this number is included implicitly as the dimension of Λ matrix, (T+1)(T+1), as well as the dimension of $\vec{X} = (x_0, \ldots, x_T)$ and $\vec{Y} = (y_0, \ldots, y_T)$. To summarize, we do the following: (1) Derive marginal distributions of peak and volume, e.g., $p(\vec{X})$ and $p(\vec{Y})$. (2) Determine which variable is more important as the constraint in the model, e.g., derive (20) and (21) for the same characteristic and compare with its frequency distribution. (3) Follow the Burg-Jaynes algorithm and determine the spectrum of the process, as well as how many Lagrange multipliers are significant. For example, if the diagonal term of Λ matrix is dominant (λ_0), assume independency and proceed with the model simplification.

3.3 Bivariate Distribution of Flood Peaks and Volumes

Let us consider the processes of peak X and volume Y. Flood hydrographs may be of the form such that the number of peaks is greater than that of the volumes for partial duration series. For the annual flood series, the number of peaks and volumes will always be the same. In the theory of bivariate processes, the variable that carries the most information from X(t) to Y(t) and vice-versa is cross-covariance $\gamma_{XY}(k)$ often available as $c_{XY}(k)$ or the sample cross-covariance from the data:

$$c_{XY}(k) = \begin{cases} \frac{1}{n+1} \sum_{t=0}^{n-k} (x_t - \bar{x})(y_{t+k} - \bar{y}), \; k = 0, 1, \ldots, +m & (23.1) \\[2ex] \frac{1}{n+1} \sum_{t=0}^{n+k} (y_t - \bar{y})(x_{t-k} - \bar{x}), \; k = 0, -1, \ldots, -m & (23.2) \end{cases}$$

(23.1) is the case when X_t is behind Y_t which is impossible for the case of peak and volume. Thus, we are left with (23.2) or the case when the process of the volume Y_t is always behind the peak X_t. Further, define information constraint as:

$$A = (n + 1)\ c_{XY}(k)$$

$$= \sum_{t=0}^{n+k} (y_t - \bar{y})\ (x_{t-k} - \bar{x}),\ k = 0, -1, \ldots, -m$$

or

$$A = \sum_{t=0}^{n-k} (x_{t+k} - \bar{x})\ (y_t - \bar{y}),\ k = 0, 1, \ldots, m \tag{24}$$

The joint entropy using (2.2) is:

$$H(\vec{X},\vec{Y}) = - \int_{\Omega}\int p(\vec{X},\vec{Y}) \log p(\vec{X},\vec{Y}) d\vec{x}d\vec{y} \qquad 25)$$

where the random variables X and Y are random row vectors:

$$\vec{X} = (x_0, \ldots, x_n) \text{ and } \vec{Y} = (y_0, \ldots, y_n)$$

The pdf that gives the maximum entropy is:

$$p(\vec{X},\vec{Y}) = \frac{1}{Z(\Lambda)} \exp[-\Lambda A] \tag{26}$$

where Λ is the lower diagonal matrix of Lagrange multipliers:

$$\Lambda = \begin{bmatrix} \lambda_0 & 0 & 0 & 0 & 0 \\ \lambda_{-1} & \lambda_0 & & & 0 \\ \cdot & & & & \\ \cdot & & & & \\ \lambda_{-m} & & \lambda_0 & & 0 \\ 0 & & & \lambda_0 & 0 \\ 0 & 0 & \lambda_{-m}\cdots\lambda_{-1} & & \lambda_0 \end{bmatrix} \tag{27}$$

of the (n+1)x(n+1) dimension and $Z(\Lambda)$ is the partition function defined as:

$$Z(\lambda_{-m}, \ldots, \lambda_0) = \int_{\Omega}\int d\vec{x}d\vec{y} \exp(-\Lambda A) \tag{28}$$

$p(\vec{X},\vec{Y})$ gives the pdf of the particular series of peaks and volumes:

$$p(X_1=x_1,\ Y_1=y_1;\ X_2=x_2,\ Y_2=y_2;\ \ldots;\ X_n=x_n,\ Y_n=y_n)$$

Since these probabilities are often extremely small, it is interesting to see the probability of the occurrence of the particular pair (x_i, y_i). For that case the bivariate distribution becomes:

$$p(x_i, y_i) = \frac{1}{Z(\lambda)} \exp[-\sum_{k=0}^{m} \lambda_k\, x_i\, y_i] \tag{29}$$

The pair (x_i, y_i) is assumed to have the same time of occurrence t_i.

However, a significant simplification occurs when the process of peaks and volumes is assumed to have small correlation at different lags and high correlation at the same lag. In that case,

$$\gamma_{XY}(k) \cong \gamma_{XY}(0) = \mathrm{Cov}(X,Y)$$

and Λ matrix becomes diagonal with λ_0 significant. The pdf for maximum entropy is (26) where

$$\Lambda A = (x_0-\bar{x},\ x_1-\bar{x},\ \ldots,\ x_n-\bar{x}) \begin{bmatrix} \lambda & 0 \\ 0 & \lambda \end{bmatrix} \begin{bmatrix} y_0-\bar{y} \\ y_1-\bar{y} \\ \vdots \\ y_n-\bar{y} \end{bmatrix} \tag{30}$$

The partition function is easily determined as:

$$\ln Z(\lambda) = -\sum_{j=0}^{n} \ln g_j + \text{constant}$$

$$\ln Z(\lambda) = -[\ln g_0 + \ldots + \ln g_n] = -[\ln(g_0 \ldots g_n)]$$

$$= -\ln |\Lambda|$$

$$\lambda = \frac{1}{\mathrm{Cov}(X,Y)}, \quad \text{and} \quad Z = \frac{1}{|\Lambda|} \tag{31}$$

where $|\Lambda|$ stays for determinant of Λ. Thus,

$$p(\vec{X},\vec{Y}) = |\Lambda| \exp[-\frac{1}{\gamma_{XY}(0)} \sum_{i=0}^{n} (x_i-\bar{x})\ (y_i-\bar{y})] \tag{32}$$

Since Λ is a diagonal matrix, $|\Lambda| = \lambda^{n+1}$, where n+1 is the number of countable occurrences. For purposes of testing the probability of only one pair (x_i, y_i), we obtain a simplified univariate case:

$$|\Lambda| = \lambda$$

and

$$p(x_i, y_i) = \frac{1}{\gamma_{XY}(0)} \exp\left[-\frac{(x_i-\bar{x})(y_i-\bar{y})}{\gamma_{XY}(0)}\right] \tag{33}$$

These expressions are equivalent to those of univariate case of peaks and volumes under the same assumptions:

$$p(x_i) = \frac{1}{r_k} \exp\left[-\frac{(x_i-\bar{x})(x_{i+k}-\bar{x})}{r_k}\right]$$
$$p(y_i) = \frac{1}{r_k} \exp\left[-\frac{(y_i-\bar{y})(y_{i+k}-\bar{y})}{r_k}\right] \tag{34}$$

This equivalency shows that the bivariate distribution approach is more general.

However, a serious disadvantage occurs in frequency domain. Since the cross-correlation matrix is not symmetric any more as the autocorrelation matrix r is in the univariate case, the Levinson-Burg algorithm cannot be used. Nevertheless, one can examine series dependency using the relationship between cross-spectrum $\Gamma(f)$ and cross-covariance $c_{XY}(k)$:

$$\Gamma(f) = \sum_{k=-\infty}^{+\infty} c_{XY}(k) \exp(i2\pi fk), \text{ for } |f| \leq 0.5 \tag{35}$$

and for any $c_{XY}(k)$.

Cross-covariance $\hat{\gamma}_{XY}(k)$ derived as a function of Lagrange multipliers is equivalent to (14.1) and cross-spectrum is found to be:

$$\hat{\Gamma}(f) = \frac{\gamma_{XY}(0)}{(\cos 2\pi f + i \sin 2\pi f)} \tag{36}$$

The real-value component of this cross-spectrum can be compared with the integrated cospectrum (Jenkins and Watts, 1968, p. 367) or its normalized estimator in order to test correlations at different frequencies.

Interesting theoretical cases can be discussed for the distributions in (26) characterized by (30). For long continuous record of flooding, the series can be separated into hydrological years where each year may have a couple of multiple peak hydrographs with X_t observations more than Y_t. In order to preserve the correlation for the same lag, all the peaks of a multiple hydrograph are associated with its volume. However, dependency is then introduced at different lags. We can still preserve diagonal Λ matrix but λ's on the diagonal change. For example, if we have "a" peaks more than volumes, the constraint $\gamma_{XY}(0)$ will be dominant until X_t-Y_t are of the same number, $\gamma_{XY}(1)$ when the difference is 1, etc.

The "transformed" matrix will have different eigenvalues, λ_1, ..., λ_a. By deriving each one from the partition function $Z(\lambda)$ it can be shown that each $\lambda_k \cong \text{constant}/\gamma_{XY}(k)$, where constants do not differ significantly from 1. In conclusion, any flood series of peaks and volumes can be represented sufficiently well by the distribution defined in (30).

3.4 Some Extensions

Univariate distributions of peaks and volumes can be improved by knowing the bivariate distribution p(X,Y). In practice we may want to derive the distribution of volumes knowing only the peaks. Then

$$p(\vec{Y}|\vec{X}) = \frac{p(\vec{X},\vec{Y})}{p(\vec{X})} \tag{37}$$

according to Bayes' formula. We can also compare $p(\vec{Y}|\vec{X})$ with $p(\vec{Y})$ to see if there really is improvement in the model fitting by introducing the new variable X_i. The same reasoning applies to distribution of $p(\vec{X}|\vec{Y})$.

For the equal number of peaks and volumes, neglecting occurrence of multiple peak hydrographs, we can approximate a single peak hydrograph by a triangular shape whose base is the flood duration d. Then:

$$d = 2\left(\frac{y}{x}\right)$$

The pdf of duration for known y and x is determined as the pdf for the ratio of two random variables:

$$p(d) = 2f\left(\frac{y}{x}\right) = 2\int_{-\infty}^{+\infty} |y|\ p(yd,y)dy \tag{38}$$

Since y is always positive, (38) can be simplified as:

$$p(d) = 2\int_{0}^{\infty} y\ p(yd,y)dy \tag{39}$$

using (30) and (31) as the bivariate p(x,y) or its time-univariate substitute.

4. APPLICATION

Various components of the model have been tested on the same data. Thirty years of daily mean discharges were available from the USGS for the Spring Creek (Louisiana). The low threshold value X_0 = 100 ft^3/sec was chosen so that in every year (even very dry) at least one flood was available. In this way, the model behavior can be judged more thoroughly. In total 326 peaks and associated volumes were found or taking into account the multiple peak hydrographs there were 28 peaks more than volumes which is approximately one per year.

The frequency histograms of peaks and volumes are presented in figures 1 and 2. Both peaks and volumes can be represented equally well by the exponential distribution in (21) where the mean $\bar{X}(t)$ and $\bar{Y}(t)$ are used as the constraints in the entropy model. However, there is also the possibility that the process of the peaks can be represented by skewed Gaussian distribution in (20) taking var[X(t)] as the constraint in the entropy model. Both cases were tested for both peaks and volumes and the results are presented in figures 3 and 4. Prediction was much better with the mean than variance for both cases.

To further improve prediction of flood volumes p(Y) we used (37), and at the same time tested whether the bivariate distribution of peak and volume in (30) with cov(X,Y) as the constraint performed satisfactorily. As seen in figure 5, the discrepancy in frequency distribution is in the range of 0 to 2%.

Testing in frequency domain was also done. First, mean monthly discharges were used as the input to the entropy model (8) and the series was generated in frequency domain (figure 6). The model fitted well the observed spectra indicating strong correlation at lag 12 or frequency 0.08 which is also shown in the high value of λ_k at that lag (Table 1). Since the available series was sufficiently long, 20 Lagrange multipliers were used in Λ matrix.

TABLE 1. Lagrange multipliers of the entropy model for mean monthly discharges.

k	0	1	2	3	4	5	6	7	8	9	10	11
$\lvert\lambda_k\rvert$	1.33	0.14	0.01	0.07	0.06	0.01	0.01	0.02	0.02	0.01	0.03	0.005

This table can be interpreted similar to partial autocorrelation function in time series models while introducing "knowledge" in the model about time series. The most significant information is introduced at lags 0, 12 and 1. The processes of flood peak and volume were tested in the same way while each occurrence was treated as one lag (figures 7 and 8) regardless of their times of occurrence. Except for the lag 0, both series showed increased frequency at lag 18 which corresponded to the time equivalent of 13-14 months. The increased magnitude of Lagrange multipliers confirmed our observations, although

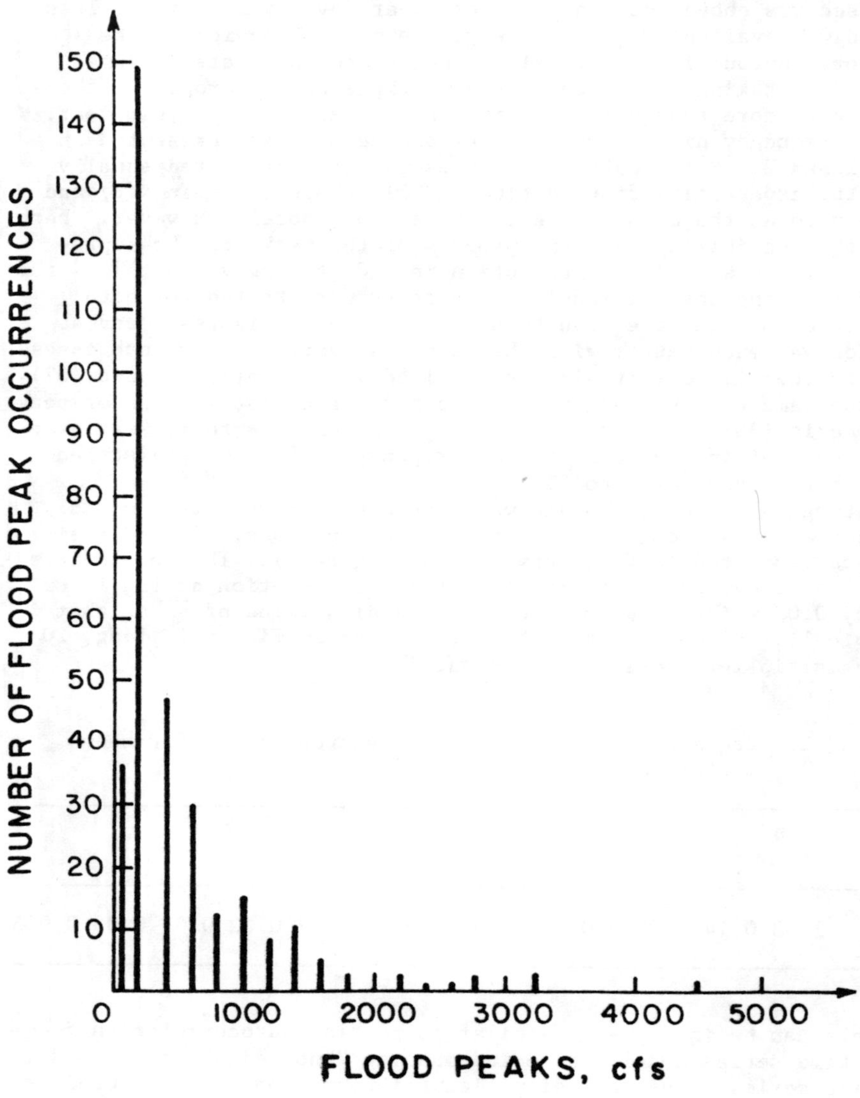

Figure 1. Histogram of flood peaks for Spring Creek (Louisiana) for the period 1956-1985.

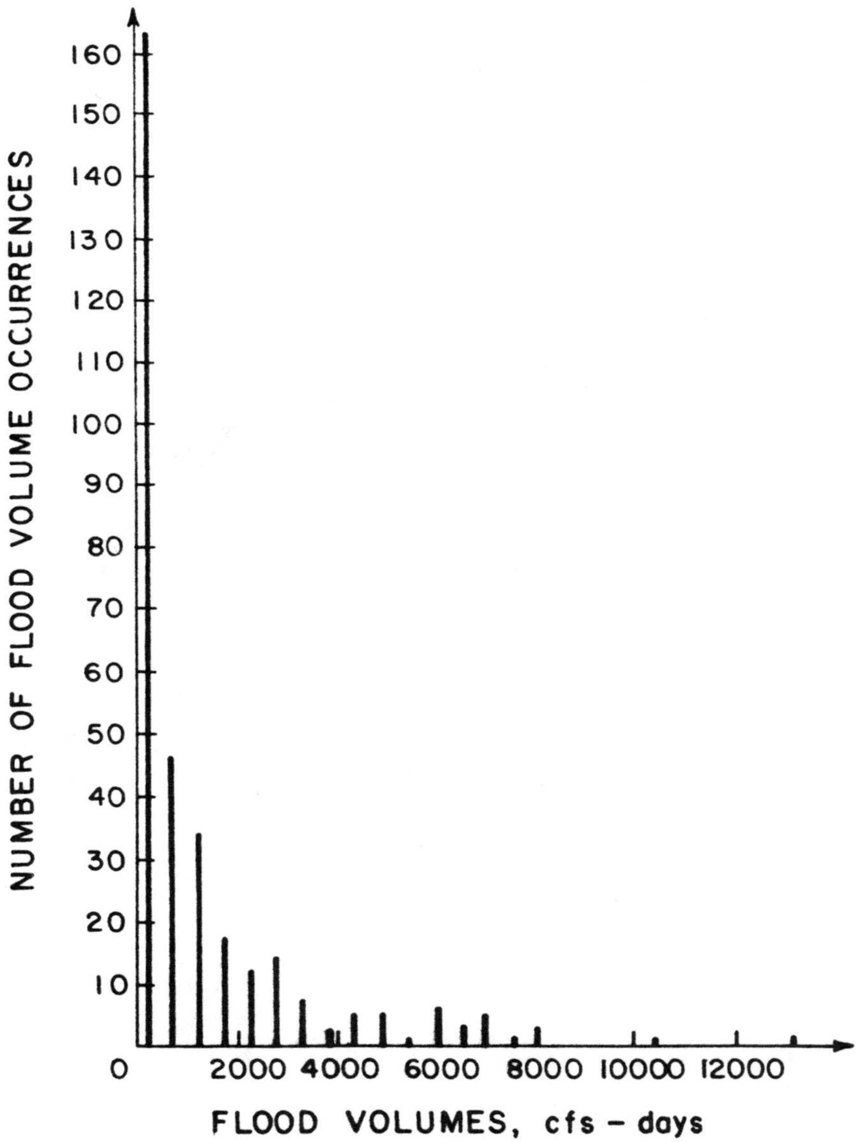

Figure 2. Histogram of flood volumes for Spring Creek (Louisiana) for the period 1956-1985.

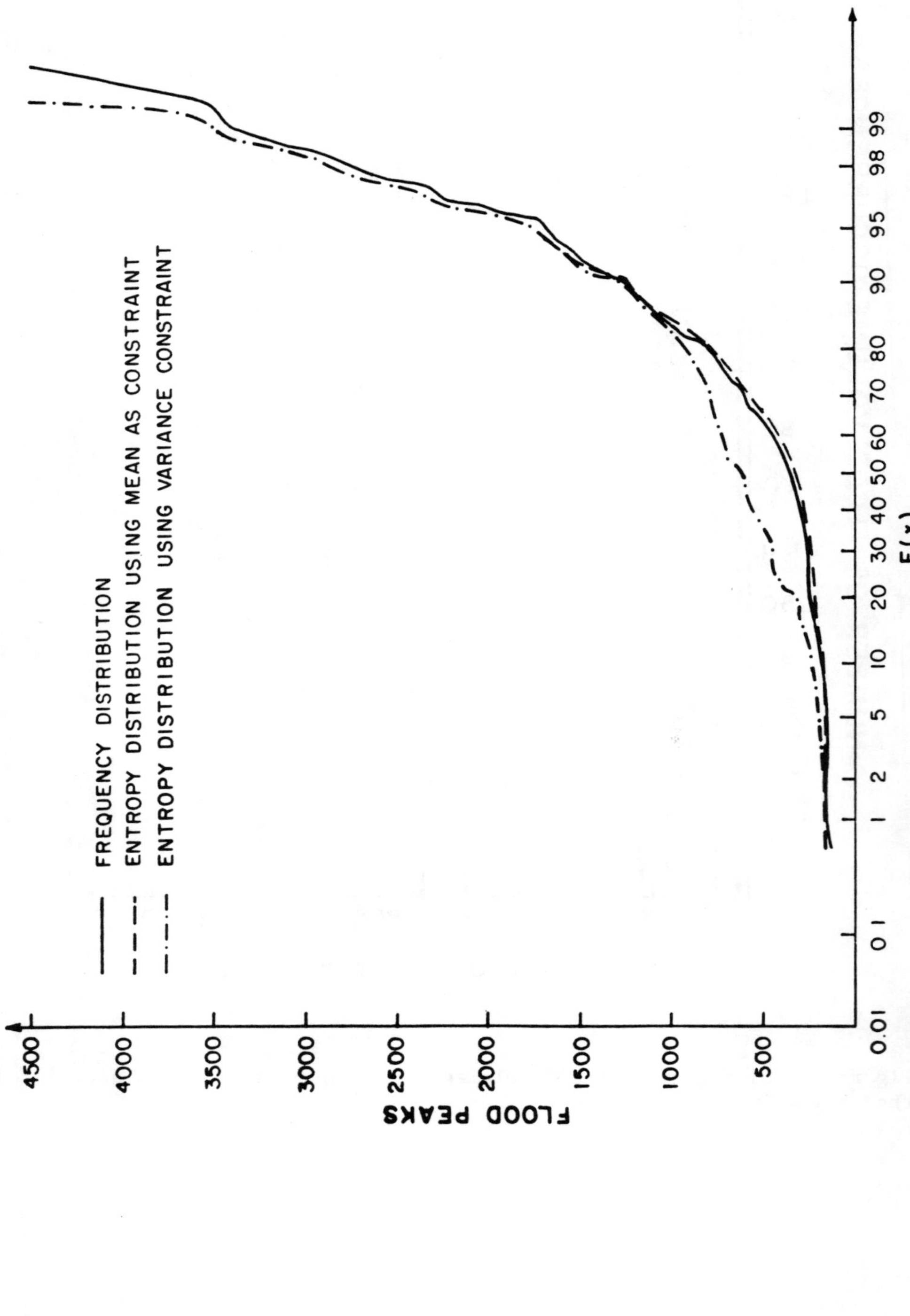

Figure 3. Cumulative distribution of flood peaks.

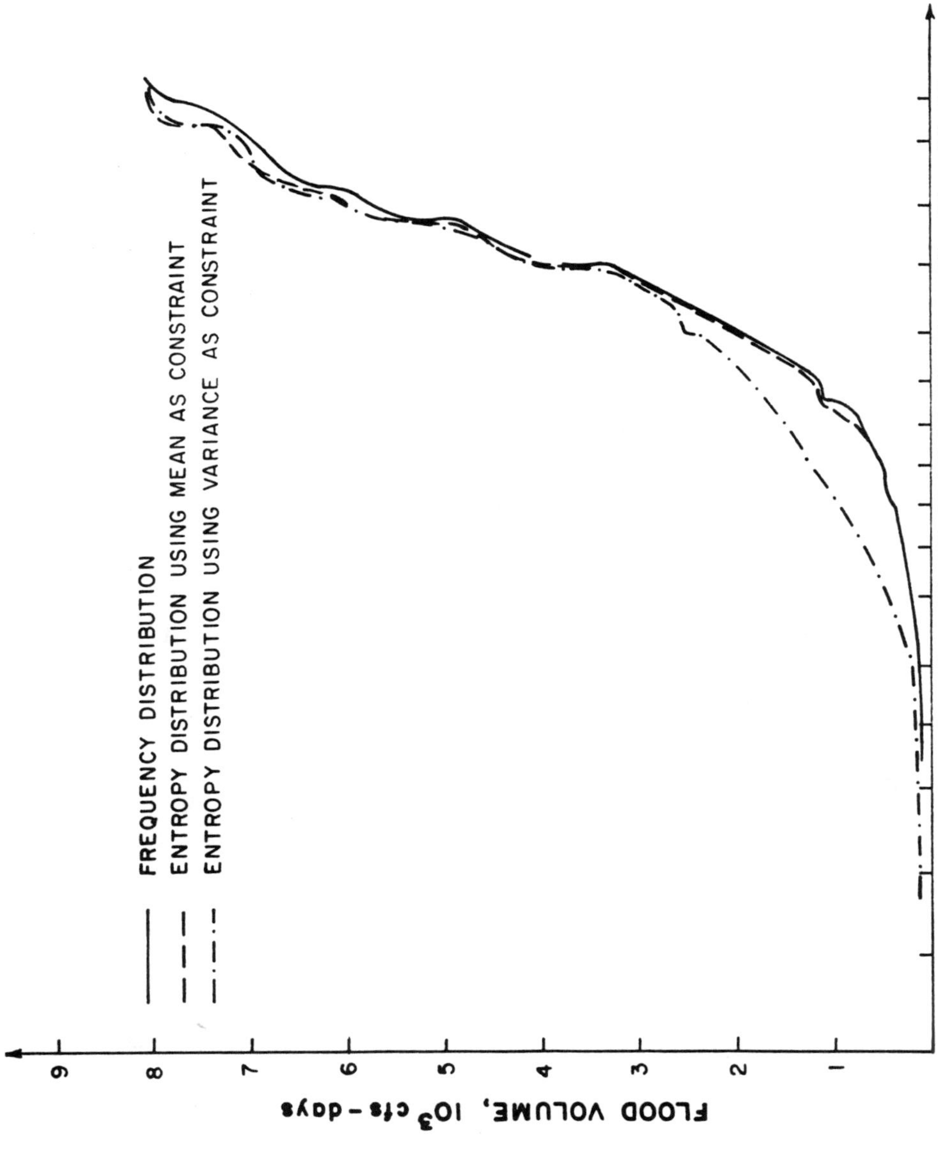

Figure 4. Cumulative distribution of flood volumes.

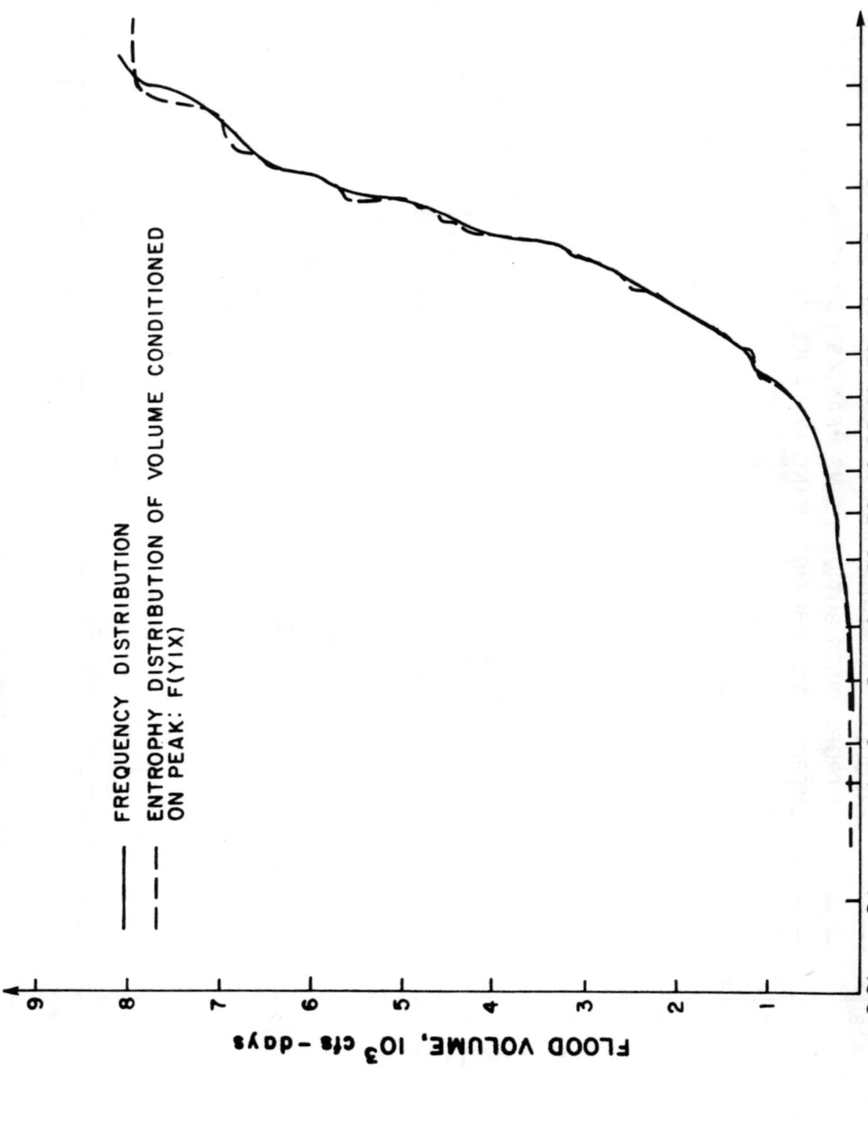

Figure 5. Cumulative distribution of flood volumes conditioned on peaks.

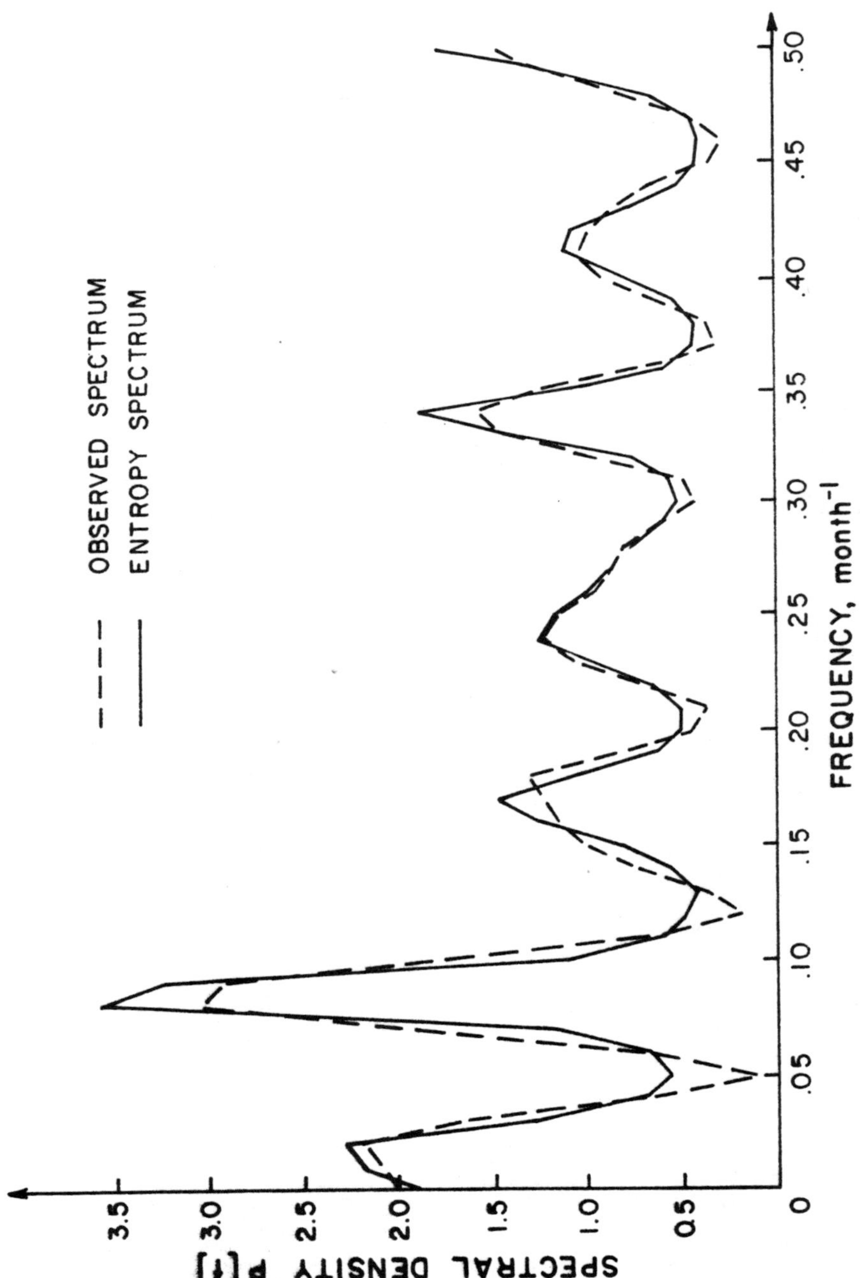

Figure 6. Spectral representation of monthly mean discharges.

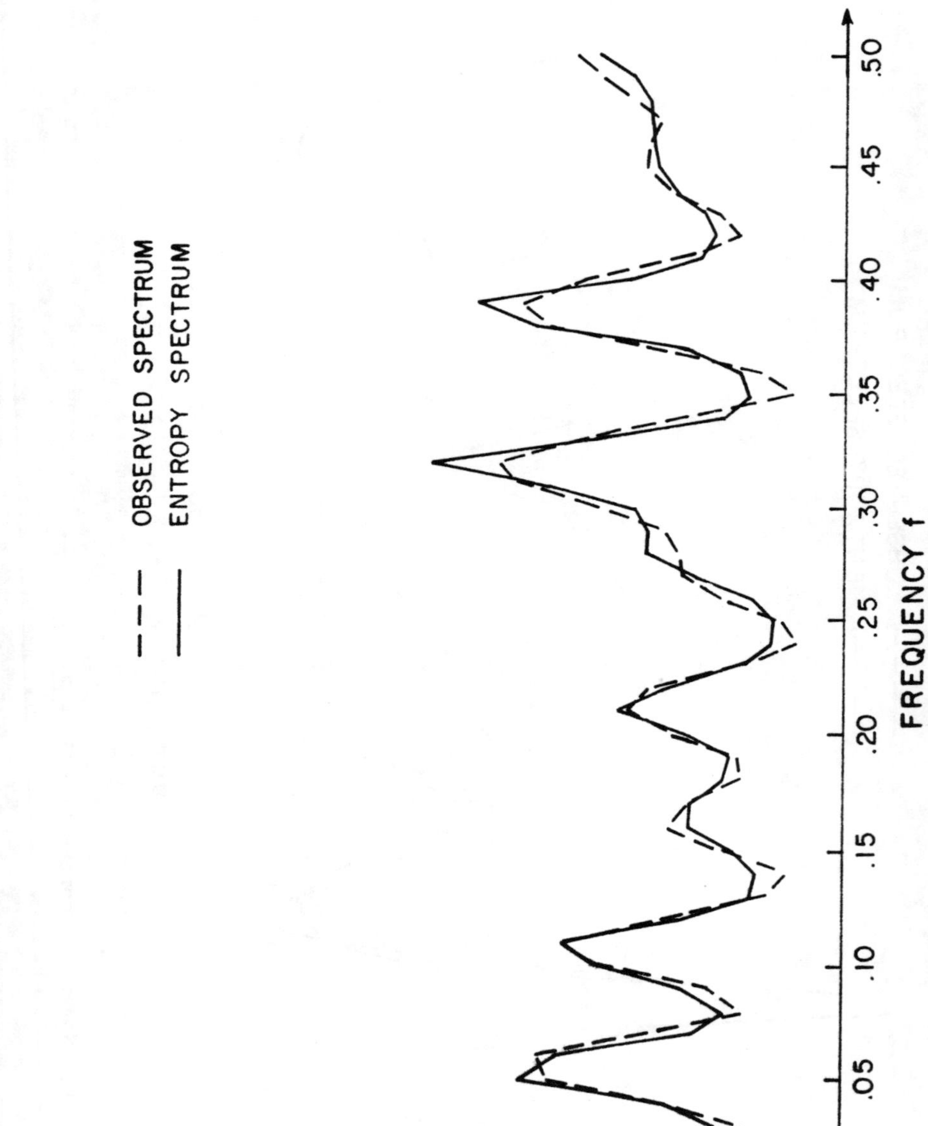

Figure 7. Spectral representation of flood peaks.

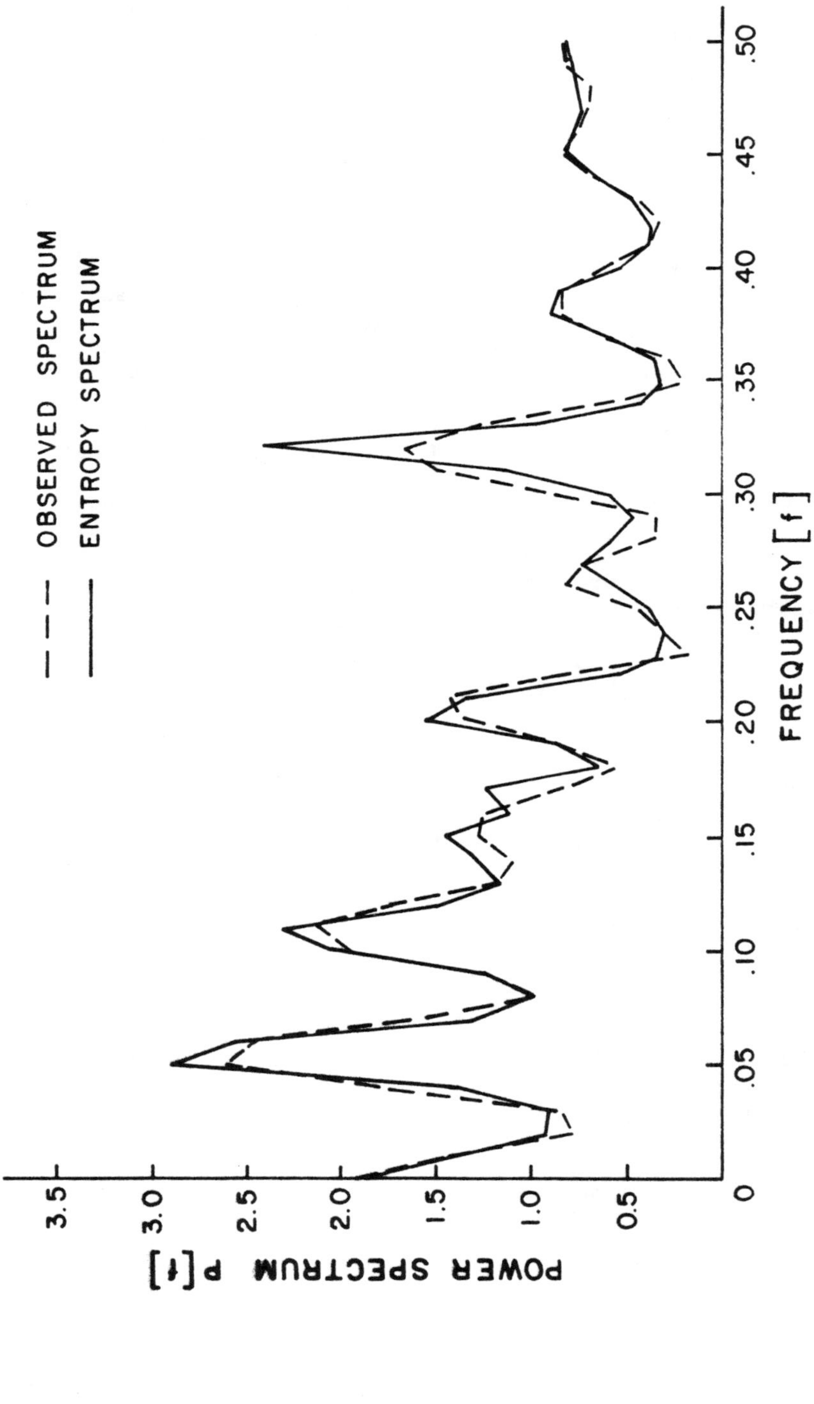

Figure 8. Spectral representation of flood volumes.

λ_{18}(peaks) ≅ 0.19 << λ_1(peaks) = 1.21, and λ_{18}(volumes) ≅ 0.12 << λ_1(volumes) = 1.38.

These tests show that the approximation of Λ matrix with the diagonal Λ for the statistical inference was valid for univariate distributions $p(\vec{X})$ and $p(\vec{Y})$.

Finally, tests for the distribution (30) were conducted for the multivariate case. Given the information on (X_i, Y_i) for 30 years or 326 occurrences, we computed joint probability for a particular year. The probabilities were small but the comparison among the years showed in which years the floods with really small probability or "unexpected" would occur. The results are presented in Table 2.

5. CONCLUSIONS

The most important features of our model are: 1. The entropy concept is shown to be general for flood prediction. Given certain information it specializes in most of the existing models: Gumbel distribution for annual series, exponential distribution for partial duration series in representing flood peaks, etc. 2. Multivariate entropy model of flood characteristics is not confined to i.i.d. assumption. However, for practical purposes, the independency of the associated peaks and volumes (X_i, Y_i) for various occurrences is reasonably reliable. 3. The multivariate model can be used to improve prediction of the single flood characteristics through the use of conditional distribution, as shown for the distribution of flood volume. 4. The entropy model with its parameter λ_k is applicable to statistical inference yielding more general results than some of what has been previously reported and to representing the process in frequency domain. 5. The parameters of the model λ_k play the same role as the partial autocorrelation function in time series models. 6. The model has sufficient flexibility, since under certain assumptions, it reduces to simpler forms easily applicable in practice.

6. ACKNOWLEDGEMENTS

This study was supported in part by funds provided by the Louisiana Water Resources Research Institute through the U.S. Department of Interior, Geological Survey, under the project, "A Multivariate Stochastic Analysis of Flood Magnitude, Duration and Volume."

7. REFERENCES

Amorocho, J. and Espildora, B., 1973. 'Entropy in the assessment of uncertainty in hydrologic systems and models.' Water Resources Research, Vol. 9, No. 6, pp. 1511-1522.

TABLE 2. Joint occurrence probability of peaks and volumes in a particular year.

Year	p(X,Y)	Year	p(X,Y)
1956	2.52×10^{-7}	1971	2.14×10^{-7}
1957	0.28×10^{-7}	1972	3.74×10^{-8}
1958	2.38×10^{-12}	1973	3.87×10^{-14}
1959	0.411×10^{-7}	1974	3.03×10^{-6}
1960	0.30×10^{-7}	1975	4.00×10^{-15}
1961	0111×10^{-7}	1976	1.33×10^{-10}
1962	6.39×10^{-11}	1977	3.62×10^{-7}
1963	2.28×10^{-7}	1978	7.96×10^{-8}
1964	3.70×10^{-7}	1979	1.102×10^{-7}
1965	0.69×10^{-7}	1980	4.45×10^{-8}
1966	5.96×10^{-7}	1981	6.68×10^{-16}
1967	4.55×10^{-9}	1982	4.91×10^{-13}
1968	1.12×10^{-12}	1983	2.67×10^{-35}
1969	2.34×10^{-10}	1984	8.68×10^{-31}
1970	5.34×10^{-7}	1985	7.90×10^{-12}

Ashkar, F. and Rouselle, J., 1982. 'A multivariate statistical analysis of flood magnitude, duration and volume.' in: V. P. Singh (editor), Statistical Analysis of Rainfall and Runoff, published by Water Resources Publications, Littleton, Colorado.

Borgman, L. E., 1963. 'Risk criteria.' Journal of Waterways, Harbors and Coastal Engineering Division, ASCE, Vol. **89**, pp. 1-35.

Box, G. E. P. and Jenkins, G. M., 1976. Time Series Analysis, Forecasting and Control. Holden Day, San Francisco, California.

Burg, J. P., 1975. 'Maximum entropy spectral analysis.' Unpublished Ph.D. Dissertation, Stanford University, University Microfilms, pp. 75-25, 499.

Gupta, V., Duckstein, L. and Peebles, R. W., 1976. 'On the joint distribution of the largest flood and its time of occurrence.' Water Resources Research, Vol. 12, No. 2.

Jaynes, E. T., 1957. 'Information theory and statistical mechanics, Part I.' Physical Review, Vol. 106, pp. 620-630, Part II, Vol. 108, pp. 171-191.

Jaynes, E. T., 1982. 'On the rationale of maximum entropy methods.' Proceedings of IEEE, Vol. 70, No. 9, pp. 939-952.

Jenkins, G. M. and Watts, D. G., 1968. Spectral Analysis and its Applications. Holden Day, San Francisco, California.

Kavvas, M. L., 1982. 'Stochastic trigger model for flood peaks: 1. Development of the model, 2. Applications of the model.' Water Resources Research, Vol. 18, No. 2, pp. 383-411.

Kavvas, M. L. and Delleur, J. W., 1975. 'The stochastic and chronological structure of rainfall sequences: applications to Indiana.' Tech. Report 57, Water Resources Research Center, Purdue University, West Lafayette, Indiana.

Kavvas, M. L., Cervantes, J. E. and Delleur, J. W., 1983. 'A cluster model for flood analysis.' Water Resources Research, Vol. 19, No. 1, pp. 209-224.

North, M., 1980. 'Time dependent stochastic model of floods.' Journal of Hydraulics Division, ASCE, Vol. 106, No. HY5, pp. 649-665.

Shane, R. M. and Lynn, W. R., 1964. 'Mathematical model for flood risk evaluation.' Journal of the Hydraulics Division, ASCE, Vol. 80, pp. 1-20.

Shannon, C. E., 1948. 'A mathematical theory of communication.' Bell System Tech. Journal, Vol. 27, pp. 379-423.

Singh, V. P. and Krstanovic, P. F., 1985. 'A stochastic model for sediment yield.' Paper presented at the IVth International Hydrology Symposium, "Multivariate Analysis of Hydrologic Processes," Colorado State University, Fort Collins, Colorado.

Singh, V. P. and Krstanovic, P. F., 1986a. 'A stochastic model for sediment yield and phosphorous loading.' Submitted to Journal of Environmental Quality.

Singh, V. P. and Krstanovic, P. F., 1986b. 'A multivariate stochastic flood analysis using entropy.' Completion Report, Louisiana Water Resources Research Insttiute, Louisiana State University, Baton Rouge, Louisiana.

Singh, V. P., Singh, K. and Rajagapol, A. K., 1985. 'Application of the principle of maximum entropy (POME) to hydrologic frequency analysis.' Completion Report, Louisiana Water Resources Research Institute, Louisiana State University, Baton Rouge, Louisiana.

Skilling, J. and Gull, S. F., 1984. 'The entropy of an image in 'inverse problems.' SIAM-AMS Proceedings, Vol. 14, pp. 167-189.

Sonuga, J. O., 1972. 'Principle of maximum entropy in hydrologic frequency analysis.' Journal of Hydrology, Vol. 27, pp. 177-191.

Sonuga, J. O., 1976. 'Entropy principle applied to the rainfall-runoff process.' Journal of Hydrology, Vol. 30, pp. 81-94.

Todorovic, P. and Rouselle, J., 1971. 'Some problems of flood analysis.' Water Resources Research, Vol. 7, No. 5, pp. 1144-1150.

Todorovic, P. and Zelenhasic, E., 1970. 'A stochastic model for flood analysis.' Water Resources Research, Vol. 6, pp. 1641-1648.

Todorovic, P. and Woolhiser, D., 1972. 'On the time when extreme flood occurs.' Water Resources Research, Vol. 8, No. 6, pp. 1433-1438.

Todorovic, P., 1978. 'Stochastic models of floods.' Water Resources Research, Vol. 14, No. 2, pp. 345-356.

MULTIVARIATE PARTIAL DURATION SERIES IN FLOOD RISK ANALYSIS

Francisco Nunes Correia
Director-General for Natural Resources
Secretary of State for the Environment and Natural Resources
Av. Almirante Gago Coutinho, 30
P-1000 Lisboa, PORTUGAL

ABSTRACT. Flood frequency and risk analysis play a crucial role in the design of many hydraulic structures and water resources systems. Traditionally only the annual flood peak discharges are considered in the flood risk assessment process. An alternative approach is the use of partial duration series. This approach has proved to be particularly suitable when short records are available. In this method not only the annual maxima are considered but all the peaks over a given threshold. When using this method it is very attractive to extend the results presented in the literature to include the flood volume and flood duration in the risk assessment process, instead of considering only the flood peak discharges. In this paper this extension is attempted and its merits are discussed. Data from twelve portuguese river basins are used to document the applicability of multivariate partial duration series and its limitations.

1. INTRODUCTION

Flood frequency and risk analysis plays an important role in the design and operation of many hydraulic structures and water resources systems. For many decades hydraulic practice has concentrated on fitting a statistical distribution to the series of annual maxima floods. This approach provides a very limited assessment of flood characteristics. In fact, risk is strictly formulated as the probability that the annual maximum peak discharge exceeds a given design value. Questions such as for how long the design value is exceeded and how frequently it may occur every year are completely excluded from this classic formulation.

In recent years Todorovic (1970), Todorovic and Zelenhasic (1970), Todorovic and Rousselle (1971), Cunnane (1973), NERC (1975), Taesombut and Yevjevich (1978), Ashkar and Rousselle (1982), and Correia (1983 and 1985), among others authors, successfully tried a different approach based on partial duration series. This approach takes in to consideration all values above a given threshold. A bivariate stochastic

V. P. Singh (ed.), Hydrologic Frequency Modeling, 541–554.

model is considered to describe the number of floods per year and the maximum discharge associated with every flood.

Commonly, the method based on partial duration series is used to estimate the distribution of the annual maximum discharge. However, there is a much larger potential to be explored in this method, namely a multivariate characterization of flood risk. Flood volume, flood duration and flood peak are the most important characteristics of floods to be considered in the design of water resources systems.

The basic purpose of this paper is to present a formulation based on partial duration series to the study of flood volumes and flood durations, under certain assumptions, and to illustrate the aplicability of the proposed methodology in Portugal.

2. MULTIVARIATE PARTIAL DURATION SERIES IN FLOOD RISK ASSESSMENT

A flood event is defined as a sequence of discharge values above a given truncation level. The peak discharge, the duration and the volume of the flood are denoted by X, Y and Z, respectively.

There is a much stronger dependence between flood volume and duration and flood volume and peak than between duration and peak. For this reason peak and duration will be studied first and volume will be considered later.

If flood peak, X, and duration, Y, are independent, the distribution function of X and Y is given by

$$H\ (x,\ y) = \int_0^x \int_0^y h\ (x)\ h\ (y)\ dx\ dy \tag{1}$$

in which h (x) and h (y) are the marginal density functions of flood peak and flood duration.

The joint distribution function of X and Y when the time interval of one year is considered may be given by

$$F\ (x,\ y) = \sum_{k=0} [H(x,\ y)]^k \cdot P\ (E_k) \tag{2}$$

where $P\ (E_k)$ is the probability that k flood events occur in one year. F (x,y) represents the probability that in one year x and y are both not exceeded:

$$F\ (x,\ y) = P\ (X \leqslant x,\ \ Y \leqslant y)$$

If the number of flood events per year is Poisson distributed Eq. (2) may be written as

$$F\ (x,\ y) = e^{\lambda H(x,\ y)\ -\lambda} \tag{3}$$

If the marginal distribution function of flood peak and duraction are both exponential, with parameters α and β respectively, Eq. (1) becomes

$$H(x, y) = (1 - e^{-\alpha x})(1 - e^{-\beta y}) \tag{4}$$

Substituting Eq. (4) in Eq. (3) we get

$$F(x, y) = \exp[\lambda(1 - e^{-\alpha x})(1 - e^{-\beta y}) - \lambda] \tag{5}$$

In these expressions the, parameters λ, α and β are easily estimated for any given truncation level. In fact λ is the average number of annual flood events, α is the inverse of the mean flood peak discharge and β is the inverse of mean flood duration.

The probability that at least one of the values (x, y) is exceeded in one year is equal to 1 - F(x, y) and the return period associated with this event is:

$$T(x, y) = \frac{1}{1 - F(x, y)} \tag{6}$$

The return period associated with the exceedance of both x and y in one year is given by

$$T'(x, y) = \frac{1}{1 - F_M(x) - F_M(y) + F(x, y)} \tag{7}$$

In this expression $F_M(x)$ and $F_M(y)$ represent the marginal distributions of x and y being exceeded in one year. These marginal distributions can be computed by

$$F_M(x) = \exp(-\lambda e^{-\alpha x}); \quad F_M(y) = \exp(-\lambda e^{-\beta y}) \tag{8}$$

Independence between flood peak and duratiou may not be realistic, principally when the truncation level in low and the annual number of flood events is relatively high.

If X and Y are not independent their joint distribution function is

$$H(x, y) = \int_0^x \int_0^y h(x \mid y)\, h(y)\, dx\, dy \tag{9}$$

If we assume that the relationship between X and Y is modeled adequately by a linear model, the value of X is given by X = AY + B + , where the residual ε is assumed to be normally distributed with zero mean and variance σ^2. The conditional density function $h(x \mid y)$ is also normal with variance σ^2 ande mean $\mu(X) = AY + B$.

If the flood duration is exponentially distributed the joint density function is:

$$h\ (x, y) = \frac{1}{\sqrt{2\pi}\ \sigma} e^{-\frac{[x-\mu(y)]^2}{2\sigma^2}} \beta e^{-\beta y} \qquad (10)$$

This function can be numerically integrated to get H (x, y) for any pair of values x and y. Once (x, y) is obtained, Eq.(3) provides the nonexceedance probability for these values and Eq. (6) or (7) the corresponding return period.

The problem of assessing the flood volume must be adressed now. It is common that the flood volume associated with a flood event of duration Y and peak discharge X, can be computed, with reasonable accuracy, as the area of the triangle with base Y and height X:

$$Z = \frac{XY}{2} + e \qquad (11)$$

where the residual e is assumed to be normally distributed with zero mean and variance σ_e^2. Then, it is possible to consider the (1 - α) confidence interval for the flood volume Z, associated to any pair (X, Y)

$$Z \in \left[\frac{XY}{2} - \sigma_e\ \phi^{-1}\left(\frac{\alpha}{2}\right), \frac{XY}{2} + \sigma_e \phi^{-1}\left(\frac{\alpha}{2}\right)\right] \qquad (12)$$

If the assumptions of Eq. (12) are supported by data, it is possible to establish in a quite simple way the confidence interval of any size for the flood volume associated with this pair of values.

3. APPLICATION OF MULTIVARIATE PARTIAL DURATION SERIES IN PORTUGAL

Twelve hydrometric stations, representing very distinct hydrologic conditions, were selected. Some characteristics of these stations are presented in Table 1.

The results of the test of hypothesis relative to the Poisson distribution and serial independence are presented in Table 2 for average number of exceedances, λ , equal to one, two and four. To test the Poisson distribution a gaussian approximation was used and then the chi-square statistics were computed (Sukhatme, 1938). The probabilities that the value of the chi-square statistics are exceeded, 1-P(χ^2), are presented in Table 2. These probabilities can be taken as a goodness-of-fit criteria and very low values of 1 - P (χ^2) must imply the rejection of the hypothesis.

With respect to the serial independence the lag-one autocorrelation coefficient, r, was computed. The test of Anderson, described by Salas et Al. (1980), was used to test the hypothesis of r being equal to zero. The probability that $|r|$ is exceeded, if the hypothesis is true, is

Table 1 - Basic Characteristics of the Hydrometric Stations used in this Study

RIVER	HYDROMETRIC STATION	AREA (Km^2)	ALTITUDE (m)	Nº OF YEARS OF DATA
Beça	Cunhas	338	199	29
Corgo	Ermida	258	120	21
Sabor	Qtª Laranjeiras	3 464	111	14
Paiva	Castro Daire	286	445	28
Alva	Pte. Mucela	666	70	29
Simonte	Louçainha	4	640	17
Tera	Pavia	610	224	17
Degebe	Amieira	1 454	107	31
Pena	Vale da Ursa	80	15	25
Corona	Moinho do Bravo	218	27	25
Odelouca	Mte. dos Pachecos	386	51	16
Arão	Pereira	20	80	21

Table 2 - Goodness-of-fit of the Poisson Distribution and Test of the Hypothesis that the Lag-one Autorrelation Coefficient is Zero

RIVER	λ	Poisson distribution %	INDEPENDENCE Peak	Duration	Volume
Beça	1.0	15.42	14.	39.	17.
	2.0	98.33	53.	00.	02.
	4.0	7.02	29.	01.	04.
Corgo	1.0	78.30	14.	73.	04.
	2.0	7.02	24.	19.	14.
	4.0	13.21	73.	25.	18.
Sabor	1.0	25.78	01.	71.	41.
	2.0	94.16	41.	11.	20.
	4.0	24.48	68.	52.	23.
Paiva	1.0	68.79	03.	26.	13.
	2.0	11.65	42.	67.	63.
	4.0	40.32	53.	33.	44.
Alva	1.0	48.51	60.	46.	74.
	2.0	62.32	64.	58.	79.
	4.0	82.54	71.	45.	68.
Simonte	1.0	32.44	50.	23.	57.
	2.0	1.85	24.	31.	25.
	4.0	92.07	81.	18.	61.
Tera	1.0	42.33	28.	39.	62.
	2.0	47.22	52.	80.	25.
	4.0	45.25	42.	05.	42.
Degebe	1.0	34.06	12.	17.	31.
	2.0	27.69	06.	16.	14.
	4.0	87.08	86.	79.	36.
Pena	1.0	62.50	46.	82.	69.
	2.0	48.63	58.	56.	76.
	4.0	43.67	72.	79.	90.
Corona	1.0	15.07	43.	19.	52.
	2.0	61.35	11.	19.	28.
	4.0	8.27	67.	89.	38.
Odelouca	1.0	0.50	40.	54.	70.
	2.0	4.71	61.	43.	58.
	4.0	74.81	03.	07.	32.
Arão	1.0	2.00	12.	33.	24.
	2.0	0.50	87.	23.	67.
	4.0	12.81	74.	00.	00.

presented in Table 2.

The results of the test of hypothesis relative to the distribution of flood peaks, durations and volumes, are presented in Table 3. The exponential and two parameter gamma distribution were used and the chi-square statistics were computed. The probabilities of exceedance, 1-P (χ^2) are presented in Table 3, and can be taken as a goodness-of-fit criteria. The average values for λ = 1.0, 2.0 and 4.0, when all hydrometric stations are considered, are presented in Table 4, as well as the number of rejections if a 95% confidence level in considered. The average values presented in this table are relative to the non-rejected situations.

From these tables it can be concluded that the Poisson and independence hypothesis are very well supported by data and there is no reason to prefer two parameter gamma distribution instead of the exponential distribution. It can be also concluded that the truncation level, and the corresponding average number of flood events per year, has not much influence in the number of rejections for the Poisson and independence hypothesis. For the exponential hypothesis a low truncation level implies s larger number of rejections.

Two aditional hypothesis have to be analysed. The first one is the linear dependence between flood peaks and durations. The second is the use of Eq. (11). In Table 5 the linear correlation coefficients relating flood peaks and durations, and observed volumes and volumes computed by Eq. (11), are presented. With respect to the linear dependence between peaks and durations a quite erratic behaviour can be observed. However, there is a general trend corresponding to an increase of the correlation coefficient with the average number of flood events.

Plots of peaks versus durations confirmed this idea and the independence hypothesis is rather general for high trucation levels. For low trucation levels the degree of dependence and the use of a linear model must be carefully investigated. With respect to the use of Eq. (11), very large correlation coefficients were computed in all cases. This fact shows that when flood peak and duration are given, the flood volume can be computed with a rather small uncertainty.

In Correia (1985) the detailed study of two hydrometric stations, namely River Corgo at Ermida and River Tera at Pavia, is done. Some of the results of that paper are presented here. The truncation levels that were considered and the corresponding average number of annual exceedances, mean flood peak, duration and volume, are presented in Table 6. The marginal distributions of annual maximum flood peak and flood duration are presented in Table 7 and 8, respectively. With respect to the joint distributions of annual maximum peaks and durations some results are displayed on Figures 1 through 4.

Table 3 - Goodness-of-fit of the Exponential and Gamma Distributions

RIVER	λ	EXPONENTIAL DISTRIBUTION			GAMMA DISTRIBUTION		
		Peak	Duration	Volume	Peak	Duration	Volume
Beça	1.0	16.52	16.48	17.62	20.80	12.40	0.50
	2.0	11.11	90.90	0.50	9.20	0.70	14.40
	4.0	2.56	4.77	0.50	7.30	21.20	0.90
Corgo	1.0	72.54	0.50	69.57	32.20	39.80	0.80
	2.0	76.09	60.85	87.57	61.80	29.40	0.70
	4.0	90.78	55.78	21.92	65.00	41.40	5.20
Sabor	1.0	72.91	44.22	0.50	2.60	2.70	0.50
	2.0	22.05	54.18	5.67	1.80	69.20	0.50
	4.0	0.50	29.12	0.50	3.40	1.20	2.20
Paiva	1.0	4.01	32.15	92.54	25.00	82.00	0.50
	2.0	38.69	19.81	2.12	3.40	13.70	7.50
	4.0	21.96	79.62	0.50	8.20	25.50	32.20
Alva	1.0	17.79	25.05	27.59	6.20	86.80	0.50
	2.0	0.50	0.50	9.68	49.10	44.00	0.60
	4.0	0.50	2.01	0.50	10.00	4.20	11.20
Simonte	1.0	53.06	0.50	68.80	0.50	65.50	7.40
	2.0	92.99	26.71	40.81	0.50	72.60	0.50
	4.0	34.97	1.56	55.51	0.50	65.50	1.50
Tera	1.0	80.06	89.25	24.45	43.30	39.30	0.50
	2.0	84.90	38.65	18.17	37.30	43.00	0.50
	4.0	66.44	60.01	0.50	45.60	45.80	6.50
Degebe	1.0	86.29	44.00	35.01	34.00	52.00	6.30
	2.0	74.56	85.65	0.50	92.40	10.50	28.10
	4.0	0.50	0.69	0.50	0.50	13.30	4.30
Pena	1.0	24.25	0.50	14.24	11.40	2.30	0.50
	2.0	59.35	65.48	26.56	0.50	17.20	0.50
	4.0	0.50	19.57	36.44	11.20	68.00	3.40
Corona	1.0	33.80	70.42	66.28	28.10	53.00	0.50
	2.0	77.59	0.50	23.30	52.20	47.00	0.50
	4.0	27.71	15.27	31.62	21.60	2.50	20.10
Odelouca	1.0	34.29	59.32	67.45	65.50	36.10	0.50
	2.0	68.76	85.04	92.17	65.50	46.60	21.50
	4.0	23.20	49.21	7.71	14.80	3.10	31.90
Arão	1.0	21.88	71.01	0.83	13.50	47.80	0.50
	2.0	43.01	82.30	23.80	0.50	25.90	0.50
	4.0	0.50	45.66	52.21	0.50	0.50	0.50

Table 4 - Average Values of Goodness-of-fit Criteria and Number of Rejections

HYPOTHESIS		λ = 1.0		λ = 2.0		λ = 4.0	
		Fit-ness	Reject.	Fit-ness	Reject.	Fit-ness	Reject.
Poisson	Nº. of flood events	42.32	2	43.33	3	44.29	0
Independence	Peak	31.88	2	43.60	0	65.17	1
	Duration	43.50	0	38.40	1	43.10	2
	Volume	43.33	1	47.49	1	45.20	2
Exponential Distribution	Peak	46.67	2	59.01	2	44.18	6
	Duration	50.21	3	60.96	2	44.28	4
	Volume	48.36	2	36.41	3	34.24	6
Gamma Distribution	Peak	28.00	2	52.50	5	22.96	4
	Duration	51.47	2	41.91	2	40.08	5
	Volume	6.85	10	17.88	8	17.85	6

Table 5 - Linear Correlation Coefficients of Flood Peaks versus Durations and Observed versus Computed Volumes

RIVER	FOOD PEAKS VS. DURATIONS			OBSERVED VS. COMPUTED		
	λ =1.0	λ =2.0	λ =4.0	λ =1.0	λ =2.0	λ =4.0
Beça	0.18	0.30	0.47	0.95	0.96	0.94
Corgo	0.03	0.08	0.30	0.95	0.95	0.92
Sabor	0.29	0.31	0.54	0.98	0.97	0.94
Paiva	0.32	0.52	0.57	0.97	0.97	0.96
Alvão	0.12	0.30	0.31	0.98	0.97	0.94
Simonte	0.24	0.17	0.33	0.97	0.92	0.95
Tera	0.15	0.11	0.37	0.93	0.92	0.91
Degebe	0.11	0.26	0.34	0.96	0.95	0.95
Pena	0.17	0.15	0.35	0.93	0.95	0.94
Corona	0.25	0.34	0.26	0.94	0.96	0.96
Odelouca	0.04	0.48	0.64	0.97	0.93	0.95
Arão	0.24	0.19	0.05	0.97	0.97	0.99
MEAN VALUES	0.18	0.27	0.38	0.96	0.95	0.95

Table 6 – Truncation Levels, Average Number of Annual Exceedances and Mean Flood Peak, Duration and Volume

HYDROMETRIC STATION	TRUNCATION LEVEL (m^3/s)	λ	MEAN VALUES Peak (m^3/s)	Duration days	Volume ($10^6 m^3$)
Corgo at Ermida	110.	1.0	48.45	1.70	5.58
	72.	2.0	51.21	2.40	7.63
	32.	4.0	54.19	5.15	13.19
Tera at Pavia	86.	1.0	74.89	1.69	10.14
	51.	2.0	54.50	2.21	8.23
	6.7	4.0	55.27	7.27	14.16

Table 7 – Marginal Distribution of Annual Maximum Peak Discharge, in m^3/s

HYDROMETRIC STATION	λ	RETURN PERIOD (YEARS) 5	25	50	100	500	1000
Corgo at Ermida	1.0	182.67	264.97	299.05	332.88	411.05	444.66
	2.0	184.31	271.29	307.31	343.07	425.69	461.22
	4.0	188.41	280.45	318.57	356.41	443.84	481.43
Tera at Pavia	1.0	198.33	325.54	378.22	430.51	551.34	603.28
	2.0	170.52	263.10	301.43	339.48	427.42	465.22
	4.0	160.22	260.10	298.98	337.57	426.75	465.08

Table 8 – Marginal Distribution of Annual Maximum Flood Duration, in Days

HYDROMETRIC STATION	λ	RETURN PERIOD (YEARS) 5	25	50	100	500	1000
Corgo at Ermida	1.0	2.55	5.44	6.63	7.82	10.56	11.74
	2.0	5.26	9.34	11.03	12.70	16.58	18.24
	4.0	14.86	23.61	27.23	30.83	39.14	42.71
Tera at Pavia	1.0	2.53	5.41	6.59	7.77	10.50	11.67
	2.0	4.85	8.60	10.16	11.70	15.26	16.80
	4.0	20.98	33.33	38.45	43.52	55.25	60.29

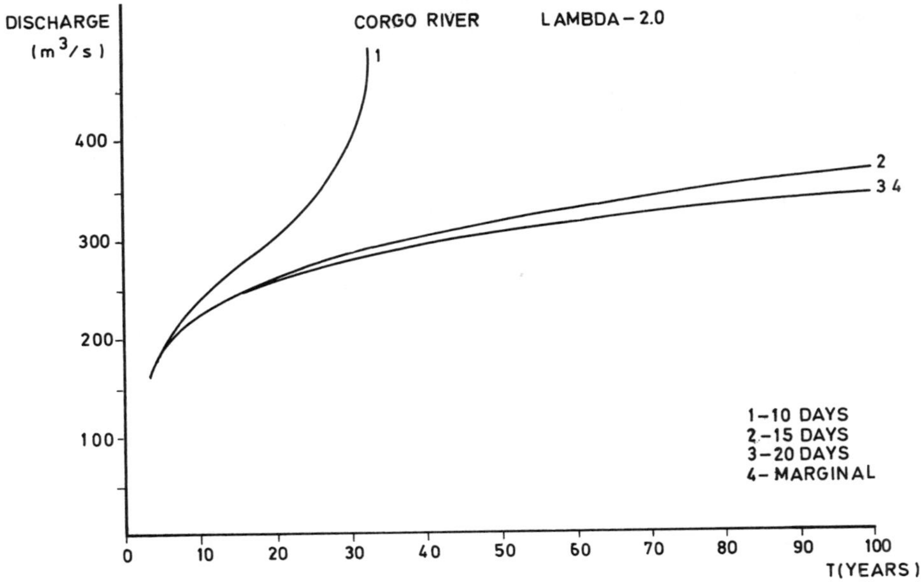

Fig. 1 – Peak discharges for several flood durations assuming independence between peaks and durations, at river Corgo

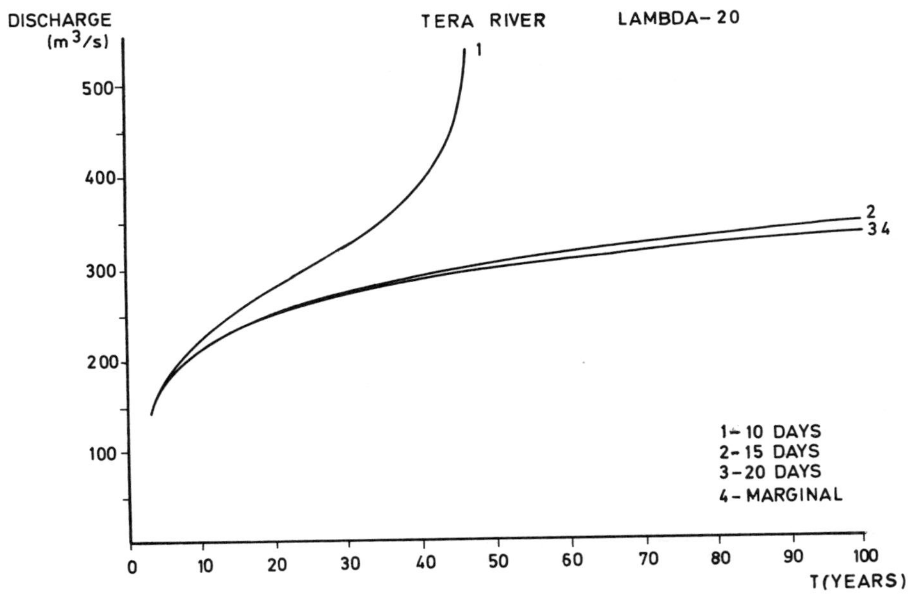

Fig. 2 – Same as above, at river Tera

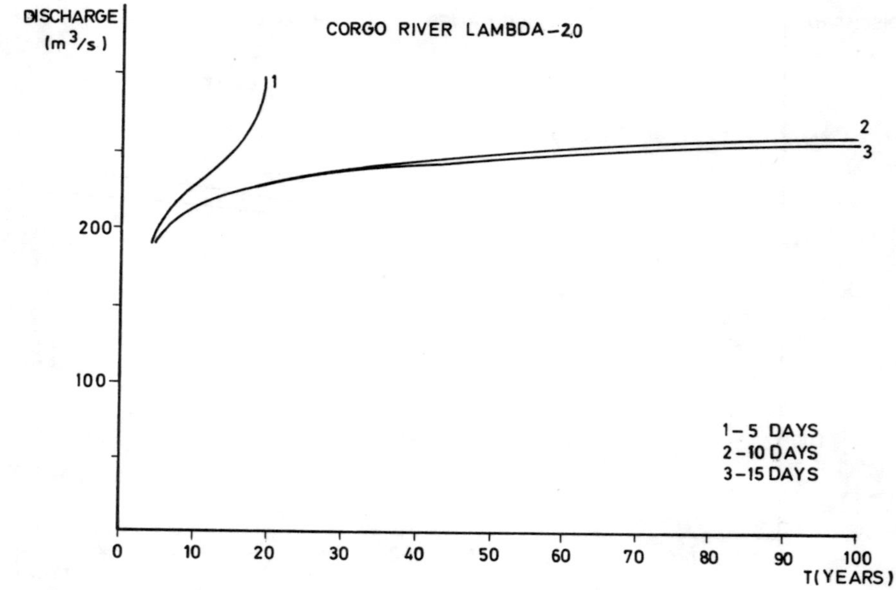

Fig. 3 - Peak discharges for several flood durations assuming a linear dependence between peaks and durations, at river Corgo

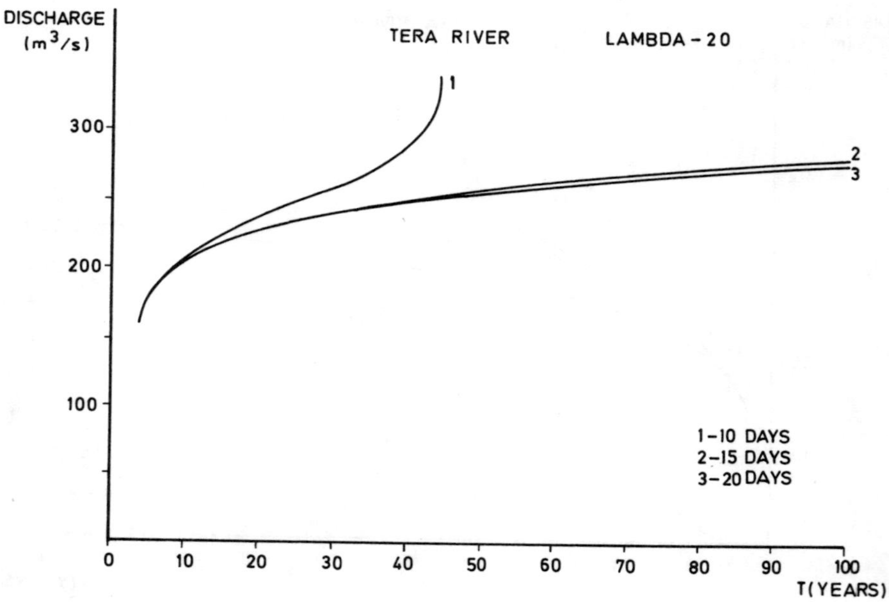

Fig. 4 - Same as above, at river Tera

4. CONCLUSIONS

The use of partial duration series provided an useful framework for the multivariate characterization of flood peak, duration and volume. A much better assessment of flood frequency and risk can be done based on this approach.

The basic hypothesis relative to the statistical distributions of annual number of excedances, flood peaks, durations and volumes, as well as the hypothesis of serial independence, can be accepted in most situations in Portugal. However, some additional research is needed before final conclusions can be formulated with respect to the general use of multivariate partial duration series.

ACKNOWLEDGEMENTS

Dr. Maria Alzira Santos, researcher at LNEC, is gratefully acknowledged for the fruitful discussions on the topic of this paper.

This research was done under Project 62/11/8490 financed by PIDAC (Central Administration Investment and Development Plan). Data was provided by Direcção-Geral dos Recursos e Aproveitamentos Hidráulicos.

REFERENCES

ASHKAR, F.; J. ROUSSELLE 1982 - A Multivariate Statistical Analysis of Flood Magnitude, Duration and Volume. In Statistical Analysis of Rainfall and Runoff, edited by V. Singh, Water Resources Publications, Littleton, Colorado.

CORREIA, F.N. 1983 - Métodos de Análise e Determinação de Caudais de Cheia. Laboratório Nacional de Engenharia Civil, Lisboa

CORREIA, F.N. 1985 - Engineering Risk using Multivariate Partial Duration Series, Nato Advanced Study Institute on Engineering Reliability and Risk in Water Resources, University of Arizona, Tucson, Arizona

CUNNANE, C. 1973 - A Particular Comparison of Annual Maxima and Partial Duration Series Methods of Flood Frequency Prediction. Journal of Hydrology, nº 18, North-Holland Publishing Company, Amsterdam

NERC, 1975 - Flood Studies Report. Natural Environment Research Council, London

TAESOMBUT, V.; V. YEVJEVICH 1978. Use of Partial Duration Series for Estimating Distribution of Maximum Annual Flood Peak. Hydrology Paper nº 97, Colorado State University, Fort Collins, Colorado

TODOROVIC, P. 1970 - On some Problems Involving Random Number of Random Variables, Annuals of Mathematical Statistics, Vol. 41, nº 3

TODOROVIC, P.; E. ZELENHASIC 1970 - A Stochastic Model for Flood Analysis. Water Resources Research, Vol. 6, nº 6, American Geophysical Union, Washington, D.C.

TODOROVIC, P.; J. ROUSSELLE 1971 - Some Problems of Flood Analysis. Water Resources Research, Vol. 7, nº 5, American Geophysical Union, Washington, D.C.

SALAS, J.; J. DELLEUR; V. YEVJEVICH; W. LANE 1980 - Applied Modeling of Hydrologic Time Series. Water Resources Publications, Littleton, Colorado

ANOTHER LOOK AT THE JOINT PROBABILITY OF RAINFALL AND RUNOFF

C. T. Haan and B. N. Wilson
Agricultural Engineering Department
Oklahoma State University
Stillwater, Oklahoma

ABSTRACT. Considerable discussion exists in the literature dealing with the relationship between the return period of a runoff event and the return period of the rainfall that produced the runoff. In most instances the assumption is made that the two return periods are the same when a design flow is estimated. This assumption has been widely criticized but continues to be used because of the lack of a suitable alternative. Differences in the return period of runoff and rainfall can be attributed to variation in the antecedent conditions that exist on a catchment at the time of a storm. Thus a procedure that can simultaneously account for rainfall probability and antecedent condition probability is needed.

This paper develops a procedure for estimating the magnitude of a flow event for a given return period that incorporates the joint probability of rainfall depths and antecedent soil water conditions. Rainfall probabilities are determined from the Extreme Value distribution, and soil water probabilities are based on 22 years of data from an experimental watershed near Stillwater, OK. The resulting flow estimates for a given return period are compared with similar estimates based on the assumption of equality of return periods for rainfall and runoff.

INTRODUCTION

Hydrologists and engineers are concerned with estimating runoff volumes and peaks from rainfall. Generally the assumption is made that the computed runoff has the same frequency of occurrence as the rainfall used in the computation. It is this assumption that is frequently discussed and criticized.

Schaake, Geyer and Knapp (1967) investigated the rational equation as a transformation from rainfall frequencies to peak flow frequencies. They found generally good agreement with a tendency for the C coefficient to increase as the return period increased. Hjelmfelt (1980) in a similar study based on runoff volumes found the curve number techniques of the Soil Conservation Service (1972) to be

V. P. Singh (ed.), Hydrologic Frequency Modeling, 555–569.

a reasonable transformation from rainfall frequencies to runoff frequencies with the possible exception of arid watersheds.

Reich (1970) studied extreme value statistics of several watersheds in Pennsylvania. He found that extreme rains in the spring are more closedly related to annual peak flows than are annual maximum rainfalls. No usable relation between extreme value statistics for rainfall and flood flows could be found. Larson and Reich (1972) studied data from some of the same Pennsylvania watersheds used by Reich (1970). They state that " . . . individual rainfall events do not produce a peak flow of the same recurrence interval and that a wide variation exists". Based on a rank correlation diagram relating peak runoff and precipitation and a frequency distribution of rank differences, they concluded (1) ranks of annual floods and annual storm rainfalls have a central tendency toward equality; (2) for individual years, a wide but symmetrical variation from this central tendency exists; and (3) for design purposes, the assumption that a design storm of a certain recurrence interval will produce a peak flow of the same recurrence interval is appropriate.

From Larson and Reich's (1972) study, the relation between recurrence intervals for rainfall and runoff can be visualized by recognizing that for a given rainfall the resulting runoff may vary according to some probability distribution with the recurrence interval of the expected value of runoff from a given rainfall equal to the recurrence interval of the rainfall.

The purpose of this paper is to present a methodology for computing runoff frequencies from the probabilistic behavior of rainfall and other factors contributing to runoff. The procedure is illustrated using two common hydrologic techniques and assumed, but reasonable, probability distribution functions. The methodology is not, however, limited to these hydrologic techniques nor to the particular probability distributions used. The method can be generalized to consider several random variables in addition to rainfall. In the end the methodology allows the probability distribution of runoff to be calculated from random variations in rainfall and other factors affecting runoff.

RUNOFF VOLUME

The U.S. Soil Conservation Service (1972) uses the relationship

$$Q = (R - 0.2S)^2/(R + 0.8S) \qquad R > 0.2S \tag{1}$$

to estimate runoff volume, Q, from rainfall volume, R, and a parameter S. Equation (1) can be rearranged to give

$$S = 5R + 10Q - 10\sqrt{Q^2 + 1.25\ RQ} \tag{2}$$

In applying equation (1), a correction is generally applied to S if the 5-day antecedent precipitation is outside an "average" range.

Antecedent moisture condition I (AMC I) is drier than the average, AMC II, while AMC III is wetter than AMC II.

For the period July 1951 through October 1972, 151 storms having 24-hour rainfall volumes in excess of 1.00 inch where studied on a grassland watershed north of Stillwater, OK. Of these storms 118 could be classed as AMC I storms, 15 as AMC II, and 18 as AMC III. Gray et al. (1982) in a study of runoff in the Kentucky - Indiana area also found that 80-90% of the time during the growing season the watersheds were in AMC I.

For each of the 151 storms, the observed rainfall and runoff were used to compute S from equation (2). The means and standard deviations of R, Q, and S and their intercorrelations are shown in Table I. The overall mean and standard deviation of S for the 151 storms were 3.10 and 2.68 inches, respectively.

In agreement with other studies (Sharma, Gander, and Hunt, 1980; Hjelmfelt, Kramer, and Burwell, 1981), S was found to follow a lognormal distribution within each AMC class. Figure 1 shows the lognormal plot of S for the entire 151 events. Perhaps another distribution would better fit this set of data; however, it was decided to follow the lead of earlier investigators and use the lognormal distribution.

A common procedure for estimating the T-year runoff volume is to use the T-year rainfall volume and an average antecedent condition in equation (1). The antecedent condition as reflected in the parameter S can be considered a random variable. For a given R, the expected runoff volume would then be a random variable as well. Table II contains the resulting flow estimates considering each AMC separately and using the overall average S. In these calculations the 100 yr - 24 hr and 2 yr - 24 hr rainfall volumes were taken as 8.6 and 3.75 inches, respectively, which is appropriate for Stillwater, OK. Volumes for other return periods were determined from the Extreme Value Type I distribution with the parameters of the distribution estimated by the procedure shown in the Appendix.

It is possible to calculate the pdf of the random variable Q based on equation (1) and the joint pdf of R and S using the method of derived distribution (see Appendix). The data in Table I indicate little correlation between R and S so their joint pdf may be taken as the product of the individual pdfs of R and S.

Following the procedure shown in the appendix for derived distributions, define the variable Y to be

$$Y = R \tag{3}$$

Table I. Means, Standard Deviations and Correlations on R, Q, and S

AMC	#Events	Mean (in)			Std dev (in)			Correlations		
		R	Q	S	R	Q	S	R&Q	R&S	Q&S
I	118	1.79	0.49	3.47	0.81	0.57	2.83	0.42	0.46	-0.53
II	15	2.06	0.94	2.70	1.18	1.24	2.60	0.79	0.05	-0.52
III	18	1.67	1.09	1.06	0.85	0.87	1.45	0.87	-0.05	-0.47

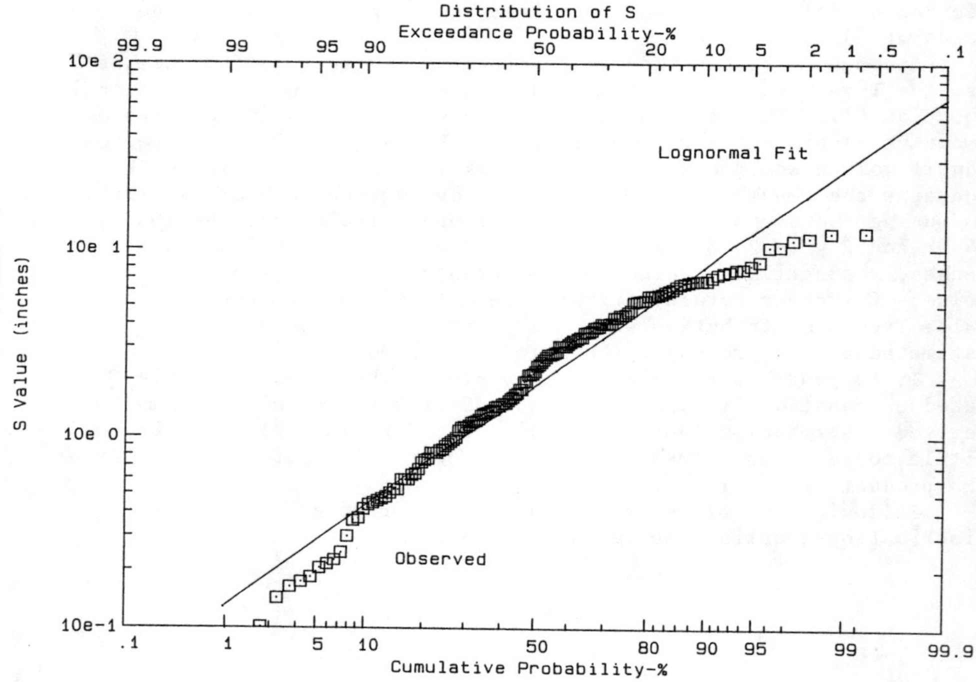

Figure 1. Log normal plot of S

Thus equations (1) and (3) constitute equations (a15) of the appendix. The joint pdf of Q and Y is given by

$$p_{Q,Y}(q,y) = p_{R,S}(r,s) \mid J(r,s/q,y) \mid \tag{4}$$

$$= p_R(r)\, p_S(s) \mid J(r,s/q,y) \mid \tag{5}$$

The determinant of the Jacobian is given by

$$|J(r,s/q,y)| = |\partial r/\partial q \;\; \partial s/\partial y - \partial s/\partial q \;\; \partial r/\partial y|$$

From equation (3) it can be seen that $\partial s/\partial y$ is 0 and $\partial r/\partial y$ is 1. Therefore

$$|J(r,s/q,y)| = 10 \left[1 \; \frac{2q + 1.25r}{2(q^2 + 1.25\ rq)^{1/2}} \right] \tag{6}$$

Equation (a10) gives the lognormal distribution for S as

$$p_S(s) = \exp(-(\ln s - \mu)^2/2\sigma^2) \; / \; s\sqrt{2\pi\sigma^2} \tag{7}$$

where μ and σ^2 are the mean and variance of the natural logarithms of S.

Extreme rainfall is often described using the Extreme Value Type I distribution (Hershfield, 1961). From equation (a1), the Extreme Value Type I distribution of rainfall is given as

$$p_R(r) = \exp\{-(r - \beta)/\alpha - \exp[-(r - \beta)/\alpha]\}/\alpha \tag{8}$$

The pdf of Q and Y is now obtained by substituting equations (2) and (3) into (7) and (8) and then combining equations (6), (7), and (8) in equation (5) giving the joint pdf of Q and the defined variable Y. The univariate pdf of Q is found by integrating out Y.

$$p_Q(q) = \int_q^{\infty} p_{Q,Y}(q,y)dy \tag{9}$$

The cumulative distribution function of Q is given by

$$P_Q(q) = \int_o^q p_Q(q)dq \tag{10}$$

The T-year Q is the value of q* satisfying

$$1/T = 1 - P_Q(q^*) \tag{11}$$

The results of applying equation (11) are shown in Table III. Numerical integration was used for the integrations required by equations (9) and (10). Comparing Tables II and III, it can be seen that equation (11) tended to estimate greater flow volumes for the same return period than equation (1). It also appears that the estimates tend to converge as the return period gets shorter.

Table IV contains the annual maximum observed flows for the storms analyzed. A frequency plot of the data is shown in Figure 2 along with the estimates based on the overall average S and equations (1) and (11). As is generally true when comparing frequency estimates, definite statements cannot be made. However, the estimates based on equation (11) are reasonable approximations to the data at high return periods.

PEAK FLOWS

The rational equation continues to be a popular method of estimating peak flows. The return period of the peak flow computed from the rational equation is generally taken as equal to the return period of the rainfall used in the computation. The rational equation is given by

$$Q = C\ I\ A \tag{12}$$

where Q is the peak flow, I is the rainfall intensity for the desired frequency and proper duration, A is the drainage area and C is a runoff coefficient. The coefficient C must reflect many hydrologic factors including antecedent conditions. If C is calculated from equation (12) using data from several observed storm events, C will be found to be a random variable reflecting otherwise unquantified variation in the hydrologic conditions of the drainage area.

Using the method of derived distributions, the pdf of Q can be determined. Following the same steps as used in the previous section on runoff volume, define

$$K = C \tag{13}$$

The joint distribution of Q and K is then given by

Table II. Flow Volumes for Various Return Periods Calculated from Equation (1).

T (yrs)	R (in)	Flow Volume Q (inches)			Overall
		AMCI	AMCII	AMCIII	
100	8.60	5.49	6.04	7.45	5.74
50	7.60	4.60	5.11	6.46	4.83
25	6.75	3.85	4.33	5.63	4.07
10	5.75	3.00	3.43	4.65	3.16
5	5.00	2.38	2.78	3.92	2.56
2	3.75	1.43	1.74	2.72	1.57

Table III. Flow Volumes for Various Return Periods Calculated from Equation (11)

T (yrs)	Flow Volume Q (inches)			Overall
	AMCI	AMCII	AMCIII	
100	6.95	7.40	8.02	7.10
50	5.84	6.28	7.10	6.01
25	4.94	5.38	6.24	5.12
10	3.84	4.26	5.15	4.02
5	2.99	3.40	4.28	3.16
2	1.70	2.05	2.94	1.85

Table IV. Annual Maximum Observed Flow Volumes (inches)

Year	Flow	Year	Flow
1951	3.57	1962	1.42
1952	1.76	1963	1.01
1953	1.07	1964	0.89
1954	0.53	1965	0.31
1955	2.16	1966	0.43
1956	0.01*	1967	2.16
1957	3.05	1968	1.47
1958	0.03	1969	1.67
1959	4.79	1970	1.41
1960	1.41	1971	1.44
1961	1.36	1972	0.74

*assigned - actual value was 0.00

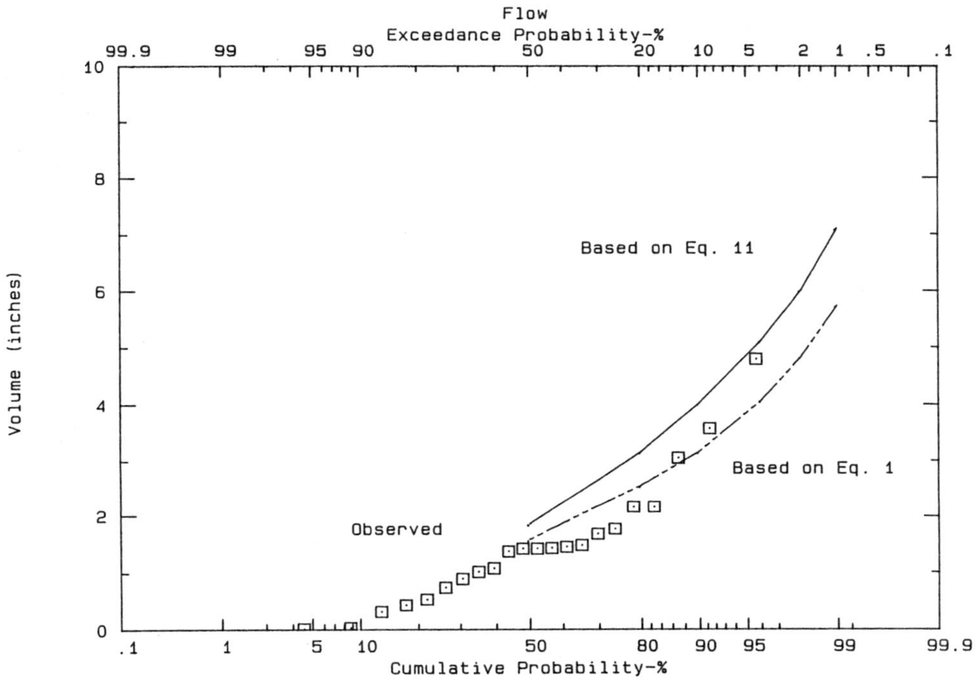

Figure 2. Frequency plot of observed and calculated data.

$$p_{Q,K}(q,k) = p_C(c)\ p_I(i)\ \ J(c,i/q,k) \tag{14}$$

assuming independence of C and I. Operations on equations (12) and (13) produce

$$|J(c,i/q,k)| = 1/CA$$

Since C is bounded by 0 and 1, a Beta distribution (equation a4 in Appendix) is used to describe $p_C(c)$. The Extreme Value Type I distribution is used to describe $p_I(i)$. Thus

$$p_{Q,K}(q,k) = p_C(c)\ p_I(i)/CA \tag{15}$$

where $p_C(c)$ and $p_I(i)$ are the Beta distribution and Extreme Value Type I distribution, respectively. The univariate pdf of Q is given by

$$p_Q(q) = \int_K p_{Q,K}(q,k)dK \tag{16}$$

Since C = 0 maps to K = 0, C = 1 maps to K = 1, I = 0 maps to Q = 0, and I = maps to Q = , the limits on the integration are from 0 to 1. Making the substitutions C = K and I = Q/KA, we have

$$p_Q(q) = \int_o^1 p_C(k)\ p_I(q/kA)/kA\ dk \tag{17}$$

Again the T-year peak flow can be computed from equations (10) and (11) where Q now represents the peak flow and $p_Q(q)$ comes from equation (17).

The results of this analysis are shown in Table V. Again the estimates from equation (11) are larger than the estimates from equation (12) with convergence noted at lower return periods. This is in agreement with the earlier observation of Schaake, Geyer and Knapp (1967) that C may be larger for storms with greater return periods.

DISCUSSION

The joint distribution approach presented here considers all combinations of rainfall values and antecedent conditions in a joint probability approach to estimate flow frequencies. As a result of allowing the possibility of large rains and wet antecedent conditions, it is possible for estimated flows for large return periods to be greater than similar estimates that treat the antecedent condition as a constant and use return period rainfall directly. Obviously the procedures discussed in this paper require greater study before any sweeping conclusions can be made. The joint probability, derived distribution approach provides a mechanism for estimating flows of various return periods that incorporates rainfall probabilities and random variation in other factors governing runoff.

Table V. Peak Flow Estimates from Equation (11) and (12)

T (yrs)	(iph)	Average C Value 0.3	0.5	0.7
		Q(cfs)		
100	3.50	1.05/1.69	1.75/2.14	2.45/2.69
50	3.17	0.95/1.46	1.59/1.89	2.22/2.43
25	2.84	0.85/1.24	1.42/1.67	1.99/2.15
10	2.39	0.72/0.96	1.20/1.35	1.67/1.78
5	2.04	0.61/0.73	1.02/1.10	1.43/1.42
2	1.50	0.45/0.40	0.75/0.73	1.05/1.04

The estimate from equation (12) is the first entry in each column. Thus for C = 0.3 and T = 100, equation (12) estimated a peak flow of 1.05 and equation (11) a flow of 1.69.

ACKNOWLEDGEMENTS

Approved as Journal Article No. 4942 of the Oklahoma Agricultural Experiment Station.

REFERENCES

Gray, D.D., Katz, P.G., deMonsabert, S.M. and Cogo, N.P. 1982. 'Antecedent Soil Moisture Condition Probabilities'. Journal Irrigation and Drainage Division, American Society of Civil Engineers, Vol. 108(IR2), pp. 107-114.

Haan, C.T. 1977. Statistical Methods in Hydrology. Iowa State University Press, Ames, Iowa.

Hahn, G.J. and Shapiro, S.S. 1967. Statistical Models in Engineering. John Wiley and Sons, New York.

Hershfield, D.M. 1961. Rainfall Frequency Atlas of the United States. U.S. Weather Bureau Technical Paper 40. U.S. Department of Commerce, Washington, D.C.

Hjelmfelt, A.T., Jr. 1980. 'Empirical Investigation of Curve Number Technique'. Proceedings American Society of Civil Engineers, Vol. 106(HY9), pp. 1471-1476.

Hjelmfelt, A.T., Jr., Kramer, L.A. and Burwell, R.E. 1981. 'Curve Numbers as Random Variables'. in: V.P. Singh (Editor), Rainfall-Runoff Relationships. Proceedings International Symposium on Rainfall-Runoff Modeling. May 18-21, 1981, Mississippi State University. Published by Water Resources Publications, Littleton, Colorado.

Hughes, W.C. 1977. 'Peak Discharge Frequency from Rainfall Information'. Proceedings American Society of Civil Engineers, Vol. 103(HY1), pp. 39-50.

Larson, C.L. and Reich, B.M. 1972. 'Relationship of Observed Rainfall and Runoff Recurrence Intervals'. in: Schulz, E.F., Koelzer, V.A. and Mahmood, K. (Editors), 1973, Floods and Droughts, pp. 34-43, Water Resources Publications, Fort Collins, Colorado.

Reich, B.M. 1970. 'Flood Series Compared to Rainfall Extremes'. Water Resources Research, Vol. 6(6), pp. 1655-1667.

Schaake, J.C., Jr., Geyer, J.C. and Knapp, J.W. 1967. 'Experimental Examination of the Rational Method'. Proceedings American Society of Civil Engineers, Vol. 93(HY6), pp. 353-370.

Sharma, M.C., Gander, G.A. and Hunt, C.G. 1980. 'Spatial Variability of Infiltration in a Watershed'. Journal of Hydrology, Vol. 45, pp. 101-122.

Soil Conservation Service. 1972. Hydrology, Section 4, National Engineering Handbook. Soil Conservation Service, U.S. Department of Agriculture, Washington, D.C.

APPENDIX

Extreme Value Type I

The Extreme Value Type I (EVI) distribution is given by (Haan, 1977)

$$p_X(x) = \exp\{-(x - \beta)/\alpha - \exp[-(x - \beta)/\alpha\}/\alpha \tag{a1}$$

where α and β are parameters of the distribution. The expected value of the EVI is given by

$$E(X) = \beta + \gamma\alpha \tag{a2}$$

where γ is Euler's constant. The variance is given by

$$Var(x) = \pi^2\alpha^2/6 \tag{a3}$$

The T-year frequency factor for the EVI is given by

$$K_T = \sqrt{6/\pi^2}\ \{\gamma + \ln\ln[T/(T - 1)]\} \tag{a4}$$

By using the relationship

$$X_T = E(X) + K_T \sqrt{Var(X)} \tag{a5}$$

where X_T is the value of X with a return period of T and using equations (a2-a4), it can be shown that

$$E(X) = 0.95\ X_2 + 0.05\ X_{100} \tag{a6}$$

$$Var(X) = 0.092\ (X_{100} - X_2)^2 \tag{a7}$$

$$\alpha = 0.236\ (X_{100} - X_2) \tag{a8}$$

$$\beta = X_{100} - 1.086\ (X_{100} - X_2) \tag{a9}$$

Thus by specifying the 100-year and 2-year value, the pdf given by equation (a1) is completely specified. Similar relationships could be developed using any 2 return periods.

The Log Normal Distribution

The log normal distribution (LN) is given by (Haan, 1977)

$$p_X(x) = \exp\,[-(\ln x - \mu)^2/2\sigma^2]/x\sqrt{2\pi\sigma^2} \tag{a10}$$

where the parameters μ and σ^2 are estimated as the mean and variance of the natural logarithms of X.

Beta Distribution

Haan (1977) gives the Beta distribution as

$$p_X(x) = x^{\alpha-1}\,(1-x)^{\beta-1}/B(\alpha,\beta) \tag{a11}$$

$$0 < x < 1\ ;\alpha\ ,\ \beta > 0$$

where $B(\alpha,\beta)$ is the Beta function related to the gamma function by

$$B(\alpha,\beta) = \Gamma(\alpha)\Gamma(\beta)/\Gamma(\alpha+\beta) \tag{a12}$$

Hahn and Shapiro (1967) give the moment estimation for the parameters as

$$\hat{\beta} = (1-\bar{x})\,[\bar{x}(1-\bar{x}) - s^2]/s^2 \tag{a13}$$

and

$$\hat{\alpha} = \bar{x}\,\beta/(1-\bar{x}) \tag{a14}$$

Derived Distributions

If $p_{X,Y}(x,y)$ is the joint probability density function (pdf) of the random variables X and Y and U and V are functions of X and Y given by

$$U = u(X,Y) \text{ and } V = v(X,Y) \tag{a15}$$

then under certain conditions (Haan, 1977), the joint pdf of U and V is given by

$$p_{U,V}(u,v) = p_{X,Y}(u,v)\ |J(x,y/u,v)| \tag{a16}$$

where J(·) is the Jacobian of the transformation between U,V and X,Y computed as the determinant of the matrix given by

$$|J(x,y/u,v)| = \begin{vmatrix} \partial x/\partial u & \partial x/\partial v \\ \partial y/\partial u & \partial y/\partial v \end{vmatrix} \tag{a17}$$

The univariate pdf of U is given as the marginal distribution of U computed as

$$p_U(u) = \int_v p_{U,V}(u,v)dv \tag{a18}$$

taking care that the proper limits on v are used.

A BIVARIATE FLOOD MODEL AND ITS APPLICATION

B. Sackl and H. Bergmann
Institute for Hydromechanics, Hydraulics and Hydrology
Mandellstraße 9/I
A - 8010 Graz
Austria

ABSTRACT. In this paper, a bivariate model will be presented, using flood peak and flood volume of direct runoff as the two characteristic variables of a flood event. The model can also be applied to other bivariate problems in hydrology.

The bivariate normal distribution was chosen as the parent bivariate distribution function, which seems to be suitable for the practical user. It is necessary to transform the marginal distributions of both samples into normal distributions. The theoretical bi-normal distribution is fitted and tested by using the equi-lines of probability density function (ISO-PDF-lines).

The model offers various possibilities of probability interpretation. ISO-PDF-lines show a range within which a certain percentage of events lie. The conditional distributions of one variable can be computed for a constant value of the other variable, which shows the probability by which the variable is reached or exceeded for the other fixed variable. A useful calculation is given by the so called "probability of quadrants". For instance, the "upper-right probability" of a certain pair of values defines the probability of events with both characteristics being higher than those of the presumed values.

Applying this method to the planning of flood protection projects, design flood events of chosen probabilities are determined by a certain equi-line of quadrants probability.

1. INTRODUCTION

Hydrological phenomena like precipitation or runoff always appear as multivariate events. Therefore they are characterized by numerous parameters. Conventional statistical methods only regard the frequency of a certain characteristic, like the peak discharge of a flood event. Therefore, by univariate frequency analyses no statemants about the frequency of the whole event can be made. For many problems in water management it is not satisfying to know only the frequency of single characteristics. It will be more important to know the probability of the whole event. For example, by regarding the analyses of the runoff

V. P. Singh (ed.), Hydrologic Frequency Modeling, 571–582.

process, this fact will be of importance for the planning of flood retention basins. In such cases it will be more favourable to use a multivaraite statistical model, which describes the hydrological process i its entirety. But on the one hand, the expense for a multivariate will usually exceed the possibilties of practical application, on the other hand , the fit of a theoretical distribution function is only possible with a sufficient large sample size. For instance, if a sample 20 points is available for a univariate anlyses, 20^2 = 400 points will be necessary for a bivaraiate analyses to achieve the same reliability . For the above reasons a bivariate model will be described in this paper.

2. DATA ERRORS

The more complex and expensive a model is, the more requirements will be expected from input data and boundary conditions. The most complex model will be of no use if the input data are unreliable or incorrect. Frequent error sources, which may arise with runoff analyses are for instance three types of extrapolation errors.

2.1. Stage runoff curve

The highest runoff values measured at most water gauges do not exceed 30 to 40% of a flood with an exceedance of 1%. That will be about the runoof peak of a flood with an exceedance of 50%. No matter which method is used, the extrapolation of the stage runoff curve will be unreliable (fig.1).

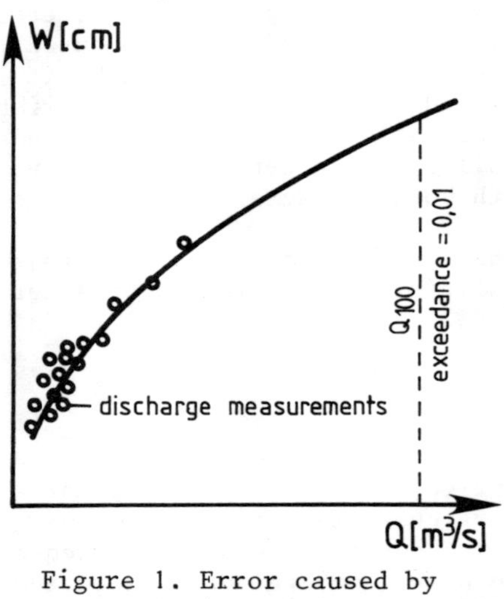

Figure 1. Error caused by extrapolation of stage runoff curve.

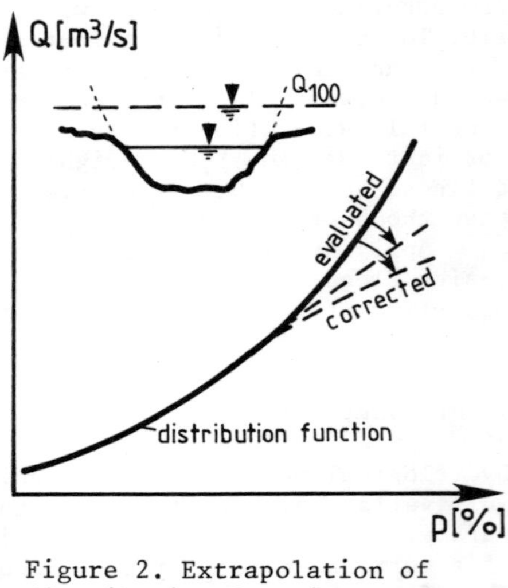

Figure 2. Extrapolation of distribution function, error caused by retention effect.

2.2. Distribution function

The extrapolation of theoretical distribution functions will be unreliable, because the time of observation is too short in most cases. In comparing investigations at the water gage Köln at the river Rhein, which has been observed since 1825, several series of 30 years were evaluated. The flood of a return period of 100 years, resulting from these series, ranged between 7000 m^3/s and 13000 m^3/s (Liebscher, 1970). Therefore it will be favourable to use partial series as sample for bivariate analyses. Usually the minimum annual flood of the one or the other variable is chosen as threshold.

2.3. Retention effect

Quite frequently you will find a sample with floods that do not overtop or hardly overtop the banks, but probabilities of floods will have to be determined, which show extensive overtoppings with a more or less sigificant retention effect. As the runoff peak is diminished by the retention, the extrapolation tail will consequently have to be corrected towards lower values. This will be of importance especially for the planning of flood retention basins (fig.2).

3. CHARACTERISTICS OF FLOOD EVENTS

The most significant characteristics of a flood event (fig.3) are the runoff peak of the total runoff (Q_s), the runoff peak (Q_d) and the volume (V_d) and duration (T) of the direct runoff, base flow and time parameters , which describe the time evolution of the event.

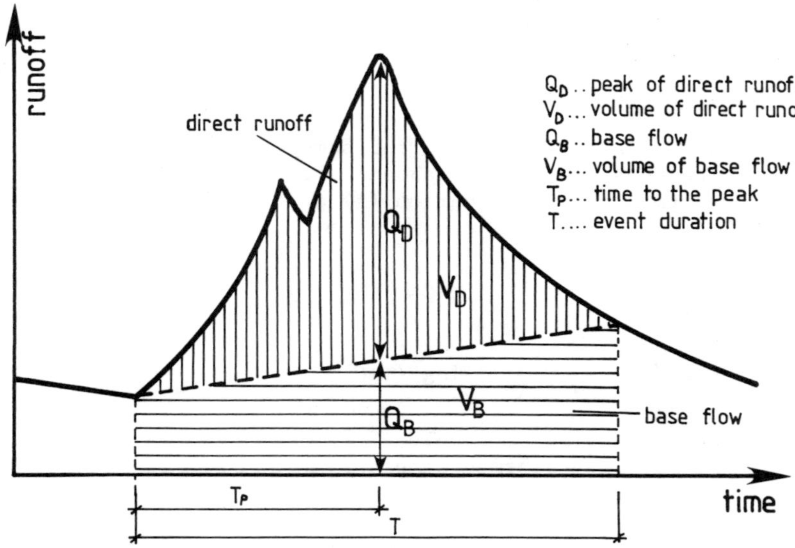

Figure 3. Characteristic values of a flood event.

4. BIVARIATE NORMAL DISTRIBUTION

The volume V_d and runoff peak Q_d are the variables X and Y of the bivariate model. Every event is shown as a pair of values in the X-Y plane (fig.4). For frequency analyses the bi-normal distribution will be used.

Marginal distributions are the univariate distribution functions of the single values (fig.4). The application of the bi-normal distribution assumes normal marginal distributions. Usually this is not the case with the characteristics Q_d and V_d. Therefore, a transformation of the variables will be necessary.

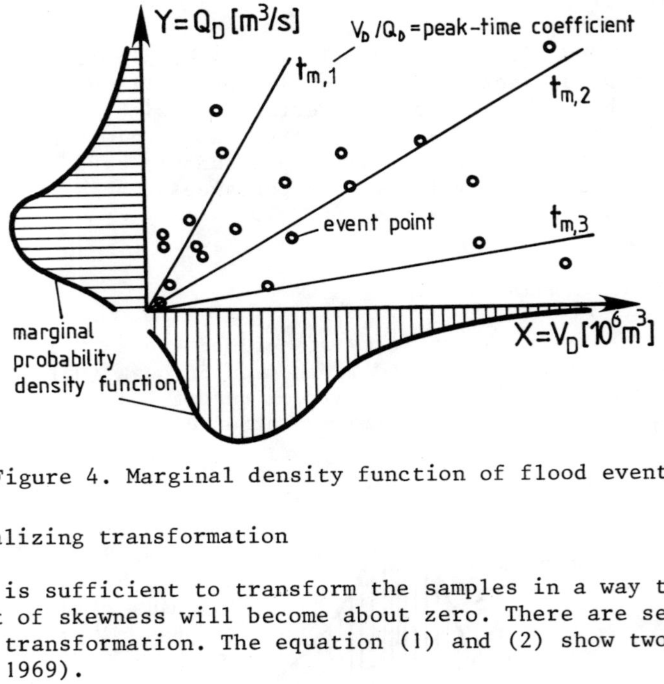

Figure 4. Marginal density function of flood events.

4.1. Normalizing transformation

Usually it is sufficient to transform the samples in a way that the coefficient of skewness will become about zero. There are several methods of transformation. The equation (1) and (2) show two examples (Schröder, 1969).

$$X' = \left(\frac{X-X_{min}}{\Delta X}\right)^C \qquad (\emptyset < C \leqslant 1) \qquad (1)$$

$$X' = \log\left(1+C\,\frac{X-X_{min}}{\Delta X}\right) / \log(1+C) \qquad (2)$$

in which X = original sample
X'= transformed sample
X_{min}= minimal value of sample
ΔX= sample range
C = transformation coefficient, evaluated iteratively

To be exact, it would be necessary to normalize the excess of the samples (excess = 3 for normal distribution), but it is of minor importance and it will only be advisable with a sufficient large sample size. Normal distributed marginal distributions follow from a bi-normal distributed sample, but this will not be the case conversely (Dyck, 1980). But nevertheless, the inverse conclusion will be permissanle with good approximation.

4.2. Bi-normal density function

For the determination of the bi-normal density function it will be suitable to compute with standardized variables (3), whereby the correlation coefficient r remains as the only parameter.

standardized variables:

$$u = (X-X_m) / s_x \qquad v = (Y-Y_m) / s_y \tag{3}$$

bi- normal density function:

$$g(u,v;r) = \frac{1}{2\pi\cdot(1-r^2)^{0.5}} \exp\left(\frac{u^2-2ruv+v^2}{2\cdot(r^2-1)}\right) \tag{4}$$

in which X_m, Y_m = means
s_x, s_y = standard deviations

Figure 5 shows a spatial describtion of a bi-normal density function.

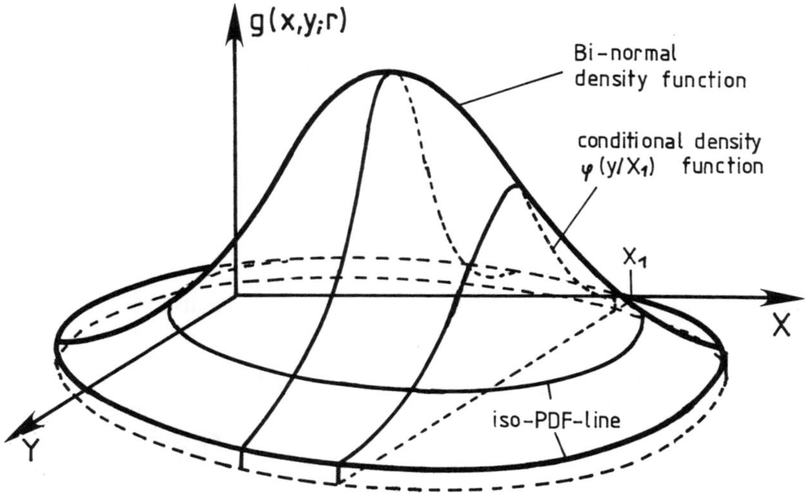

Figure 5. Bi- normal density function

<u>Conditional distributions.</u> Vertical cuts through the density function in a right angle to the X-axes result in conditional distributions of Y for a fixed value X_1 = const. (5) and vice versa (6). These conditional distributions are again univariate normal distributions.

$$\varphi(Y|X_1) = \frac{1}{(2\pi)^{0.5} s_y (1-r^2)^{0.5}} \exp -(\frac{1}{2s_y^2(1-r^2)}\{Y - -[Y_m + r\frac{s_y}{s_x}(X_1 - X_m)]^2\}) \quad (5)$$

$$\varphi(X|Y_1) = \frac{1}{(2\pi)^{0.5} s_x (1-r^2)^{0.5}} \exp \ (\frac{1}{2s_x^2(1-r^2)}\{X - -[X_m + r\frac{s_x}{s_y}(Y_1 - Y_m)]^2\}) \quad (6)$$

From conditional distributions systems of curves are derived, which determine certain probabilities. Figure 6 shows an example for $\varphi(Y|X_1)$. The 50%-line or mean-line describes the most probable value of Y for a certain value X_1 = const.

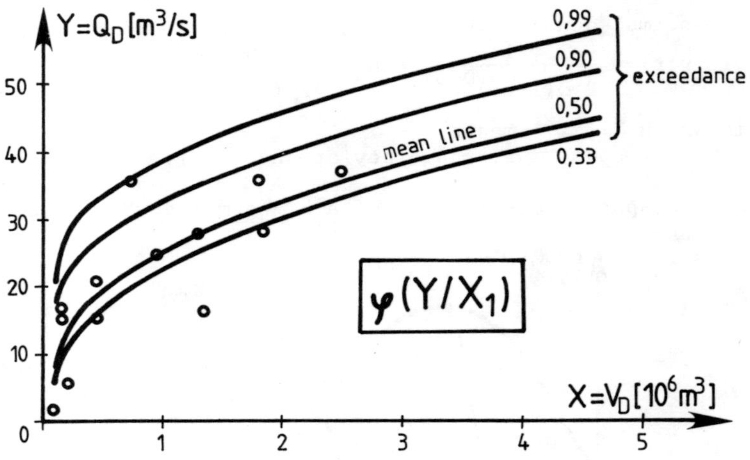

Figure 6. Conditional distributions $\varphi(Y\ X_1)$

Equi-probability density function lines: Horizontal cuts through the bi-normal density function (fig.5) result in lines of equal probability density function (short: ISO-PDF-lines). With the normalized variables, using the bivariate normal distribution , a system of similar, concentric ellipses is obtained (Stange, 1971), as it is shown in figure 7.

ISO-PDF-lines offer an appropriate possibility to test the goodness of fit of the bi-normal distribution to the sample. In the first place, this can be done by counting the number of values within a certain ellipse, and secondly, by visual control to be sure that the values are distributed uniformly to the axes of the ellipses (fig.7). Applying the generally used transformation method (1), a good fit is achieved in most cases. The correlation coefficient r of both characteristics, which is won by flood analyses, is usually positive. By retransformation of the elliptic curves to the original system of coordinates, shell-like

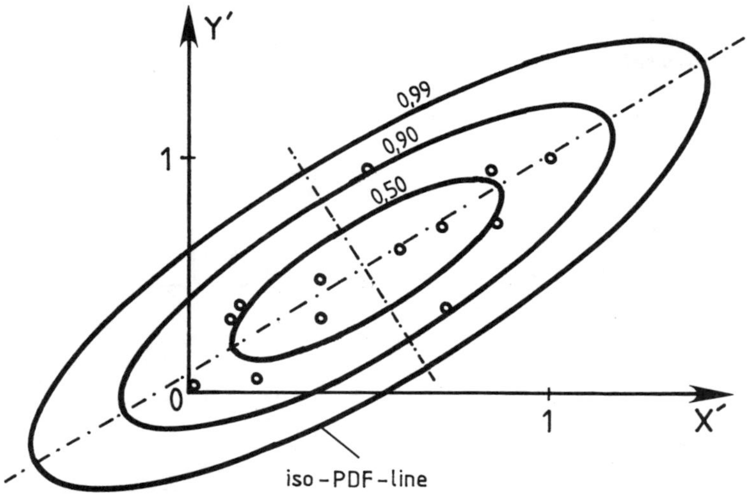

Figure 7. ISO - PDF - ellipses (transformed system)

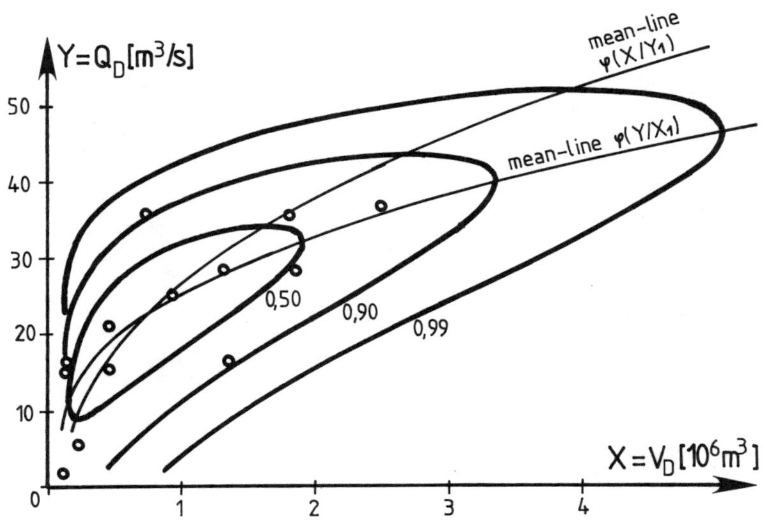

Figure 8. ISO - PDF - lines (original system)

4.3. Bi-normal distribution function

In bivariate models there is no direct analogy to the distribution function of univariate models, because there are infinite possibilities to delimitate an integration range in the X-Y plane. A simple integration is only possible in special cases as it is shown in figures 9 and 10.

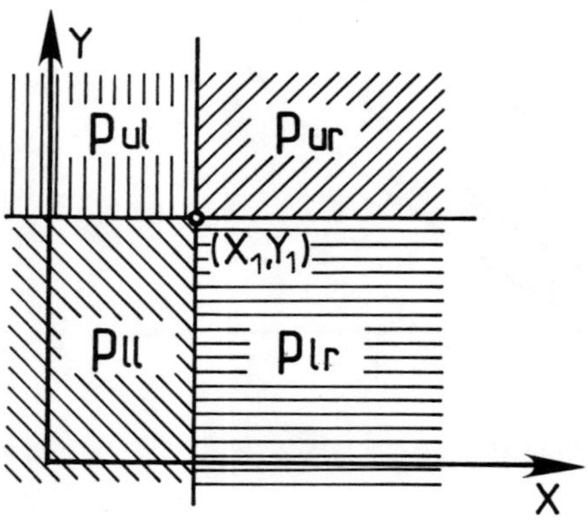

Figure 9. Delimitation of X-Y-range to bivariate distribution function, "probability of quadrants".

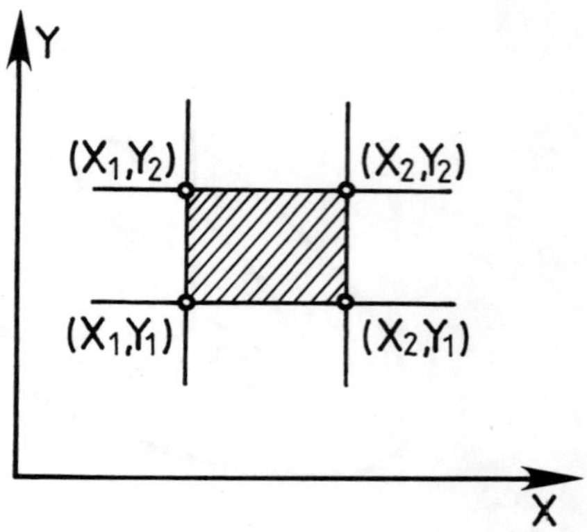

Figure 10. Delimitation of X-Y-range, rectangular random-range.

Probability of quadrants : In univariate statistics there is a clear rank order of values, but in bivariate statistics it will not always be possible to determine a pair of values (event point) to be higher or lower than another one. Figure 9 shows four possibilities of rank order for the pairs of values of the X-Y plane with the following "probabilities of quadrants":

(a) upper right quadrant $(X > X_1,\ Y > Y_1)$:

$$p_{ur} = \int_{X_1}^{\infty} \int_{Y_1}^{\infty} g(X,Y;r)\ dXdY \qquad (7)$$

(b) upper left quadrant $(X < X_1,\ Y > Y_1)$:

$$p_{ul} = \int_{-\infty}^{X_1} \int_{Y_1}^{\infty} g(X,Y;r)\ dXdY \qquad (8)$$

(c) lower left quadrant $(X < X_1,\ Y < Y_1)$:

$$p_{ll} = \int_{-\infty}^{X_1} \int_{-\infty}^{Y_1} g(X,Y;r)\ dXdY \qquad (9)$$

(d) lower right quadrant $(X > X_1,\ Y < Y_1)$:

$$p_{lr} = \int_{X_1}^{\infty} \int_{-\infty}^{Y_1} g(X,Y;r)\ dXdY \qquad (10)$$

It will be sufficient to compute the probability p_{ll} (8) for a pair of values (X_1,Y_1), from which the other three probabilities of quadrants (7),(9),(10) can be derived, as shown in equations (11),(12) and (13).

$$p_{ur} = p_{ll} - p(h) - p(k) + 1 \qquad (11)$$

$$p_{ul} = p(h) - p_{ll} \qquad (12)$$

$$p_{lr} = p(k) - p_{ll} \qquad (13)$$

where $h = \frac{X-X_m}{s_x}$, $k = \frac{Y-Y_m}{s_y}$

and p(h), p(k).....single non-exceedance of h and k

The functions p(h) and p(k) are calculated by using an approximation (error $\epsilon < 1.5 \cdot 10^{-7}$) (Abramowitz and Stegun, 1972). For solution of the double integral p_{ll} (8), tables and nomograms are found in literature (Müller , Prellberg, 1978), although it will be more advisable to work with an algorithm for computer programs (Donnely, 1978).

Step by step the lines of equal probability of quadrants can be computed, wherefrom systems of curves are obtained, which are shown in the figures 11 to 14. Most important are the lines of equal upper-right-probability (ISO-p_{ur}-lines) (fig.12). They are used, for instance, for computing design hydrographs, respectively design volume curves for the dimensioning of a flood retention basin.

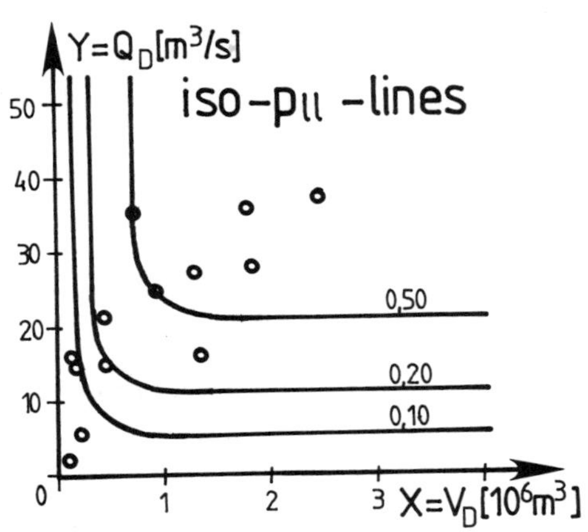

Figure 11. Equi-lines of "lower left quadrant-probability"

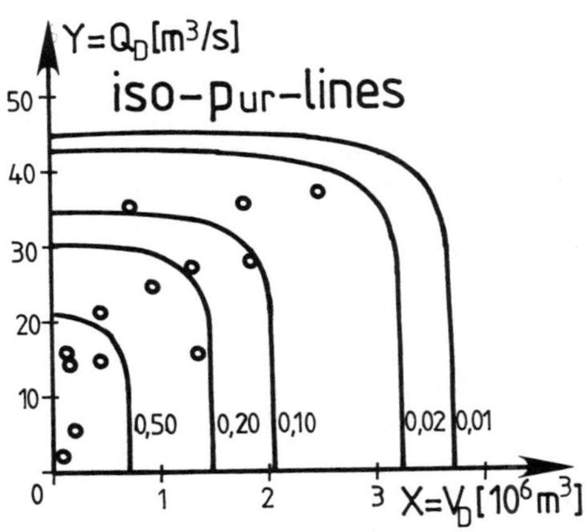

Figure 12. Equi-lines of upper right quadrant-probability"

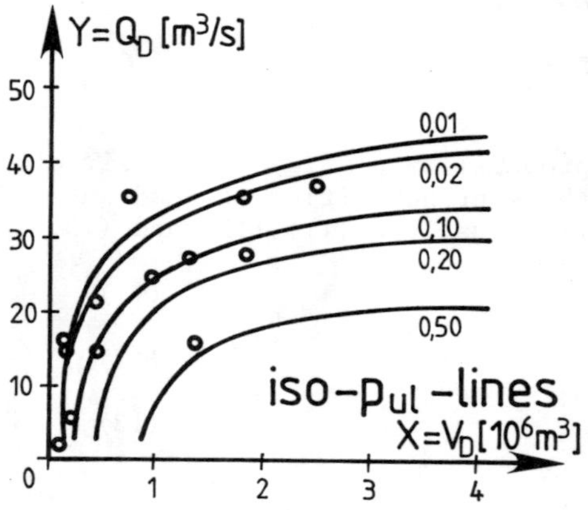

Figure 13. Equi-lines of "upper left quadrant-probability"

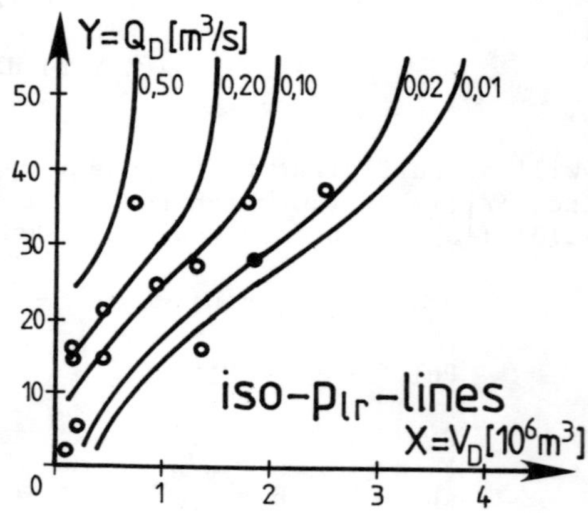

Figure 14. Equi-lines of "lower right quadrant-probability"

5. DESIGN FLOOD EVENTS

For every problem in water management a probability appropriate to the respective risk, has to be defined for the determination of design flood events.

5.1. Design hydrographs

A design hydrograph, according to a flood event (pair of values Q_d, V_d) on a certain ISO-p_{ur}-line, is determined by means of a standard hydrograph of a certain catchment area. For different ranges of flood peak-time -coefficients (t_m-values) several types are determined empirically (Bergmann and Sackl, 1985). For the determination of design hydrographs a number of pairs of values Q_d, V_d on a ISO-p_{ur}-line will be selected, by which standard hydrographs are transformed into design hydrographs. An example is shown in figure 15. The boundary values $t_{m,min}$ and $t_{m,max}$ are empirically determined (Bergmann and Sackl, 1985). Events on a straight line through the origin have the same quotient V_d/Q_d and therefore a constant value t_m.

By selection of certain t_m-values and determination of the intersection point of the appropriate straight lines with a certain ISO-p_{ur}-line, the pairs of values Q_d, V_d are obtained, which are useful for design. Finally, by these values the appropriate standard hydrograph is transformed into the design hydrograph.

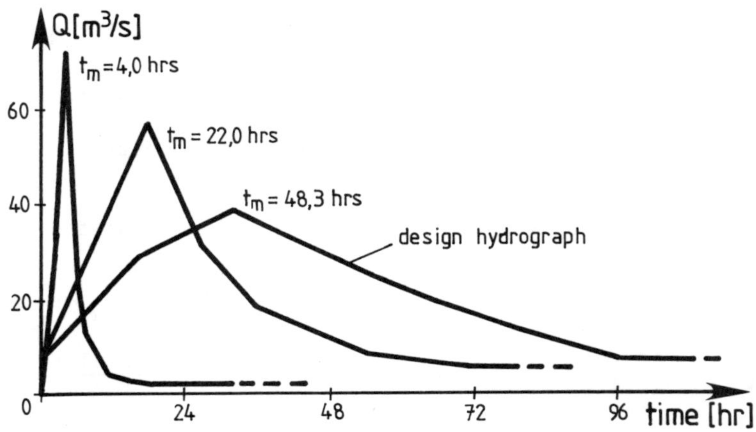

Figure 15. Design hydrographs for p_{ur} = 0.01 at the water gauge Oberwart at the river Pinka/Austria (annual maximum series 1971 - 1983)

5.2. Design volume curves

By integration of a design hydrograph from the flood peak downwards the appropriate design volume curve is computed. By this method a system of volume curves is obtained from design hydrographs, determined by a certain ISO-p_{ur}-line. The envelope curve of these volume curves can be used for the design of a flood retention basin with a given critical flow (fig. 16).

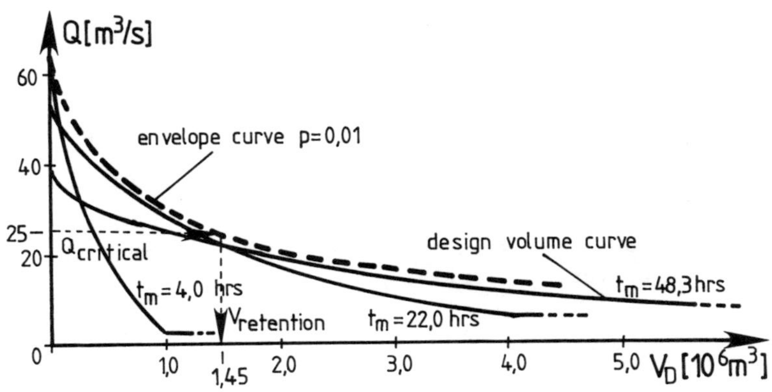

Figure 16. Design volume curves for p_{ur} = 0,01 at the water gauge Oberwart at the river Pinka/Austria (annual maximum series 1971 - 1983)

6. CONCLUSIONS

By comparing the univariate and bivariate method the advatages of the bivariate method are :

- The method is more adequate to the multivariate character of flood events.
- The variability of design flood events allows a better selection of the most crucial case according to a special water management problem.

This paper is part of more extensive investigations about the same topic.

7. REFERENCES

Abramowitz, M., Stegun, I.A. 1972. Handbook of Mathematical Functions, Dover Publications INC., pp.932, New York.

Bergmann, H., Sackl, B. 1985.'Twodimensional Frequency Analysis of Flood Water Flow and its Application in Water Management' in:Preprinted Proceedings of 21th Congress of the IAHR in Melbourne, Vol.3, pp. 521-526, The Institution of Engineers, Australia.

Donelly, T.G. 1978.'Bivariate Normal Distribution (S15)', Algorithm 462. in: Collected Algorithms from ACM, New York.

Dyck, S. 1976. Angewandte Hydrologie, Teil 1, Verlag für Bauwesen, Berlin.

Liebscher, H.J. 1970.'Hochwasserwahrscheinlichkeit', Deutsche gewässerkundliche Mitteilungen, Vol. 14/4, pp. 103-110, Koblenz.

Müller, P.H., Prellberg, D. 1978.'Methodik zur Berechnung zweidimensionaler Normalverteilungen', Mitteilungen des Leichtweiss-Institutes für Wasserbau der Technischen Universität Braunschweig, Vol. 62, Braunschweig.

Schröder, R. 1969.'Häufigkeitsanalyse hydrologischer Daten', Technischer Bericht Nr.5 des Institutes für Hydraulik und Hydrologie der Technischen Hochschule Darmstadt, Darmstadt.

Stange, K. 1971. Angewandte Statistik, 2. Teil, Mehrdimensionale Probleme, Springer Verlag, Berlin.

ANALYSIS AND SIMULATION OF THREE-COMPONENT FLOODS IN THE OHIO RIVER BASIN

Tiao J. Chang
Assistant Professor
Department of Civil Engineering
Ohio University
Athens, Ohio 45701-2979

ABSTRACT. Flood series including three components, namely, the occurrence time, the discharge volume, and the flood duration, are formed from observed daily streamflows. The intensity function describing the flood occurrence rate is derived assumming the nonhomogeneous Poisson process. A fitted equation is applied to the estimated intensities by the least squares method. The Thinning method based on controlled deletion of points in a Poisson process uses the fitted intensity function to generate the flood occurrence times which are expressed by a sequence of time intervals. The procedure of the marked point process is introduced to link the flood duration and the discharge volume, which are modeled, respectively, by exponential distributions, to the flood occurrence time. Therefore the simulated result consists of three-component flood series including the occurrence time, the flood duration and the discharge volume, which can be used for environmental pollution protection and flood-control purposes. The statistical comparisons made between the generated and the observed series are reasonably agreeable.

INTRODUCTION

Floods as runoff extremes are damaging natural resources with which humanity has to live and struggle since the floodplain is a desirable location for man and his activities. Therefore, it is important to study the flood behavior so that the damage can be reduced to an acceptable level. Flood data is usually listed either in the form of an annual flood series or a partial flood series. The concept of partial flood series has proved to be useful in many facets of flood analysis (Todorovic, 1978). Mathematical methods for determining the distribution of flood occurrences have been presented in many previous papers (Shane and Lynn, 1964; Todorovic and Zelenhasic, 1970; Todorovic and Woodhiser, 1972; Gupta, et.al., 1976; North, 1980). The present paper is concerned with stochastic flood models which consist of flood occurrences, discharge volume, and flood duration based on partial flood series. The aim here is to present the construction of this three-

V. P. Singh (ed.), Hydrologic Frequency Modeling, 583–594.

component flood series, which can be modeled by the marked point process, and the success of using the Thinning method for simulations.

In order to provide a perspective for this presentation, a brief review of recent developments in the direction shall be discussed. Shane and Lynn (1964) developed a probability model based on the time independent Poisson process in the analysis of base-flow flood data. Kirby (1969) defined floods as exceedances in a sequence of randomly spaced Bernoulli trials. The times between exceedances, at sufficiently small exceedance probabilities, approach those implied by trials from a Poisson process. Todorovic and Zelenhasic (1970), who developed the theory of flood analysis based on partial duration series, maintained that the magnitude of exceedances are independent of the number of exceedances. Gupta et.al. (1976) constructed the expression for the joint distribution function of the largest flood peak and its time of occurrence. North (1980) proposed a nonhomogeneous Poisson distribution for flood occurrences with independent discharge volumes. None of the previous work addresses the combined study of the flood occurrence, the flood volume, and the flood duration, which are discussed in the following.

DEFINITIONS

Adopting the partial duration series (PDS), a flood event is constructed by considering a continuous streamflow hydrograph as shown in Figure 1 and a certain tracer level (e.g. $\bar{Q} + S$; $\bar{Q}$ is the observation mean and S is the standard deviation). Then a flood event means an event formed from the time of the hydrograph up-crossing the tracer level to that of the hydrograph down-crossing the tracer level. Given a starting time 0, the occurrence time for a flood event is the time from the starting time 0 to that of the up-crossing of the event, i.e. t_i as shown in Figure 1. The difference between the occurrence time of a flood event and that of the following event is defined as the interval time of these two flood events expressed as

$$x_i = t_i - t_{i-1} \tag{1}$$

where $t_o = 0$. The time between the up-crossing and the down-crossing of a flood event is defined as the flood duration time, D_i. A flood duration series is formed by $\{D_i, i \geq 1\}$. The total discharge of the hydrograph between the up-crossing and the down-crossing of a flood event is called the flood volume as the shaded area shown in Figure 1. $\{V_i, i \geq 1\}$ form the flood volume series. Therefore, a three-component flood series means all the flood events associated with their flood occurrence times, flood volumes, and flood duration forming a discrete series, $\{N(t), t \geq 0\}$, which can be modeled by the point process.

FLOOD INTENSITY FUNCTION

Admitting the nonhomogeneous Poissonian behavior of a flood series, the flood intensity function is defined in the following. Let $\{N(t); t>0\}$ be a non-negative integer-valued series governed by a Poisson process with independent increments, then the characteristic function of $\{N(t) ; t>0\}$ can be expressed,

$$\begin{aligned} \phi(u) &= E[Z^n N(t)] \\ &= \sum_{n=0}^{\infty} Z^n \, P[N(t) = n] \\ &= e^{m(t)\,(e^{iu}-1)} \end{aligned} \tag{2}$$

where E: expectation; $|Z|<1$; P: probability. If $\{N(t); t\geq 0\}$ has stationary increments, then

$$\begin{aligned} m(t) &= E[N(t)] \\ &= \mu t \end{aligned} \tag{3}$$

where μ is a constant. It is called the homogeneous Poisson process. If equation (3) fails to hold, it is defined as a nonhomogeneous Poisson process. Assume that m(t) is continuous and differentiable, then

$$\mu(t) = \frac{d}{dt} m(t) \tag{4}$$

where $\mu(t)$ is named the flood intensity function.

The classical way to estimate the flood intensity function is to form a histogram from all possible sums of contiguous times between flood events in the observed series (Cox and Lewis, 1966). Assume that n flood events occcur at time $t_1, t_2, \ldots \ldots t_n$ measured from the starting time 0. Neglecting the time from 0 to t_1, one can construct the 1/2 n(n-1) sums of contiguous time intervals, t_2-t_1, t_3-t_2, ... t_n-t_{n-1}. Dividing (0, To), where To is the end time of the observation, into equal time intervals of length, τ, a histogram can be formed by counting the number of those sums in each interval $(m\tau, m\tau+\tau)$ and dividing by $n\tau$, where n is assigned as m = 0, 1, 2, ... , M, i.e. the number of the counting interval. Analytically it can be expressed by

$$\nu(t) = \frac{1}{n} \sum_{i=1}^{n-1} \sum_{j=1}^{n-1} \delta(t_{i+j} - t_i - t) \tag{5}$$

where $\delta(t)$ denotes the Dirac function at time t. In this paper an additional smoothing factor is used to obtain the unbiased estimation, $\nu(t)$.

$$\hat{\nu}(t) = \frac{To}{(To - m\tau - \frac{1}{2}\tau)\tau} \int_{m\tau}^{(m+1)\tau} \nu(t)\, dt \tag{6}$$

The τ is taken as the mean value of observed time intervals and m takes 50 in order to assure that there is at least one occurrence in each histogram interval.

THINNING METHOD

Several methods have been developed for generating nonhomogeneous processes. First, it can be derived from equation (2) as the time-scale transformation of a Poisson process by the inverse of the continuous and non-decreasing, m(t). One can define its inverse function, $m^{-1}(u)$. For u>0, $m^{-1}(u)$ is the smallest value of t satisfying $m(t)\geq u$. Then the stochastic process, $\{M(u);\ u\geq 0\}$ is defined by

$$M(u) = N\{m^{-1}(u);\ u\geq 0\} \tag{7}$$

This is a Poisson process with

$$E[M(u)] = u \tag{8}$$

Consequently, $\{M(u);\ u\geq 0\}$ is a homogeneous Poisson process with an intensity function equal to 1. Therefore, one can generate the Poisson process with intensity, 1, and transforms the time-scale by the inverse function. Unfortunately, it is not always possible to obtain the inverse function. For instance, the flood intensity function usually has a periodical cycle which cannot be expressed by the one-to-one inverse function.

A second method to generate the nonhomogeneous Poisson process with an integrated intensity function, m(t), is based on generating the time intervals between points individually (Lewis, 1978). Suppose $T_1 = t_1$, $T_2 = t_2$, ... , $T_k = t_k$, with $T_1 < T_2 < T_3$ . . . . . $< T_k$ and that the time interval to the next point, $T_{k+1} - T_k$, is independent of t_1, t_2, ... , t_{k-1} with a distribution function

$$F(x) = 1 - \exp\,[-m(t_k + x) - m(t_k)] \tag{9}$$

where exp: exponential function. Thus, it is possible to find the inverse distribution function $F^{-1}(u_k)$ and to generate $T_{k+1} - T_k$ by $F^{-1}(u_k)$, where u_k is a uniform random number in [0, 1]. It is, however, necessary to take into account of all the time intervals which have different parameters and different forms in their distribution function.

The third way, assuming the nonhomogeneous Poisson process to occur in a fixed time interval $[0, T_o]$, can reduce the generation to simulating a Poisson number by order statistics from a fixed density function. Conditionally given the occurrence number, $N(t) = n$, the observed n events in [0, To] have their occurrence times $t_1, t_2, t_3, \ldots t_n$ governed by the following distribution function

$$F(t) = \{m(t)-m(0)\} / \{m(To)-m(0)\}, \tag{10}$$

where $0<t\leq To$. This method was criticized because of requiring too much memory during the interval-by-interval generation (Lewis and Shedler, 1978).

The Thinning method based on controlled deletion of points in a Poisson process is used in this study (Lewis and Shedler, 1979). Considering a nonhomogeneous Poisson process, $\{N(t)^*; t\geq 0\}$, with intensity function, $\mu^*(t)$, in a fixed interval [0, To], the number of events, $N^*(t')$, has a Poisson distribution with the parameter

$$U_0^* = M^*(t') - M^*(0), \tag{11}$$

in the interval [0, t'], where $M^*(t)$ is the integrated intensity function corresponding to the intensity function. The intensity function, $\mu^*(t)$, is taken to be a constant, μ_0^*, which is the maximum of $\mu^*(t)$ in the interval [0,t'], so that the time interval to the next event $t_{k+1} - t_k$,

is obtained by generating and cumulating exponential random numbers $E_{1,k}$, $E_{2,k}$ . . . . . until for the first time

$$U_{j,k} \leq \mu^*(t_k + E_{1,k}^* + E_{2,k}^* + \ldots + E_{j,k}^*) / \mu_o^*, \tag{12}$$

where $\{U_{j,k}\}$ are independent uniform random numbers on (0, 1). This process is continued until the specified number of events is convered or until the specified time t' is exceeded. The Thinning method can efficiently handle the simulation of a nonhomogeneous Poisson process whose intensity function has a fixed cycle plus trend (Lewis and Shedler, 1979).

SIMULATION OF FLOOD OCCURRENCE TIME

Daily streamflows in the Ohio River Basin are used to form three-component flood series based on the tracer level, $\bar{Q} + S$; where $\bar{Q}$ is the mean value of observed daily flows and S is the standard deviation. Equation (6) is applied to estimate the unbiased flood intensity function of the flood series as two examples show in Figures 2 and 3. It is desirable to represent the flood intensity function, which appears to have a periodical cycle with trend, as a function of time in order to use the Thinning simulation method. The nonlinear least squares method is applied to find the best-fit equation based on the observed period and trend.

$$f(x) = \exp\,[a_0 + a_1 x + a_2 \sin x + a_3 \cos x], \tag{13}$$

where $x = \frac{2\pi t}{365}$; t: days; a_0, a_1, a_2, a_3 : constant.

Table 1 gives two examples of fitted flood intensity functions from the Ohio River Basin.

To illustrate the applicability of the Thinning simulation method, consider a nonhomogeneous Poisson process whose intensity function as shown in Table 1. The same number of observed flood events are generated to compare with the observed values where the flood occurrence times are represented by a sequence of time intervals, $\{x_i,\ i \geq 1\}$. Table 2 shows the comparison of their mean values between the observed and the generated, which is very consistent.

SIMULATION OF THREE-COMPONENT FLOOD

The mean flood duration, D, and mean flood volume, V, are respectively obtained from observed sequences $\{D_i\}$ and $\{V_i\}$. The flood duration is generated by the exponential distribution function as follows:

$$F_D\,(d) = 1 - \exp\,[\,\frac{-d}{D}] \tag{14}$$

The flood volume is simulated in a similar way of its distribution function.

$$F_V(\nu) = 1 - \exp\,[\frac{-\nu}{V}] \tag{15}$$

Tables 3 and 4 compare the mean values of observed flood duration and volume with those of the generated and show that they are somewhat agreeable.

The marked point process is applied to link the flood duration and volume to the flood occurrence time. A marked point process is a point process with an auxiliary variable, called a mark, associated with each point (Snyder, 1975). Following the previous definitions to start from an original time, t_o, N(t) is the occurrence number at time t; t_i is the time to the ith event from t_o. Each occurrence time, t_i, in the basic point process has associated with a mark, y_i, which is a random variable from a marked space, Y; where Y is a denumerable collection, $\{y_i\}$.

Assume that $\{N(t),\ t \geq t_o\}$ is a nonhomogeneous Poisson process with an intensity function as given in Equation (13) and that $\{y_i,\ i \geq 1\}$ is a sequence of mutually independent, identically distributed random variables such as the generated values from Equations (14) and (15). Then the marked point process is termed a compound process. In this study the $\{y_i\}$ is also taken to be independent of $\{N(t),\ t \geq t_o\}$, where y_i represents either the flood duration or the flood volume. Therefore,

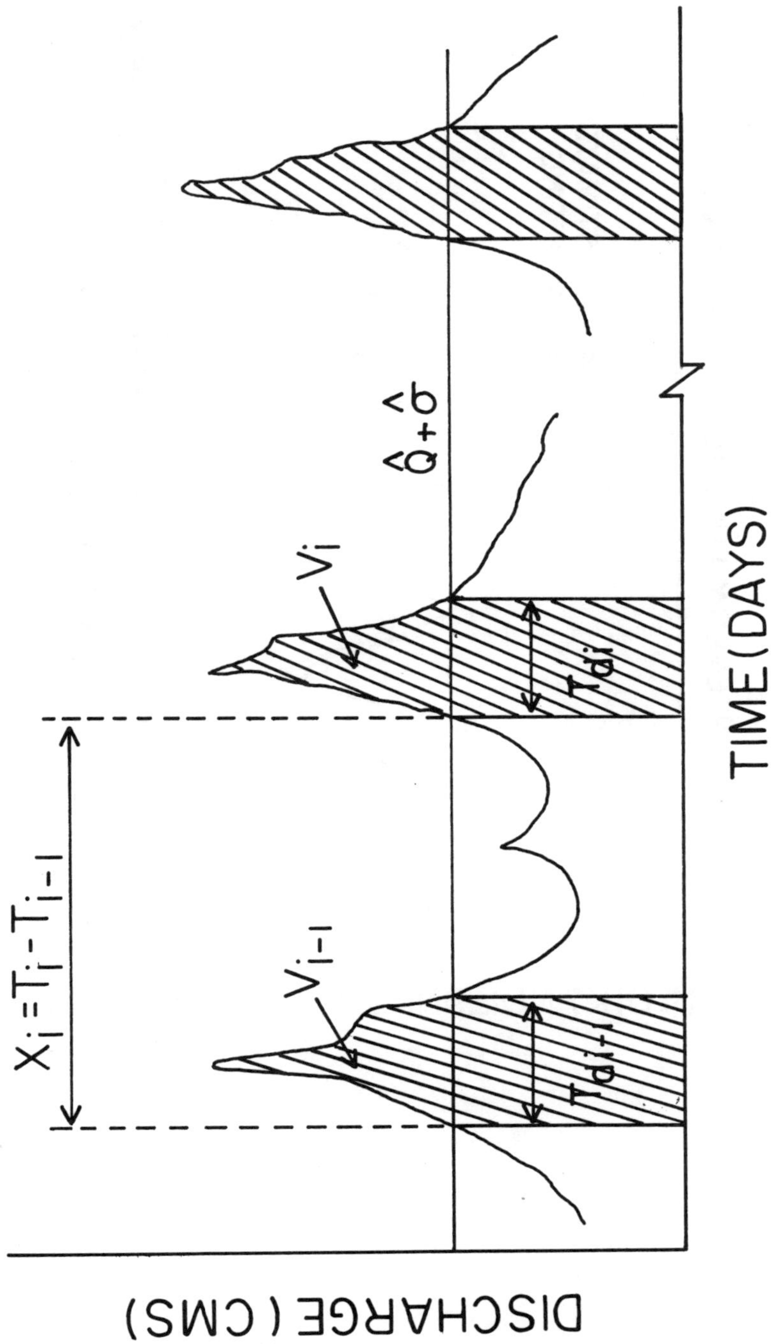

FIGURE 1. SCHEMATIC FLOODS

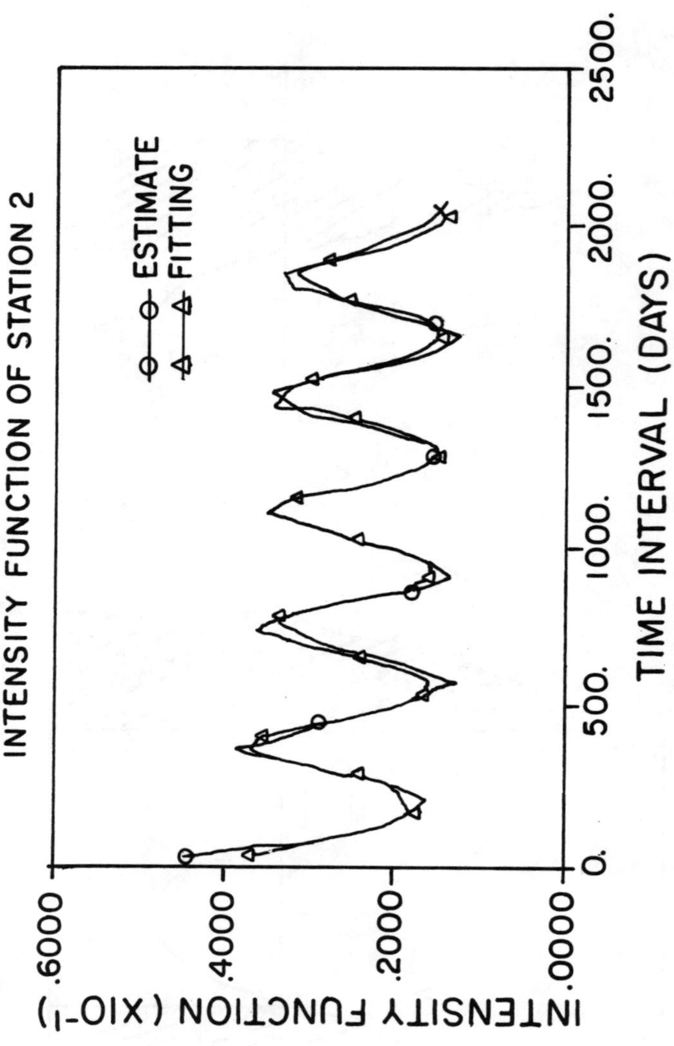

FIGURE 2. MAD RIVER NEAR SPRINGFIELD, OHIO

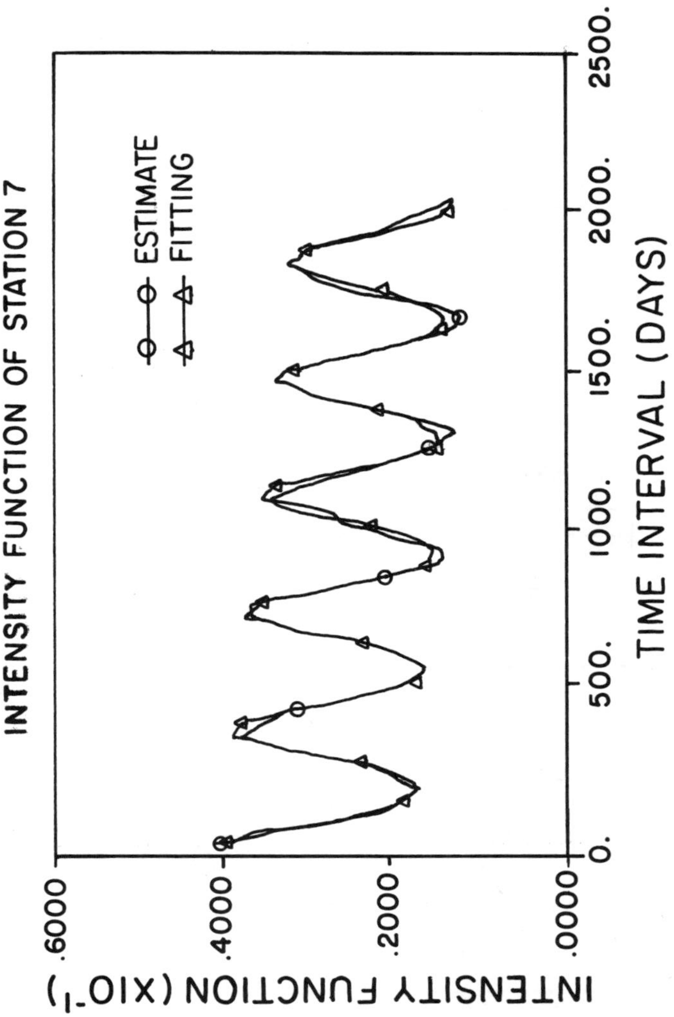

FIGURE 3. SALAMONIE RIVER AT DORA, INDIANA

Table 1. Fitted Flood Intensity Functions, $f(x^*)$

Sta. No.	No. of Years Data	Fitted Flood Intensity Function
2^{**}	56	exp [-3.676 - .004x + .131sinx + .398 cosx]
7^{***}	45	exp [-3.627 - .007x + .137sinx + .414 cosx]

*: $x = \frac{2\pi}{365} t$, t : days

**: Mad River near Springfield, Ohio

***: Salamonie River at Dora, Indiana

Table 2. Comparison of Mean Time Interval

Sta. No.	No. of Events	Mean Time Interval (days) Observation	Simulation
2	485	40.1	42.2
7	393	40.4	41.8

Table 3. Comparison of Mean Flood Duration

Sta. No.	No. of Events	Mean Flood Duration (days) Observation	Simulation
2	485	2.3	2.4
7	393	3.4	3.4

Table 4. Comparison of Mean Flood Volume

Sta. No.	No. of Events	Mean Flood Volume (cfs-day) Observation	Simulation
2	485	5,569	5,648
7	393	11,833	11,902

they can be randomly marked to the flood occurrence time, t_i, to form the three-component flood series.

CONCLUSIONS

Based on the definition in this paper, the three-component flood series obtained from the data in the Ohio River Basin appear that the flood intensity functions have a fixed cycle as well as trends, which can be fitted by the proposed equation. The Thinning method illustrates its high capability and efficiency in simulating the flood occurrences. The method of the marked point process successfully link the flood duration and volume to the flood occurrence time to form three-component flood series.

ACKNOWLEDGEMENTS

Helpful discussions with Dr. J. W. Delleur of Purdue University and the encouragement from Dr. T. R. Robe of Ohio University are gratefully acknowledged. The author also would like to thank Pam Stettler for her assistance in typing this paper.

REFERENCES

Cox, D.R. and P.A.W. Lewis, 1966, The Statistical Analysis of Series of Events, Methuen & Co. Ltd., London; John Wiley and Sons Inc., New York.

Gyupta, V.K., L. Duckstein, and R.W. Peebles, 1976, 'On the Joint Distribution of the Largest Flood and Its Time of Occurrence,' Water Res. Res., Vol. 12, No. 2, pp. 295-304.

Kirby, W., 1969, 'On the Random Occurrence of Major Floods,' Water Res. Res., Vol. 5, No. 4, pp. 778-784.

Lewis, P.A.W. and G.S. Shedler, 1976, 'Simulation of Nonhomogeneous Poisson Processes with Log Linear Rate Function,' Biometrika, 63,3, pp. 501-505.

Lewis, P.A.W. and G.S. Shedler, 'Simulation of Poisson Processes with Degree-Two Exponential Polynominal Rate Function," Operational Research, Vol. 27, No. 5, pp. 1026-1040, 1979.

Lewis, P.A.W. and G.S. Shedler, 'Simulation of Nonhomogeneous Poisson Processes by Thinning,' Naval Research Logistics Quarterly, 26(3), pp. 403-413, 1979.

North, M., 'Time-Dependent Stochastic Model of Floods,' J. of Hydraulic Division, ASCE, Vol. 10b, No. Hy5, pp. 649-665, 1980.

Parzen, Z., Stochastic Processes, Holden-Day, 1962.

Shane, R.M. and W.R. Lynn, 'Mathematical Model for Flood Risk Evaluation,' J. of Hydraulic Division, ASCE, Vol. 90, No. Hy6, pp. 1-20, 1964.

Snyder, D.L., Random Point Processes, John-Wiley & Sons, Inc., New York, 1975.

Todorovic, P. and Z. Zelenhasic, 'A Stochastic Model for Flood Analysis,' Water Res. Res., Vol. 6, No. 6, pp. 1641-1648, 1970.
Todorovic, P. and D.A. Woolhiser, 'On the Time When the Extreme Flood Occurs,' Water Res. Res., Vol. 8, No. 6, pp. 1433-1438, 1972.
Todorovic, P., 'Stochastic Models of Floods,' Water Res. Res., Vol. 8, No. 6, pp. 1433-1438, 1972.
Todorovic, P., 'Stochastic Models of Floods,' Water Res. Res., Vol. 14, No. 2, pp. 345-356, 1978.

A PROBABILISTIC MODEL FOR FLOODING DOWNSTREAM OF THE JUNCTION OF TWO RIVERS

J.A. Raynal
Engineering Graduate Studies Division
Universidad Nacional Autonoma de Mexico
Cd. Universitaria, 04510 Mexico, D.F.
Mexico

and

J.D. Salas
Hydrology and Water Resources Program
Department of Civil Engineering
Colorado State University
Fort Collins, Colorado 80523
U.S.A.

ABSTRACT. Flood frequency analysis downstream of the junction of two rivers, when flood information is available upstream of the confluence is the main subject of this paper. For this purpose, it is assumed that the joint distribution function of the floods at two upstream gaging stations follows a bivariate extreme value distribution. The distribution at the junction is obtained by the well-known convolution integral which gives the distribution of the sum of two random variables. The model is applied successfully to an actual case study.

1. INTRODUCTION

The problem of how to model extreme flooding downstream of the junction of two rivers, when flood information is available upstream of the confluence, has been tackled by Linsley and Franzini (1979) and Salas (1980). In these studies, the assumptions of either complete dependence or complete independence between the floods at the available sites were made. Furthermore, their approaches were based on empirical schemes.

The use of convolutions of random variables in hydrology was first suggested by Gumbel (1961). The main application that Gumbel made was concerning the probability distribution of the difference of two independent extreme random variables for the purpose of identifying which of the differentiable models for bivariate extreme value distributions, best fits the actual data of two samples. The selected distribution was the well-known Logistic distribution. The use of such bivariate distributions for flood frequency analysis has been outlined by Salas et al (1983).

The use of convolutions to construct a probability model that approximately describes the frequency distribution of flooding

V. P. Singh (ed.), Hydrologic Frequency Modeling, 595–602.

downstream of the junction of two rivers, is explored in this paper. The practical application of the model is shown through of a specific case study.

2. BIVARIATE EXTREME VALUE DISTRIBUTIONS

The general form of the logistic model for bivariate extreme value distributions for the maximum is, Gumbel (1960):

$$F(x,y,m) = \exp\{-[(-LnF(x))^m + (-LnF(y))^m]^{1/m}\} \qquad (1)$$

where m is the association parameter, F(u) is the marginal extreme value distribution of u, and $m \geq 1$.

The six possible distribution functions suggested by Gumbel (1962), reduces to three by using the general extreme value (GEV) and the Gumbel distributions, as marginals of the bivariate model. Thus, three forms of bivariate GEV distributions are (Raynal, 1985):

a). Type 11 Bivariate Extreme Value Distribution (BEV11) or Bi-gumbel distribution,

b). Type 12 Bivariate Extreme Value Distribution (BEV12) or Gumbel-GEV distribution, and

c). Type 22 Bivariate Extreme Value Distribution (BEV22) or Bi-general Extreme Value Distribution.

The corresponding cumulative distribution and density functions derived from eq. (1) are given in the appendix.

3. MODEL FOR FLOOD FREQUENCY ANALYSIS DOWNSTREAM OF THE JUNCTION OF TWO RIVERS

It is assumed that flood data is available at two gaging stations located upstream of the confluence of two rivers, and the corresponding bivariate distribution is known or can be fitted based on empirical data. A model can be constructed for the distribution function of the sum of two random variables, (Woodroofe, 1975):

$$F(t) = \int_{\infty}^{t} \int_{\infty}^{\infty} f(t-s,s)\, ds\, dt \qquad (2)$$

where t is the sum of the random variables.

For any of the bivariate extreme value distribution functions shown in the appendix of this paper, eq. (2) can be evaluated numerically by using any of the existing bidimensional numerical integration computer codes, like the one proposed by Stroud (1971). Thus, the model can be easily used and applied as depicted in the analysis of the case study presented in the next section.

4. CASE STUDY

In order to show the applicability of the proposed methodology to a real case, two gauging stations located upstream of the junction of two rivers were selected. The stations chosen were the Bear River and the Dry Creek as they appear in Salas (1980). Since the two stations are close to the junction, there was no need to transfer the information to a site closer to the junction.

The bivariate distribution which was best fitted to the available flood data was the BEV11 distribution. Thus, the model for flood frequency analysis downstream of the junction of the two rivers is:

$$F(t) = (1/\alpha_1\alpha_2) \int_{-\infty}^{t} \int_{-\infty}^{\infty} e^{-m(t-s-u_1/\alpha_1)} \; e^{-m(s-u_2/\alpha_2)} \;\; x$$

$$x \; [\; e^{-m(t-s-u_1/\alpha_1)} + e^{-m(s-u_2/\alpha_2)} \;]^{(1/m)-2} \;\; x$$

$$x \; \{\; m - 1 + [e^{-m(t-s-u_1/\alpha_1)} + e^{-m(s-u_2/\alpha_2)}]^{1/m} \;\} \;\; x$$

$$x \;\; \exp \{\; -[e^{-m(t-s-u_1/\alpha_1)} + e^{-m(s-u_2/\alpha_2)}]^{\;1/m}\} ds\; dt \qquad (3)$$

where u_i and α, i = 1,2 are the location and scale parameters of the bivariate model.

The parameters of the bivariate distribution function were estimated by the maximum likelihood method (Raynal, 1985). The estimates obtained for the given data were: u_1 = 134.32 m^3/s, α_1 = 116.71 m^3/s, u_2 = 48.62 m^3/s, α_2 = 50.58 m^3/s, and m = 2.6976. Then, eq.(3) was integrated numerically.

Table I gives a comparison of frequency of flooding at the junction of Bear River near Weatland and Dry Creek near Weatland by various procedures. Column (3) was obtained by adding the flows which actually occurred at the same time and performing the frequency analysis on the sums. Column (4) was derived by using eq. (3) as referenced above. Column (5) is a frequency curve for the sums of the two floods computed on the assumption of complete dependence, i.e. floods of equal probability are assumed to occur simultaneously and the total flood at the junction is the sum of the floods on the two streams having the same exceedance probability. Column (6) assumes complete independence so that the probability of any two floods occurring simultaneously is the product of the probability that they will occur independently. These different flood probabilities are also shown in Fig. 1. Usually for relatively large watersheds one would expect that the probability of flooding at the junction of the streams would fall between the curves of complete independence and complete dependence. However, in the particular case studied herein, in which the streams are small with a correlation coefficient between the floods of about 0.86 (corresponding to m = 2.70), the probability of flooding at the junction of the streams lies very close to the curve of complete dependence.

Table I. Comparison of Frequency Flooding at the Junction of Two Streams by Different Procedures.

Order	Probability $P(T \leq t)$ (%)	From Simultaneous Flood Data (m^3/s)	From Eq. (3) (m^3/s)	Assuming Complete Dependence (m^3/s)	Assuming Complete Independence (m^3/s)
1	4.76	15.26	5.31	-3.33	-37.11
2	9.52	35.11	49.49	39.88	-1.97
3	14.29	36.90	79.91	71.60	23.03
4	19.05	58.90	105.10	98.35	43.62
5	23.81	64.00	128.16	122.51	61.84
6	28.57	78.44	149.65	145.23	78.64
7	33.33	130.54	170.56	167.19	94.60
8	38.09	147.24	191.25	188.86	110.09
9	42.86	152.06	212.06	210.67	125.43
10	47.62	162.45	233.36	236.88	140.81
11	52.38	199.64	255.43	255.87	156.47
12	57.14	337.54	278.76	280.04	172.68
13	61.90	428.00	303.84	305.25	189.70
14	66.66	471.19	330.44	333.91	207.91
15	71.43	474.02	361.36	365.17	227.83
16	76.19	489.82	396.21	400.78	250.10
17	80.95	528.11	437.78	442.96	275.95
18	85.71	532.78	490.40	495.69	307.53
19	90.48	609.66	564.08	568.07	349.74
20	95.23	705.66	683.36	687.90	417.26

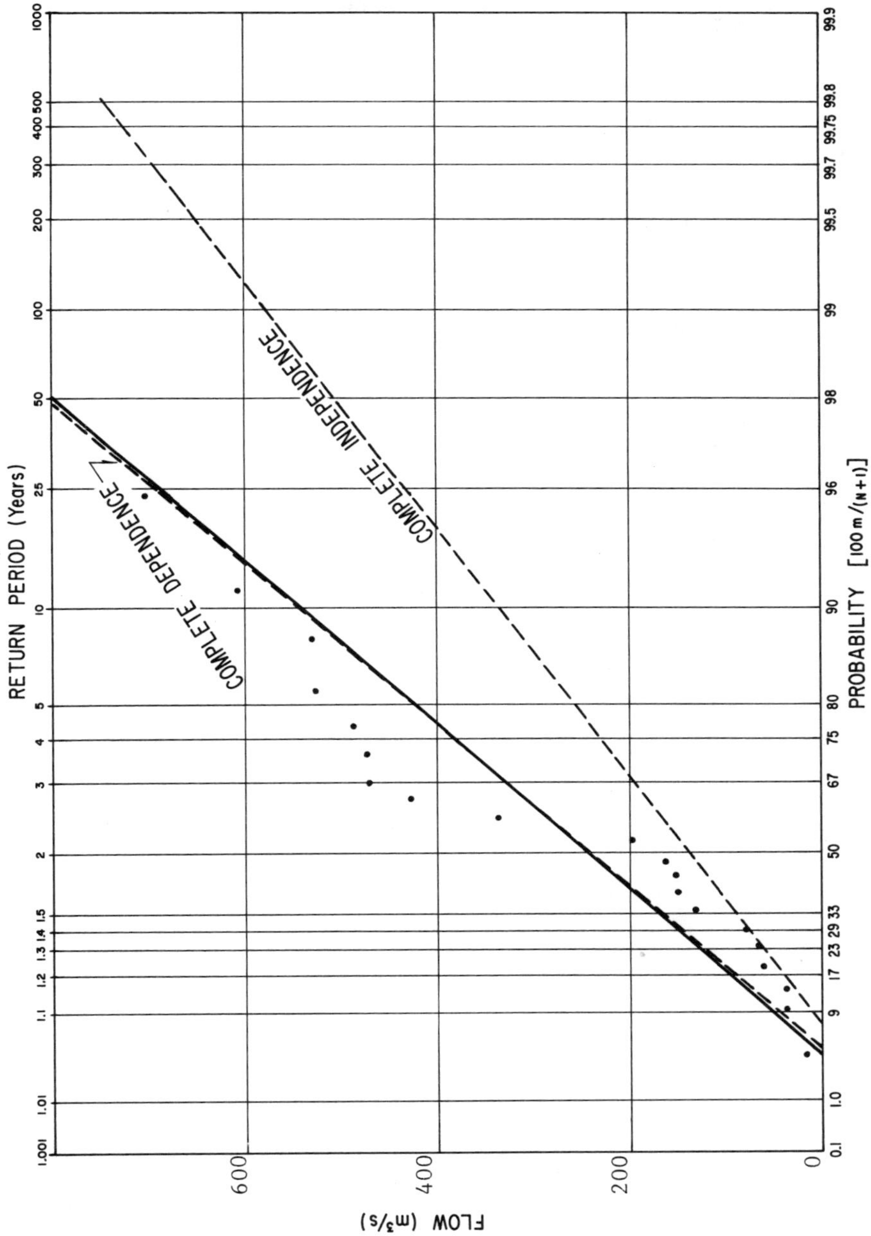

Figure 1 Frequency Curves of the Annual Maximum Flood T at the Junction of Bear River near Weatland and Dry Creek near Weatland.

5. CONCLUSIONS

A probabilistic model has been presented as an approximation to the frequency distribution of flooding downstream of the junction of two rivers, when information is available at two points upstream of the confluence. The model assumes that the floods at these two points are jointly distributed. In this study, such joint distribution was assumed to be the bivariate GEV distribution.

The model was applied to an actual case and showed to be a good approximation. It is expected that for gauging stations with longer period of recorded flows, the results will improve significantly. The final conclusion of this study is that bivariate models may be quite useful for a number of problems related to flood frequency distributions in general and frequency distribution at the junction of two streams in particular.

6. ACKNOWLEDGEMENTS

The authors wish to acknowledge the support of the U.S. National Science Foundation project entitled "Prediction of Flood Hazards Based on Joint Distribution Functions", Grant ECE-8612307, and the Engineering Graduate Studies Division of the Universidad Nacional Autonoma de Mexico. The support of CONACYT (Mexico) for the graduate studies of the first author is gratefully acknowledged.

7. REFERENCES

Gumbel, E.J. 1960. Distributions des Valeurs Extremes en Plusieurs Dimensions, Publications de L'Institute de Statistique. **Vol. 9,** pp 171-173.

Gumbel, E.J. 1961. Sommes et Differences de Valeurs Extremes Independantes. Computes Rendus Academie De Sciences. **Vol. 253**, pp 2838-2839.

Gumbel, E.J. 1962. Statistical Theory of Extreme Values (Main Results). Chapter 6. A.S. Sarhan and B.G. Greenberg (editors), Constribution to Order Statistics. pp 59-93, John Wiley & Sons.

Linsley, R.K. and Franzini, J.B. 1979. Water Resources Engineering, 3rd Edition, McGraw-Hill Book Co., New York, pp 125-126.

Raynal Villasenor, J.A. 1985. Bivariate Extreme Value Distributions Applied to Flood Frequency Analysis. Ph. D. Dissertation, Civil Engineering Department, Colorado State University, Fort Collins, Colorado.

Salas, J.D. 1980. Transfer of Information to Improve Estimates of Flood Frequencies in: T.G. Sanders (editor) Hydrology for Transportation Engineers, pp 634-642, U.S. Department of Transportations Federal Highway Administration, Offices of Research and Development, Implementation Division (HDV-21).

Salas, J.D. Rueda, E. and Raynal, J., 1983. Bivariate Gumbel Distribution for Flood Frequency Analysis. Paper Presented at the AGU Fall Meeting, San Francisco, California.

Stroud, A.H. 1971. Approximate Calculation of Multiple Integrals Prentice Hall, pp 200-205.

Woodroofe, M. 1975. Probability with Applications, McGraw-Hill Book Co. New York, pp 179-190.

APPENDIX A

CUMULATIVE DISTRIBUTION AND DENSITY FUNCTIONS FOR BIVARIATE EXTREME VALUE DISTRIBUTIONS (LOGISTIC MODEL)

A.1 BEV11 Distribution.

$$F(x,y) = \exp\{ -[e^{-m(x-u_1/\alpha_1)} + e^{-m(y-u_2/\alpha_2)}]^{1/m} \}$$

$$f(x,y) = (1/\alpha_1\alpha_2)\, e^{-m(x-u_1/\alpha_1)}\, e^{-m(y-u_2/\alpha_2)} \cdot$$

$$\cdot\ [e^{-m(x-u_1/\alpha_1)} + e^{-m(y-u_2/\alpha_2)}]^{(1/m)-2} \cdot$$

$$\cdot\ \{ m-1 + [e^{-m(x-u_1/\alpha_1)} + e^{-m(y-u_2/\alpha_2)}]^{1/m}\} \cdot$$

$$\cdot\ \{ -[e^{-m(x-u_1/\alpha_1)} + e^{-m(y-u_2/\alpha_2)}]^{1/m} \}$$

A.2 BEV12 Distribution

$$F(x,y) = \exp\{ - [e^{-m(x-u_1/\alpha_2)} + (1-(y-u_2/\alpha_2)\beta_2)^{m/\beta_2}]^{1/m}\}$$

$$f(x,y) = (1/\alpha_1\alpha_2)\, e^{-m(x-u_1/\alpha_1)}\, [1-(y-u_2/\alpha_2)\beta_2]^{(m/\beta_2)-1} \cdot$$

$$\cdot\ \{ e^{-m(x-u_1/\alpha_1)} + [1-(y-u_2/\alpha_2)\beta_2]^{m/\beta_2} \}^{(1/m)-2} \cdot$$

$$\cdot\ \{ m - 1 + [e^{-m(x-u_1/\alpha_1)} + (1 - (y-u_2/\alpha_2)\beta_2)^{m/\beta_2}]^{1/m}\} \cdot$$

$$\cdot\ \exp\{ - [e^{-m(x-u_1/\alpha_1)} + (1 - (y-u_2/\alpha_2)\beta_2)^{m/\beta_2}]^{1/m} \}$$

A.3 BEV22 Distribution.

$$F(x,y) = \exp\{ - [(1-(x-u_1/\alpha_1)\beta_1)^{m/\beta_1} + (1-(y-u_2/\alpha_2)\beta_2)^{m/\beta_2}]^{1/m} \}$$

$$f(x,y) = (1/\alpha_1\alpha_2)\, [1-(x-u_1/\alpha_1)\beta_1]^{(m/\beta_1)-1}\, [1-(y-u_2/\alpha_2)\beta_2]^{(m/\beta_2)-1} \cdot$$

$$\cdot\ \{[1-(x-u_1/\alpha_1)\beta_1]^{m/\beta_1} + [1-(y-u_2/\alpha_2)\beta_2]^{m/\beta_2} \}^{(1/m)-2} \cdot$$

$$\cdot\ \{ m-1 + [(1-(x-u_1/\alpha_1)\beta_1)^{m/\beta_1} + (1-(y-u_2/\alpha_2)\beta_2)^{m/\beta_2}]^{1/m}\} \cdot$$

$$\cdot\ \exp\{ - [(1-(x-u_1/\alpha_1)\beta_1)^{m/\beta_1} + (1-(y-u_2/\alpha_2)\beta_2)^{m/\beta_2}]^{1/m} \}$$

CONCURRENT FLOODING PROBABILITIES

Charles D. Morris
Associate Professor
Dept. of Civil Engineering
Univ. of Missouri-Rolla
Rolla, Missouri 65401

Lloyd Chris Wilson
Ph.D. Candidate
Dept. of Civil Engineering
Univ. of Missouri-Rolla
Rolla, Missouri 65401

ABSTRACT. Concurrent flooding, the flooding at a location due to two or more causative factors, has not been adequately addressed in the literature. The estimation of stage probabilities is important to hydrologic engineers and becomes complex where a location can be inundated due to flooding of two or more rivers. A probability concept of concurrent flooding is developed to determine flooding probabilities on a tributary river where the stage-flow relationship on the tributary is affected by backwater due to a flooding event on a nearby main river. A systematic approach for determining concurrent flooding probabilities call the "critical combination" method was developed and applied to the Meramec River near its confluence with the Mississippi River in Missouri. A "critical combination" is a tributary river flow and main river stage pair that could combine to cause a yearly peak stage at a point on the tributary. The stage resulting from each critical combination, computed using a backwater program, produces an annual stage series to which a probability distribution can be fitted. This procedure was used in the Meramec-Mississippi case and shown to be a valid approach.

INTRODUCTION

The estimation of stage probabilities is of importance to hydrologic engineers and becomes complex where a location can be inundated due to flooding of two or more rivers. The term concurrent flooding is used to describe the flooding on a tributary river where the stage-flow relationship on the tributary is affected by backwater due to a flooding event on a nearby main river. Since each river responds to its own unique physical impetus and is therefore governed by a different probability relationship, a correct probabilistic approach must be found to interrelate them.

A review of published literature reveals a lack of information on the subject of concurrent flooding, except for a general statement of the problem (Linsley, Kohler and Paulus, 1975), and some proposed non-probabilistic (Federal Insurance Administration, 1977) and pseudo-

V. P. Singh (ed.), Hydrologic Frequency Modeling, 603–614.

probabilistic approaches (Hydrologic Workshop, undated, and Tetra Tech, Inc., 1978). These two methods are summarized and compared in Wilson, 1980. Total probability theory is used to arrive at stage- frequency relationships at locations where concurrent flooding can occur and is applied to the lower Meramec River near its confluence with the Mississippi River in Missouri.

There are an infinite number of combinations of tributary river flow and main river stage that can produce any given stage on the tributary river in the area of the confluence of two rivers. These combinations can be plotted and shown as conditional rating curves since the stage at any point is a function of both flow and the main river stage. From probability theory, the total probability is defined as the probability of the union of all possible ways that the result can occur. Total probability theory requires the probability of exceedence for a given stage on the tributary river to be the summation of the probabilities of the occurrence of all the events that can possibly cause that stage to be exceeded.

CRITICAL COMBINATION METHOD

In general it is desired to find the stage-frequency relationship, including consideration of concurrent flooding at specific points along a tributary river near its confluence with a main river. A particular point location on the river can experience flooding due to:

1. Flood flows on the tributary river.
2. Flood stages on the main river that back up into the tributary river.
3. Any combination of (1) and (2).

Therefore, the probability structures of both tributary river flows and main river stages or flows must be somehow combined to find the correct stage frequency function at a particular location. Two methodologies were developed to compute the correct probabilities in a concurrent flooding area. One approach is to use a bi-variate probability distribution that describes the probability structure of both tributary and main river flows (Calise, 1981). This paper describes a more straight-forward technique which still satisfies the total probability theory.

A methodology is developed to calculate concurrent flooding probabilities that is theoretically sound, accurate, and yet computationally efficient. The basis of the critical combination method is a period-of-record stage simulation. However, the method is unique because it optimally selects those combinations of tributary flow and main river stage which may have caused the peak annual stages at any point along the tributary river in the area of the confluence. This approach is attractive because the required backwater calculations will be systematically chosen to avoid unneeded calculations and thereby reduce expense. Also, since the method utilizes traditional frequency analyses, it is straightforward, thus giving the engineer the opportunity to more easily interpret the results.

Four separate steps are required in the application of this method:

1. Selection of the "required critical combinations," which are the extreme combinations of daily tributary river flows and main river stage.
2. Definition of the set of conditional rating curves which define a three dimensional "rating surface".
3. Simulation of stages at the point of interest for selected extreme events from the period of record.
4. Frequency analyses of the annual stage series.

Combination Selection. The most straight-forward yet time consuming approach for determining concurrent flooding probabilities would be to simulate a daily stage record at the point of interest using a step backwater computer program and the recorded daily main river stages and tributary river flow (stage-flow combinations). The annual series of peak stages at any point on the river may easily be analyzed using the above calculated step backwater elevations. This method uses the same simulation concept, except it avoids the simulation of combinations that 1) could not produce peak stages, 2) have magnitudes insignificantly different from each other, or 3) have not occurred in previous months or years. This is accomplished by creating intervals of main river stages and tributary flows into which recorded data can be grouped.

A matrix is thus created containing combinations of main river stage and tributary river flow, hereafter referred to as a stage-flow matrix. It consists of a number of equivalently sized stage and flow intervals whose bracket sizes are chosen to result in nearly equal effects on the computed water surface elevations.

An algorithm was developed to select only those combinations "critical combinations" (elements in the matrix) which could have caused an extreme stage at any of the points along the area of the confluence for every year during the period of record. The selection process is made for each year of the period of record, and then the results are combined to give period of record results.

This selection procedure uses a disqualification process deleting those combinations which hydraulically could not have caused the annual maximum stage at a particular point on the river. This disqualification process is applied by finding the annual peak flow on the tributary river and then finding the corresponding stage on the main river for any particular year of the record. If there occurs more than one day in which the flow falls in the peak flow interval, one chooses that combination which has the higher corresponding stage interval on the main river. This is done because the combination with a higher stage for the same interval will obviously give a higher resultant stage on the tributary river in the area near the confluence.

Further, for the same year one finds the annual peak stage on the main river and the corresponding flow on the tributary. If there occurs more than one day in which the stage falls in the peak stage interval that combination which has the higher corresponding flow on the tributary should be chosen. These two peak combinations are now called "critical combinations" for that year, and they bound a critical region in the stage-flow matrix which includes combinations of stage and flow that potentially could combine to give a higher resultant stage than either of the two endpoints. This critical region does not include (disqualifies) flows less than the flow corresponding to the peak combination ("critical combination") produced by the maximum stage that occurred on the main river. Since, these flows would produce stages less than that produced by the peak combination. Likewise, all stages less than the stage corresponding to the "critical combination" associated with the maximum flow that occurred on the tributary river are disqualified. Since they have no possibility of causing a water surface elevation greater than that created by the corresponding maximum value they should be disqualified.

In addition only those critical combinations that have actually occurred or simulated to occur are systematically selected by examining combinations of tributary flow and main river stage. Those critical combinations which occur, along with the end-points are termed "critical combinations". These "critical combinations" can further be reduced in number by disqualifying any combination of tributary flow less than the maximum that occurred in each stage interval. Likewise, any combination of stage less than the maximum that occurred in that tributary flow interval should be disqualified. After the set of "required critical combinations" for each year is determined and combined for the entire period of record, a total set of "required critical combinations" are known for each of the points of interest.

Stage Interpolation. One way to reduce the number of backwater computations would be to increase the flow and stage interval sizes and repeat the critical combination selection process. In order to substantially reduce the number, it has been found that very large interval sizes were needed. For example, to achieve a reduction of 50 percent in the number of backwater calculations, interval sizes of 3 feet and 6500 cfs were required for the Mississippi River stages and Meramec River flows respectively. This is undesirable because of the loss of resolution for the high outlying values which is of particular importance in the later frequency distribution fit of stages. It is preferable to use the smallest practical interval sizes and then select from those required combinations a small number of points for which backwater computations will be made and from which the other needed points will be estimated using a stage interpolation routine.

For each of the points along the river in the confluence where a stage frequency relationship is desired, the stage corresponding to each of the total "required critical combinations" must be computed. This can be done using any backwater model. However, if small stage and flow intervals are chosen, the number of required backwater computations might be large. Therefore, it is desirable to reduce the number of combinations for which profiles must be computed.

A set of conditional rating curves may be computed and expressed as a three dimensional "rating surface". These are applicable when the flow-stage relationship at a point is affected by downstream stages on the main river. It should be recognized that the X-Y plane (horizontal) of the three dimensional rating surface contains the stage-flow matrix discussed previously, while the Z dimension represents resultant tributary stages. This concept of a rating surface permits the evaluation of the stages given the concurrent tributary river flow and main river stage data. The critical combination method allows the definition of only the elements on the 3-D rating surface that might be important in finding the annual stage series. By combining these two concepts, flow profile determination for only a small number of the total "required critical combinations" may be made. These will be used to estimate the remaining stages based on the shape of the rating surface.

Once the number of required critical combinations is reduced to a reasonable number so that backwater runs can be made, the stages for the remaining combinations must be estimated, that is, the rating surface must be more completely defined. The planer interpolation method was used because the estimated rating surface will always pass through the evenly spaced known points and the method is easy to use. The three close points on the rating surface for which backwater runs were made are used to estimate the stage at a critical combination for which a run was not made. This interpolation or extrapolation process must be repeated for each unknown critical combination as well as each point of interest due to the uniqueness of the rating surfaces. Another possible technique that could be used but was not employed in this study, is the calculation of a two-dimensional spline function.

Extreme Events and Frequency Analysis. Once all of the total "required critical combinations" have had stage estimates either calculated, interpolated or extrapolated for all points of interest on the tributary river, the annual stage series can be synthesized. The employed procedure processes the simultaneously recorded daily data on the tributary and main rivers one year at a time. The peak annual stage is the highest stage that results from the simultaneous occurrence of any of the "critical combinations" of main river flows and tributary river stages. This process is repeated for each year in the record, at each point of interest.

The annual series of stages can be treated as a stationary time series of independent random variables. The behavior of data can be modeled by the use of established theoretical probability distributions, such as the Log Pearson Type III distribution. Computer programs were developed to perform the critical combination selection process, data reduction, and planer interpolation.

APPLICATION TO THE MERAMEC RIVER

The Meramec River Basin is in the eastern portion of the State of Missouri and joins with the Mississippi River about 12 miles (19 km) south of St. Louis. The basin covers about 3,980 square miles (10,308 km). Near its confluence with the Mississippi River, the river is rather sinuous, flat, and slowly flowing with a wide flood plain. Many areas in this region are urbanized and significant economic damage occurs due to floods on the Meramec and/or Mississippi Rivers.

Tributary River Data. The stream gage closest to the mouth of the Meramec River is the United States Geological Survey (U.S.G.S.) gage number 0701900, situated near Eureka, Missouri. It is located at river mile 34.1 (54.87 km), as measured from the confluence of the two rivers. This gage has a contributing drainage area of 3,788 square miles (9,811 km) and has continuous data from October 1921 to present, the first twelve years collected from a nonrecording gage and the remaining from a water-stage recorder. The records have been labeled "good", meaning that it is estimated that about 95 percent of the daily discharges are within 10 percent of the true values. As is common for U.S.G.S. gages, the Eureka site was chosen because it was believed to be outside the influence of backwater from the Mississippi River. This assumption was checked by computing a profile for two extreme tailwater conditions and expected stages at Eureka were found to differ by only 0.05 feet (.02 m) due to backwater effects. Therefore, the stages at Eureka are not substantially affected by Mississippi River water surface elevations.

Main River Data. Daily stage elevations on the Mississippi River at the mouth of the Meramec River were required for the proposed method. A recording stage gage maintained by the St. Louis District Corps of Engineers was used to estimate this data since no gage is located at the desired point. Gage 0179A records stage and is located about 18.8 miles (30.3 km) upstream of the Meramec River (U.S. Army Engineer District, 1970). Due to alterations by both man and nature in the river, channel geometry over the past century and the relationship between stage and flow has been changing. To correct for this, a synthetic daily time series was developed by converting the tabulated daily discharges from 1922 to 1946 to river stages using a present day rating curve.

To arrive at the necessary time series of stages at the mouth of the Meramec River, many Mississippi River profiles were examined and because little variance was observed, an average drop of 9.12 feet in water surface elevation between the gage and the confluence of the two rivers was adopted. Since only the data subsequent to January 1922 was available on the Meramec at the time of the study, the combined period of record for the Meramec and Mississippi Rivers included water years 1923 to 1979.

Data Characteristics. The question of dependence between Meramec River daily flows and Mississippi River daily stages was investigated. The null hypothesis of independence was rejected in favor of the alternate hypothesis, dependence, using a two-tailed Student's "t" statistic with a 99 percent level of significance.

A backwater model with both accuracy and general acceptance was desired. The Corps of Engineers HEC-2 model was chosen. Furthermore, this program has successfully been used extensively by the St. Louis District C.O.E. and others in the lower Meramec River Basin. All input data to the HEC-2 model was supplied by the St. Louis District Corps of Engineers, with Manning's "n" values calibrated to reflect high discharge conditions. A number of checks have been made by the St. Louis District to verify the results of the hydraulic model.

Mean daily stage data was available for the period of record at a

U.S. Weather Bureau gage located near Valley Park, Missouri, at river mile 22.1 (35.6 km). Although the historical stages can be used as a check, the accuracy and thus the applicability of the data is suspect. During most of the period of record, a landfill adjacent to the gaging site has progressively encroached upon the river channel. This encroachment has probably changed the flow-stage characteristic at the gage site compromising the validity of the data and its subsequent frequency analysis.

In the Meramec and Mississippi River cases, since only mean daily flows were readily available, a procedure was investigated to relate the daily values to peak instantaneous values on the Meramec River (Boughton, 1976). There is a high degree of correlation, with a least squares correlation coefficient of 0.998, between the mean daily flows and corresponding peak daily flows as recorded at Eureka from years 1922 to 1979. It is also necessary to find the relationships between mean daily stages and peak stages for the Meramec River. There again is a high degree of correlation with a coefficient of 0.991. Therefore, a regression relationship was used to relate the mean values to the peak values. Further, since the Mississippi River rises and falls very slowly, there is practically no difference between the peak and the mean stage values in a day. Due to a lack of stage data on the Meramec River between its mouth and Eureka, it is assumed that the multiplier for mean to peak stage elevations varies linearly with distance along the channel.

An approximation was made by using recorded Eureka flowrates as the discharges along the entire lower portion of the Meramec River. Although no major natural tributaries join the Meramec River below Eureka and the basin is rather narrow near the mouth, there is some inflow into the channel along the lower reach. This was neglected for this study. Due to the vastly different times of concentration, the peak annual stages are not likely to be substantially affected.

CRITICAL COMBINATION METHOD RESULTS

The proposed methodology previously presented was used to find stage frequency relationships at three points along the Meramec River, at river miles 4.42 (7.11 km), 10.64 (17.12 km), and 22.03 (35.45 km). Also, the interpolation procedures developed to reduce computer expenditures were tested for adequacy.

The Meramec River flow interval and Mississippi River stage interval should be selected to give congruent accuracy. The change in stage was selected to be 1 foot per interval, which corresponds on the average to a flow of approximately 2,500 cfs (70 cms) on the Meramec River according to the most recent rating curve at the Eureka gage. A total of 166 "required critical combinations" were found for these flow and stage intervals. However, rather than decreasing this number by selective interpolation (as discussed earlier), HEC-2 runs were made for each of the 166 required pairs of data. This was done so that the results of all of these runs could serve as a baseline to test the accuracy of the interpolation routine and perform a sensitivity analysis.

Accuracy of Interpolation Routine. Various values for multiples (M) of the assigned intervals for discharge and stage in the critical combination matrix were chosen. This was done to determine the sensitivity of the rating surface estimation to various levels of screening of the critical combination matrix. For each of the three locations of interest in the study, the rating surfaces were defined by HEC-2 runs for these selected combinations which were always from the original set of 166 runs made previously. The remaining points were estimated using the planer interpolation routine.

These partially interpolated rating surfaces are then compared to the rating surfaces that were defined by the 166 HEC-2 calculations. The number of points selected from the total "required critical combinations" matrix for which HEC-2 runs would be required was found to decrease exponentially as (M) increased. The deviations of the interpolated water surface elevations from the HEC-2 defined water surface elevations for each of the "required critical combinations" at each of the three points in the area of the confluence under study were computed. The error of the planer interpolation routine is dependent on the shape of the rating surface and hence the location of interest. The overall accuracy of the analysis is limited by the least accurately defined rating surface.

It is important to note that the day of peak stage for any one year was not always identical for different locations on the river. Several different years were examined including 1943, 1945, and 1946. These were selected because of the divergence of the peak days. The following was concluded from examination of these data:

1. The peak annual stage at any point does not necessarily occur on the day of the peak tributary flow or peak main stage, but rather on the day of worst combination. This affirms the critical combinations approach.

2. The peak stages for various points may occur on different days of the year. This shows the continuously variable degrees of influence exerted by the tributary river and main river inputs on the area of confluence stages.

3. A daily analysis is necessary to accurately model this physical phenomena.

Frequency Analyses and Results. The accuracy of the interpolation procedure in estimating the rating surface was previously discussed. A more useful sensitivity analysis of the stage frequency estimates can be conducted by repeating the annual maximum stage simulation for different values of (M) and re-computing the Log Pearson Type III frequency estimates. The variation of the frequency-elevation estimates as a function of (M) for each of the three locations are shown in Tables 1, 2, and 3. At all points there were insignificant deviations for an (M) value of 2 and a maximum of only 0.2 feet (.06 m) for (M) equal to five. It is clear that very little accuracy is sacrificed after performing the interpolation procedure and making only 47 backwater calculations, which is a 72 percent reduction over the original 166 required.

Log Pearson Type III distribution was fitted to the annual series of stages at Valley Park on the Meramec River. This fit and the computed peak stages at nearby river mile 22.03 are shown in Table 4.

CONCLUSIONS

A new, rigorous approach to concurrent flooding frequency has been developed. This methodology was then used to evaluate flood stage probabilities on the lower Meramec River. Finally, the accuracy and sensitivity of the new method were evaluated.

The critical combination method was developed in response to the inadequacies of existing approaches to the problem of concurrent flooding. Its advantages include:

1. Freedom from dependence considerations.
2. Systematic, and therefore relatively cost efficient selection of backwater computational runs.
3. Use of traditional frequency analysis.

The economy of the critical combination approach was further increased by a select planer interpolation scheme. By comparing the results of the method using different degrees of screening, it was found that for the Meramec-Mississippi River system, a significant reduction in backwater computations can be achieved with little loss of accuracy.

Comparison of the critical combination method at river mile 22.03 to recorded Valley Park stages showed a basic consistency, however, the poor stage data does not allow a definite conclusion to be drawn. The critical combination method needs to be performed on other stream confluences with measured historical stage data in order to test its general applicability for determining the stage-frequency in concurrent flooding regions.

ACKNOWLEDGEMENTS

This work was performed at the Institute of River Studies, University of Missouri-Rolla, and was funded by the St. Louis District Corps of Engineers. The authors are indebted to Messrs. Gary Dyhouse, Ed Rothwell, and Bob Barkau of the St. Louis Corps of Engineers, Tim F. Wright of the Mathematics Department, U.M.R., S. John Calise, Institute of River Studies, U.M.R., and K. Randall Jones, Law Engineering Testing Co.

TABLE 1

TABLE SENSITIVITY OF STAGE FREQUENCY ESTIMATES-RIVER MILE 4.42

Value for (M)	T-Year Stage Estimates (Ft. Msl. (M Msl.)) 10	50	100
1	410.1 (125.00)	415.3 (126.58)	417.2 (127.16)
2	410.1 (125.00)	415.3 (126.58)	417.2 (127.16)
3	410.1 (125.00)	415.3 (126.58)	417.2 (127.16)
4	410.1 (125.00)	415.4 (126.61)	417.3 (127.19)
5	410.1 (125.00)	415.4 (126.61)	417.3 (127.19)
6	410.1 (125.00)	415.4 (126.61)	417.2 (127.16)
7	410.1 (125.00)	415.3 (126.58)	417.1 (127.13)
8	410.1 (125.00)	415.5 (126.64)	417.3 (127.19)
9	410.2 (125.03)	415.3 (126.58)	417.1 (127.13)
10	410.1 (125.00)	415.2 (126.55)	417.0 (127.10)
11	410.1 (125.00)	415.3 (126.58)	417.1 (127.13)
12	410.1 (125.00)	415.3 (126.58)	417.1 (127.13)
14	410.2 (125.03)	415.3 (126.58)	417.1 (127.13)
16*	410.2 (125.03)	414.9 (126.46)	416.5 (126.95)
18*	410.4 (125.09)	414.6 (126.37)	416.0 (126.80)

*Some values could not be interpolated because of low density.

TABLE 2

SENSITIVITY OF STAGE FREQUENCY ESTIMATES-RIVER MILE 10.64

Value for (M)	T-Year Stage Estimates (Ft. Msl. (M Msl.)) 10	50	100
1	412.4 (125.70)	416.5 (126.95)	417.9 (127.38)
2	412.4 (125.70)	416.5 (126.95)	417.9 (127.38)
3	412.4 (125.70)	416.4 (126.92)	417.8 (127.35)
4	412.4 (125.70)	416.4 (126.92)	417.8 (127.35)
5	412.4 (125.70)	416.4 (126.92)	417.8 (127.35)
6	412.5 (125.73)	416.5 (126.95)	417.8 (127.35)
7	412.4 (125.70)	416.3 (126.89)	417.5 (127.25)
8	412.6 (125.76)	416.5 (126.95)	417.8 (127.35)
9	412.7 (125.79)	416.2 (126.86)	418.0 (127.41)
10	412.5 (125.73)	416.2 (126.86)	417.3 (127.19)
11	412.7 (125.79)	416.8 (127.04)	418.2 (127.47)
12	412.7 (125.79)	416.5 (126.95)	417.8 (127.35)
14	412.9 (125.85)	417.2 (127.16)	418.7 (127.65)
16*	412.8 (125.82)	417.0 (126.10)	418.4 (127.53)
18*	413.9 (126.16)	418.1 (127.44)	419.5 (127.86)

*Some values could not be interpolated because of low density.

TABLE 3

SENSITIVITY OF STAGE FREQUENCY ESTIMATES-RIVER MILE 22.03

Value for (M)	T-Year Stage Estimates (Ft. Msl. (M Msl.)) 10	50	100
1	422.4 (128.75)	426.8 (130.09	428.2 (130.52)
2	422.4 (128.75)	426.8 (130.09)	428.2 (130.52)
3	422.4 (128.75)	426.7 (130.06)	428.1 (130.49)
4	422.5 (128.78)	426.6 (130.03)	428.0 (130.45)
5	422.5 (128.78)	426.6 (130.03)	428.0 (130.45)
6	422.4 (128.75)	426.4 (129.97)	427.7 (130.36)
7	422.5 (128.78)	426.4 (129.97)	427.6 (130.33)
8	422.6 (128.81)	426.6 (130.03)	427.9 (130.42)
9	422.6 (128.81)	426.8 (130.09)	428.1 (130.48)
10	422.6 (128.81)	426.6 (130.03)	427.9 (130.42)
11	422.6 (128.81)	427.1 (130.18)	428.5 (130.61)
12	422.6 (128.81)	427.0 (130.15)	428.4 (130.58)
14	422.7 (128.84)	427.7 (130.36)	429.3 (130.85)
16*	422.9 (128.90)	428.2 (130.52)	430.0 (131.06)
18*	424.9 (129.51)	431.2 (131.43)	433.3 (132.07)

*Some values could not be interpolated because of low density.

TABLE 4

PEAK DAILY STAGE FREQUENCY DISTRIBUTION AT VALLEY PARK - COMPUTED AS COMPARED TO MEASURED

Return Period (Yr)	Critical Combinations (Ft. Msl. (M Msl.))	Recorded (Ft. Msl. (M Msl.))
10	422.4 (128.75)	421.4 (128.44)
25	425.1 (129.57)	424.6 (129.42)
50	426.8 (130.09)	426.5 (130.00)
100	428.2 (130.52)	428.4 (130.58)

REFERENCES

Boughton, W.C. March, 1976. 'Flood Estimation from Short Records,' Journal of the Hydraulic Division, ASCE, Vol. 102, No. HY3, pp. 241-254.

Calise, S. John. 1981. 'The Probability of Concurrent Flooding Using a Bivariate Probability Distribution,' thesis presented to the University of Missouri, at Rolla, Missouri, in partial fulfillment of the requirement for the degree of Master of Science.

Federal Insurance Administration. October, 1977. 'Guidelines and Specifications for Study Contractors,' pp. 2-12.

Hydrologic Probabilities Workshop VII, Coincident Frequencies. Unpublished Document of U.S. Army Corps of Engineers, Hydrologic Engineering Center.

Linsley, R., Kohler, M., and Paulus, J. 1975. Hydrology for Engineers, 2nd Edition, McGraw-Hill, New York, pp. 353-356.

Tetra Tech, Inc. 1978. 'Coastal Flooding Storm Surge Model, Part 1, Methodology,' pp. 5-1 - 5-5.

U.S. Army Engineer District, St. Louis, Corps of Engineers. 1970. 'Stages and Discharges of the Mississippi River and Tributaries in the St. Louis District,' p. 24.

Wilson, Lloyd C. 1980. 'The Probability of Concurrent Flooding on the Meramec and Mississippi Rivers,' thesis presented to the University of Missouri, at Rolla, Missouri, in partial fulfillment of the requirement for the degree of Master of Science.

BIVARIATE ANALYSIS OF CONCURRENT FLOODING

C.D. Morris
Associate Professor
Department of Civil Engineering
University of Missouri-Rolla
Rolla, Missouri 65401

S.J. Calise
Planning & Project Manager
West Coast Regional
Water Supply Authority
Clearwater, Florida 33519

ABSTRACT. This study investigates the probability of concurrent flooding near and around the confluence of the Meramec and Mississippi Rivers in Missouri. A new general approach called the "bivariate probability method" was developed as a result of this investigation.

The "bivariate probability method" utilizes the fit of a bivariate probability distribution in order to describe the probability relationship between tributary river flow and main river stage. A backwater program is used for the conversion of the tributary river. The method also implements the theory of total probability so that all flow/stage combinations which could have created a given stage on the tributary river are accounted for in the probability computations.

An important aspect of the proposed methodology is the applicability to basins which have limited periods of record. This application is accomplished by using relationships developed between the short-term gage and nearby longer-term gage to estimate the parameters required to fit the bivariate probability distribution.

1. INTRODUCTION

Hydrologic engineers have been faced with the problem of predicting stage frequencies of rivers and streams. The problem of predicting stage frequencies becomes doubly complex in regions where concurrent flooding can occur. A location which may be inundated by the flooding on two or more rivers is defined as being a concurrent flooding region. Hydrologic engineers have typically estimated stage frequencies by first fitting a selected probability distribution to the recorded river flow data and then proceed to predict stages from a known stage-flow relationship, i.e., a rating curve, or by using Manning's formula (U.S. Water Resources Council, 1977; Chow 1964). This procedure clearly fails at locations where the stage-flow relationship is affected by more than one river. The main reason this procedure fails is the natural fact of hydraulic dependence between the two rivers which creates a zone that is affected by the conditions of both rivers involved. This zone, known as the concurrent flooding region requires special consideration when determining stage frequency relationships. Thus, the probability structures

V. P. Singh (ed.), Hydrologic Frequency Modeling, 615–632.

of both the tributary river flows and the main river stages must be enmeshed to find the correct stage frequency relationship at any point in the concurrent flooding region. There have been attempts at solving this particular problem, yet each attempt suffers its own shortcomings and seems to evade the obvious bi-functional relationship exists between the tributary river flows and the main river stages. The purpose of this paper is to present a new methodology can better define stage frequency relationships within concurrent flooding regions by incorporating this bi-functional relationship (Calise, 1981).

Development of a New Methodology

Since there has been only a minimal amount of research directed towards solving the probability aspects of concurrent flooding, the review of existing approaches is somewhat limited. The most recent being an investigation by Wilson (1980) which thoroughly discusses the probability theory that has been applied to concurrent flooding in the past, along with a discussion of the limitations and drawbacks of each approach.

Since there are an infinite number of combinations of tributary river flow and main river stage that can produce concurrent flooding, several of the existing approaches attempt to apply the law of total probability. From probability theory, the law of total probability is defined as the probability of the union of all possible ways that the result can occur. The most complete and general application of this total probability approach would require the use of a joint probability density function, also referred to as a bivariate probability density function, to describe the probability of all the possible ways a flood stage can occur. Haan (1977) gives a brief discussion of a bivariate probability function along with a simplified example of its application at the confluence of two rivers. Rochberg (1976) uses a joint density function in order to probabilistically describe the flowrates at the confluence of two rivers, but had to make the assumption of simultaneous occurrence for the prediction of concurrent flooding when using an annual series. In other words, even though the two rivers might flood in the same year they do not necessarily flood on the same day within the year which is implied when using annual probabilities for the prediction of concurrent flooding. This use of annual data inherently produces unreliable estimates of flood frequencies.

To solve the "duration" problem, that is, to insure that the events are being statistically related as actually occur at relatively the same time, historical data representing same-time events (i.e. same-time of occurrence) will be used. If the case arose where a basin was so small that the use of daily data would be a poor representation of the same-time events, the sampling interval should be made less than one day. Also, a check of the travel time from where the data is recorded to the region of interest is a must in order to insure that the events will affect the region at relatively the same time (i.e. same-time effect). If the travel time is greater than one day, say 36 hours, the concurrent flooding region should be divided into separate zones. Under the prescribed conditions, Zone I would be analyzed using the actual unadjusted tributary river flow data, while Zone II would be analyzed with a one day lag period adjustment being applied to the tributary river flow data. In order to fully understand the data requirements, one must keep in mind the two previously discussed basic concepts of same-time of occurrence and same-time effects.

To solve the problem of probabilistic dependence, the use of a discrete bivariate (of joint) probability function, represented as a conditional probability matrix, was investigated. The trivariable relationship is just an extension of a typical two-dimensional rating surface which will be mathematically described in order to best define the relationship between the three given variables. Using this tri-variable relationship as described by the bivariate stage matrix, one can then represent flow/stage combinations resulting in equal tributary stages for any point within the confluence area. Envision the plan view of the bivariate stage matrix having been inscribed with contour lines representing the resulting tributary stages. Using the law of total probability coupled with the laws of conditional probability, the daily probability of a certain stage being exceeded can be computed. This utilization of these fundamental laws of probability for a range of stages allows a daily stage-probability histogram to be developed for any location.

This histogram which describes the probability of daily stages must be converted to an annual series distribution in order to estimate the conventional 2, 5, 10, 50, or 100 year flood events. Grigoriu (1978) developed a method based on crossing theory that uses mean daily flows to predict extreme flows. This analysis seems to be valid only when predicting events with recurrence intervals greater than 100 years and is very sensitive to the selection of a probability distribution. Mejia (1971) used the broken line process as a hydrologic model for generating multivariate sequences exhibiting the Hurst phenomenon. This analysis uses mean daily flows for determining monthly maxima and computes annual probabilities by applying the union rule of probability. Both of the approaches previously described for converting from daily to annual probabilities must make the assumption of serial independence between daily events. Both authors state that the assumption of serial independence is not as crucial as one might initially believe. Thus, for the developed methodology the union rule of probability will be implemented in order to transform daily probabilities to the desired annual probabilities.

Upon review of the proposed methodology, it was decided that instead of using a bivariate probability function (discrete case), the fit of a bivariate probability density function (continuous case) would be more appropriate. There are two main reasons why the fit of a bivariate density function is best suited for this approach. First, from an application point of view, it is much simpler and does not suffer significant loss of accuracy in areas of limited records due to the small sample sizes. The use of a bivariate probability distribution broadens the application of this approach to basins with limited records. Second, from a theoretical point of view, it describes all possible events that can occur and therefore is not restricted only to historical occurrences. In other words, even though the historical data is used to describe or fit the bivariate probability density function, a probability has been associated with all events, even the ones that have not yet occurred.

Since the developed methodology incorporates daily data to statistically fit the selected bivariate probability distribution, the sample size has been increased tremendously. It follows that the sample estimates of the distribution parameters will likely reflect a better estimate of the population values. This application to short-term records is one of the most significant aspects of this newly developed bivariate probability approach, the other being the association of probability to all events.

METHOD

The method presented herein to calculate concurrent flooding probabilities is the most general, yet complete probabilistic approach developed to date. It is based upon the premise that the dependent relationship between tributary river flows and main river stages can be properly described by a bivariate probability density function. The method assumes independence between sampled events in order to fit the bivariate probability distribution and to transform the computed daily probabilities to annual probabilities. The bivariate approach incorporates the existing recorded data, in this case daily records, which had a combined effect upon flood stages at the point of interest lying within the tributary concurrent flooding region. This method is computationally straightforward and easy to perform using a computer and is attractive because of its applicability to basins with very limited records.

There are four separate steps required in the application of the bivariate probability method:

1. Selection and fit of a bivariate probability distribution.
2. Defining the bivariate stage matrix with backwater calculations and interpolation.
3. Computation of daily probabilities using a numerical integration approach which applies the laws of total probability.
4. Transformation of daily probabilities to annual probabilities using the union rule of probability.

Bivariate Probability Distribution

Using a bivariate probability distribution as a means of probabilistically describing the dependent relationship between daily tributary river flows and main river stages, the foundation of the four methodology is established. This probabilistic tool can easily account the complexities involved when dealing with two dependent random variables, as is the case with concurrent flooding.

Selection of distribution

As a result of a literature search investigating the possible choices for a bivariate probability distribution, the bivariate "normal" distribution was selected. Most of the literature combined discussions and applications of the bivariate normal distribution indicating its common use and acceptability. From a theoretical point of view the section of the bivariate normal distribution is substantiated from the central limit theorem.

The bivariate normal distribution requires five parameters in order to uniquely describe this bivariate relationship: 1) average daily main river stage, 2) variance of daily main river stage, 3) average daily tributary river flowrate, 4) variance of daily tributary river flow, and 5) correlation coefficient between daily tributary river flow and main river stage. It should be noted that the five parameters that have been defined are characterizing the populations values.

Estimation of parameters

The concept of using sample estimates to describe the unknown parameters must not be forgotten. Depending upon the reference, a reliable sample of events should contain a minimum of 20 to 40 sample values. A sample containing less than 20 values should be used with caution and is not recommended because of its lack of reliability in estimating population parameters.

When dealing with hydrologic data, especially streamflow data, the shortage of data is common while the over-abundance of data is rare. When using an annual series, one should collect at least 20 years of data (i.e., 20 sample values) before ever attempting to fit a probability distribution. Referring now to the bivariate approach, which in this case uses a daily series, the sample size is much larger when compared to the same 20 years of annual data. Obviously, the daily series is not an independent random sample, but in the strictest sense neither is the annual series. Admittedly the annual series exhibits less serial dependence than does the daily series, but does not necessarily contain as many independent observations. The equivalent numbers of "independent" samples for estimating the mean and variance of a first order Gauss-Markovian series can be obtained respectively from an approximate expression (Siddiqui, 1962; Kendall and Stuart, 1967).

Therefore, if this approximate expression is applied to a record of 20 values with a lag one auto-correlation coefficient of 0.80, instead of having 20 sample values to estimate the mean and variance of an annual series, there are 811 and 1602 equivalent sample values respectively, when using a daily series. It follows for a given record of a daily series, more "independent" observations are generally available to estimate the mean and variance than compared to an annual series. The concept of probabilistic independence is central to the successful application of probability theory. From a purely practical point of view, the analysis of many probabilistic models would become hopelessly complex if engineers were unwilling to adopt the assumption of independence of certain random variables in a number of key situations. Flow gages must lie outside of the concurrent flooding region in order to insure hydraulic independence. In other words, the tributary gage should not be influenced by the backwater effects of the main river.

As discussed in a previous section, one must insure that the time of the data selected, be it hours, days or weeks, does reflect concurrent events. The concepts of same-time occurrence and same-time effect should be represented by the data. For the analysis, the data should be segregated into periods of assumed stationarity in order to satisfy the assumption that the selected probability distribution does not change with respect to time. When using a daily series, the data will more than likely contain inherent yearly cycles corresponding to periods of high and low flows. Commonly, seasonal periods or monthly periods are selected to satisfy the stationarity requirements. In order to rigorously analyze or dissect the data into its underlying trends or cycles, an in-depth time series analysis should be performed upon the raw data.

Bivariate stage matrix

For a given location where the stage frequency relationship is desired, the stages corresponding to the different flow/stage combinations must be determined. This can be accomplished by the use of a computerized backwater profile model. Since there are an infinite number of flow/stage combinations which could produce a single given stage at a given location, the number of backwater computations would theoretically be infinity. Obviously, this is undesirable and warrants investigation if the proposed methodology is going to be practicable. In order to make the proposed methodology useful to the practicing engineer, it is necessary to formulate a procedure which will reduce the number of backwater computations to a practical number.

1. Rating surface concept. Since the stage at given location is a function of main river stage, x, and tributary river flow, y, the tri-variable relationship could be expressed by the equation:

$$\text{stage} = f(x,y)$$

This bivariate stage function (three-dimensional surface) could vary with each variable and is just an extension of the typical two-dimensional rating curves used to transform a recorded stage to a flowrate. Thus, one might refer to the three-dimentional surface as a rating surface. Given any combination of main river stage and tributary river flow, the resulting stage would be known. This elementary rating surface concept will be the bridge used to connect a given stage at a given location to a corresponding bivariate probability.

2. Rating surface limits. Since the data will be segregated into separate periods with each period contributing its own stage probability to the overall annual stage probability, it follows that the "high flow" or "high stage" periods will contribute the most probability to the annual stage probabilities. Realizing that the procedure of this analysis is to attach annual probabilities to a range of stages at a particular location in the concurrent flooding area, the rating surface need only to be defined for the given range of stages. In estimating range of stages for which the rating surface should initially be defined, the statistics computed for each period should be used to select a high flow or high stage period.

It should be pointed out that since the flows and stages are generally correlated, high flow periods will tend to coincide with high stage periods. Then using a high flow or a high stage period as a reference, one should extend the rating surface outward (increasing) four or five standard deviations from each of its estimated mean values. This rule of thumb should be used as a guide for an initial starting range. One can always extend the range further if needed.

3. Spline interpolation. Having determined the range for which the rating surface will be defined, the next step is to describe it mathematically. Keeping in mind that the goal is to minimize the number of backwater computations while at the same time sufficiently describing the rating surface, an interpolation approach was selected. For instance, if specified flow/stage combinations were known from backwater computations, then one could estimate the unknown flow/stage combinations by interpolating. The interpolation routine selected for this incorporates the use of a spline curve fitting technique. The spline fit is a unique fit in that the fitted curve passes through every given data point and fits a cubic relationship between the data points.

Since an interpolation technique is being used for the description of the continuous rating surface, the rating surface will herein be referred to as a bivariate stage matrix. Referring back to the three-bivariate stage matrix, it is evident that the two-dimensional spline interpolation will have to be carried out a number of times in order to be implemented into the analysis. The spline interpolated curves would be fit to approximate the conditional relationship between the tributary river stages at a given location for a given tributary river flow and main river stages. Thus, separate conditional curves described by the bivariate stage matrix correspond to the tributary river flowrates with each curve containing unique cubic fits. From these fitted continuous curves, one can now perform spline interpolations between the main river stage spline interpolations and the tributary river flows for given main river stages. An investigator can now fill in the so-called grid system of the bivariate stage matrix at a given location on the tributary river. Accordingly, the interpolation process can be repeated several times in both directions in order to describe as fine a grid system as needed or desired. This interpolatory process using spline functions is computerized and accessible as subroutine "IBCIEU" from the International Mathematical and Statistical Library, Inc. (IMSL Library, 1979).

4. Selection of points. Once the boundaries of the bivariate stage matrix have been initially set, the next step is to choose the flow/stage combinations for which the backwater program will actually compute the resulting stages at a given location on the tributary river. These selected points become the "known" points from which the previously discussed spline interpolation process will originate. The number of points selected is both a function of economics and overall accuracy. Using economics as the guide for determining the initial number of backwater runs, the ultimate deciding factor for the number of runs should be accuracy.

For example, once the flow/stage combinations determine the initial boundaries of the bivariate stage matrix, one can select the initial set of "known" grid points. These points become known by making the corresponding backwater runs. Using these points, proceed to fill in the grid system utilizing the spline interpolation routine. To check the relative accuracy of the spline interpolation, compute some backwater runs at points which are furthest from the initial set of points, and compare them to the interpolated values. If the values do not match up within the accuracy of the backwater model, the grid system is then redefined with a larger set of known points and the process is repeated. The larger set of known points should certainly contain the initial set so that cost may be minimized while accuracy is being maximized.

Daily Exceedance Probability Computations

In order to determine the daily probability of exceeding a given stage S, all the flow/stage combinations that can cause S to occur must be accounted for. This concept is recalled as being the law of total probability, which is the backbone of the proposed methodology.

The exceedance stage probability will be computed as the complement of the comulative stage probability.

$$P(S(trib) > S) = 1 - P(S(trib) \leq S)$$

where:

S(trib) = random variable associated with tributary river stage

S = stage at point on the tributary river for which the probability will be associated

To properly calculate the cumulative stage probability, one must find the probability of a flow/stage combination resulting in a stage less than or equal to S. To determine the cumulative stage probability, one can probabilistically use a numerical integration approach along a single S contour curve.

Illustrated in general terms, given a function g(X,Y,Z = constant), the bivariate cumulative probability distribution could be implemented to probabilistically integrate g(X,Y,Z = constant). The bivariate cumulative probability distribution F(S,Y) is described by the following general equation:

$$F(X,Y) = P(X \leq x \text{ and } Y \leq y) = \text{double integral of } f(x,y)\,dx\,dy$$

where: X and Y are continuous random variables of the bivariate probability distribution f(x,y).

By applying a fundamental property of cumulative probabilities, one can estimate the probability of being within the single interval. This property of cumulative probabilities is described by the following equation:

$$P\,(x(1) \leq X \leq x(2) \text{ and } Y \leq y(avg)) = F(x(2),\, y(avg)) - F(x(1),y(avg))$$

This application could be performed several times along the curve to probabilistically integrate the total probability region as follows:

$$P(Region) = \sum_{i=1}^{n} P\,(\text{being within } i(th) \text{ interval})$$

where: n = number of intervals

Therefore, the total probability of Z being less than or equal to g(X,Y,Z = constant) is the summation of P(Region).

There is another way to probabilistically integrate the volume of Region which is computationally more efficient. The approach involves the use of conditional probabilities as applied by the total probability theorem. First, a good understanding of what is meant by a conditional probability is necessary. The probability of an event B occurring when it is given that some event A has occurred is called a conditional probability and is denoted by P(B/A). The symbol P(B/A) is usually read "the probability of B, given A." Understanding the concept of conditional probability, it is now appropriate to introduce the total probability theorem which can be visually represented by constructing a Venn diagram. If E(1), E(2), E(3),..., E(n) represents a set of mutually exclusive and collectively exhaustive events, one can determine the probability of another event A from the following:

$$P(A) = \sum_{i=1}^{n} P(A/E(i)) \cdot P(E(i))$$

Accordingly, the probability for total probability region is represented by the following:

$$P(S(trib) \leq S) = \sum_{i=1}^{n} P[Q \leq q(avg)/S'@M(S'(i)] \cdot P[S'@M(S'(i))]$$

where:

n = number of stage intervals between $\bar{S}$ and S(limit)
S(trib)= random variable associated with tributary river stage
S = specific stage at point R on the tributary river for which the probability will be associated
Q = random variable associated with tributary flow
Q(avg) = averaged tributary flow for i(th) main river stage interval
S = random variable associated with main river stage
M(s(i)) = i(th) stage interval for main river stage

The quantity $P[Q \leq q(avg)\ S@M(s(i)]$ will be computed by transforming the conditional distribution of Q, given S@M(s(i)), to a marginal normal distribution. Stated in general terms, if X and Y are two random variables distributed n(M(1),SD(1)) and n(M(2),SD(2)) respectively, where M,SD, and V is the mean, standard deviation, and variance, and R is the correlation coefficient of X and Y, it can be shown (Hogg, 1970) that the conditional distribution of X, given Y = y, is the normal distribution with:

mean = M(1) + R[V(1)/V(2)][y-M(2)]

and

variance = SD $(1 - R^2)$

The quantity P[S@M(s(i))] is easily computed using cumulative probabilities of the marginal probability distribution of S which is normally distributed.

P[S@M(s(i)) = F(S(2)) - F(S(1))

The IMSL subroutines "MDNOR" and "MDBNOR" are utilized for computing the cumulative probabilities of the normal and bivariate normal distributions, respectively.

Annual Exceedance Probability Computations

In order to arrive at the desired annual probabilities, the computed daily probabilities must undergo two probability transformations. For illustrative purposes throughout this section, the assumed stationary periods will be represented by monthly periods. One must first transform the daily probabilities for each month to monthly probabilities, and secondly, transform these monthly probabilities into an annual probability. Each transformation is accomplished by applying the union rule of probability.

Daily to monthly transformation

Stated in general terms, the union rule of probability is represented by the following equation and illustrated:

$$P(A \cup B) = P(A) + P(B) - P(A \cap B)$$ where U is union and I is the intersection

and

where: A and B represent specific events.

This union probability is interpreted as the probability that "at least one" of the events, A or B, will occur. If events A and B are independent, P(A I B) = P(A) P(B) and the previous equation becomes:

$$P(A \cup B) = P(A) + P(B) - P(A)\,P(B)$$

The computed daily probability, P(S(trib) > S), can be interpreted as the probability that stage S will be exceeded on any day chosen at random during the given month. For transforming this probability from a daily time frame to a monthly time frame, the union rule is expanded and represented by the following expression:

$$P(Z(1) \cup Z(2) \cup Z(3) \cup \ldots \cup Z(m)) = P\left[\bigcup_{i=1}^{m} Z(i)\right]$$

where:

m = number of "independent" observations for the given month

Z(i) = daily event associated with exceeding stage S(trib) on the i(th) day of the given month

This monthly probability is interpreted as the probability that the given stage S(trib) will be exceeded at least once during the given month.

The reader is reminded that the assumption of serial independence between m daily events is still made. To account for this daily dependence, m possibly could be estimated as n(eq) by applying the expressions used for determining the equivalent number of independent observations as presented in earlier section. The concept of determining an n(eq) could possibly be investigated even further in order to arrive at a specific expression for estimating a bivariate equivalent number of independent observations. The approach is complicated further when attempting to apply a so-called n(eq) expression by the determination of the auto-correlation coefficient, at a particular point in the concurrent flooding region. The dilemma of estimating R(1) is not easily solved due to the fact that R(1) varies throughout the flooding region. Obviously, R(1) for a given point is a function of the tributary river and the main river auto-correlation coefficients. Being unable to resolve this dilemma, m was then chosen to be equal to the number of days in the given month.

The application of De Morgan's theorem (Copi, 1978) simplifies the computations of the union probabilities into a form that is much easier to perform and may be represented by the following equation:

$$P\left(\bigcup_{i=1}^{m} Z(i)\right) = 1 - P\left(\mathrm{SUM}_{i=1}^{m} Z(i)^{c}\right)$$

where:

$$P\,(Z(i)^{c} = 1 - P\,(Z(i))$$

Then, since P (Z(1)) = P (Z(2)) = P (Z(3)) = ... = P (Z(m)), it follows that

$$P\,(Z(1)^{c} = P\,(Z(2)^{c}) = P\,(Z(3)^{c} = \ldots = P\,(Z(m)^{c}).$$

This equality of the assumed serially independent daily probabilities further simplies the union computations to the following expression:

$$P\left(\bigcup_{i=1}^{m} Z(i)\right) = 1 - \mathrm{SUM}_{i=1}^{m} P\,(Z(i)^{c}) = 1 - P\,(Z(i)^{c})^{m}$$

The resulting probability should be expressed as, P (S(trib) > S month), and is recalled as being the probability of exceeding the given stage S at least once during the given month. This probability transformation is performed for each month and thus produces 12 unique conditional probabilities, corresponding to each month, for the exceedance of S.

Monthly to annual transformation

With a good understanding of the preceding section, one can similarly apply the union rule again in order to estimate the desired annual exceedance probability:

$$P\,(Z(1) \cup Z(2) \cup \ldots \cup Z\,(j) \cup \ldots \cup Z(12)) = 1 - \prod_{j=1}^{12} P\,(Z(j)^{c})$$

where:

Z(1), Z(2), ..., Z(j) = daily event associated with exceeding stage S(trib) on any day during the given month j.

$Z(j)^{c}$ = compliment of Z(j)

This completes the second probability transformation and results in the desired annual exceedance probability which is interpreted as "the probability of exceeding stage S on any day during any given year." In other words, it is the probability that the given stage S will be exceeded at least once during the given year.

APPLICATION OF BIVARIATE PROBABILITY METHOD

The application of the bivariate probability method will proceed in a manner to serve a dual purpose. First, the period of record containing the entire period of record will be analyzed, then separate 5-year periods contained within the period of record will be analyzed as the limited periods of record. The analysis of the 5-year periods will serve as the basis for evaluating the applicability of the bivariate method to basins with limited periods of record. The concurrent or region studied is located in Missouri at the confluence or junction of the Meramec and Mississippi Rivers. These two rivers, which have a favorable amount of recorded data, will form the basis of applying the bivariate probability method.

Data and Assumptions

The data base was selected to be one day so that the requirements same-time occurrence and same-time effect would be satisfied. This time base selection process was documented previously, and also substantiated the selection of a daily time frame for this particular application.

Tributary river data

The Meramec River Basin is in the mid-eastern portion of the state of Missouri and joins with the Mississippi River about 12 miles (19 km) south of St. Louis City. The basin covers about 3,980 square miles (10,308 sq. kms) in the foothills of the Ozark Mountains and is affected both by frontal storms of several days duration and storms of short duration, and high intensity thunderstorms. The Meramec River itself is rather sinuous, flat, and slow moving near its confluence with the Mississippi River where there is a wide flood plain. Much of the area in this region is urbanized, therefore, as a result of floods on the Meramec and/or Mississippi Rivers, there is significant economic damage incurred during periods of flooding.

The stream gage closest to the mouth of the Meramec River is United States Geological Survey (USGS) gage number 07019000, located near Eureka, Missouri. It is located at river mile 34.19 (55.01 km), which is about 3 miles (4.83 km) downstream from the junction with the Big River. The gage has a drainage area of 3,788 square miles (9,811 sq. kms) and continuous data from October 1921 to present, with the first twelve years collected from a nonrecording gage and the remaining from a water-stage recorder. These records have been labeled "good", meaning that it is estimated that approximately 95 percent of the daily discharges are within 10 percent of the true values. The Meramec River streamflow data at Eureka consisting of mean daily flows was compiled and distributed on magnetic computer tape by the USGS Water Resources Division, Reston, Virginia.

As is common practice for USGS gages, the Eureka gage site was chosen because it was thought to be beyond the backwater influence of the Mississippi River. Wilson (1980) verified this assumption using the hydraulic model developed by the Hydrologic Engineering Center of the U.S. Army Corps of Engineers, HEC-2, by showing insignificant backwater effects for a range of flow/stage conditions.

Main river data

Daily stage elevations of the Mississippi River at the confluence of the Meramec River are required for the proposed method. A recording stage gage maintained by the St. Louis District Corps of Engineers was used to estimate the data since no gage is located at the required point. Gage 0179A records the flow that drains about 701,000 square miles (1,815,590 sq. kms) of the midwestern United States, and is located at the foot of Market Street, St. Louis, which is about 18.8 miles (30.25 km) upstream of the Meramec River (U.S. Army Engineer District, 1970).

The natural flow on the river is affected to some extent by the numerous irrigation and navigation structures on the Missouri and upper Mississippi Rivers. Since some attenuation occurs due to these structures, the data could be considered somewhat nonstationary. Of more importance is the fact that the channel geometry has been altered by both nature and man over the past century (Deckerd, 1979). This causes a changing relationship between stage and flow at St. Louis and perhaps causing a nonstationarity in the recorded stages. In order to account for this undesired phenomenon, a synthetic daily time series was developed by converting the tabulated daily discharges from 1922 to 1946 to river stages using a present day rating curve.

To arrive at the necessary daily time series of stages at the mouth of the Meramec River, many Mississippi River flow profiles were examined and because little variance was observed, an average drop of 9.12 ft. (14.67 km) in water surface elevation between the gage and the confluence of the two rivers was adopted. Since only the data to January 1922 was available at the time of the study, the combined period of record for the Meramec and Mississippi Rivers included water years 1923 to 1979.

Nearby river data

The nearby stream gage selected as the long-term gage is located on the Meramec River at river mile 149.4 (240.38 km) near Steelville, Missouri. The gage, USGS gage 0701300, has a drainage area of 781 square miles (2,002.8 sq. kms) and has continuous data from October 1922 to present, with the first twelve years being collected from a non-recording gage and the remaining data being collected from a water-stage recorder. The period of record, 1923-79, was chosen to correspond to the tributary river gage located at Eureka which is being investigated as the limited period of record gage. The records have also been labeled "good". The Meramec River streamflow data at Steelville of mean daily flows was also compiled and distributed on magnetic computer tape by the USGS Water Resources Division.

Hydraulic model

In order to best define the bivariate stage matrix, a backwater model with both accuracy and general acceptance was desired. The Corps of Engineers' HEC-2 model was selected because of its extensive use by the St. Louis District and others in the lower Meramec River basin for flood insurance studies and other hydrologic studies. Furthermore, this model was chosen because it is considered by many to be the present state-of-the-art for steady-state modeling.

All input data to HEC-2 was supplied by the St. Louis District of Engineers, with the Manning's "n" values calibrated to reflect flooding conditions (i.e., high discharges). A number of verifying checks have been made by the St. Louis District to justify the results of the hydraulic model. For high flows, the model was calibrated to high water marks observed during 1979 for which the tolerance was about 0.1 ft. (0.03 m) in the stage estimated. For lower flows, the model was checked at Valley Park (river mile 22.1) (35.56 km) using Eureka flows and observed stages less than within-bank flow conditions. For discharges between 1,000 and 20,000 cfs (28.32 and 566.40 cu m/sec), the tolerance was found to be about 0.5 ft. (0.15 m). For the proposed applications the HEC-2 model is considered to be sufficiently accurate.

Although no major natural floodways flow into the Meramec River below Eureka, and the basin is rather narrow near the mouth, there still would be some inflow to the channel along its lower reach. The assumption of constant flow as opposed to spatially varied flow will tend to give slightly smaller flow volumes, but due to the vastly different times of concentration, the effects upon the tributary river stages will be insignificant.

Another limitation to this method involves the use of a steady flow analysis. It assumes that the Meramec flowrate occurs for an indefinite period of time, i.e., that equilibrium exists. However, for various flow/stage combinations, the approximate travel time for an element of water from Eureka to the mouth of the Meramec River is 15 hours. With this travel time in mind, and knowing that the base of a typical flow hydrograph for this basin is relatively small, it is possible that the phenomenon of unsteady-gradually varied flow exists. Therefore, the storage and acceleration-attenuation that may occur in these cases often the downstream stages considerably less than those estimated from a steady flow analysis (Chow, 1959). The question of dependence between the Meramec River daily flows and the Mississippi River daily stages was investigated by Wilson (1980) using two different procedures. The chi-square test and the Students t test both implied that the two rivers were correlated on a daily basis at the 99 percent level of significance. The proposed bivariate probability method is equipped to handle the dependence which is inherent and expected between the two rivers near their confluence.

Bivariate Probability Method Results

The bivariate probability method will be applied to the Meramec River at Valley Park, Missouri. The four basic steps required for the application of the method will be performed and discussed accordingly.

Parameter estimates

In order to satisfy the stationary requirements of the data for utilization of a probability distribution, the daily data was segregated into monthly periods. These monthly periods are assumed to contain stationary processes but have not been rigorously investigated by a thorough time series analysis.

For this analysis the Meramec flowrates were transformed to better represent a normal distribution (i.e., normalized) by the use of a simple log-transformation. This common transformation implies that the Meramec flowrates are assumed to be log-normally distributed. The log-transformation seemed to adequately transform the raw flowrate data. No transformation was applied to the Mississippi stages. It should be pointed out that it is always theoretically possible to determine a function f(X) which will transform the skewed distribution into a normal one.

Comparing the transformed statistics to the period of record statistics, the linear transformations seem to be transforming the 5-year sample statistics of the Meramec River to reflect that of the 57 year population estimates. This transforming procedure is the link required to "best" estimate the population parameters needed to reliably apply the bivariate method to basins with short periods of record.

Selection of points

The points selected for the description of the bivariate stage matrix were chosen to represent the desired stages S. It is only necessary to define the stage matrix for those resulting stages which are to be related to any occurrence frequency. It is natural for some months to have a higher probability of producing concurrent flooding than others. Since the analysis is concerned with flooding, the extreme events associated with a "high probability" month will be used as a guide for the selection of the initial stage matrix. From the statistical analysis, the month of May was selected since it is a good representation of a "high flow" and "low stage" period. Using the mean values for this month, 3.47342 (approximately 3000 cfs) (84.96 cu m/sec) and 387.4 ft. MSL (118.08 m), the limits of the stage matrix were extended outward (increasing) approximately five standard deviations from each mean value. An initial total of 50 backwater runs were performed in order to thoroughly describe the outlined region of the stage matrix.

Spline interpolation

The original stage matrix will be "filled in" using a spline routine and the original 50 resulting computed stages. The coarseness of the grid is governed by the desired accuracy of the probability integration which is performed using the interpolated grid points.

For this analysis, the intervals for both axis of the original 10 x 5 stage matrix were subdivided by a factor of five to produce, by interpolation, a "filled in" 46 x 21 stage matrix. This "filled in" coarseness is anticipated to be satisfactory for the probability computations. To verify the spline interpolation routine, the points furthest from the "known" points were all checked for accuracy. As described earlier, these center points are more likely to have the interpolatory error than those points nearer the "known" points.

Computation of annual exceedance probabilities

For the computation of the annual exceedance probabilities, the program was developed and combined the daily probability integration along with the two probability transformations into one program. The relative accuracy of the selected interval size (i.e., coarseness) of the bivariate stage matrix was checked and verified. The annual exceedance probability results were tabulated and include the entire period of record analysis along with the eleven separate 5-year periods. The 5-year estimates correspond well to the period of record estimates given the very limited sample records. The dispersion of the limited sample records tends to increase as one moves away from the confluence.

Since the data used for the analysis consisted of mean daily values, instead of the typically used peak values, the probability events of mean daily stages have been associated with annual exceedance probabilities. To arrive at the typically computed peak daily probabilities from mean daily probabilities, the relationship developed by Wilson (1980) using regression techniques is implemented. These appropriate multipliers or correction factors are to be used in order to relate the mean daily stages to the peak daily stages for the points of interest on the Meramec River. The preceding multipliers were applied to the frequency vs. mean daily stages to estimate the frequency vs. peak daily stages at Valley Park. The resulting estimates are presented in Table 1 along with the recorded historical frequency vs. peak daily stages (U.S. Army Engineer District, 1970).

CONCLUSIONS

A new approach to concurrent flooding frequency analysis has been developed. This new methodology was applied at Valley Park on the lower Meramec River to estimate the probability of flood stages. The following general conclusions are made as a result of this particular investigation into the probability of concurrent flooding on the Meramec River near the confluence with the Mississippi River:

1. The developed methodology is unique in its approach and application of probability theory in regions of concurrent flooding.
2. It appears that this methodology has general applicability to regions that are subject to concurrent flooding.
3. This method estimates the annual exceedence probabilities of concurrent flooding events.
4. The method seems to be adequate for estimating concurrent flooding exceedence probabilities for basins with limited periods of record.

TABLE 1. PEAK DAILY STAGE FREQUENCY AT VALLEY PARK - COMPUTED AS COMPARED TO MEASURED

Return Period (Yr)	Bivariate Analysis (Ft. Msl. (M Msl.))	Recorded (Ft. Msl. (M Msl.))
10	421.5 (128.47)	421.4 (128.44)
25	424.3 (129.32)	424.6 (129.42)
50	426.2 (129.90	426.5 (130.00)
100	428.5 (130.60)	428.4 (130.58)

ACKNOWLEDGEMENTS

This work was performed at the Institute of River Studies, University of Missouri-Rolla, and was funded by the St. Louis District Corps of Engineers. The authors are indebted to Messrs. Tom Lovelace, Gary Dyhouse, Bob Barkau, Ed Rothwell, Charles Bader and Dennis Stevens of the St. Louis Corps of Engineers, and C. Lloyd Wilson of the Institute of River Studies, U.M.R.

REFERENCES

Calise, S.J. 1981. The Probability of Concurrent Flooding Using A Bivariate Probability Distribution. Thesis presented to the University of Missouri, at Rolla, Missouri in partial fulfillment of the requirement for the degree of Master of Science.

Chow, V.T. (Editor in Chief) 1964. Handbook of Applied Hydrology - A Compendium of Water-Resources Technology, McGraw-Hill, New York.

Chow, V.T. 1959. Open Channel Hydraulics, McGraw-Hill, New York.

Copi, I.M. 1979. Introduction to Logic, MacMillan Publishing Co., New York.

Deckerd, H.L. 1979. An Analysis of Selected Hydraulic Features on the Mississippi and Atchafalaya Rivers in Louisiana. Thesis presented to the University of Missouri, at Rolla, Missouri in partial fulfillment of the requirement for the degree of Master of Science.

Grigoriu, M. 1979. On the Prediction of Extreme Flows , Proceedings, International Symposium on Risk and Reliability in Water Resources, I, II, June 26-28.

Hann, C.T. 1977. Statistical Methods in Hydrology, The Iowa State University Press, Ames, Iowa.

Hogg, R.V., Craign, A.T. 1970. Introduction to Mathematical Statistics, MacMillan Company, New York.

IMSL Library. 1979. International Mathematical and Statistical Library, Inc. Houston, Texas.

Kendall, M.G. and Stuart, A. 1979. The Advanced Theory of Statistics, Hafner Publishing Company, 3rd ed., III, New York.

Mejia, A.J.M. 1971. On the Generation of Multivariate Sequences Exhibiting the Hurst Pheonomenon and Some Frequency Analysis. Thesis presented to Colorado State University, at Fort Collins, Colorado in partial fulfillment of the requirement for the degree of Doctor of Philosophy.

Rochberg, R. 1979. Analysis of Flood Damage at the Confluence of Two Rivers. Preliminary report submitted to U.S. Army Corps of Engineers, St. Louis District, Missouri.

Siddiqui, M.M. 1962. Some Statistical Theory for the Analysis of Radio Propagation Data , Journal of Research of the National Bureau of Standards, LXVI, 5.

U.S. Army Engineer District, St. Louis, Corps of Engineers. 1970. Stages and Discharges of the Mississippi River and Tributaries in the St. Louis District, p. 24.

U.S. Water Resources Council. 1977. Guidelines for Determining Flood Flow Frequency'.

Wilson, Lloyd C. 1980. The Probability of Concurrent Flooding on the Meramec and Mississippi Rivers. Thesis presented to the University of Missouri, at Rolla, Missouri in partial fulfillment of the requirement for the degree of Master of Science.

SUBJECT INDEX

AUTHOR INDEX